VIII
							2 He 4.003
		III	IV	V	VI	VII	
		5 B 10.81	6 C 12.011	7 N 14.007	8 O 15.9994	9 F 18.998	10 Ne 20.18
I	II	13 Al 26.98	14 Si 28.09	15 P 30.97	16 S 32.06	17 Cl 35.453	18 Ar 39.95
28 Ni 58.71	29 Cu 63.55	30 Zn 65.37	31 Ga 69.72	32 Ge 72.59	33 As 74.92	34 Se 78.96	35 Br 79.90	36 Kr 83.80
46 Pd 106.4	47 Ag 107.87	48 Cd 112.40	49 In 114.82	50 Sn 118.69	51 Sb 121.75	52 Te 127.60	53 I 126.90	54 Xe 131.30
78 Pt 195.09	79 Au 196.97	80 Hg 200.59	81 Tl 204.37	82 Pb 207.2	83 Bi 208.98	84 Po (210)	85 At (210)	86 Rn (222)

63 Eu 151.96	64 Gd 157.25	65 Tb 158.93	66 Dy 162.50	67 Ho 164.93	68 Er 167.26	69 Tm 168.93	70 Yb 173.04	71 Lu 174.97
95 Am (243)	96 Cm (245)	97 Bk (247)	98 Cf (249)	99 Es (249)	100 Fm (255)	101 Md (256)	102 No (254)	103 Lw (257)

AN INTRODUCTION TO
CHEMICAL PRINCIPLES

JACK E. FERNANDEZ   ROBERT D. WHITAKER

University of South Florida, Tampa

AN INTRODUCTION TO
CHEMICAL
PRINCIPLES

MACMILLAN PUBLISHING CO., INC.
NEW YORK

COLLIER MACMILLAN PUBLISHERS
LONDON

A portion of this material has been adapted from *Modern Chemical Science,* copyright © 1971 by Jack E. Fernandez.

MACMILLAN PUBLISHING CO., INC.
866 Third Avenue, New York, New York 10022

COLLIER-MACMILLAN CANADA, LTD.

Library of Congress Cataloging in Publication Data

Fernandez, Jack E   (date)
    An introduction to chemical principles.

    1. Chemistry.   I. Whitaker, Robert D., joint author.   II. Title.
    QD31.2.F48   540   74-80
    ISBN 0-02-337070-X

Printing: 1 2 3 4 5 6 7 8   Year: 5 6 7 8 9 0

To

ALBERT
BARRY
CAMPBELL
DALLAS
JACK
MARINA
RICHIE
RUDY
RUTH
SYLVIA

PREFACE

First year college chemistry courses have come to mean many different things to different people. The diverse backgrounds of entering college students and their varied purposes in taking the course have created problems for instructors and textbook writers alike. As teachers with combined experience of over thirty years, we have long been interested in these problems. This textbook is our attempt to find a solution. Its originality lies more in the integration of ideas, principles, and theories than in a new selection of topics. This book is designed to serve that wide range of students who come to their first year college chemistry course with only an average or no high school chemistry background. Additionally, there is sufficient versatility of material so that this text can meet the needs of the students who will follow their first course with courses in organic and analytical chemistry as well as those students taking first year chemistry as a terminal course.

The text is divided into four broad areas:

    Part 1. Elementary Ideas
    Part 2. Chemical Structure
    Part 3. Chemical Dynamics
    Part 4. Chemistry in the Service of Man

The fundamentals of chemistry are covered in Parts 1, 2, and 3. Part 4 is a special topics section that can be utilized, at the discretion of the instructor, either as collateral material as the underlying principles are developed or as independent study for the better student. The section can even be eliminated entirely without loss of critical principles.

The objective of our approach is to emphasize relationships among important chemical principles. Students who have had no previous course in chemistry should experience little difficulty in grasping the fundamental concepts because mathematical sophistication is not a goal nor is it assumed on the part of the student.

We have tried to make the coverage truly *general.* We have therefore included some concepts that are not usually covered in a freshman text. These are included, not because they are new or unusual, but rather because they aid in understanding and in giving the student a better idea of the scope of chemistry. Some such concepts are molecular symmetry and stereochemistry, polymer structure, reaction mechanisms, nuclear and chemical technology, and chemicals that affect life. By the same token, we have omitted some concepts traditionally covered where our experience has convinced us that they are neither essential nor helpful to the student's grasp of fundamental principles and facts.

Three concepts omitted are *gram atomic weight, equivalent weight* and its associated concentration term *normality,* and *electronegativity.* We feel that these ideas are more confusing than helpful to the beginning student. If one takes the broad view of the mole concept, then surely the idea of a *mole of atoms* need not be obscured by the introduction of gram atomic weight. Equivalent weight was an interesting early solution to the historic problem of obtaining relative atomic masses. However, the use of equivalent weights in our opinion creates unnecessary confusion in dealing with the far more important concept of atomic weight. Normality, while a convenient concentration term in the analytical laboratory, is certainty a nonessential for beginning chemistry. The concept of electronegativity appears to be passing out of the vocabulary of research chemists. Although it does possess considerable explanatory utility, beginning students are often too prone to seize upon electronegativity as the ultimate explanation for all phenomena. We, therefore, have used the directly measurable quantitative properties of ionization energy and electron affinity to develop the qualitative idea of the relative ability of an atom to attract electrons without reference to the pseudo-quantitative concept of electronegativity.

In all numerical examples the correct numbers of significant figures and proper units have been employed. Significant figures and unit analysis are not discussed in great detail but the instructor can emphasize them if he chooses. Both of these topics along with fairly complete discussions of other mathematical skills are treated in Appendix A for easy reference and assignment.

Our approach in Part 1 is to begin with common experience and proceed slowly in an experimental route through the introductory concepts and language of chemistry. Our aim in the first three chapters is to define some chemical terms and to develop the most fundamental ideas upon which atomic theory will be based. The next three chapters develop kinetic molecular theory through a study of gases and phase changes. The remaining chapters of Part 1 present a gradual development of atomic theory culminating in its application to chemical bonding. Throughout Part 1, we have attempted to provide as far as possible an integration of these concepts. An approach found in many other books is to present kinetic molecular theory, atomic structure, stoichiometry, periodicity, and bonding as separate topics to be mastered independently. We feel that each of these is understood best in terms of the others. For this reason we have attempted to develop the concepts of periodicity, atomic structure, and bonding in a parallel fashion, beginning with the operational viewpoint and leading to the theoretical. Moreover, these same topics are carried over to Parts 2 and 3 where their development and application continue along with the introduction of new concepts.

Part 2 begins at Chapter 13 with a subject that is both fundamental to molecular considerations and interesting to students—stereochemistry. This subject is a logical extension of atomic theory and bonding and forms one of the pillars upon which subsequent study of molecular structure rests. The most recent stereochemical conventions have been adopted.

Chapters 14 through 18 comprise a study of chemical structure that is divided into covalent, metallic, and ionic bond types. Within each of these bond types, the role of stereochemistry is examined as a means of relating structure to observable properties. Such ideas as physical properties, chemical energy, and reactivity can thus be correlated with structure. Of course, periodicity and atomic and bonding theory provide the basis.

The general scheme of Part 3 is to present equilibrium as the fundamental way to view chemical changes. The introductory chapter on chemical equilibrium serves to link the generalized concept of chemical equilibrium with simple types of physical equilibria. Le Chatelier's principle is presented and is shown to be a necessary consequence of the existence of an equilibrium constant. Separate chapters cover electrolytic dissociation, oxidation-reduction, acid-base reactions, and equilibria important in analytical chemistry. Quantitative treatments of weak acids and bases, buffer systems, and multiple equilibria are considered in detail in several sections of Chapter 22, but these sections may be deleted in terminal courses without loss of continuity.

Chemical kinetics is presented in Chapters 24 and 25. The striking differences among reaction rates and the effect of temperature serve as the starting point. The idea of mechanism is then discussed as a logical explanation of reactions and their rates using the kinetic molecular paradigm. Finally, reaction rate theory is developed from Arrhenius' early ideas of activation and is refined by a consideration of collision theory.

Chapter 25 uses collision theory as the basis to investigate various chemical reactions from a mechanistic point of view. Some examples are presented to show how chemists "tailormake" molecules. The importance of an understanding of reaction mechanism in this effort is emphasized.

Chapter 26 is a brief introduction to the chemistry of the most important organic functional groups. It serves to prepare the way for the next chapter, which is an attempt to deal with life as a complex chemical system. The emphasis is on how this chemistry differs from simpler chemistry. Life chemistry is thus shown to be different in organization rather than in principle.

The final chapter of this section treats nuclear reactions. Of special importance in this chapter is the fascinating story of the scientific revolution that culminated in the practical use of atomic energy.

Several points that are felt to be novel should be stressed. First, in dealing with chemical dynamics, organic chemistry is not treated separately from other "chemistries." In keeping with the idea that this book is a truly "general" textbook, reactions traditionally reserved to an organic section are presented where they serve to illustrate an important principle or merely to show some interesting chemistry. Likewise, coordination compounds are treated simply as additional examples of the variety of molecular geometry. Descriptive chemistry is not isolated from principles because principles in chemistry are not developed in isolation but, rather, are shown to spring from a consideration of "real" chemistry and "real" reactions.

Part 4 is an attempt to deal with the interface between the science of chemistry and its application to modern life. Much has been said and written in recent years about the need for relevance in science courses. Our response to this concern is to present some chemical topics that bear directly on man and over which man has increasingly to make direct value judgments. These topics need not be deferred to the end of the course. Instead they may be assigned at any time after the necessary groundwork had been laid.

A suggested sequence for the use of Part 4 as collateral material is as follows:

Section in Part 4	Can Be Used Any Time Following Chapter
29-1 Nylon	16 Polymers
29-2 Nylon Starting Materials	26 Organic Chemicals
29-3 Sulfur	3 Pure Substances
29-4 Sulfuric Acid	19 Chemical Equilibrium
29-5 Iron from Its Ore	11 Periodic Properties of the Elements
29-6 Sodium and Aluminum	21 Oxidation and Reduction
29-7 Electrolytic Copper Refining	21 Oxidation and Reduction
29-8 Cryogenics	6 Kinetic Molecular Theory
29-9 Cryogenic Liquids	6 Kinetic Molecular Theory
29-10 Synthetic Diamonds	6 Kinetic Molecular Theory
30 Chemicals That Affect Life	26 Organic Chemicals
31 Origin of Life	16 Polymers
32 Nuclear Technology	28 Nuclear Reactions
33 Energy	6 Kinetic Molecular Theory

The mathematical manipulations of Appendix A are designed to allow the student to acquire on his own the manipulative skills necessary for beginning chemistry. Students with only minimal backgrounds in algebra should be able to handle the mathematics in the text after a thorough review of this appendix.

Appendix B gives rules for the systematic naming of organic compounds and coordination compounds.

Appendix C—"Qualitative Inorganic Analysis"—is a treatment of the traditional "qual scheme." The analysis of some common anions and cations is presented to serve several purposes: (1) To afford a convenient means for the student to acquire a working knowledge of some descriptive chemistry, (2) to provide a direct interaction between text material and laboratory practice, and (3) to emphasize the application of equilibrium theory. This section presents both theory and actual laboratory directions. This material is suitable for use at any time after Chapter 23.

We wish to acknowledge Ginger Weir for typing the manuscript.

J. E. F.

R. D. W.

CONTENTS

10

ELECTRONIC THEORY OF ATOMIC STRUCTURE

11

PERIODIC PROPERTIES OF THE ELEMENTS

12

CHEMICAL BONDING

PART 2

CHEMICAL STRUCTURE

13

STEREOCHEMISTRY—THE SHAPES OF MOLECULES

17

METALLIC STRUCTURES

18

STRUCTURES OF IONIC COMPOUNDS

25

CHEMICAL REACTIVITY

26

ORGANIC CHEMICALS

27

THE CHEMISTRY OF LIFE

PART 4

CHEMISTRY IN THE SERVICE OF MAN

APPENDIX

A

MATHEMATICAL MANIPULATIONS

APPENDIX

B

NOMENCLATURE

APPENDIX

C

QUALITATIVE ANALYSIS FOR COMMON CATIONS AND ANIONS

ELEMENTARY
IDEAS

CHEMISTRY—A PROLOGUE

Over 50 centuries ago the inhabitants of the Nile valley began to preserve for posterity their view of the world. In the 30 centuries that followed, several civilizations rose to heights of achievement that in many ways rival our own. During the six centuries before the Christian era, the Greeks introduced a new subject to civilization—philosophy. Although preceeding peoples had employed the secrets of nature to their advantage in the form of technology, it was left to the Greeks to begin the process of reducing the countless phenomena of our universe to a comprehensible body of ideas. Even though they did not perform experiments, the Greeks had conceived the spirit of science.

The remaining 20 centuries of our history have seen many upheavals. But during this time science was undergoing its inexorable gestation. During all the ups and downs of these centuries science grew, sometimes almost imperceptibly, but always toward a fuller, more consistent, and more comprehensive understanding of nature.

Now, science has come to its full power. The scientific method has achieved the status of faith. Man has merely to will, and science can do. Or so it seems. The growth of science has brought about many side effects. Some of these—pollution, overpopulation, wars of indescribable horror—threaten mankind itself. It is as though man has created, with great patience and skill, a giant slave who has suddenly become uncontrollable. Perhaps the analogy will be thought overdramatic but few will think it untrue.

Science, to serve man truly, must be under his control. This control can be realized only if man understands science. It is with this end in mind that this book is written—to make the science of chemistry accessible to all citizens of a free society, both future scientists and others. Failure of our citizenry to be aware of the nature, scope, power, and limitations of science could be disastrous.

3

1-2. CONTEMPORARY CHEMISTRY

Let us turn now to the science of chemistry. Chemical problems confront us daily in modern society. As consumers we are deluged with a never ending array of new products as well as confusing but important assertions of vital concern to everyone. Some recent arguments have concerned fluoridation of drinking water, detergents that persist undegraded in sewage systems, food additives, and drugs. In addition, a problem that offers dire threats is environmental pollution. Regardless of the source, pollution of the environment is primarily a chemical problem. Another problem that an affluent society must face is nutrition. What kinds of foods do we require, and how can we best produce them?

In addition to these obvious chemical problems, chemistry plays a behind-the-scene role that is not often perceived by the average citizen. An example of such a role is the use of chemicals as raw materials in the production of common goods, such as plastics, paint finishes, metals, fuels, and lubricants that go into the manufacture of automobiles. All of these products stem from the chemical industry. Over the past century the chemical industry has grown from an embryonic state to a Goliath. Along with this growth have come problems, many of them social, including employment and taxation.

1-3. ENERGY AS A CHEMICAL RESOURCE

Still another product of chemistry is energy. The affluence of modern society can be measured in terms of the amount of energy that it consumes. Figure 1–1 shows the correlation between the per capita income of several nations and their energy consumption. Those countries that have very low energy consumptions have very low per capita incomes. Those with very high energy consumptions have high per capita incomes and the relationship seems to be roughly linear. While no cause-and-effect relationship is claimed, the graph suggests that affluence can be measured roughly by the energy consumption of a nation. The consumption of energy in the United States has increased some 30-fold since 1850. At that time wood supplied more than 90% of all the energy consumed. By 1900 the dominant fuel was coal, which accounted for over 70% of the total energy expended in the United States. By 1950 coal had been replaced by oil and natural gas as the leading energy sources. By 1970 nuclear energy accounted for 0.3% of our energy. It is estimated that by the year 2000 nuclear energy will account for approximately 50% of the energy produced in the United States. Thus, for the next several decades at least, chemical fuels will continue to provide an important source of energy in the world.

1-4. CHEMISTRY AND HEALTH

Since the earliest recorded history man has grappled with diseases in many ways. One of the oldest ways has been the use of chemical products. Today the search for solutions to the health problems of mankind continues through attempts to devise new chemical substances. Although some of the ancient diseases of mankind, such as the plagues of the middle ages, polio, and many childhood diseases

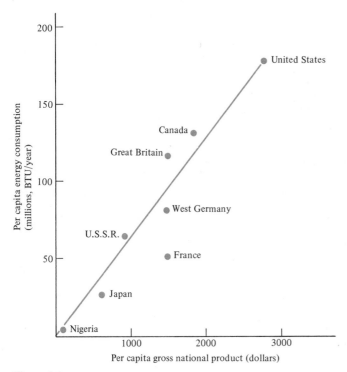

Figure 1-1

have been effectively eradicated, many, such as heart disease, cancer and the common cold, are still of concern. In fact some, such as heart disease, are on the increase. Solutions for these scourges have not yet been found, although it is hoped that the science of chemistry coupled with its sister sciences will soon furnish answers.

1-5. SOURCES OF CHEMICAL RAW MATERIALS

The accelerating depletion of natural resources that results from increasing population and affluence demands new sources of raw materials. The oceans probably afford the greatest reservoir of natural chemical resources in the world (Table 1-1). In addition to mineral resources, current studies offer the hope that one day the oceans will provide a bountiful harvest of food for man's use.

1-6. CHEMISTRY AS AN INTELLECTUAL ACHIEVEMENT

The discussion in previous sections relates essentially to chemical products—substances that affect our lives. But what of the intellectual achievements of chemistry? The science of chemistry has helped us to understand the workings of the universe, the sources of energy, life, and other glimpses into the workings of nature. The great theoretical superstructure that has been constructed piece by piece over the past centuries has led to an intellectual colossus that explains

Table 1-1*. Chemical
Elements Present in a
Cubic Mile of Sea Water

Element	Tons
chlorine	89,400,000
sodium	49,700,000
magnesium	6,000,000
sulfur	4,200,000
calcium	1,900,000
bromine	310,000
manganese	47
copper	47
iron	94
lead	19
silver	1.4
vanadium	1.4
gold	0.028

*Courtesy of Professor Dean F. Martin, University of South Florida.

the substances around us in terms of the unobservable and almost unimaginably small particles called atoms.

Further, the structure of these atoms has been elucidated to reveal electrons, protons, neutrons, and still smaller particles. This enormous body of knowledge forms an intellectual synthesis not surpassed by the greatest endeavors of the ancient Greek philosophers or the renaissance artists and architects. But of even more importance are the methods of science that extract from unyielding nature her deepest secrets. Although more difficult to appreciate than a work of art, chemical science is one of the great achievements of man.

1-7. SOCIAL IMPLICATIONS OF CHEMISTRY

Although pollution of the environment has received popular attention in recent years, it has been a problem since the beginning of factories. The widespread use of coal brought many pollution problems even as early as the nineteenth century.

The importance of this and other problems to life on our planet has brought scientists out of their cloistered laboratories into the arena of public life. No longer can scientists afford to ignore the results of their labors. They now must confront questions of ethics and morality while engaged in their research.

The profound problems brought about by science cannot be solved by scientists alone. The public must enter the scientific arena as well in helping to make the important decisions concerning the applications of science to society. Thus the study of chemistry has become a mandatory part of a liberal education.

2

CLASSIFICATION OF MATTER AND ENERGY

As we view nature, many regularities reveal themselves. It is the study of these regularities that comprises the activity called science.

One of the most common regularities in nature is the occurrence of substances in three physical states—solid, liquid, and gas. Most substances can occur in more than one state, and many can occur in all three states under the proper conditions of temperature and pressure. A substance in its **solid** state possesses not only a definite volume, but also a definite shape. These two characteristics define the solid state. A **liquid** exhibits a fixed volume but has no definite shape of its own. Instead, it assumes the shape of its container. A **gas** has neither shape nor volume of its own. When placed into a container, a gas will expand or contract to fill the container completely regardless of its size or shape.

2-2. MEASUREMENT OF TEMPERATURE

Temperature is a useful measure of the heat intensity of an object. Temperature is also useful because it allows the determination of the direction of heat flow. It is common experience that heat always flows spontaneously from a hotter body to a colder one. Therefore, by knowing the temperature of two objects that are in contact, we can know the direction in which heat will flow.

The temperature scale most commonly used in scientific work is the Celsius (formerly called centigrade) scale; the freezing point of water determines 0°C and the boiling point determines 100°C. On the Fahrenheit scale, the freezing point of water is 32°F and the boiling point is 212°F (Figure 2–1).

Figure 2–1 shows that for every degree on the Celsius scale, there are 1.8 degrees on the Fahrenheit scale. Using this simple observation one can convert

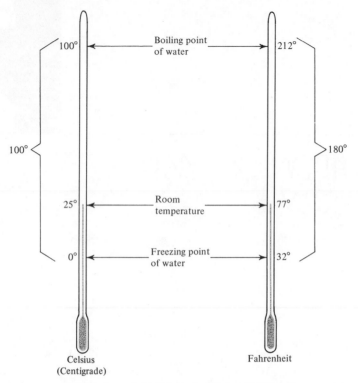

Figure 2-1. Comparison of the Celsius and Fahrenheit thermometer scales.

a temperature change of $x°C$ to degrees Fahrenheit by multiplying by 1.8. In order to arrive at the correct actual temperature in degrees Fahrenheit, however, one must add 32 to the calculated degrees Fahrenheit because $0°C$ corresponds to $32°F$. Thus the equation for converting degrees Celsius to degrees Fahrenheit is

$$°F = (1.8 \times °C) + 32$$

where F and C are the Fahrenheit and Celsius temperatures, respectively.

2-3. METRIC SYSTEM

Before continuing our discussion it is important to have an understanding of the units of measurement that scientists commonly employ. The system of units of measurement in common use in the United States at the present time is known as the English system. It is based on the units, inch (or foot), pound, and second. At the present time only the United States continues to use this system. All scientists, including Americans, use the metric system which is based on the units, meter for length, kilogram for mass, degree Kelvin* for temperature, and second for time. The most common units used by chemists, however, are grams, centimeters, and seconds. These units and their English equivalents are defined in Table 2-1.

8   *The Kelvin temperature scale will be discussed in Section 4-3.

Table 2-1. Metric System Units and Their English Equivalents

	Metric	*English Equivalent*
Length	meter (m)	39.37 in.
	centimeter (cm)	0.3937 in.
	millimeter (mm)	0.03937 in.
	Ångstrom (Å) ($1\ \text{Å} = 10^{-8}$ cm)	3.937×10^{-9} in.
Mass	kilogram (kg)	2.203 lb.
	gram (g)	0.002203 lb.
	gram (g)	0.03524 oz.
	454 g	1.00 lb.
	28.38 g	1.00 oz.
Volume	liter (l)	1.06 qt.
	cubic centimeter	
	(cc or cm^3)	0.00106 qt.
	(1 milliliter (ml) = 1 cm^3)	

When solving problems the student is strongly urged to label all quantities with their units and to apply the same mathematical operations to the units as to the numerical quantities. This procedure is important for several reasons: First, concepts are best understood in terms of their units. A length of 5 has no meaning, whereas a length of 5 in. or 5 m has. Secondly, by performing all mathematical operations on both the quantities and their units, the student is better able to find errors. For example, in converting 5 lb into gram units, do you multiply by 454 or divide by 454? It is easily seen that dividing gives the incorrect units

$$\frac{5.0\ \text{lb}}{454\ \text{g/lb}} = 0.011\frac{\text{lb}^2}{\text{g}}$$

On the other hand, multiplying gives

$$5.0\ \text{lb} \times 454\frac{\text{g}}{\text{lb}} = 2270\ \text{g}$$

This technique is amplified in Appendix A.

One might question why we employ so many different units for the same kinds of measurements. Table 2-1 lists four different metric length units—meter, centimeter, millimeter, and Ångstrom. These units are necessary to avoid the inconvenience that arises from measuring large lengths in very small units or of measuring small quantities in large units. For example, it is cumbersome to measure body weight in grams because the average man (170 lb) weighs 77,180 g. Instead the kilogram may be used—77.2 kg.

Still another, and more important, reason for the use of the proper unit is to preserve the appropriate significance in the numbers used to represent a certain quantity. For example, the weight of 77,180 g, involving five figures, implies that the weight of the man was measured to a very high degree of accuracy. The weight of 77.2 kg, on the other hand, contains three figures and implies a lower degree of accuracy. In this case the insignificant figures in the first number cannot be dropped because they are required to place the decimal point.

When multiplying or dividing numbers you should pay careful attention to significant figures. The rule to follow is to report your answer using a number of significant figures equal to those in the *least* significant of the numbers multiplied. Thus the product of the numbers

$$123 \times 42 \times 12345 = 68,329,575$$

should be expressed in no more than two significant figures and must therefore be expressed in exponential notation, usually called **scientific notation**

$$6.8 \times 10^7$$

In scientific notation, numbers are always expressed as the product of a number between 1 and 10 and the number 10 raised to an integral power. Table 2–2 lists some exponentials and their more common equivalents.

When adding or subtracting numbers, the same care must be taken to retain the correct number of significant figures. In the example that follows, note that the smaller numbers are lost because they are smaller than the probable uncertainties in the larger numbers.

$$
\begin{array}{r}
123 \\
3456 \\
21.09 \\
\underline{0.0006} \\
3600.0906
\end{array}
$$
  may be expressed as 3600

In the case of addition, the sum of the units column is a significant figure if each number added was significant in the units column. The same is true for the tens, hundreds, thousands, and so on. Therefore, in the above addition, the correct sum has four significant figures and is given unambiguously as

$$3.600 \times 10^3$$

Table 2–2.   Scientific Notation

Number	Exponential Form	Meaning
1	10^0	1
10	10^1	10
100	10^2	10×10
1000	10^3	$10 \times 10 \times 10$
10000	10^4	$10 \times 10 \times 10 \times 10$
100,000	10^5	$10 \times 10 \times 10 \times 10 \times 10$
1,000,000	10^6	$10 \times 10 \times 10 \times 10 \times 10 \times 10$
0.1	10^{-1}	$1/10$
0.01	10^{-2}	$1/(10 \times 10)$
0.001	10^{-3}	$1/(10 \times 10 \times 10)$
0.0001	10^{-4}	$1/(10 \times 10 \times 10 \times 10)$

These topics are further expanded in Appendix A. The student who is not familiar with these ideas is urged to practice by working out the problems in Appendix A.

2–4. PHYSICAL CHANGES AND PHYSICAL PROPERTIES

Changes of state are common occurrences, especially with water. Heating converts water from its liquid state to its gaseous state. The term **vapor** is used when describing a gas that is normally a liquid under the conditions observed. Thus gaseous water present in a room at ordinary temperatures is called water vapor, whereas the oxygen in the same room is called oxygen gas.

Any change in the physical state of a substance is termed a **physical change.** Physical changes do not involve a change in the chemical composition of a substance but only in its physical state or condition. Examples are conversion from a liquid to a gas, from solid to liquid, and the grinding or smashing of a solid.

Melting and Freezing □ An important physical change is melting, or its reverse, freezing. Melting and freezing are fascinating phenomena. If water is cooled, the temperature can be observed to drop until a temperature of 0°C is reached. At this point, crystals of ice begin to form, but continued removal of heat from the water does not result in a temperature drop. Instead, as long as liquid water and ice are simultaneously present, the temperature remains constant at 0°C. Only after all the water has frozen does the temperature again begin to drop. This phenomenon is described by the curve in Figure 2–2. The same thing is observed if the process is reversed. Because this phenomenon is normally observed with most pure substances, the melting point (or freezing point, since these are the same) can be taken as a physical property that can be used to identify pure substances.

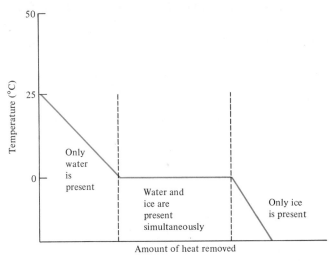

Figure 2–2. Curve representing the cooling (removal of heat) of a quantity of water initially at room temperature.

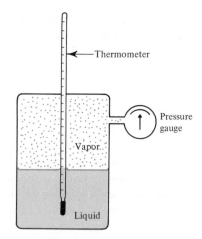

Figure 2-3. Apparatus for determining the vapor pressure of a pure liquid.

Vapor Pressure □ The boiling point of a pure substance is similarly a useful physical property for identification purposes. During the process of boiling (discussed later in this section) a constant temperature is maintained. For this reason, it takes just as long to boil an egg whether the water boils very vigorously or very slowly. In either case the same temperature is attained.

A further word must be added about the phenomenon of boiling. A liquid may evaporate—form its vapor—at any temperature below its boiling point. It is a common occurrence that a container of water can evaporate to dryness while at room temperature. Boiling may be distinguished from evaporation by the introduction of a new physical property—the **vapor pressure.**

If a pure liquid is placed in a container (Figure 2–3) from which all the air is removed, it can be shown that at a given temperature a fixed pressure can be read on the gauge. This pressure is called the vapor pressure, and is given in units of millimeters of mercury (mm Hg) where atmospheric pressure equals 760 mm Hg (Chapter 4). It can further be observed that the vapor pressure of a pure liquid recorded in this way is fixed for each temperature. Figure 2–4 shows the variation of vapor pressure with temperature for benzene. The vapor pressure results from the evaporation of liquid into the open space and the reverse process of condensation of vapor to reform liquid. In the apparatus of Figure 2–3, these two processes are rapidly balanced so that a definite vapor pressure exists for every liquid. Like the melting and boiling points, vapor pressure is therefore a useful physical property.

Boiling □ Let us return to the phenomenon of boiling. When the temperature of a liquid rises to the point at which the vapor pressure of the liquid just exceeds the pressure of the atmosphere, then evaporation is so rapid that it begins to occur not only at the liquid surface but also in the body of the liquid. At this point, bubbles of vapor form and rise to the surface where they escape into the air. When this situation occurs, the liquid is said to boil. We may also define boiling point in terms of vapor pressure: the **boiling point** is the temperature at which the vapor pressure of a liquid is equal to the external pressure. In the case of benzene shown in Figure 2–4, the boiling point is 80°C.

The data of Figure 2–4 also suggest that, if the atmospheric pressure changes, then the boiling point changes also. Thus, at a pressure of 700 mm Hg, the boiling

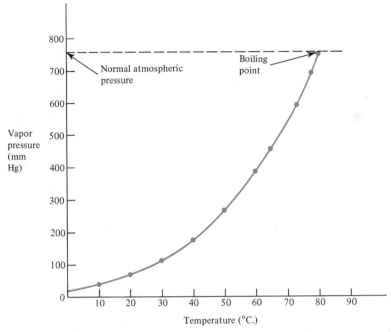

Figure 2-4. Curve showing the variation of vapor pressure of benzene versus its temperature.

point of benzene is 78°C. A practical example of this effect occurs when one cooks at high altitudes where the atmospheric pressure is so low that the boiling point of water is significantly reduced.

Density ☐ Another useful and common physical property is **density.** Density is the weight of a specified volume of substance. More precisely, the density of a substance is defined as the weight of a given quantity of the substance divided by its volume, or

$$\text{density} = \frac{\text{weight}}{\text{volume}}$$

The density may be expressed in any convenient weight and volume units. Most often, scientists use the units grams per milliliter (g/ml)

■ **Example 2-1**

What is the density of benzene if 15 ml weighs 13.2 g?

☐ *Solution*

$$\text{density} = \frac{\text{weight}}{\text{volume}} = \frac{13.2 \text{ g}}{15 \text{ ml}} = 0.80 \text{ g/ml}$$

Table 2-3. Physical Properties of Some Common Substances

Substance	Melting Point (°C)	Boiling Point,* °C	Vapor Press.† (mm Hg)	Density† (g/ml.)	Solubility†	
					In Water	In Benzene
sodium chloride	801	1413	0	2.165	soluble	insoluble
copper	1083	2300	0	8.94	insoluble	insoluble
sucrose (sugar)	186	(decomposes)	0	1.57	very soluble	insoluble
naphthalene (moth balls)	80	217.9	1	1.15	insoluble	soluble
aspirin	143	(decomposes)	0	—	insoluble	insoluble
cholesterol	149	360 (decomposes)	0	1.07	insoluble	soluble
urea	132	(decomposes)	0	1.34	soluble	insoluble
phenol	43	182	<1	1.07	soluble	soluble
isooctane (gasoline)	−107	99.3	50	0.69	insoluble	soluble
sulfur	113	445	1	2.0	insoluble	slightly soluble
ethyl alcohol	−114	78.5	50	0.79	soluble	soluble
water	0	100	23.8	1.0	soluble	insoluble
benzene	5.5	80	100	0.88	insoluble	soluble

*At ordinary atmospheric pressure
†At room temperature (25°C)

Solubility ☐ A different kind of physical property is **solubility.** The solubility of a substance is defined as the weight of the substance that will dissolve in a given quantity of a pure liquid at a given temperature. Usually the substance dissolved is called the **solute** and the liquid in which it dissolves is called the **solvent.** For our purposes, solubility may be given in the approximate terms—soluble or insoluble. A substance is said to be insoluble in a given solvent if a single small crystal, the size of a grain of sand, will not dissolve in 1 ml of the solvent. A soluble substance will normally be defined as one that dissolves to the extent that table salt does in water (about 35 g/100 ml of water at room temperature).

Solubility is an important physical property. It allows the separation of mixtures into pure components. Decisions can frequently be made about the nature of a solute from knowledge of the kinds of solvents in which it will dissolve.

Two convenient solvents for purposes of classifying substances are water and benzene. Benzene is a hydrocarbon (contains only the elements carbon and hydrogen). Water and benzene are mutually insoluble. They are convenient because many substances are soluble either in water or in benzene. Rarely is a substance soluble in both benzene and water.

Table 2-3 lists some common substances and their melting points, boiling points, vapor pressures, densities, and solubilities in water and benzene.

2-5. MIXTURES AND SOLUTIONS

Section 2-4 dealt with the physical properties of pure substances. But substances in nature rarely occur in pure form. We shall now examine the ways in which matter occurs in nature. We begin our classification with mixtures. A **mixture** may be defined as a sample of matter that contains more than one identifiable component. Most of the common substances found in nature are mixtures, such as sea water, earth, milk, and blood. It is not always easy, however, to distinguish a mixture from a pure substance on the basis of appearance alone. Sea water, for example, looks exactly like pure distilled water, although it contains a great number of dissolved substances. Its taste, of course, is quite different and allows the distinction to be made.

Mixtures are distinguished by the fact that they can be separated into their pure components by ordinary nonchemical means. A mixture of sand and water, for example, is a **heterogeneous** mixture, that is, contains two visually identifiable components—a solid and a liquid. Sand and water can be separated easily by filtration. A mixture of salt and water is a **homogeneous** mixture because the salt dissolves and a clear solution results. This mixture, therefore, cannot be as readily separated because a salt solution easily passes through a filter. Separation may be carried out in this case by taking advantage of the great difference between the boiling points of salt and water. Water boils at 100°C, whereas salt boils above 1400°C. Thus the separation is effected by heating the solution until all the water has boiled away and only salt remains. A solution of sugar and water presents still another problem because heating this homogeneous mixture to a temperature that will boil the water away causes a chemical reaction to occur—the sugar chars and forms caramel—that destroys the sugar. In this case other methods must be employed.

So far we have seen that there are two kinds of mixtures—homogeneous and heterogeneous. These two kinds may be distinguished in still another way: homogeneous mixtures have the same composition throughout, whereas heterogeneous mixtures vary in composition within the mixture itself. Homogeneous mixtures are usually called **solutions.**

A common feature of mixtures is that they have variable compositions. A salt solution may contain a very small amount of solute or a very large amount and still be considered a salt solution. The same holds for heterogeneous mixtures.

We can now summarize the properties of a mixture as follows: (a) A mixture contains more than one identifiable pure substance, (b) it can be separated into its pure components by ordinary physical (nonchemical) means, and (c) it may vary in the proportions of its components. Figure 2–5 illustrates the distinction between solutions and other mixtures.

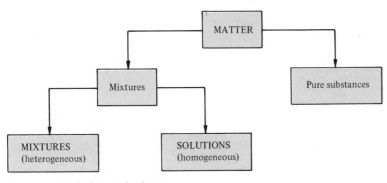

Figure 2–5. Solutions and mixtures.

2-6. COMPOUNDS AND ELEMENTS

Some substances are homogeneous but are not mixtures or solutions. These are pure substances that have definite and invariable compositions and fixed properties. Examples of such materials are salt, sugar, and water. Ordinary table salt—sodium chloride—contains exactly 39.3% sodium and 60.7% chlorine regardless of its source or method of manufacture. Sugar—sucrose—contains 42.2% carbon, 6.4% hydrogen, and 51.4% oxygen. Still another example of such a substance is water which consists of 11.1% hydrogen and 88.9% oxygen. Such substances differ from mixtures in a new and important way: the properties of sodium chloride are entirely different from those of sodium metal, which is very reactive, and chlorine gas, which is very poisonous. Similarly, sugar does not resemble its constituents—carbon, hydrogen, and oxygen—nor does water resemble its constituents—hydrogen and oxygen. Substances that are not mixtures but can be decomposed into simpler substances by chemical means and have definite compositions are called **compounds.**

Still another type of pure substance exists that is not a mixture. Such substances are also pure but cannot be decomposed into simpler substances by any ordinary means either physical or chemical, and are called **elements.**

Figure 2-6 is a diagram that summarizes the above categories of substances.

2-7. CHEMICAL CHANGE

Since man's earliest history he has been intrigued by the changing world about him. One of the most obvious things in the world, and in fact the universe, is change. All things change: the seasons, the positions of the stars in the sky, plants, and animals. Organisms are born, grow to maturity, age, and then die. Rivers and lakes freeze in the winter and thaw in the spring. The leaves change color. Dead organisms decay. These and many other changes are around us every day. Chemistry deals with many of these changes.

But which changes can we call **chemical changes** and which can we not call chemical changes? The melting of ice is certainly not a chemical change nor is

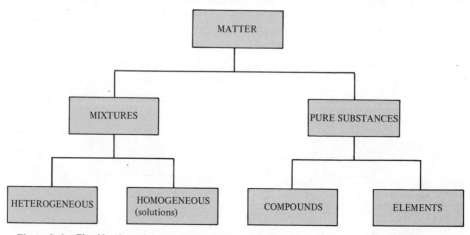

Figure 2-6. Classification of matter into mixtures, solutions, compounds, and elements.

the boiling of water. Decaying of flesh and the change in color of the leaves must certainly be chemical changes. The reason for this statement is that in a chemical change the actual chemical constitution of the substance is different after the change has occurred. For example, when ice melts, liquid water is formed. Liquid water, however, has the same chemical composition as ice. In fact, only its physical state has changed. Similarly when water boils the vapor that is produced has the same chemical composition as the liquid—the same percentage hydrogen and percentage oxygen. When wood burns, on the other hand, the composition changes. Wood, which is a complex mixture of many kinds of substances including cellulose and various oils, becomes transformed upon combustion into carbon dioxide and water. Carbon dioxide and water are certainly different substances from those that were present in the wood. Hence we say that the burning of wood is a chemical change.

To summarize, a chemical change occurs only when the original substance ceases to exist and in its place new substances are produced with entirely new sets of physical and chemical properties.

2–8. CONSERVATION LAWS

In 1775, Antoine Lavoisier proclaimed the law of the conservation of matter which states that in any chemical change matter can neither be created nor destroyed. This law today is deeply ingrained into the thinking of all scientists. It is difficult to imagine a time when the law of the conservation of matter was not accepted. It was probably the failure to recognize this law, however, that held back the progress of chemistry for so many centuries.

Lavoisier's statement was made in support of the idea that combustion involves the combination of the combustible substance with oxygen from the air. Thus, although wood burns to form only a small amount of ash (and therefore undergoes an apparent weight loss) if one could capture the gaseous products that are produced, one would find that all of the apparent weight loss can be accounted for. In terms of an equation

$$\text{wood} + \text{oxygen} \longrightarrow \text{carbon dioxide} + \text{water} + \ldots$$

The weights of wood and oxygen consumed are exactly equal to those of carbon dioxide plus water plus the other products formed.

The subject of chemical equations will be taken up in Chapter 3. It should only be noted here that the law of the conservation of matter allows the chemist to account *quantitatively* for all the substances involved in a chemical reaction. For example, in the above process, if 10 g of wood is consumed in burning and 25 g of total products is formed—carbon dioxide, water, and so on—then one knows immediately that 15 g of some other material must have been consumed also. The other substance in this case is oxygen.

Another conservation law was established in the nineteenth century—the law of the **conservation of energy.** This law states that energy can neither be created nor destroyed but can only be changed from one form to another. Here again the conservation law is convenient because it allows an accurate accounting of energy changes that occur during physical and chemical changes.

It may be noted that the advent of atomic energy has necessitated a change in these laws. They have been replaced by a combination conservation law—the law of the **conservation of mass-energy:** the sum of mass and energy cannot change. This will be taken up in our discussion of nuclear energy. In all *chemical* reactions, however, the old laws of conservation of matter and conservation of energy hold exactly.

2-9. ENERGY

Energy is a term familiar to everyone. Certainly the expressions *energetic* and *full of energy* describe people who are always "on the go," ever ready to do more work. Actually, our intuitive feelings about energy carry a fairly accurate description of its scientific meaning which is generally stated to be: the ability or capacity to perform work. Work, in its scientific meaning, is the mathematical product of force and the distance through which that force acts.

$$\text{work} = \text{force} \times \text{distance}$$

Various forms of energy are recognized, but three forms in particular will be of interest in the present chapter—potential energy, kinetic energy, and heat energy.

(1) **Potential energy** (PE) is energy possessed by an object by virtue of its *position* or *configuration* relative to some reference point. For example, an object of mass m, poised above the surface of the earth at a height, h, possesses a potential energy given by

$$PE = mgh \qquad (2\text{-}1)$$

where g is the acceleration due to gravity. According to Newton's second law of motion, the force required to accelerate an object of a given mass is equal to the mathematical product of the mass, m, and the acceleration, a: $F = ma$. The object in our original example is said to possess **potential energy** because of its ability to perform work if released and allowed to fall to earth. The work which could be done would simply be

$$\text{work} = \text{force} \times \text{distance}$$

but force $= mg$ and distance $= h$ so

$$\text{work} = mgh$$

which is simply the potential energy of the object. However, if the object were released and allowed to fall, at the instant of striking the earth, the object would have lost all of its potential energy, since $h = 0$ at the surface of the earth. In fact, as the object falls, its potential energy decreases (as h decreases), yet we say it has the ability to expend energy at the moment of striking the earth by an amount equal to the original potential energy of the object. If the potential energy is lost as the object falls, where does it go? The answer is that, as an object falls, its potential energy is converted into kinetic energy.

(2) **Kinetic energy** (KE) is energy possessed by an object by virtue of its *motion* relative to some reference point. To calculate the kinetic energy we can return to our example above. At any point during the fall of the object, the kinetic energy of the object will be equal to the work done by gravity on the object

$$\text{work} = \text{force} \times \text{distance}$$

and

$$\text{force} = ma$$

but the acceleration of the object, $a = $ velocity at a given time/time of fall, so

$$a = v/t$$

where v is the velocity at time t; therefore,

$$\text{force} = mv/t$$

Also,

$$\text{distance} = \text{average velocity} \times \text{time}$$

so

$$\text{distance} = vt/2$$

where v is the velocity at any time. (The factor 2 in the denominator is necessary to convert the velocity at any time into average velocity because the object started from rest and acceleration is constant.) Thus, finally

$$\text{work} = mv/t \times vt/2$$
$$= \tfrac{1}{2}mv^2$$

or

$$\text{KE} = \tfrac{1}{2}mv^2 \qquad\qquad (2\text{-}2)$$

The above exercise should not be considered a rigorous derivation of the kinetic energy equation because the assumed conditions of an object starting from rest and undergoing constant acceleration are not at all necessary. Kinetic energy is always calculated using equation (2-2) regardless of how the object is put in motion. This little exercise, however, does give some insight into how the original potential energy of the object is converted into kinetic energy as the object falls and, at the point of impact, all of the potential energy would have been converted into kinetic energy. As the object strikes the earth, it would perform an amount of work equal to the kinetic energy and, once again, a conversion of energy from one form to another would have occurred. Perhaps our object may strike a nail in a board and do work on the nail in driving it into the wood (Figure 2-7).

Finally, however, we must consider the ultimate fate of the work done on the nail. Where does this energy go? The answer is that the work done on the nail is initially stored in the nail as random molecular energy which is eventually transferred to the surroundings as heat energy.

(3) **Heat energy** is that form of energy that is transferred between two objects by virtue of a temperature difference alone. The conversion of the work done

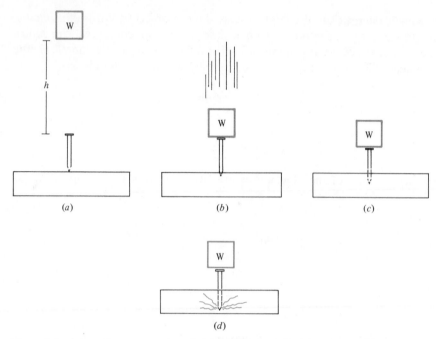

Figure 2-7. Schematic representation of the driving of a nail with a weight (W). (a) Energy of the weight is in the form of potential energy. (b) At the point of striking the nail, the potential energy has been converted into kinetic energy. (c) The kinetic energy is converted into work done on the nail increasing the random molecular energy of the nail and raising its temperature. (d) Random molecular energy of the nail is dissipated as heat into the surroundings.

on the nail into random molecular energy would be reflected by a rise in the temperature of the nail. Anyone who has ever driven a nail and touched the head of the nail can attest to the temperature rise produced by doing work on the nail. Cooling occurs as heat energy is transferred from the nail to the cooler surroundings. The temperature of the surroundings is thereby increased as heat is absorbed and stored as random molecular energy in the surroundings. A given amount of heat absorbed by a substance will produce a temperature rise in the substance that depends on its mass and the identity of the substance.

$$\text{heat} = ms\Delta t \tag{2-3}$$

where m is the mass, s is the specific heat, and Δt is the change in temperature. Specific heat must, in general, be measured for a particular material although its calculation from theory is often possible. The **specific heat** of a substance is the amount of energy required to increase the temperature of 1 g of the material by 1 °C. The most common unit for s is calories/gram-degree (cal/g °C), although heat, as a form of energy, may be expressed in any energy unit. A commonly used unit of energy when dealing with heat is the calorie (cal); 1 **cal** is defined as the amount of energy necessary to raise the temperature of 1 g of water from 14.5 °C to 15.5 °C.* Table 2-4 lists some common substances and their specific heats.

*A larger energy unit commonly used in chemistry is the kilocalorie (kcal), which is equal to 1000 cal.

Table 2-4. Specific Heat
of Some Substances

Substance	s (cal/g °C)
acetic acid	0.47
aluminum	0.22
benzene	0.41
copper	0.092
gold	0.031
silver	0.056
water (liquid)	1.0
water (ice)	0.49

Finally, before leaving the subject of energy, one very important special form of potential energy must be mentioned. Fuels like gasoline contain a form of potential energy called **chemical energy.** When chemical energy is released, by burning in this case, kinetic energy is produced in the form of heat and light; if the release is rapid enough, an explosion ensues, which is a form of mechanical energy. Chemical energy is potential energy because it is stored in the substance and is not experienced until it is released. Chemical energy is a very important subject in the study of chemistry and we shall have many opportunities to refer to it in the chapters that follow.

PROBLEMS

1. Define in clear concise terms: solid, liquid, and gas.
2. What is the difference between a gas and a vapor?
3. What distinction can be made between the terms mixture and solution?
4. How do compounds and mixtures differ in **(a)** their composition, **(b)** their properties?
5. Distinguish between the terms compound and element.
6. Distinguish between the terms physical change and chemical change.
7. Classify the following changes as physical or chemical.
    (a) the effect of a cold front in causing rain
    (b) a forest fire
    (c) the healing of a wound
    (d) respiration
    (e) digestion
    (f) rusting of iron
    (g) fading of colors in time
    (h) fermentation of grape juice to produce wine
    (i) distillation of wine to produce brandy
    (j) the effects of fertilizers on plants
    (k) the drying of clothes when hung in the sun
    (l) boiling an egg
    (m) cutting a piece of wood

21

8. Referring to Figure 2–1, suggest why the removal of heat energy from a mixture of ice and water at 0°C does not result in a temperature drop.

9. Following are vapor pressure data on water at various temperatures. Plot the data and answer the following questions.

Temperature (°C)	Vapor Pressure (mm Hg)
0	4.6
10	9.2
20	17.5
30	31.8
40	55.3
50	92.5
60	149
70	234
80	355
90	526
100	760

(a) At what temperature is the vapor pressure of water equal to 400 mm Hg?
(b) At what temperature does water boil at an elevation of 14000 ft above sea level (1 mile = 5280 ft) using the following data:

Elevation Above Sea Level (miles)	Pressure (mm Hg)
0.0	760
0.6	675
1.2	598
1.6	563
2.5	467
3.1	411

10. Given that atmospheric pressure is uniformly 14.7 lb/in.^{2}, calculate the total atmospheric force on the entire surface of the earth. Assume that the earth is a sphere of diameter = 7000 miles.

11. Assuming that the atmosphere extends to an altitude of 50 miles above the surface of the earth, calculate the average density of the atmosphere.

12. A certain substance has a density of 2.4 g/ml. What is the weight of 25 ml of this substance?

13. How much volume would 10 g of the substance in problem 12 occupy?

14. A rectangular solid that measures 1.5 in. by 2.0 in. by 3.0 in. weighs 250 g. What is its density? (Note that 1 in. = 2.54 cm.)

15. Suggest how the following pairs of substances could be separated. (Refer to Table 2–3.)

(a) benzene and water   **(d)** cholesterol and aspirin
(b) copper powder and sodium chloride   **(e)** cholesterol and isooctane
(c) urea and phenol

16. Of what utility are conservation laws?
17. Distinguish between the terms energy and work.
18. Distinguish between the terms potential energy and kinetic energy.
19. What two forms of potential energy does a can of gasoline placed on a high shelf possess?
20. What is the kinetic energy of a $\frac{1}{4}$-lb ball as it leaves a pitcher's hand if its velocity is 100 miles/h as it leaves.
21. Describe the energy changes that occur when a pendulum is allowed to swing freely.
22. How many calories of heat energy are required to raise the temperature of (a) 100 g of gold by 75°C? (b) 100 g of acetic acid by 75°C? (c) 100 g of aluminum by 75°C? (d) Disregarding cost and snob appeal, would you suggest gold or aluminum as the construction material for a coffee cup? Explain.
23. Convert 50°C to degrees Fahrenheit.
24. Write an equation to show how one could convert temperature in degrees Fahrenheit to degrees Celsius.
25. Perform the following conversions using the appropriate equations.
    (a) 25°C to °F   (c) 100°C to °F
    (b) 100°F to °C   (d) 98.6°F to °C
26. At what temperature are the values of degrees Fahrenheit and degrees Celsius equal?

PURE SUBSTANCES—ELEMENTS AND COMPOUNDS

3-1. INTRODUCTION

Two major problems that had to be solved before chemistry could flourish as a science were those of **purity** and **identity.** Even at the present time, work on the nature or properties of any new substance must wait until it is established to be a single material and that it is of high purity.

In the previous chapter a compound was defined as a pure substance that (a) can, under the proper conditions, be decomposed into simpler substances, (b) has its own set of properties that are not merely the summation of the properties of its constituent elements, and (c) has a definite composition. Elements were defined in the same way except that they cannot be decomposed into simpler substances.

It is the objective of this chapter to explore the meaning and implications of these definitions and to see how these properties lead to the idea of atoms and molecules.

3-2. DECOMPOSITION OF COMPOUNDS

One of the most straightforward examples of the decomposition of a compound into its elements is the electrolysis of water. This is accomplished by allowing a direct electric current to pass through water in the manner shown in Figure 3-1. During the electrolysis, gaseous oxygen collects in the tube over the positive electrode and gaseous hydrogen collects in the tube over the negative electrode. The two gases are formed in the ratio 2 volumes of hydrogen to 1 volume of oxygen. This volume relationship will be discussed in a later chapter.

The first time that this process was carried out, the experimenter demonstrated with it that water is not an element because it can be dissociated into two simpler

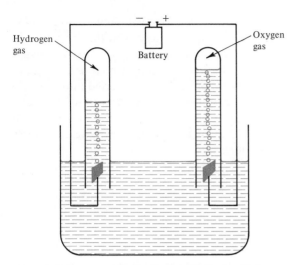

Figure 3-1. Apparatus for the electrolysis of water. Hydrogen gas collects in the tube over the negative electrode and oxygen gas collects in the tube over the positive electrode. At the beginning of the process, both tubes are filled with water.

substances. The experimenter was Henry Cavendish and the time was the late eighteenth century. Another result of this reaction is that, regardless of the source of the water, the weights of hydrogen and oxygen produced from a given amount of water are always the same.

■ **Example 3-1**

If 18 g of water is electrolyzed, 2.0 g of hydrogen and 16 g of oxygen are formed. How many grams of each would form if 50 g of water were electrolyzed?

□ *Solution*

$$\% \text{ hydrogen in water} = \tfrac{2.0}{18} \times 100 = 11.1\%$$

$$11.1\% \text{ of } 50 \text{ g} = (0.111)\,(50) = 5.6 \text{ g of hydrogen}$$

$$50 \text{ g} - 5.6 \text{ g} = 44 \text{ g oxygen}$$

The student should recall that the conversion of water into hydrogen and oxygen is a chemical change because the properties of water are totally different from those of hydrogen or oxygen.

3-3. PERCENTAGE COMPOSITION

Usually it is not so simple to decompose compounds directly into their elements. Instead, compounds are usually decomposed into other simpler compounds. For example, all compounds of the element carbon may be burned to form carbon dioxide if an excess of oxygen is provided. Similarly, if the compound also contains hydrogen, then burning in a large amount of oxygen results in the formation of water. These compounds represent the ultimate combustion products of carbon and hydrogen, respectively. Even though hydrocarbons (compounds containing only the elements carbon and hydrogen) cannot easily be converted into the pure elements, their compositions may be determined by combustion to form carbon dioxide and water. Because the percentage of carbon in carbon dioxide and the percentage of hydrogen in water are well known, the percentage of these elements in the original hydrocarbon can be determined.

■ **Example 3-2**

The complete combustion of 10.0 g of the hydrocarbon isooctane yields 30.9 g of carbon dioxide and 14.2 g of water. Given that carbon dioxide contains 27.3% carbon and 72.7% oxygen, and the previous percentages for water, calculate the percentages of carbon and hydrogen in the original hydrocarbon.

□ *Solution*

$$(30.9 \text{ g carbon dioxide}) \times 0.273 = 8.44 \text{ g carbon in } 30.9 \text{ g carbon dioxide}$$

$$(14.2 \text{ g water}) \times 0.111 = 1.58 \text{ g hydrogen in } 14.2 \text{ g water}$$

$$\% \text{ carbon in isooctane} = \frac{8.44}{10.0} \times 100 = 84.4\%$$

$$\% \text{ hydrogen in isooctane} = \frac{1.58}{10.0} \times 100 = 15.8\%$$

$$\text{Total} = 100\%$$

■ **Example 3-3**

In the previous example, the composition of carbon dioxide was given. Assuming that carbon dioxide cannot be decomposed directly into its elements, suggest how one might determine the composition experimentally.

□ *Solution*

By simply burning a known quantity of pure carbon in an excess amount of oxygen, the weight of gaseous carbon dioxide produced affords the percentage composition as the following typical experiment shows:

Weight of carbon burned	10.0 g
Weight of carbon dioxide produced	36.6 g

Since all of the carbon was converted into carbon dioxide

$$\% \text{ carbon} = \frac{10.0}{36.6} \times 100 = 27.3\%$$

By difference

$$\% \text{ oxygen} = 100.0 - 27.3 = 72.7\%$$

3-4. LAWS OF CHEMICAL COMPOSITION

Before beginning our discussion of atomic theory—the central theme that underlies all of modern chemistry—it is worthwhile to sketch briefly the kinds of evidence that led to the concept. Ideas as important as the atom do not germinate in a vacuum but are nurtured by need.

The most far-reaching idea regarding chemical composition is the law of conservation of matter, enunciated by Lavoisier in the latter quarter of the eighteenth century. He stated that, in all chemical transformations, **matter is neither created nor destroyed;** that is, the same amount of matter is present after the chemical transformation as was present before (Section 2–8).

At about the same time, another law was formulated—the **law of definite proportions.** According to this law a compound always contains its constituent elements in exactly the same proportions by weight regardless of its history or source. Thus water, whether obtained from a well or by the combustion of hydrogen, can always be decomposed to yield exactly 11.1% hydrogen and 88.9% oxygen. The essential point of this generalization is that there is no variation whatever in the composition of a particular compound.

The significance of this law should not be glossed over. Most of our everyday experiences lead us to believe otherwise, so the law of definite proportion is actually contrary to preconceived notions. Contrary examples from common experiences are cooking and concrete mixing. In either case, the ingredients are usually exactly specified; however, a little more or less sand in concrete or milk in a cake will not ruin the final product. Or, more to the point, these excess quantities will be incorporated and not be found left over. In the case of a chemical combination, on the other hand, if a mixture of hydrogen and oxygen containing more than 11.1% hydrogen is allowed to react, the excess hydrogen will be found unreacted along with the water that was formed.

3-5. DALTON'S ATOMIC THEORY

By 1800, these startling observations including the laws of conservation of matter and definite proportions had led to serious confusion about the nature of matter. The solution was proposed by the English chemist, John Dalton, in 1805. According to him these laws could be explained readily by assuming that elements consist of atoms that are indestructible and eternal. All atoms of a given element are the same and differ in weight from atoms of all other elements. The law of conservation of matter was thus rationalized by the fact that the atoms of which all matter is made are indestructible. The law of definite proportions is rationalized by the fact that molecules of a compound are formed by the combination of a definite fixed number of each kind of atom. Because all atoms of a given element have the same weight, there could be no variation in the percentage composition of a compound.

Dalton's atomic theory, being a good scientific theory, allowed certain predictions to be made that, if verified, could add weight to the theory. If the predictions were not verified, the theory could thus be shown to be false.

One of the results of his theory was the establishment of a scale of atomic weights. If atoms of the different elements differ in their weights, then the weights of equal numbers of atoms would be a measure of the relative weights of single atoms.

An example of this principle is the substance hydrogen chloride. Hydrogen chloride contains 2.74% hydrogen and 97.26% chlorine. Thus, assuming that the molecule of hydrogen chloride consists of one atom of each element, it follows that the relative atomic weights of hydrogen atoms to chlorine atoms are in the ratio 2.74 to 97.26 or 1 to 35.5. Since hydrogen is the lightest element, its relative atomic weight was assigned a value of 1. The molecular weight of the hydrogen chloride molecule is then simply the sum of the atomic weights of its atoms, namely, 36.5.

Dalton's table of atomic weights was thus the first fruit of his new theory. Although there were inaccuracies in his values, they paved the way for much more research that eventually led to the presently accepted values. The atomic weights of some common elements are shown in Table 3–1 below.

3-6. CHEMICAL SYMBOLS

By now it has probably become obvious that dealing with chemical names is cumbersome. In order to overcome this problem, chemists developed a system

Table 3-1. Some Common Elements and Their Symbols

Element	Symbol	Atomic Weight	Element	Symbol	Atomic Weight
hydrogen	H	1.008	chlorine	Cl	35.5
helium	He	4.00	argon	Ar	39.9
lithium	Li	6.94	potassium	K	39.1
beryllium	Be	9.01	calcium	Ca	40.1
boron	B	10.8	iron	Fe	55.8
carbon	C	12.01	nickel	Ni	58.7
nitrogen	N	14.01	copper	Cu	63.5
oxygen	O	16.00	arsenic	As	74.9
fluorine	F	19.0	bromine	Br	79.9
neon	Ne	20.2	silver	Ag	107.9
sodium	Na	23.0	tin	Sn	118.7
magnesium	Mg	24.3	iodine	I	126.9
aluminum	Al	27.0	gold	Au	197.0
silicon	Si	28.1	mercury	Hg	200.6
phosphorus	P	31.0	lead	Pb	207.2
sulfur	S	32.1			

of designating the chemical elements by a shorthand notation that uses the first letter or two of the name of the element. Symbols of the elements that have already been mentioned are C for carbon, O for oxygen, H for hydrogen, and Cl for chlorine. These symbols represent more than simply the identity of the element; they represent also a fixed amount of the element. In general, the symbol can represent one atom of the element. Later we shall see that the symbol more often represents a fixed number of atoms. Since each element consists of atoms, all of which are the same, then the chemical symbols represent an exact weight of the element. Here then is another meaning of chemical symbols—**the chemical symbol represents an exact weight of the element.** Table 3–1 lists some of the more common elements with their symbols and atomic weights. The student is urged to memorize these symbols as early as possible but not the atomic weights.

3-7. CHEMICAL FORMULAS AND EQUATIONS

Chemical formulas employ the symbols described in Section 3-6 to provide a shorthand version of the composition of chemical compounds. These symbols should become part of your working vocabulary, and they can be easily remembered after repeated use during your studies. Numerical subscripts following elemental symbols in a formula represent the ratio of atoms of those elements that are present in the compound. Thus the formula H_2O states that water consists of hydrogen and oxygen atoms in the ratio of 2 to 1. Similarly, the formula for sucrose (table sugar) is $C_{12}H_{22}O_{11}$ which means that the elemental ratio is 12 carbon atoms to 22 hydrogen atoms to 11 oxygen atoms.

A chemical equation is also a form of shorthand that states, in terms of formulas, the facts of a chemical reaction. Thus the equation

$$H_2O \longrightarrow H_2 + O_2$$

states that water forms hydrogen and oxygen. The arrow shows the direction in which reaction occurs, and is read "yields" or "reacts to form." The above is not a correct equation, however, since the quantities on both sides of the arrow are not equal, or **balanced.** To make it a true equation, each element has to occur on both sides in exactly the same amount. Thus, in the above unbalanced equation, there are two oxygen atoms on the right of the arrow and only one on the left. The situation is corrected by the use of coefficients.

$$2\,H_2O \longrightarrow 2\,H_2 + O_2$$

It should be noted that in balancing an equation, only coefficients can be added since these indicate the number of formula units of the substance. In no case can one change the subscripts for to do so would be to change the elemental composition and therefore the identity of the substance.

Many substances, for example, water, exist in discrete units called **molecules.** In the case of water, the molecule is a combination of one oxygen atom with two hydrogen atoms, and the formula H_2O represents one molecule of water just as H represents one atom of hydrogen. Other substances, however, do not exist as discrete molecular units, and instead are composed of vast networks of atoms.

■ **Example 3-4**

Balance the equation $Fe + Cl_2 \longrightarrow FeCl_3$.

☐ *Solution*

Because there are two chlorine atoms on the left side and three on the right side, they can be balanced at six on each side as follows

$$Fe + 3\,Cl_2 \longrightarrow 2\,FeCl_3$$

Multiplying the iron by 2 on the left side now balances the equation.

$$2\,Fe + 3\,Cl_2 \longrightarrow 2\,FeCl_3$$

Sodium chloride is an example of such a nonmolecular substance. The formula, NaCl, does not represent an individual *unit* of sodium chloride because there are no such discrete units. Instead, in cases like this, the formula represents simply the atomic ratio. Another example is calcium chloride, where the formula, $CaCl_2$, tells us that this compound consists of calcium atoms and chlorine atoms in the ratio of 1 to 2.

The distinction between molecular and nonmolecular substances will be discussed in Chapter 12.

3-8. AVOGADRO'S NUMBER—MOLES

From the ideas of atoms and relative atomic weights it is just a short step to an idea that is central to the understanding of chemistry. The chemist describes chemical reactions in terms of formulas and equations that represent how atoms and molecules react. For example, the decomposition of water to form hydrogen and oxygen is given by the equation

$$2\,H_2O \longrightarrow 2\,H_2 + O_2$$

The immediate point is simply that one obviously cannot deal with one or two atoms at a time. The equation is thus an abstraction of events at the atomic level.

Actually, very large numbers of atoms are involved in even the smallest perceivable samples of matter. For convenience, then, the following conventions have been adopted: A **mole** of a substance is an amount of the substance equal to its formula weight expressed in a suitable weight unit. The formula weight is obtained by taking the sum of the atomic weights in the formula of a substance. Thus the formula weight of H_2O is $(2 \times 1) + (1 \times 16) = 18$, because the atomic weights of hydrogen and oxygen are 1 and 16, respectively. The formula weight of O_2 is 32, and thus 32 g of O_2 constitutes 1 gram-mole (g-mole) of O_2. However, one can also speak of a gram mole of O atoms, which is 16 g of oxygen.

Table 3–2. Weights of Various Moles

Substance	Formula	Relative Formula Weight	Gram-Mole	Pound-Mole
water	H_2O	18	18 g	18 lb.
hydrogen	H_2	2	2 g	2 lb.
chlorine	Cl_2	71	71 g	71 lb.
sodium chloride	NaCl	58.5	58.5 g	58.5 lb.

Thus, a **gram-mole** is the quantity equal to the formula weight in gram units, the **ton-mole** is the quantity equal to the formula weight in ton units. This idea is expanded in Table 3–2, which gives the weights of various mole quantities for various substances. To avoid confusion, the student should recognize that the definition of the mole is purely arbitrary and depends on the value of the weight unit employed. However, in most applications the gram-mole is used, and it will be the one used throughout this book.

The importance of the mole, of course, lies in the fact that 1 mole of any substance contains the same number of formula units as 1 mole of any other substance. For this reason, one can speak of the decomposition of 2 moles of water to form 2 moles of hydrogen and 1 mole of oxygen, and now one is speaking of measurable amounts of materials in the correct molecular ratios in which they react chemically.

Since, as stated above, 1 g-mole of any substance contains the same number of formula units as 1 g-mole of any other substance, the next question to be asked is how many formula units there are in 1 mole of a substance. At this time we shall not delve into the experimental method of determining this number, but will simply state it to be 6.02×10^{23} formula units per gram-mole. This is often called **Avogadro's number** after the Italian chemist who contributed greatly to our understanding of the concept of the molecule (see Chapter 4). Obviously, Avogadro's number applies only to *gram*-moles. Other mole units will contain different numbers of formula units.

■ Example 3–5

How many molecules are there in 1 g of water?

□ *Solution*

Formula weight = 18, therefore 18 grams contains 6.02×10^{23} molecules. 1 g contains $\frac{1}{18}$ of Avogadro's number of molecules, which equals

$$\frac{6.02 \times 10^{23} \text{ molecules/g-mole}}{18 \text{ g/g-mole}} = 3.3 \times 10^{22} \text{ molecules/g}$$

Exponential numbers such as Avogadro's number are much too large to be visualized. In order to give the student some appreciation of its magnitude,

6.02×10^{23} rice grains would fill enough railroad boxcars to form a line to the sun and back 1000 times! For this reason, any perceivable sample of matter is truly statistically valid.

3-9.   CALCULATION OF PERCENT COMPOSITION—SIMPLEST FORMULA

A direct consequence of Dalton's atomic theory is the ease of calculation of the percentage composition of a compound from its formula. Taking the simple compound water, for example, the formula weight, obtained from the sum of the weights of the atoms, is 18.

$$2 \, H = 2 \times 1.0 = 2.0$$

$$1 \, O = 1 \times 16.0 = \underline{16.0}$$
$$\text{Formula weight (H}_2\text{O)} = 18.0$$

The percentage composition can now easily be obtained from this formula weight and the atomic weights of hydrogen and oxygen.

$$\% \text{ O in H}_2\text{O} = \frac{16.0}{18.0} \times 100 = 88.9\%$$

$$\% \text{ H in H}_2\text{O} = \frac{2.0}{18.0} \times 100 = 11.1\%$$

■ **Example 3-6**

Calculate the percentage composition of sulfuric acid (H_2SO_4).

□ *Solution*

The formula weight obtained from the sum of the atomic weights of all the atoms presents is

$$2 \, H = 2 \times 1.0 = 2.0$$
$$1 \, S = 1 \times 32.1 = 32.1$$
$$4 \, O = 4 \times 16.0 = \underline{64.0}$$
$$\text{H}_2\text{SO}_4 \text{ formula weight} = 98.1$$

The percentage of each element present is then obtained

$$\% \text{ H} = \frac{2.0}{98.1} \times 100 = 2.0\%$$

$$\% \text{ S} = \frac{32.1}{98.1} \times 100 = 32.7\%$$

$$\% \ O = \frac{64.0}{98.1} \times 100 = 65.3\%$$

$$\text{Total} \quad \overline{100.0\%}$$

Chemists usually obtain analyses of actual chemical substances in terms of percentage composition (Sections 3–2 and 3–3). Working from these percentage compositions, it is possible to arrive at a formula which is called the **simplest formula** or **empirical formula.** The simplest formula is a useful datum because it affords the precise ratio of the atoms of each element present in a compound. For example, the simplest formula of the sugar, glucose, is CH_2O. The actual molecular formula of glucose (a molecular substance) is $C_6H_{12}O_6$, which means that the simplest formula must be multiplied by six to obtain the molecular formula. The uses of simplest formulas will be described below.

Simplest formulas are easily obtained from percentage compositions by the following procedure. Given that the unknown compound has the following percentage composition:

40.0% carbon
6.7% hydrogen
53.3% oxygen

In a 100.0-g sample of the substance, the above percentages become weights of the elements present. Dividing each weight by the atomic weight of that element gives the number of gram-moles of that element.

carbon: 40.0 g ÷ 12.0 g/mole = 3.33 C g-moles
hydrogen: 6.7 g ÷ 1.0 g/mole = 6.7 H g-moles
oxygen: 53.3 g ÷ 16.0 g/mole = 3.33 O g-moles

If it is not obvious that the atoms are present in the ratio, C to H to O = 1 to 2 to 1, the student can simply divide each of the numbers by the smallest— 3.33—to obtain the smallest ratio.

Note that any formula in which the ratio of C to H to O is 1 to 2 to 1 will have the same percentage composition; namely, $C_2H_4O_2$, $C_3H_6O_3$, $C_6H_{12}O_6$, and so on.

The correct molecular formula can be obtained only if the molecular weight of the compound is known. If, in the above example, the molecular weight is known to be 180, then the molecular formula is found to be six times the simplest formula because the latter adds up to a **simplest formula weight** of 30. The molecular formula is thus related to the simplest formula in the way shown

$$\text{simplest formula} = CH_2O$$

$$\text{molecular formula} = C_nH_{2n}O_n$$

where $$n = \frac{\text{molecular weight}}{\text{simplest formula weight}}$$

33

CH. 3 PURE SUBSTANCES—ELEMENTS AND COMPOUNDS

3-10. NAMING OF COMPOUNDS

Certain compounds have individual names that give no clue to their chemical identity. Water (H_2O) and ammonia (NH_3) are two common examples. However, most compounds are named according to definite rules which relate the name to the chemical composition of the substance. Many simple compounds consist of a metallic atom or group of atoms and a nonmetallic atom or group of atoms. The metallic part is named first followed by the nonmetallic part. The subscripts employed in writing the formulas of compounds depend on the relative combining capacity of the atoms involved. The origin and nature of the variation of combining capacities of the elements will unfold in the following chapters. For the present, the student should simply become familiar with the names of the more common metallic and nonmetallic atoms and groups so that the correct name can be given to a formula. Some of the metallic and nonmetallic elements have been listed in Table 3–1. Table 3–3 lists some of the most common elements along with groups of atoms that react as a unit and thus have special names. All of the species in Table 3–3 have a fixed combining capacity and, hence, there is but one possible compound for each metal-nonmetal combination. The student should memorize Table 3–3.

The combination of a metallic species and a nonmetallic species can be named directly by combining the names given in Table 3–3. For example, note the similarity of names in the following groups of compounds.

NaOH	sodium hydroxide
$Mg(OH)_2$	magnesium hydroxide
$Al(OH)_3$	aluminum hydroxide
Ag_2O	silver oxide
CaO	calcium oxide
Al_2O_3	aluminum oxide

Table 3–3. Names of Some Important Atoms and Groups of Atoms

Metallic Species			
Al	aluminum	Li	lithium
NH_4	ammonium	Mg	magnesium
Ba	barium	K	potassium
Be	beryllium	Ag	silver
Ca	calcium	Na	sodium

Nonmetallic Species			
Br	bromide (bromine)	OH	hydroxide
CO_3	carbonate	I	iodide (iodine)
Cl	chloride (chlorine)	NO_3	nitrate
ClO_3	chlorate	O	oxide (oxygen)
CrO_4	chromate	PO_4	phosphate
F	fluoride (fluorine)	S	sulfide (sulfur)

The student should practice naming the combinations of the metallic and nonmetallic species listed in Table 3–3. The situation is more complex when more than one compound can form between a metal and nonmetal, for example, $FeCl_2$ and $FeCl_3$. Methods of differentiating between these two chlorides of iron and similar compounds are discussed in Chapter 21. Likewise, most nonmetallic elements form more than one compound with each other, for example, CO (carbon monoxide) and CO_2 (carbon dioxide), SO_2 (sulfur dioxide) and SO_3 (sulfur trioxide). Systematic rules for naming similar as well as more complex compounds are also discussed in Chapter 21. Appendix B gives rules for the systematic naming of organic compounds and complicated inorganic species called coordination compounds.

PROBLEMS

1. Calculate the percentage composition of each of the compounds below from the data given.

Substance	Analysis
sodium chloride (NaCl)	46 g Na; 71 g Cl
iron oxide (Fe_2O_3)	55.85 g Fe; 24 g O
hydrogen peroxide (H_2O_2)	4 g H; 64 g O
sucrose ($C_{12}H_{22}O_{11}$)	144 g C; 22 g H; 176 g O

2. In each of the compounds of problem 1, calculate the number of grams of one element that would combine with 10 g of the other; for example, how many grams of sodium combine with 10 g of chlorine? How many grams of chlorine combine with 10 g of sodium?

3. Given the following percentage compositions, calculate the relative atomic weights of all the atoms involved. Assume that the formulas are correct and that hydrogen has a relative weight of 1.

Substance	Percent Composition
HCl	2.74% H; 97.26% Cl
NaCl	39.3% Na; 60.7% Cl
H_2O	11.1% H; 88.9% O
NaOH	57.5% Na; 40.0% O; 2.5% H
NaBr	22.3% Na; 77.7% Br
CO_2	27.3% C; 72.7% O
CH_4	75.0% C; 25.0% H
H_2S	5.9% H; 94.1% S

4. The complete combustion of 7.8 g of a compound containing only the elements carbon and hydrogen resulted in the formation of 26.4 g of carbon dioxide and 5.4 g of water. Calculate the percentage composition of the original compound.

5. How many grams of water would be formed by the complete combustion of 30 g of hydrogen?

6. The complete combustion of 4.6 g of a compound containing only the elements carbon, hydrogen, and oxygen resulted in the formation of 8.8 g of carbon dioxide and 5.4 g of water. Calculate the percentage composition of the original compound.

7. Calculate the formula weights of each of the following:

(a) H_2 (e) H_2O (i) H_2S
(b) NaOH   (f) SO_2 (j) H_2SO_4
(c) HCl   (g) Fe_3O_4 (k) H_3PO_4
(d) CO_2 (h) NaCl   (l) $Ca(OH)_2$

8. Calculate the number of moles and the number of molecules in each of the following molecular substances.

(a) 2 g H_2 (d) 35.5 g Cl_2
(b) 18 g H_2O (e) 10 g O_2
(c) 20 g H_2

9. Calculate the number of formula units in (a) a pound-mole of any substance, (b) a ton-mole of any substance, (c) a milligram-mole of any substance.

10. Balance the following equations.

(a) $Na + H_2S \longrightarrow Na_2S + H_2$
(b) $H_2O + Na \longrightarrow NaOH + H_2$
(c) $F_2 + H_2O \longrightarrow HF + O_2$
(d) $H_2O_2 \longrightarrow H_2O + O_2$
(e) $CH_4 + O_2 \longrightarrow CO_2 + H_2O$
(f) $K + Cl_2 \longrightarrow KCl$

11. In Section 3–4, the example was cited of hydrogen chloride which contains 2.74% hydrogen and 97.26% chlorine. Assuming the formula to be HCl, the relative atomic weights were calculated to be H = 1 and Cl = 35.5. Calculate the relative atomic weights if the formula were (a) H_2Cl, and (b) HCl_2. (Note that the correct formula *is* HCl.)

12. Suppose that the atomic weight of oxygen were changed, by agreement, to 48.0. What would be the formula weight of CO_2?

13. Calculate the percentage composition of each of the following compounds: (a) CO_2, (b) H_2O_2, (c) $C_6H_{11}O_5$, (d) TiO_2, (e) C_5H_{10}.

14. Calculate the formula weights of each of the following compounds: (a) C_2H_4, (b) CH_2O (c) $Ba(OH)_2$, (d) $(NH_4)_2SO_4$, (e) $AlCl_3$, (f) $MgCO_3$, (g) $Ag(NH_3)_2Cl$, (h) H_3PO_4, (i) H_2SO_4.

15. A compound of nitrogen and oxygen is found to contain 25.9% nitrogen. What is its simplest formula?

16. Calculate the simplest formula of a compound if 3.0 g of this compound contains 1.2 g C, 0.2 g H, and 1.6 g O.

17. A compound is found to contain 41.0% oxygen and 59.0% sodium. Calculate the simplest formula. Give two more possible molecular formulas in addition to the simplest formula.

18. In 1819 Pierre Dulong and Alexis Petit announced that the product of the atomic weight of a solid element and its specific heat (Section 2–9) is approximately 6 cal/mole °C.

(a) Test this relationship using the elements in Table 2–4.

(b) How can specific heat measurements aid in the determination of atomic weights?

19. Name the following compounds.

(a) NH_4Br

(b) $(NH_4)_2CO_3$

(c) $Ca_3(PO_4)_2$

(d) $BaCl_2$

(e) $KClO_3$

(f) $Mg(NO_3)_2$

(g) Li_2O

(h) Ag_2CrO_4

(i) NaF

(j) K_3PO_4

(k) K_2S

(l) $BeCl_2$

(m) AlF_3

(n) AgI

(o) KOH

(p) Na_3PO_4

(q) $AgNO_3$

(r) LiCl

(s) BeO

(t) $BaCrO_4$

(u) CaS

(v) $Mg(ClO_3)_2$

(w) MgO

(x) Al_2O_3

(y) $(NH_4)_3PO_4$

(z) NaI

PROPERTIES AND BEHAVIOR OF GASES

The point in time when man became aware of the air that surrounds him is lost in antiquity. Primitive man must have wondered at the wind and rain, heat and cold that constantly kept him on his guard, but it is a large jump from these obvious phenomena to the rather sophisticated concept of a gaseous atmosphere. This chapter deals with some of the steps that led to a quantitative treatment of gases.

4-1. BASIC DEFINITIONS

A **gas** has neither a definite shape nor volume of its own but rather assumes the shape and volume of its container.

The **pressure** of a gas is the force exerted per unit area on the sides of its container.

$$\text{pressure} = \frac{\text{force}}{\text{area}}$$

Atmospheric pressure is the weight of the air per unit area and at sea level amounts to about 14.7 psi (pounds per square inch). In scientific units, this pressure is 1.01×10^6 dynes/cm². *

The **temperature** of a substance is a quantitative measure of its relative hotness and coldness, and has been defined in Section 2–2.

*A force of 1 dyne is approximately equal to the weight of an average mosquito.

One great stumbling block in our path toward understanding the atmosphere was its perennial presence. Only when a vacuum was produced could one observe what it was like not to have air present. Within a short span of time in the first half of the seventeenth century, several men were responsible for removing this stumbling block.

Galileo Galilei found that air has weight, but it was left to his pupil, Evangelista Torricelli to construct the first mercury barometer in 1643. A glass tube, sealed at one end, was filled with mercury and, with the finger held over the open end, inverted and submerged in a dish of mercury. Upon releasing the finger, the mercury only partially ran out of the tube and Torricelli concluded that the pressure exerted by the mercury column was exactly balanced by the pressure of the surrounding air on the mercury pool (Figure 4–1).

The reality of atmospheric pressure was strikingly demonstrated in 1654 by Otto von Guericke. He was the mayor of Magdeburg and like many of his contemporaries, he pursued science as a hobby, often with important results. His apparatus came to be called the "Magdeburg hemispheres": Two copper hemispheres about 2 ft in diameter were placed together to form a sphere and an airtight seal was made with oil-soaked leather rings (Figure 4–2). When the air was removed from the spheres, by means of a pump designed by von Guericke, the hemispheres could not be pulled apart by two teams of eight horses each! The wonder of this demonstration is easily explained by the pressure of the atmosphere—nearly 3 tons—holding them together. This experiment can be performed more easily but less dramatically by the use of a plumber's plunger on a flat surface.

A standard pressure unit is **1 atmosphere** (1 atm) and is defined as the pressure exerted by a vertical column of mercury exactly 760 mm high at 0°C and measured at sea level where the force of gravity is standard. Atmospheric pressure

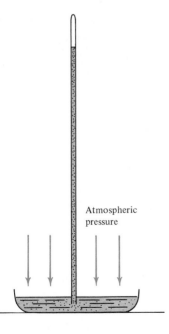

Atmospheric pressure

Figure 4-1. Torricelli's barometer made by inserting a tube filled with mercury into a dish of mercury.

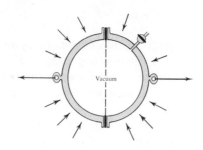

Figure 4-2. Schematic drawing of the Madgeburg hemispheres designed by Otto von Guericke. The arrows represent the pressure of the atmosphere.

may, on occasion, be equal to 1 atm, but often it is less. When a meteorological high pressure system passes over an area, atmospheric pressure sometimes rises above 1 atm. A pressure of 1 **torr** (named in honor of Torricelli) is 1/760 of 1 atm, and is therefore numerically equal to a height of the mercury column of 1 mm. Hereafter we shall employ the term torr in place of millimeters of mercury to describe the unit of pressure.

4-3. BASIC GAS LAWS

The quantitative relationship between the volume of a fixed amount of air and its pressure was first studied by Robert Boyle, one of those seventeenth century men of wealth and independence who devoted their leisure to the study of science. In 1661, he published his results, which showed that the pressure and volume of a quantity of air are inversely proportional (Figure 4-3). All of Boyle's experiments were carried out at the constant temperature of his laboratory.

$$V \propto \frac{1}{P}$$

or,

$$PV = k \quad \text{(mass and temperature constant)} \tag{4-1}$$

where k is a constant whose value depends upon the amount of gas used and its temperature. An alternate form of Boyle's law, useful for problem solving, can be stated

$$P_1 V_1 = P_2 V_2 \quad \text{(mass and temperature constant)} \tag{4-2}$$

since the PV product is always equal to the same constant. The subscripts 1 and 2 denote initial and final conditions, respectively.

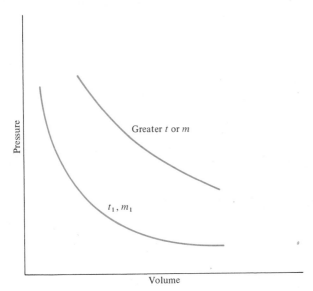

Figure 4–3. Variation of the volume of a gas with pressure at constant mass and temperature: Boyle's law.

■ **Example 4–1**

A sample of gas at a certain temperature occupies a volume of 5.00 liters at a pressure of 0.450 atm. What volume will it occupy under a pressure of 1.80 atm at the same temperature?

☐ *Solution*

$$P_1 V_1 = P_2 V_2$$

We are given that $P_1 = 0.450$ atm; $V_1 = 5.00$ liters; $P_2 = 1.80$ atm. Then solving for V_2

$$V_2 = \frac{P_1 V_1}{P_2}$$

$$= \frac{0.450 \text{ atm} \times 5.00 \text{ liters}}{1.80 \text{ atm}}$$

$$= 1.25 \text{ liters}$$

Of great significance to chemists is the fact that this equation is the same for *all* gases.

Jacques Charles, in 1787, found that all gases expand upon heating and contract upon cooling provided their pressure is maintained at the same value throughout the change. The fractional change in the volume of gases for each

41

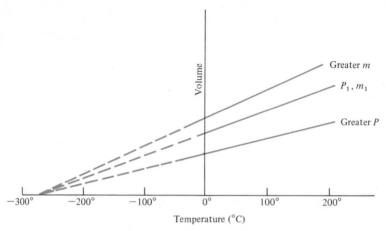

Figure 4-4. Variation of the volume of a gas with temperature at constant mass and pressure: Charles' law.

degree change in temperature amounts to approximately 1/273 of their volume at 0°C. Stated mathematically,

$$V = V_0\left(1 + \frac{t}{273}\right) \quad \text{(mass and pressure constant)} \qquad (4\text{-}3)$$

where V is the volume at the temperature, t°C, and V_0 is the volume of the particular gas at 0°C. A typical plot of volume versus temperature is given in Figure 4-4.

All gases eventually condense to a liquid at low enough temperatures. There-fore, measurements of gas volumes obviously cannot be extended to temperatures where the gas no longer exists. However, if the straight line plots of actual measurements are extrapolated until they intersect the temperature axis, they meet approximately at a common point of −273°C. A simplification of Charles' law results if a new temperature scale is defined whose zero point is placed at this point of intersection (−273°C). Thus,

$$T = t + 273$$

This scale is called the **absolute (or Kelvin) temperature scale** and is designated, °K. The effect of this definition is to translate the volume axis to the point −273 on the Celsius temperature axis as shown in Figure 4-5.

If substitution of $(T - 273)$ for t is made in equation (4-3), there results

$$V = cT \quad \text{(mass and pressure constant)} \qquad (4\text{-}4)$$

where c is a constant equal to $V_0/273$. The value of c depends on the nature of the gas, its mass, and the pressure. Since

$$\frac{V}{T} = c$$

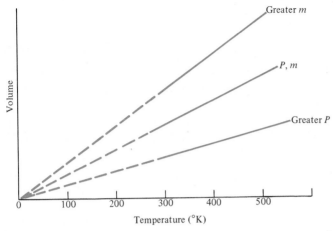

Figure 4-5. Variation of the volume of a gas at constant mass and temperature with absolute temperature (degrees Kelvin).

we may write Charles' simplified law as

$$\frac{V_1}{T_1} = \frac{V_2}{T_2} \quad \text{(mass and pressure constant)} \qquad (4\text{-}5)$$

Only absolute temperatures, of course, may be used when Charles' law is stated in this form.

■ Example 4-2

At a fixed pressure, 3.60 liters of gas is cooled from 27°C to −73°C. What is the new volume of the gas?

☐ *Solution*

$$\frac{V_1}{T_1} = \frac{V_2}{T_2}$$

We are given $V_1 = 3.60$ liters; $T_1 = 27 + 273 = 300°\text{K}$; and $T_2 = -73 + 273 = 200°\text{K}$. Then solving for V_2

$$V_2 = \frac{V_1 T_2}{T_1}$$

$$= \frac{3.60 \text{ liters} \times 200°\text{K}}{300°\text{K}}$$

$$= 2.40 \text{ liters}$$

4-4. COMBINED GAS LAW

Boyle's and Charles' laws treat changes in the condition of gases under specific restrictions. Real changes, however, often occur with simultaneous changes in the three variables of pressure, volume, and temperature. A combination of Boyle's and Charles' laws is thus desirable. Let us consider an actual change in P, V, and T for a given amount of gas to occur by two hypothetical consecutive changes, one at constant temperature and the other at constant pressure (Figure 4-6). For the constant temperature change, Boyle's law applies, and from equation (4-2)

$$V_x = \frac{P_1 V_1}{P_2} \quad \text{(mass and temperature constant)} \tag{4-6}$$

The constant pressure change is governed by Charles' simplified law, and substituting into equation (4-5) gives

$$V_x = \frac{T_1 V_2}{T_2} \quad \text{(mass and pressure constant)} \tag{4-7}$$

Since V_x has but one value, the right hand sides of equations (4-6) and (4-7) may be equated

$$\frac{P_1 V_1}{P_2} = \frac{T_1 V_2}{T_2} \quad \text{(mass constant)}$$

and rearranged to give

$$\frac{P_1 V_1}{T_1} = \frac{P_2 V_2}{T_2} \quad \text{(mass constant)} \tag{4-8}$$

Notice that Boyle's and Charles' simplified laws are now simply special cases of the combined gas law. If $T_1 = T_2$ (constant temperature), equation (4-8) becomes

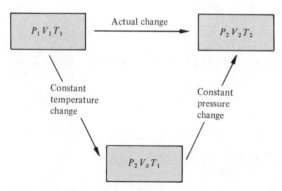

Figure 4-6. Diagram showing how an actual change in the pressure, volume, and temperature of a gas may be made to occur by two hypothetical changes—one at constant temperature and one at constant pressure.

identical with equation (4-2), and if $P_1 = P_2$ (constant pressure), equation (4-5) results. Equation (4-8) may be used to calculate changes in any one variable of pressure, volume, or temperature for simultaneous changes in the other two variables provided a constant mass of gas is considered. Such calculations are, of course, subject to the inaccuracies inherent in Boyle's and Charles' laws as will be discussed in Chapter 6.

Since the combination PV/T is invariant for a given mass of gas regardless of the specific values of pressure, volume, and temperature, the combined gas law may be written

$$\frac{PV}{T} = C \quad \text{(amount of gas constant)} \tag{4-9}$$

where C is a constant whose value depends on the identity and amount of gas.

■ **Example 4-3**

A sample of gas occupies a volume of 500 ml at 7°C and 0.200 atm. What pressure will the gas exert if it is confined in a volume of 1.00 liter at 107°C?

☐ *Solution*

$$\frac{P_1 V_1}{T_1} = \frac{P_2 V_2}{T_2}$$

We are given $P_1 = 0.200$ atm; $V_1 = 500$ ml; and $T_1 = 7 + 273 = 280°K$.

$$V_2 = 1.00 \text{ liter} \times 1000 \text{ ml/liter} = 1000 \text{ ml}$$

$$T_2 = 107 + 273 = 380°K$$

Solving for P_2

$$P_2 = \frac{P_1 V_1 T_2}{T_1 V_2}$$

and substituting the known values

$$P_2 = \frac{0.200 \text{ atm} \times 500 \text{ ml} \times 380°K}{280°K \times 1000 \text{ ml}}$$

$$= 0.136 \text{ atm}$$

4-5. GAY-LUSSAC AND THE COMBINING VOLUMES OF REACTING GASES

At the beginning of the nineteenth century, a relationship was observed involving the chemical combination of gases. Joseph Gay-Lussac in France observed that,

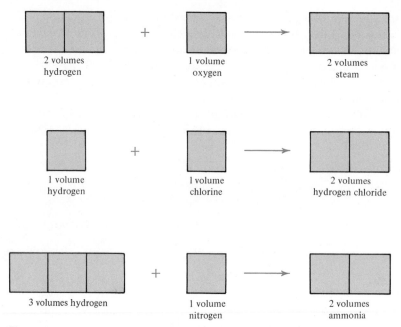

Figure 4-7. Schematic representation of Gay-Lussac's law of combining volumes. All squares represent the same volume of gas at the same temperature and pressure.

although there was no obvious relationship governing the weights of gases that combined, there was one that governed the volumes that combined. The Scot, Henry Cavendish, had determined that when hydrogen and oxygen combine chemically to form water the volumes of the two gases (measured, of course, at the same temperature and pressure) that combine are in the precise ratio of 2 to 1. Gay-Lussac investigated the volume ratios with which many different gases react and generalized these observations (Figure 4-7) in his law of combining volumes. In his own words:

> . . . the compounds of gaseous substances with each other are always formed in very simple ratios, so that representing one of the terms by unity, the other is 1, or 2, or at most 3. These ratios by volume are not observed with solid or liquid substances nor when we consider weights, . . .

4-6. AVOGADRO'S LAW

The Italian Amadeo Avogadro, by means of a brilliant intuitive leap, expressed in 1811 the underlying truth to be found in Gay-Lussac's work in these words:

> It must . . . be admitted that very simple relations . . . exist between the volumes of gaseous substances and the numbers of . . . molecules which form them. The first hypothesis to present itself in this connection, and apparently the only admissible one, is the supposition that the number of . . . molecules in any gases is always the same for equal volumes, or always proportional to the volumes . . .

Under the same conditions of temperature and pressure, therefore, the volumes of different gases are all directly proportional to the number of molecules in the gases. What Gay-Lussac was really studying, then, was simply the relative numbers of gaseous molecules that were reacting! The law of conservation of mass obviously requires that the ratio of atoms in a molecule of a substance be the same as the ratio in which they react to form that molecule.

2 volumes hydrogen + 1 volume oxygen $\longrightarrow$ 2 volumes water vapor

If it is assumed that the gases hydrogen and oxygen exist as single atoms (as Dalton had assumed), Avogadro's idea cannot be reconcilled with the facts expressed by the above equation.

$$2\ H + O \xrightarrow{\ \ \ \times\ \ \ } 2 \text{ volumes water vapor}$$

There is no way that one oxygen atom can be divided to yield two molecules of water. Consequently, Avogadro concluded that elementary gases do not exist as individual atoms but rather as molecules with an even number of atoms, for example, H_2 and O_2. The molecules can thus divide upon reaction. Furthermore, in order to conform to the law of conservation of mass and balance the chemical equation, Avogadro was forced to write the formula of water as H_2O in order to accommodate all the hydrogen atoms. Avogadro thus considered the following possibilities

$$2\ H_2 + O_2 \longrightarrow 2\ H_2O$$

$$2\ H_4 + O_4 \longrightarrow 2\ H_4O_2$$

$$2\ H_6 + O_6 \longrightarrow 2\ H_6O_3$$

because any one of them would give the observed volume ratios. At this point, Avogadro made the simplifying assumption that there are only two atoms per elementary gaseous molecule. This assumption was subsequently proved correct. The equations for Gay-Lussac's findings can thus be represented as shown in Figure 4–8.

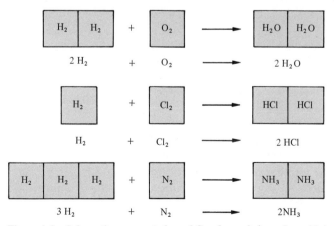

Figure 4–8. Schematic representation of Gay-Lussac's law of combining volumes as interpreted by Avogadro. All squares represent the same volume of gas at the same temperature and pressure.

Avogadro realized the tremendous import of his work. If equal volumes of different gases (at the same pressure and temperature) contain the same number of molecules, then the ratio of the masses of the two gas volumes is the same as the ratio of the masses of the individual molecules. The problem of the determination of relative molecular weights is solved because the molecular formulas of gaseous substances are determinable. For example, under the same conditions of temperature and pressure, 1 liter of ammonia is $8\frac{1}{2}$ times as heavy as 1 liter of hydrogen. If we choose to make the atomic weight of the hydrogen atom 1.0, then the hydrogen molecule must have a molecular weight of 2.0 (remember Avogadro determined that elementary gases are diatomic). The molecular weight of ammonia must therefore be $2 \times 8\frac{1}{2} = 17$. Since the formula of ammonia is NH_3, the atomic weight of nitrogen must be 14.

The concept of **gram molecular volume** is readily apparent. This is the volume of a gas measured at a specified temperature and pressure which contains sufficient molecules to give a weight (in grams) equal to the molecular weight of the gas. Moreover, the gram molecular volume is the same for all gaseous substances at a given temperature and pressure. At 1 atm pressure and 0°C (standard pressure and temperature), the gram molecular volume (GMV) is 22.4 liters. At a different temperature and pressure, the GMV has a different value but is, of course, the same for all gaseous substances. The number of molecules of a substance required to produce a weight in grams equal to the molecular weight of the substance is called **Avogadro's number** (Section 3–6). Often, the weight in grams equal to the molecular weight of a substance is called the **gram molecular weight** (GMW). One GMW has already been referred to as 1 gram-mole or simply 1 mole. The mole idea was introduced in Section 3–6, and will, as you will see, be employed throughout the study of chemistry.

4-7. IDEAL GAS LAW

Let us return to the combined gas law as expressed by equation (4–9)

$$\frac{PV}{T} = C \quad \text{(mass constant)}$$

If this equation is rearranged to solve for volume, we obtain

$$V = C\frac{T}{P} \quad \text{(mass constant)} \tag{4–9a}$$

We know from the work of Avogadro that at a fixed temperature and pressure, the volume of a gas is determined only by the number of molecules present. A simple way to express the number of molecules is through gram moles because we know that a gram mole of any substance contains Avogadro's number of molecules. We may thus write

$$C \propto n \tag{4–10}$$

where n represents the number of gram moles of the gas, that is, grams of gas ÷ molecular weight of the gas. Alternatively, we have

$$C = nR \qquad\qquad (4\text{-}10a)$$

where R is a proportionality constant. This term is called the universal gas constant because it is a true constant of nature, completely independent of the particular gas being considered. If we substitute the value of C given by equation (4-10a) into equation (4-9a) and rearrange, there results

$$PV = nRT \qquad\qquad (4\text{-}11)$$

Equation (4-11) is called the **ideal gas equation** or the **ideal gas law.** It is independent of any restriction, but it is subject to the same inaccuracies encountered with Boyle's and Charles' laws, which will be discussed in Chapter 6. An ideal gas may be defined as a gas that obeys equation (4-11). No real gas is an ideal gas under all conditions, but all gases approximate ideal behavior as temperature is increased and pressure is decreased.

■ **Example 4-4**

0.500 mole of gas exerts a pressure of 700 torr in a container of fixed volume at 27°C. Some gas is removed while cooling the container until the temperature is −23°C and the pressure is 0.461 atm. How much gas was removed?

□ *Solution*

In this problem volume is constant.

$$PV = nRT$$

Thus,

$$\frac{P}{nT} = \frac{R}{V} = \text{constant}$$

so

$$\frac{P_1}{n_1 T_1} = \frac{P_2}{n_2 T_2}$$

We were given $P_1 = 700$ torr; $n_1 = 0.500$ mole; $T_1 = 27 + 273 = 300°$K; $P_2 = 0.461$ atm × 760 torr/atm = 350 torr; and $T_2 = -23 + 273 = 250°$K. Then solving for n_2

$$n_2 = \frac{P_2 n_1 T_1}{P_1 T_2}$$

$$= \frac{350 \text{ torr} \times 0.500 \text{ mole} \times 300°\text{K}}{700 \text{ torr} \times 250°\text{K}}$$

$$= 0.300 \text{ mole}$$

49

and

$$n_1 - n_2 = 0.500 - 0.300 = 0.200 \text{ mole}$$

so that the gas removed = 0.200 mole.

The universal gas constant, R, must be determined experimentally. Rearrangement of equation (4–11) gives

$$R = \frac{PV}{nT} \qquad (4\text{–}11a)$$

Since 1 mole of any gas at 1 atm and 0°C is found to occupy a volume of 22.4 liters, substitution of these values into equation (4–11a) yields the value of R

$$R = \frac{1.00 \text{ atm} \times 22.4 \text{ liters}}{1.00 \text{ mole} \times 273°\text{K}}$$

$$= 0.0821 \text{ liter-atm/mole } °\text{K}$$

Use of other units to express volume or pressure will result in different numerical values for R. One must be very careful to use the value of R that is consistent with the pressure and volume units used in a particular problem. Table 4–1 gives the values of R expressed in several different units.

Table 4–1. Values of R in Various Units

Pressure Unit	Volume Unit	Value of R
atmosphere	liter	0.0821 liter-atm/mole °K
atmosphere	ml	82.1 ml-atm/mole °K
torr	liter	62.4 liter-torr/mole °K
torr	ml	62.4×10^3 ml-torr/mole °K
dynes per square centimeter	cm³ (ml)	8.32×10^7 erg/mole °K

The ideal gas law summarizes Boyle's law, Charles' simplified law, and Avogadro's law.

$$PV = nRT$$

1. Boyle's law. If n and T are constant, then

$$PV = \text{constant}$$

2. Charles' simplified law. If n and P are constant, then

$$\frac{V}{T} = \frac{nR}{P} = \text{constant}$$

3. Avogadro's law. The volumes of two different gases, V_1 and V_2 are given by

$$V_1 = \frac{n_1 R T_1}{P_1} \qquad V_2 = \frac{n_2 R T_2}{P_2}$$

If conditions of pressure and temperature are the same, that is, $T_1 = T_2$ and $P_1 = P_2$, then

$$\frac{V_1}{V_2} = \frac{n_1}{n_2}$$

which says that under these conditions, the volumes of the two gases are proportional to the moles or numbers of molecules of each gas. If the volumes should happen to be equal, that is, $V_1 = V_2$, then obviously the numbers of molecules in each gas would also be equal, that is, $n_1 = n_2$.

The Ideal Gas law is applicable to any mixture of nonreacting gases. However, the quantity n must be the sum of the moles of all gases present in the mixture.

Several alternate forms of the ideal gas law are useful in solving specific problems. By definition

$$n = \frac{g}{MW}$$

where g is the weight of gas and MW its molecular weight. Therefore

$$PV = \frac{gRT}{MW} \qquad\qquad (4\text{--}11b)$$

$$MW = \frac{gRT}{PV} \qquad\qquad (4\text{--}11c)$$

Also note that $g/V = d$, the density of the gas.* Equation (4–11c) may thus be expressed

$$MW = \frac{dRT}{P} \qquad\qquad (4\text{--}11d)$$

■ **Example 4-5**

Calculate the density of hydrogen at $-50°C$. and 3.25 atm.

☐ *Solution*

$$PV = nRT \quad \text{and} \quad PV = \frac{gRT}{MW}$$

---

*Unlike liquids and solids for which density is expressed as grams per milliliter, the density of a gas is expressed in the units grams per liter.

but

$$d = \frac{g}{V}$$

so

$$P = \frac{dRT}{MW}$$

or

$$d = \frac{P(MW)}{RT}$$

We were given $P = 3.25$ atm; hydrogen has MW $= 2.02$ g/mole; $R = 0.0821$ liter-atm/mole °K; and $T = -50 + 273 = 223°$K. Therefore

$$d = \frac{3.25 \text{ atm} \times 2.02 \text{ g/mole}}{0.0821 \text{ liter-atm/mole-}°\text{K} \times 223°\text{K}}$$

$$= 0.359 \text{ g/liter}$$

4-8.   DALTON'S LAW OF PARTIAL PRESSURES

Mixtures of nonreactive gases may be treated by the laws applicable to a single gas. The *total* pressure and *total* number of moles are involved, however, in any calculations regarding gas mixtures using the ideal gas law. Dalton in 1802 first stated the relationship between the total pressure of a mixture of gases and the pressure exerted by each component of the mixture.

> At constant temperature, the total pressure exerted by a mixture of gases in a definite volume is equal to the sum of the individual pressures which each gas would exert if it occupied the same volume alone.

The individual pressure that each gas exerts independently of any other gas which is present in a mixture is called the **partial pressure** of the gas. In terms of partial pressures, Dalton's law may be stated mathematically

$$P_{\text{total}} = P_1 + P_2 + P_3 + \dots \quad \text{(temperature and volume fixed)} \quad (4\text{--}12)$$

where the supscripted P's refer to the partial pressures of each gaseous component in the mixture. An example will illustrate the meaning of this statement: Suppose three 1-liter flasks at 25°C are filled, respectively, with hydrogen at 0.50 atm, carbon monoxide at 2.00 atm, and argon at 0.20 atm. If now all three gases are mixed in the same 1-liter flask at 25°C, the total pressure will be found to be 2.70 atm.

$$P = P_{H_2} + P_{CO} + P_{Ar}$$

$$= 0.50 + 2.00 + 0.20$$

$$= 2.70 \text{ atm}$$

Experimentally, it has been found that Dalton's law of partial pressures is accurate under conditions where significant deviations from the ideal gas law occur. However, at extremely high pressures, Dalton's law will not accurately predict the pressure of a mixture of gases.

Under conditions such that the gaseous components of a mixture behave ideally, a useful relationship may be derived from Dalton's law. In this case, the partial pressure of each component may be expressed using equation (4–11)

$$P_1 = \frac{n_1 RT}{V} \tag{4–11e}$$

$$P_2 = \frac{n_2 RT}{V} \tag{4–11f}$$

$$P_3 = \frac{n_3 RT}{V} \tag{4–11g}$$

$$\vdots \qquad \vdots$$

and since

$$P_{total} = P_1 + P_2 + P_3 + \cdots$$

we may write

$$P_{total} = \frac{n_1 RT}{V} + \frac{n_2 RT}{V} + \frac{n_3 RT}{V} + \cdots \tag{4–12a}$$

or

$$P_{total} = \frac{RT}{V}(n_1 + n_2 + n_3 + \cdots) \tag{4–12b}$$

because V and T are the same for each component as well as for the mixture. If we now divide equation (4–11e) by equation (4–12b)

$$\frac{P_1}{P_{total}} = \frac{n_1 RT/V}{(RT/V)(n_1 + n_2 + n_3 + \cdots)}$$

$$\frac{P_1}{P_{total}} = \frac{n_1}{(n_1 + n_2 + n_3 + \cdots)} \tag{4–13}$$

The quantity $n_1/(n_1 + n_2 + n_3 + \cdots)$ is defined as the **mole fraction** of component 1 and is symbolized by X_1. Consequently, we may rewrite equation (4–13) as

$$P_1 = X_1 P_{total} \tag{4–13a}$$

Similarly, division of equations (4–11f) and (4–11g) by equation (4–12b) leads to

$$P_2 = X_2 P_{total} \tag{4–13b}$$

53

$$P_3 = X_3 P_{\text{total}} \tag{4-13c}$$

$$\vdots \quad \vdots$$

By the use of these relationships, partial pressures can be calculated from a knowledge of the mole fraction of a particular component and the total pressure of the mixture. Conversely, the mole fraction of a component in a mixture of gases can be obtained from a knowledge of the total pressure of the mixture and the partial pressure of the component. However, the gases must each behave ideally for this relationship to be valid.

Mole fraction is a useful unit for expressing the concentration of components in a mixture because its definition requires that

$$X_1 + X_2 + X_3 + \cdots = 1 \tag{4-14}$$

■ **Example 4-6**

Calculate the pressure exerted by 8.00 g of oxygen and 14.0 g of nitrogen confined in a 10.0-liter container at 7°C. Also calculate the mole fraction of each gas.

□ *Solution*

$$PV = nRT$$

$$P = \frac{nRT}{V}$$

$$n = n_{O_2} + n_{N_2}$$

$$= \frac{8.00 \text{ g}}{32.0 \text{ g/mole}} + \frac{14.0 \text{ g}}{28.0 \text{ g/mole}}$$

$$= 0.750 \text{ mole}$$

We know that $R = 0.0821$ liter-atm/mole °K, $T = 7 + 273 = 280$°K, and $V = 10.0$ liters is given. Then

$$P = \frac{0.750 \text{ mole} \times 0.0821 \text{ liter-atm/mole °K} \times 280°\text{K}}{10.0 \text{ liters}}$$

$$= 1.72 \text{ atm}$$

$$X_{O_2} = \frac{n_{O_2}}{n_{O_2} + n_{N_2}}$$

$$= \frac{0.250}{0.750}$$

$$= 0.333$$

But

$$X_{O_2} + X_{N_2} = 1$$

so

$$X_{N_2} = 0.667$$

Alternatively, the partial pressures of the oxygen and nitrogen can be found and the total pressure obtained by adding them.

$$P_{O_2} = \frac{n_{O_2}RT}{V}$$

$$= \frac{g_{O_2}RT}{MW_{O_2}V}$$

$$= \frac{8.00 \text{ g} \times 0.0821 \text{ liter-atm/mole } °K \times 280°K}{32.0 \text{ g/mole} \times 10.0 \text{ liter}}$$

$$= 0.575 \text{ atm}$$

$$P_{N_2} = \frac{14.0 \text{ g} \times 0.0821 \text{ liter-atm/mole } °K \times 280°K}{28.0 \text{ g/mole} \times 10.0 \text{ liters}}$$

$$= 1.15 \text{ atm}$$

$$P = P_{O_2} + P_{N_2}$$

$$= 0.57 + 1.15$$

$$= 1.72 \text{ atm}$$

$$X_{O_2} = \frac{P_{O_2}}{P}$$

$$= \frac{0.575}{1.72}$$

$$= 0.334$$

■ **Example 4-7**

Under suitable conditions, ammonia is oxidized by oxygen to give nitric oxide and water.

$$4 \text{ NH}_3 + 5 \text{ O}_2 \longrightarrow 4 \text{ NO} + 6 \text{ H}_2\text{O}$$

Assume that all substances are gaseous and that volume measurements are obtained at the same temperature and pressure. If 10.0 liters of H_2O is obtained, calculate the volume of NO produced and the volumes of NH_3 and O_2 consumed.

□ *Solution*

$$10.0 \text{ liters H}_2\text{O} \times \frac{4 \text{ liters NH}_3}{6 \text{ liters H}_2\text{O}} = 6.67 \text{ liters NH}_3 \text{ consumed}$$

$$10.0 \text{ liters } H_2O \times \frac{5 \text{ liters } O_2}{6 \text{ liters } H_2O} = 8.33 \text{ liters } O_2 \text{ consumed}$$

$$10.0 \text{ liters } H_2O \times \frac{4 \text{ liters } NO}{6 \text{ liters } H_2O} = 6.67 \text{ liters } NO \text{ produced}$$

■ Example 4–8

If the reaction of Example 4–7 had been carried out at constant volume and temperature, what would the final pressure be if the NH_3 and O_2 were both initially present in the proportions required for complete reaction at a total pressure of 3.00 atm? Assume the reaction goes to completion.

□ *Solution*

$$PV = nRT$$

We are given that $V = $ constant and $T = $ constant. Therefore, since

$$\frac{P}{n} = \frac{RT}{V} = \text{constant}$$

$$\frac{P_1}{n_1} = \frac{P_2}{n_2}$$

We are also given $P_1 = 3.00$ atm and, from the coefficients in the left-hand side of the reaction given in Example 4–7, $n_1 \propto (4 + 5)$ so that $n_1 = 9K$ (where K is a proportionality constant). From the right-hand side of the reaction equation, $n_2 \propto (4 + 6)$ or $n_2 = 10K$. Solving for P_2

$$P_2 = \frac{P_1 n_2}{n_1}$$

$$= \frac{3.00 \text{ atm} \times 10K}{9K}$$

$$= 3.33 \text{ atm}$$

PROBLEMS

1. Define the following terms:
   (a) gas
   (b) pressure
   (c) temperature
   (d) ideal gas
   (e) atmospheric pressure
   (f) barometer
   (g) gram molecular volume
   (h) gram molecular weight
   (i) partial pressure

2. Consider 1 mole of a gas. Indicate the way in which the following variables would change under the influence of the various imposed conditions. Use + for increase; − for decrease; 0 for no change.

Imposed Condition	P	V	T
P is increased at constant V			
P is increased at constant T			
P and V are both increased			
P doubles and V is halved			
V is increased at constant P			
V and P are both decreased			
T increases while V decreases			
T is increased at constant P			

3. Convert 15 pounds per square inch (psi) to equivalent values in pounds per square centimeter, dynes per square centimeter, dynes per square inch, atmospheres, and torr.

4. Convert 25°F to the equivalent Celsius and Kelvin values.

5. Calculate the temperature at which the numerical values of °K and °F are equal.

6. If a barometer employing mercury indicated a pressure of 720 torr, what height of water could be supported by this pressure at 25°C?

7. Calculate the volume of a given amount of gas at 77°C and 1.0 atm if it has a volume of 2.0 liters at −23°C and 1.0 atm.

8. Air in an automobile tire has a pressure of 40 psi at 25°C. After traveling for a while, the pressure rises to 50 psi. Calculate the temperature rise. Assume the volume of the tire does not change.

9. A balloon contains 500 ml of air at 20°C and 800 torr. Calculate the pressure of the air if the balloon is heated to 40°C and the volume increases to 520 ml.

10. A sample of gas exhibits a pressure of 720 torr at a temperature of 25°C. What pressure will the gas exhibit at 50°C in the same volume?

11. Determine the molecular weight of a gas which has a density of 1.00 g/liter at 56°C and 710 torr.

12. Calculate the number of gram-moles of a gas in a bulb with a volume of 150 ml if the pressure of the gas is 1.00 atm at 27°C.

13. Calculate the pressure exerted by 1.00 mole of a gas confined to a volume of 9.86 liters at 27°C.

14. What is the molecular weight of a gas if a certain volume has a mass 2.5 times as great as an equal volume of O_2 when both volumes are measured at the same pressure and temperature?

15. 2.50 moles of an ideal gas occupies a volume of 10.0 liters at 1.00 atm and a certain temperature. How many moles of gas would occupy 20.0 liters at 2.00 atm and the same temperature?

16. Calculate the total pressure exerted by a mixture of O_2, N_2, He, and Ar in a 5.0-liter container if it has been determined that the individual pressures exerted by each of

the gases when confined individually to a 5.0-liter container at the same temperature of the mixture are as follows: $O_2 = 380$ torr, $N_2 = 1.00$ atm, He $= 760$ torr, Ar $= 0.50$ atm

17. Calculate the density of a gas if a given volume weighs 8.0 times as much as an equal volume of H_2, both volumes measured at 50°C and 0.95 atm.

18. A sample of H_2 is bubbled through water and 100 ml is collected at 20°C. The pressure exerted by the mixture is found to be 748 torr. How much H_2 is in the sample? (The vapor pressure of water in a gaseous mixture saturated with water is 18 torr at 20°C)

19. A mixture is prepared from 20.0 liters of ammonia and 35.0 liters of chlorine. These substances react according to the equation

$$2\,NH_3(g) + 3\,Cl_2(g) \longrightarrow N_2(g) + 6\,HCl(g)$$

where (g) indicates that the substance is a gas and the volumes of the gases are measured at the same temperature and pressure.
(a) What volume of HCl is produced?
(b) What is the total volume of all gases after reaction?

20. Calcium cyanamide reacts with water according to the equation

$$CaNCN(s) + 3\,H_2O(g) \longrightarrow CaCO_3(s) + 2\,NH_3(g)$$

where (s) means solid and (g) means gas. If 40.0 liters of NH_3 is produced at a certain temperature and pressure, what volume of steam was used at the same conditions?

21. Calculate the molecular weight of a gas if 0.200 g occupies a volume of 200 ml at a pressure of 0.960 atm and a temperature of 23°C.

22. Calculate the number of molecules in 16.0 g of O_2.

23. Calculate the number of hydrogen atoms in 1.0 liter of methane (CH_4) at -13°C and 1.0 atm.

24. Determine the relative densities under the same conditions of temperature and pressure of air, CO_2, O_2, CH_4, and H_2.

25. Calculate the gram molecular volume of a gas at 100°C and 0.500 atm.

26. Assume that the following reactions are carried out at constant volume and temperature, that the reactants are mixed in the proportions required by the balanced equations, and that the reactions proceed to completion. If the initial pressures were 1.0 atm in each case, what would be the final pressures, after reaction, in each case?
(a) $2\,H_2 + O_2 \longrightarrow 2\,H_2O$
(b) $H_2 + Cl_2 \longrightarrow 2\,HCl$
(c) $N_2 + 3\,H_2 \longrightarrow 2\,NH_3$

5

KINETIC THEORY OF GASES

We have seen that a productive science requires a broad, fundamental theory that can serve as a frame of reference within which to rationalize experimental observations and results and help formulate principles derived from those observations and to point the way toward future experiments. We must now consider a theory upon which much of modern physical science rests—the kinetic molecular theory.

5-1. BASIC POSTULATES

Kinetic molecular theory was first applied to gases in an effort to explain their behavior and elucidate the known, quantitative relationships embodied in the gas laws described in Chapter 4. Daniel Bernoulli, a member of an illustrious family of Swiss mathematicians, is credited with the first model of a gas as a substance composed of "very minute corpuscles, which are driven hither and thither with a very rapid motion." Bernoulli published his theory in 1738—over 60 years before Dalton's atomic theory! Bernoulli considered the pressure of a gas to be a manifestation of the countless collisions that must occur each second between the tiny particles of the gas and the walls of its container. On this basis, he showed that, at a constant temperature, the pressure and volume of a gas must be inversely proportional—a fact already arrived at empirically by Boyle. Furthermore, Bernoulli reasoned from his model that the pressure of a gas should be proportional to the average square of the velocity of the gas particles—a conclusion which was not properly appreciated for more than 100 years, when the concept of kinetic energy ($\frac{1}{2}mv^2$) was developed. It appears that his work was unknown to early kinetic theorists in the nineteenth century, and references to Bernoulli appear only after the kinetic molecular theory was firmly entrenched in the latter half of the century.

The basic assumptions of the kinetic theory of gases can be stated quite simply as follows:

1. A gas is composed of a large number of small particles of matter called molecules in ceaseless rapid motion.
2. These molecules behave like perfectly elastic spheres and their contacts with the walls of the container are also elastic. As a consequence, the total kinetic energy of the particles remains constant.
3. The free space within the walls of the container is so much greater than the total volume of the particles themselves as to allow them to move freely among each other in every direction.
4. All consideration of attractive forces is ignored, so that each particle must proceed on a straight line until it strikes against another or against the sides of the container.

On the basis of these assumptions, much of the known behavior of gases can be rationalized. In addition, it will be shown that the absolute temperature of a substance can be identified with the average kinetic energy of its molecules, so that $T \propto KE$.

■ Example 5-1

On the basis of the above postulates and the proportionality of absolute temperature and molecular kinetic energy, rationalize qualitatively (a) Boyle's law and (b) Charles' law.

□ *Solution*

Since pressure is the result of the collision of molecules against the container walls, pressure depends on (a) the energy of collision, and (b) the frequency of collision. If the volume of a gas is reduced, the same number of molecules have a shorter distance to travel between collisions, hence the collision frequency is increased, and a higher pressure results. Raising the temperature of the gas increases its kinetic energy ($\frac{1}{2}mv^2$), hence the pressure should increase. In order to keep pressure constant, the volume must be increased.

5-2. MOLECULAR VELOCITIES

An important result in the kinetic theory arises from the problem: If the molecules of a gas actually move at speeds of hundreds of meters per second at ordinary temperatures as calculated, then one would expect gases to diffuse and mix with each other very rapidly. In fact, of course, diffusion of a gas in the air, such as occurs when a gas jet is briefly opened on one side of a room, is not instantaneous, and several minutes may pass before the gas is noticed on the other side of the room. The answer lies in the fact that the chaotic motion of the molecules of a gas leads to intermolecular collisions and thus alters the paths of the molecules many times per second.

One can then pose the question: How far on the average can a molecule travel before it meets another molecule and suffers a collision? This average distance traveled between collisions is called the **mean free path** of the molecules. The mean free paths of gases under ordinary conditions are calculated to be in the range 10^{-4} to 10^{-5} cm so that it is easy to see that, although the molecules travel very rapidly, the distance traveled in any one direction is quite short. Consequently, the mixing of two gases is a relatively slow process. If a gas is released into a vacuum so that the molecules can all move more or less in the same direction, the gas rapidly fills the space, as expected from the high speeds of the molecules.

The Englishman James Clerk Maxwell can be credited with formalizing the kinetic theory. He introduced the methods of mathematical statistical analysis in treating kinetic theory and was successful in calculating some properties of gases that could be checked by experiment. For example, he used the idea of mean free path to calculate the viscosity of gases. Earlier workers had considered an average value for the velocities of the molecules. Maxwell found that the velocities of most of the molecules are clustered around the average velocity but some molecules in the assemblage always possess velocities much higher whereas some molecules have velocities much lower than the average. Furthermore, his treatment showed that with increasing temperature a higher proportion of the molecules acquire velocities considerably greater than the average and a smaller proportion have velocities much less than the average. Figure 5–1 is a graphical illustration of Maxwell's distribution law. With the experimental verification that followed the prediction made by Maxwell and others, the kinetic theory by the late nineteenth century became as widely accepted as basic atomic theory.

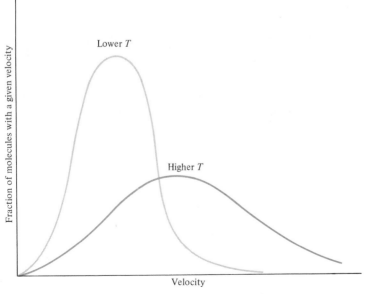

Figure 5–1. Maxwell's distribution of molecular velocities.

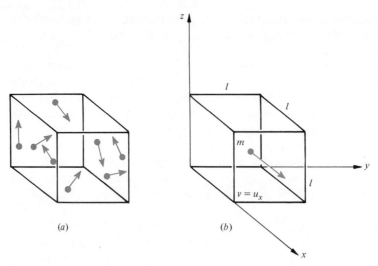

Figure 5-2. (a) N molecules moving in various directions. (b) Molecule moving along x axis with a velocity, u_x.

5-3. DERIVATION OF THE PERFECT GAS LAW

Some of the salient quantitative features of elementary kinetic theory will now be presented. The four basic assumptions given in Section 5–1 provide an adequate starting point. Consider a cube with side l containing N molecules of mass m. For the moment we concentrate on one molecule assumed to be traveling along the x direction with the velocity u_x (Figure 5–2b). When this molecule strikes the wall, it will rebound with the same velocity, but in the opposite direction, namely $-u_x$. Since momentum* has direction as well as magnitude, the change in momentum upon collision, Δp, is

$$\Delta p = mu_x - (-mu_x)$$
$$\Delta p = 2mu_x \tag{5-1}$$

The time between collisions with this wall is the time required for the molecules to travel to the back wall, rebound and return to the front wall, that is, the time required for the molecule to travel $2l$

$$\text{time} = t = 2l/u_x \tag{5-2}$$

The force exerted by the molecule at the front wall according to one form of Newton's second law of motion is

$$\text{force} = F = \Delta p/t \tag{5-3}$$

---

*Momentum is a term that describes the quantity of motion possessed by an object. For an object traveling in a straight line it is defined as the product of the mass and velocity.

Substituting equations (5-1) and (5-2) into (5-3)

$$F = \frac{2mu_x}{2l/u_x}$$

$$F = mu_x^2/l \qquad (5\text{-}3a)$$

Finally, the pressure, P', exerted by this molecule against the wall is force per unit area, and the wall area is simply l^2, so

$$P' = F/l^2 \qquad (5\text{-}4)$$

and substituting equation (5-3a)

$$P' = mu_x^2/l^3 \qquad (5\text{-}4a)$$

Equation (5-4a) is the pressure per molecule, so for N molecules, the pressure, P, would be

$$P = Nmu_x^2/l^3 \qquad (5\text{-}4b)$$

Now, of course, the molecules do not travel only along the x axis, nor along any of the axes, but rather in all possible directions Figure (5-2a). Therefore, if we consider the resolution or separation of the actual velocities of the molecules into their *component* velocities along the x, y, and z axes, the relationship between the average square of the velocity, c^2, and the average of the squares of the components of the velocity along the axes is

$$c^2 = u_x^2 + u_y^2 + u_z^2 \qquad (5\text{-}5)$$

But on the average molecules have no preferred direction, so

$$u_x^2 = u_y^2 = u_z^2 \qquad (5\text{-}6)$$

Thus, in view of equations (5-5) and (5-6), equation (5-4b) may be rewritten in terms of the **mean square velocity, c^2.**

$$P = Nm\frac{c^2}{3}/l^3$$

or

$$PV = \tfrac{1}{3}Nmc^2 \qquad (5\text{-}7)$$

where l^3 has been replaced with the volume of the container, V. Equation (5-7) is sometimes referred to as the perfect gas equation since it results from the assumption of a gas composed of mechanically "perfect" particles. Comparison with the ideal gas law as given by equation (4-11), $PV = nRT$, shows that

$$\tfrac{1}{3}Nmc^2 = nRT$$

But since

$$n = Nm/\text{MW}$$

we may write

$$\tfrac{1}{3}Nmc^2 = \frac{NmRT}{\text{MW}}$$

or

$$c^2 = \frac{3RT}{\text{MW}} \qquad (5\text{--}8)$$

Equation (5–8) may be used to obtain the **root mean square velocity** of a gas.

$$c = \sqrt{\frac{3RT}{\text{MW}}} \qquad (5\text{--}8a)$$

This velocity is simply the square root of the mean square velocity and is physically very nearly the average speed of the molecules.

Care must be taken to use the value 8.32×10^7 erg/mole °K for R in this equation in order for velocity to be expressed in units of centimeters per second. For example, an H_2 molecule is calculated to have an average velocity of 2.59×10^4 cm/sec at 0°C. It is interesting to note that, according to equation (5–8a), the velocity of gas molecules is dependent only upon their molecular weight and the absolute temperature.

The kinetic energy of 1 gram molecular weight of a gas is given by the expression

$$KE = \tfrac{1}{2}(\text{MW})c^2 \qquad (5\text{--}9)$$

because the total mass of the molecules of this amount of a substance is simply its molecular weight. If the mean square velocity in equation (5–9) is replaced by the expression provided by equation (5–8), a very interesting result is obtained

$$KE/\text{mole} = \tfrac{1}{2}(\text{MW}) \times \frac{3RT}{\text{MW}}$$

$$KE/\text{mole} = \tfrac{3}{2}RT \qquad (5\text{--}9a)$$

Equation (5–9a) says that the kinetic energy per mole for a gas is dependent only upon the absolute temperature of the gas. The identity of the gas is immaterial, and consequently we can conclude that the molecules of any two gases at the same temperature have the same kinetic energy per mole. An expression for the kinetic energy of any amount of gas is easily obtained

$$KE = \tfrac{3}{2}nRT \qquad (5\text{--}9b)$$

where n is the number of gram-moles of substance. From the way in which equations (5–9a) and (5–9b) have been derived, it would seem that they have validity only for substances obeying equation (4–11), that is, for ideal gases. Actually, however, concealed within these expressions is a far more general relationship than one limited to matter in the gaseous state (as will be shown in the next chapter).

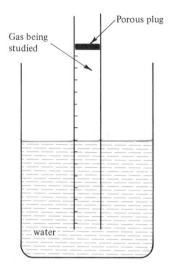

Figure 5-3. Graham's diffusion-tube to study the interdiffusion of different gases with air.

5-4. GRAHAM'S LAWS OF DIFFUSION AND EFFUSION

A remarkable series of experiments on the movement of gases into a vacuum and on the intermingling of gases under uniform conditions of pressure and temperature were carried out by Thomas Graham over the period from around 1826 to 1846. His experiments were on the interdiffusion of different gases in air. A description of his experimental method will illustrate the meaning of the term **diffusion.**

A "diffusion tube" (Figure 5–3) can be constructed by plugging one end of a calibrated glass tube with porous plaster. The open end is filled and immersed in a container of water (mercury can be used for gases which react with water). The gas to be studied is then bubbled into the tube to give the desired amount, and its volume is quickly recorded with the water levels inside and outside the diffusion tube the same. Air and the gas then diffuse across the porous plug until all the gas has diffused out of the tube and has been replaced with air. At all times, the water level in the container is adjusted to keep it equal to the water level in the diffusion tube. In this way, the pressure across the porous plug does not change. The volume of the gas that diffuses out is, of course, equal to the original volume of gas taken because it is entirely replaced with air at the end of the experiment. The volume of air that diffuses into the tube, however, is always found to be larger than the original volume of gas if air diffuses more rapidly than the gas, and smaller if the gas diffuses more rapidly than air. The amount of gas that diffuses during the time of the experiment must be proportional to its original volume, and the amount of air that diffuses into the tube must therefore be proportional to the volume of gas plus or minus the change in volume occurring in the tube during the experiment.

Average rate of gas diffusing out of the diffusion-tube: $r_{gas} \propto V_0/t$

Average rate of air diffusing into the diffusion-tube: $r_{air} \propto (V_0 + \Delta V)/t$

Here, V_0 is the original gas volume, ΔV is the change in volume (positive or negative), and t is the time of the experiment. Therefore

$$\frac{r_{\text{gas}}}{r_{\text{air}}} = \frac{V_0}{V_0 + \Delta V} \tag{5-10}$$

Graham discovered that the ratio of the rates of diffusion is inversely proportional to the square root of the ratio of the densities of the gas and air. Table 5-1 presents some data for various gases interdiffusing with air compared with the square root of the inverse ratio of the densities.

The results of Graham's work show that for the interdiffusion of two gases occurring at uniform pressure and temperature, the relationship is

$$\frac{r_1}{r_2} = \sqrt{\frac{d_2}{d_1}} \quad \text{(uniform pressure and temperature)} \tag{5-11}$$

An alternate and experimentally useful form of equation (5–11) is derived by using equation (4–11d) in a slightly rearranged form

$$d = \frac{(\text{MW})\, P}{RT}$$

which says that at constant temperature and pressure

$$d \propto \text{MW}$$

So if the molecular weight is substituted for density in equation (5–11) we find

$$\frac{r_1}{r_2} = \sqrt{\frac{(\text{MW})_2}{(\text{MW})_1}} \quad \text{(uniform pressure and temperature)} \tag{5-11a}$$

Equation (5–11a) may be used to determine the molecular weight of an unknown gas by allowing interdiffusion of the unknown gas and a reference gas (air, for example) under the conditions of the Graham experiment. The left-hand side of equation (5–11a) is thus determined experimentally and, because the molecular weight of the reference gas is known, the molecular weight of the unknown is easily calculated. If air is used as the reference gas, an "average molecular weight"

Table 5-1. Comparison of Graham's Experimental Interdiffusion Rates with the Inverse Square Root of Density

Gas	Experimental $r_{\text{gas}}/r_{\text{air}}$	Calculated $\sqrt{d_{\text{air}}/d_{\text{gas}}}$
H_2	3.83	3.79
CH_4	1.34	1.34
CO	1.01	1.02
N_2	1.01	1.02
O_2	0.95	0.95
CO_2	0.81	0.81

Table 5-2. Major Constituents of Dry Air

Gas	Mole Fraction
nitrogen	0.78
oxygen	0.21
argon	0.01

is used for it. Although air is a gaseous solution and not a compound, the composition of air is practically constant and, for calculations, may be assigned a "molecular weight" of 29.0. Table 5-2 gives the composition of air. Water vapor is a variable constituent of air and the data given are for "dry air."

The process of the **effusion** of gases into a vacuum through a hole in a thin plate was also studied by Graham (Figure 5-4). Although the quantitative expression of the relative rates of effusion of two gases is the same as that for diffusion, it should be recognized that the processes are entirely different. Even in modern times there has been some confusion between the terms diffusion and effusion.

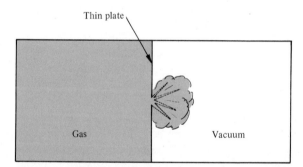

Figure 5-4. Effusion of a gas into a vacuum. Effusion—escape of a gas into a vacuum through a hole in a thin plate. Rate of effusion is inversely proportional to the square root of the molecular weight of the gas.

PROBLEMS

1. How are mean free path and rate of diffusion related?
2. Describe the dependence of the average velocity of gaseous molecules on temperature. Describe how the distribution of molecular velocities changes with temperature.
3. Distinguish between the ideal gas equation and the perfect gas equation with respect to their scientific origin.
4. Explain the term "root mean square" velocity.
5. Calculate the kinetic energy of one mole of any gas at 25°C.

6. Distinguish between the terms diffusion and effusion.

7. Use the result of kinetic molecular theory that two gases at the same temperature have the same average molecular kinetic energies to derive Graham's law of effusion.

8. The mean free path of nitrogen molecules at 27°C and 1.00 atm is 1.28×10^{-5} cm. Calculate the average number of collisions per second made by a nitrogen molecule under these conditions.

9. (a) How would the average kinetic energy of water molecules at 25°C compare with the average kinetic energy of hydrogen molecules at 25°C?

   (b) How would the average speed of water molecules compare with the average speed of hydrogen molecules at 25°C?

10. What would the temperature be if **(a)** the kinetic energy per mole of helium atoms is 840 cal? **(b)** the kinetic energy per mole of oxygen molecules is 840 cal? **(c)** the root mean square velocity of helium atoms is 3.00×10^4 cm/sec? **(d)** the root mean square velocity of oxygen molecules is 3.00×10^4 cm/sec?

11. Consider 1 mole of a gas. Indicate the way in which the following variables would change under the influence of the various imposed conditions. Use + for increase, − for decrease; 0 for no change.

Imposed Condition	P	V	T	Molecular Velocity, C	Molecular Kinetic Energy
P is increased at constant T					
V is increased at constant P					
T is decreased at constant V					
P is doubled and V is halved					

12. Calculate the relative rates of effusion of SO_2 and CH_4.

13. O_2 effuses through a small hole at a rate one fourth as great as does an unknown pure gas. Calculate the molecular weight of the unknown gas.

14. 100 ml of a certain gas, X_2 is placed in a diffusion tube. Air is allowed to diffuse into the tube, and at the end of the experiment 125 ml of air occupy the diffusion tube. Calculate the molecular weight of X_2.

15. Calculate the root mean square velocity of H_2 molecules at 0°C and at 100°C.

16. What must be the temperature of He atoms in order that their root mean square velocity be the same as that of H_2 molecules at 0°C?

17. Compare the root mean square velocities of O_2 molecules and SO_2 molecules at the same temperature.

18. Compare the root mean square velocities of a given molecular species at 27°C and at 327°C.

19. Use the data of Table 5–2 to justify the use of 29.0 as the "molecular weight" of dry air.

20. In the derivation of the perfect gas equation a cubical container was assumed for simplicity. Would you expect the same or a different final result if a different container shape were assumed? Why?

KINETIC MOLECULAR THEORY AND LIQUIDS AND SOLIDS

Boyle detected no deviations from his law. His experimental apparatus was somewhat crude, the pressure variation over which his experiments were carried out was relatively small, and air does not show deviations from Boyle's law except at high pressures or very low temperatures. It was early suspected that deviations should be found because neither Boyle's nor Charles' laws makes any allowance for the condensation of a gas.

By carrying out Boyle's experiment in a thick-walled glass capillary tube in which gases are compressed to pressures up to 3000 atm, large deviations from the gas laws were detected in the nineteenth century (Figure 6-1). The use of steel equipment in modern times has allowed even higher pressures to be achieved, and many gases have been studied to determine deviations from the gas laws. The simple gas laws apply best when the pressure is relatively low (a few atmospheres or less) and the temperature is high enough to be relatively far from the

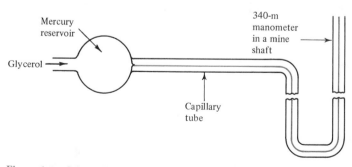

Figure 6-1. Schematic diagram of the apparatus used to study gases under pressures up to nearly 3000 atm.

69

Table 6-1. Comparison of Observed and Calculated Pressures for the Compression of Fixed Amounts of Hydrogen and Carbon Dioxide at 40°C

Gas	Initial Volume at 1.0 atm (liters)	Final Volume (liters)	Calculated Pressure Using Boyle's Law (atm)	Observed Pressure (atm)
hydrogen	26	0.38	68	71
carbon dioxide	26	0.38	68	50

condensation point of the gas. For most substances that are gaseous at ordinary room conditions, Boyle's and Charles' laws are quite accurate for pressures up to a few atmospheres and temperatures near room temperature or higher.

The degree of deviation from calculated pressures and volumes varies greatly with different substances as might be expected. Table 6–1 presents data for the experimental and calculated values of pressure for two substances of widely differing behavior. As can be seen from the data of Table 6–1, deviations of calculated values from observed values may be positive or negative and they may be relatively small (a few per cent for hydrogen over a large pressure range) or fairly large (as in the case of carbon dioxide). Normally, the more easily condensable gases show larger deviations than the permanent (difficult to condense) gases such as argon, carbon monoxide, helium, and hydrogen. Consider the boiling points of several substances at 1 atm pressure.

Gas	Boiling Point (°C)
hydrogen	−253
nitrogen	−196
carbon dioxide	−79

The lower the boiling point of a substance, the farther removed is the substance from liquid (nongaseous) behavior at higher temperature. We have seen that carbon dioxide displays much greater deviation from Boyle's law at 40°C than does hydrogen. Judging from the boiling points given above, one might conclude that nitrogen would show greater deviations from Boyle's law than hydrogen but less than carbon dioxide. This conclusion is borne out by experiment.

6-2. CONDENSED PHASES OF MATTER

It is known from experience that sufficient cooling of a gas usually results in its condensation to a liquid. A **liquid** has a definite volume but not an independent shape and simply assumes the shape of its container. Most liquids are nearly incompressible. Further cooling of a liquid eventually causes its solidification. A **solid** is a substance that does not flow; it possesses a definite shape and volume. Solids are essentially incompressible.

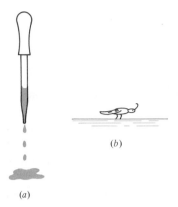

(b)

(a)

Figure 6-2. Two effects of surface tension. (a) Liquid drops are nearly spherical. (b) Light insects can walk on water.

Small particles of a liquid are observed to assume a nearly spherical shape as they drop. This effect is a consequence of the tendency of all liquids to minimize their surface area for a given volume. The tendency of a liquid to resist the expansion of its surface area is measured by its surface tension. The units of surface tension are force per unit length. This effect causes a liquid to behave as if its surface were covered by a thin, stretched membrane. Of course, it must be understood that no actual membrane covers the surface of a liquid, but the effect produced by the surface tension phenomenon causes the liquid to *act* as if there were such a covering. The higher the surface tension, the "tougher" the membrane.

Water has a relatively high surface tension (72 dynes/cm at 25°C) and, as a consequence, certain very light-weight insects can walk on the surface of ponds and lakes. Surface tension decreases as temperature increases until finally, when the liquid is converted to the gaseous state, the surface tension becomes zero.

Two types of solids are recognized—crystalline and amorphous. **Crystalline** solids are characterized by a regular geometric configuration or structure. Thus,

Figure 6-3. Crystalline solids show a variety of geometrical patterns. Left to right: calcite, pyrite, quartz, and corundum.

71

even though individual crystals of sodium chloride may look different depending upon how they are grown, the adjacent faces of all sodium chloride crystals meet at an angle of precisely 90°. **Amorphous** solids show no such regularity of structure. Glass is an example of an amorphous solid.

Most known substances are capable of existing, under appropriate conditions, in a gaseous, liquid, and solid phase. A much smaller number of compounds are also capable of existing in a fourth condition, intermediate with respect to a true solid and liquid, known as the **liquid-crystal** phase. The first recorded observation of this interesting phenomenon was made in 1888 by the Austrian botanist Fredrick Reinitzer. He noticed that the compound cholesteryl benzoate changed from a solid to a cloudy "liquid" at 145°C and remained in this condition until the temperature reached 179°C, at which point the "liquid" became clear. That the cloudy "liquid" was actually different from a true liquid phase was demonstrated shortly thereafter by the German physicist O. Lehmann, who showed that the cloudy condition was associated with a crystal like structure of the cholesteryl benzoate molecules. Lehmann suggested the appropriate name "liquid-crystal" for this phase.

Crystalline solids often have properties that vary with direction. Some crystals, for example, have different colors depending on the direction in which they are viewed. Substances displaying properties dependent on the direction in which they are measured are said to be **anisotropic,** whereas those with properties that are the same in all directions are called **isotropic.** All liquids and gases are isotropic. Many solids and liquid-crystals are anisotropic.

6-3.   EXTENSION OF KINETIC MOLECULAR THEORY TO CONDENSED PHASES

Although gases were the first state of matter to which kinetic molecular ideas were applied, it became evident in the development of the theory that condensed phases of matter could also be treated using the basic model.

Let us return to equation (5–9a), which was derived by a comparison of the perfect gas equation with the ideal gas equation. This expression can also be used to calculate the kinetic energy for molecules in the liquid or solid states. Thus the average kinetic energy of molecules in a piece of ice at 0°C is the same as the average kinetic energy of molecules in liquid water at 0°C. Indeed, the molecules of any substance in any state of aggregation possess the same average kinetic energy as do the molecules of any other substance as long as both substances are at the same temperature!

The reason for this rather startling result is that the total energy possessed by molecules is separable into various forms which are largely independent of each other. Thus, a diatomic molecule such as oxygen, O_2, may possess a given amount of energy, distributed among kinetic energy, vibrational energy associated with the chemical bond, energy related to the internal structure of the atoms, and possibly other forms of internal energy. We may write the total energy as

$$E_{total} = KE + \text{other forms of energy} \qquad (6\text{–}1)$$

The "other forms of energy" may also be split up in equation (6–1), but for the moment we are only interested in the fact that the kinetic energy of the molecules of a substance can be separated from other forms of molecular energy.

Since equation (5-9a) or (5-9b) expresses a relationship between the absolute temperature and *only* the kinetic energy of molecules, it is seen that equation (6-1) also suggests that the same relationship may exist between the kinetic energy of molecules and their absolute temperature for any substance. This general result is one of the most important deductions from kinetic molecular theory. We must be clear that unless the same *number* of molecules is being considered for two different substances, it would be inaccurate to say that the *total* kinetic energy of the molecules for the two substances is the same at the same temperature. The molecular kinetic energies *per mole* would be the same, or the *average* kinetic energies for the molecules of the two substances would be the same regardless of the total kinetic energies. Kinetic molecular theory thus gives a physical meaning to absolute temperature in terms of the kinetic energies of molecules. **Absolute zero is the point at which the molecules have no kinetic energy, that is, molecular motion has ceased.**

We may now consider the kinetic molecular interpretation of phase changes. If a gas such as steam is cooled, the average kinetic energy of the gaseous water molecules is reduced. Eventually, a temperature will be reached at which condensation into liquid water will begin, and at 1 atm this temperature is 100°C. The kinetic energy of the molecules, and hence their average speed, has been reduced to a point where the attractive forces between the molecules become important and cause them to stick together, that is, to condense.

An interesting thing is now observed as further cooling occurs. The temperature of the steam and liquid water will remain at 100°C until all of the steam has condensed to the liquid state. Yet for each gram of water condensed, 540 cal must be removed from the steam. In other words, steam at 100°C has more energy than liquid water at 100°C! This energy, which must be removed from a vapor to cause condensation (or supplied to a liquid to cause vaporization) with no change in temperature, is called the **latent heat of vaporization.** Clearly, this latent energy does not affect the kinetic energies of the molecules, because both the liquid and vapor molecules must have the same average kinetic energies at the same temperature.

An early suggestion was that latent energies changed the **potential energies** of the molecules, that is, the degree of aggregation of the molecules. Today, we use the term **entropy** to indicate the degree of molecular disorder in a system. A substance at a given temperature has a greater entropy per unit weight when it is gaseous than when it is in the liquid state, and the latent heat of vaporization, when removed from a gas as it condenses, serves to lower the entropy (reduce molecular disorder) of the substance. Of course, in boiling, the latent heat of vaporization is supplied to the liquid to convert it to a gas, thereby increasing the entropy of the substance (increasing molecular disorder). Table 6-2 lists latent heats of vaporization for a few common substances. The unusually high heats of vaporization for ammonia, alcohol, and especially, water suggest that molecules of these substances exert great attractive forces on each other. As we shall see later, the nature of these unusual attractive forces can be nicely explained.

The **latent heat of fusion** is associated with changes between the solid and liquid states. Thus, for each gram of water at 0°C that freezes to ice at 0°C, 80 cal must be removed. Again, the loss of energy serves to decrease the entropy of the water as the molecules become more ordered in the solid state. Table 6-3 lists some latent heats of fusion for a few common substances.

Table 6-2. Latent Heats of Vaporization At The Normal Boiling Point

Substance	Boiling Point (°C)	Latent Heat of Vaporization (cal/g)
ammonia	−33	327
benzene	80	94
chloroform	62	59
ethyl alcohol	78	204
ethyl ether	35	84
sulfur dioxide	−11	95
water	100	540

Table 6-3. Latent Heats of Fusion At The Melting Point

Substance	Melting Point, (°C)	Latent Heat of Fusion (cal/g)
ammonia	−78	79
benzene	6	30
hydrogen chloride	−114	13
lead	327	1240
potassium nitrate	337	28
sodium chloride	808	116
water	0	80

Entropy changes (symbolized ΔS) for phase changes of pure substances are calculated by the equation

$$\Delta S = \frac{\Delta H_l}{T} \tag{6-2}$$

where ΔH_l represents a latent heat and T is the constant absolute temperature in degrees Kelvin at which the phase change occurs. The symbol, ΔH, universally represents the heat associated with any type of process at constant pressure including chemical reactions. In all reactions, the energy or heat of reaction is always considered to be a reactant.

$$\text{reactants} + \Delta H \longrightarrow \text{products}$$

If heat is absorbed, ΔH is positive; ΔH is negative if heat is released. (This is a purely arbitrary convention that should be remembered.) Thus, if the latent heat is released, ΔH_l is negative, and if absorbed, ΔH_l is positive. The entropy change per gram of ice which melts at 0°C is

$$\Delta S = \frac{80 \text{ cal/g}}{273\,°\text{K}}$$

$$= 0.29 \text{ cal/g }°\text{K}$$

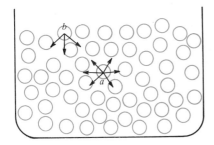

Figure 6-4. Surface tension in liquids is caused by intermolecular attractions. (*a*) A molecule in the bulk liquid is attracted equally in all directions. (*b*) A surface molecule is pulled toward the bulk of the liquid and thus resists expansion of the surface area.

and the entropy change per gram of liquid water freezing at 0°C is

$$\Delta S = \frac{-80 \text{ cal/g}}{273°\text{K}}$$

$$= -0.29 \text{ cal/g}°\text{K}$$

Primarily, the degree of intermolecular attraction determines the temperatures at which a substance condenses and solidifies. The larger the attractions, the higher the temperature at which the molecules will coalesce to a more ordered state. *We can now see the principal reason why real gases deviate from ideal behavior—intermolecular attraction.* As long as a gas is at a low pressure, the molecules remain relatively far apart and the intermolecular forces have little importance. Also, the higher the temperature, the greater the average kinetic energy; hence, the faster the molecules are traveling and the more energetically they rebound upon collision. Under these circumstances, the intermolecular forces are largely negligible. However, as pressure increases, forcing the molecules closer together on the average, and as temperature decreases, slowing down the molecules, the intermolecular forces begin to have an effect. The molecules tend to "stick" for a moment upon colliding, and the observed pressure is no longer the same as the ideal pressure. Another factor that leads to failure of the ideal gas laws arises from the fact that real molecules occupy space and do not have the total volume of the container within which to move about. At high pressure, where the density of the gas is great, the free volume in which the molecules can move may become significantly less than the volume of their container.

The surface tension of liquids is caused by intermolecular attraction. A molecule in the bulk of a liquid is subject to attractions in all directions from its neighbors. However, a molecule at the surface is attracted only downward into the liquid (Figure 6-4). Consequently, those molecules at the surface are pulled inward, and work would be required to increase the surface area of the liquid. The surface tension is a measure of the energy required to increase the surface area. The greater the intermolecular attractions, the higher the surface tension of the liquid.

Liquid-crystal phenomena usually arise in substances composed of long molecules that can align themselves preferentially in certain directions (Figure 6-5). When this alignment persists even after the bulk solid structure is destroyed, little "floating islands" with a fairly high degree of molecular order result in the liquid-crystal state.

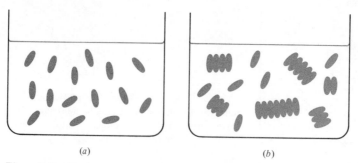

(a)   (b)

Figure 6–5. Liquid-crystals. (a) Ordinary liquids have molecules that are randomly oriented. (b) Liquid-crystals have molecules in clusters within which there is a high degree of orientation.

6–4.   KINETIC-MOLECULAR INTERPRETATION OF PHASE EQUILIBRIA

If a liquid is placed in a closed evacuated container which is maintained at a constant temperature, after a time a steady pressure will be exerted by molecules that have gone into the vapor state. The volume of the container has no effect on this steady pressure, as long as there is sufficient liquid to supply enough molecules for the vapor state. Only the temperature affects the value of the steady vapor pressure—the higher the temperature, the higher the vapor pressure. A liquid and its vapor which have reached this condition of steady vapor pressure are said to be in **equilibrium** (Figure 6–6). This constant value of the vapor pressure, dependent only upon the temperature is called the **equilibrium vapor pressure.**

The existence of an equilibrium vapor pressure is explained by our kinetic molecular picture of matter. Let us imagine the following sequence of events (Figure 6–7):

(a) Liquid initially placed in an evacuated container maintained at a constant temperature has no molecules in the vapor state.

(b) Almost immediately a few of the molecules in the liquid with high enough kinetic energies may escape into the vapor state. The necessary latent heat is supplied by the surroundings.

(c) As time passes, more molecules escape into the vapor state so that chances become great that some of the gaseous molecules will be recaptured by the liquid.

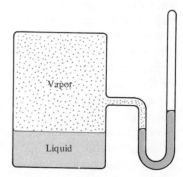

Figure 6–6. A constant vapor pressure is always reached when a liquid is placed in a closed container and held at a constant temperature.

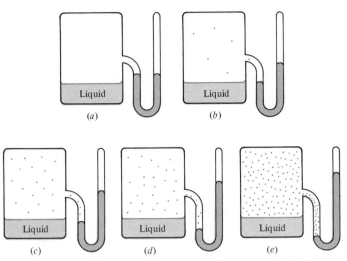

Figure 6-7. Attainment of liquid-vapor equilibrium as explained by the kinetic molecular theory. (*a*) Initially only liquid is present. (*b*) Some molecules escape to the vapor phase. (*c*) More molecules escape the liquid and some in the vapor phase are recaptured by the liquid. (*d*) Equilibrium is reached when the rates of escape and recapture are equal. (*e*) At a higher temperature, equilibrium is reestablished at a relatively higher rate of escape and recapture.

(d) After a while, the rate of escape of molecules from the liquid is exactly balanced by the rate of condensation of the gaseous molecules. The vapor pressure is now the equilibrium value and will not change as long as external conditions do not change.

(e) Elevation of the temperature causes the average kinetic energies of liquid and vapor molecules to increase equally, but the entropy of the system also increases, thus leading to increased molecular disorder and hence to more molecules in the gaseous state. Equilibrium will eventually be reestablished, but at a higher vapor pressure because a somewhat greater proportion of the molecules will be in the gaseous state.

A solid and its liquid are in equilibrium at the melting point of the solid. Again, the kinetic-molecular model pictures equilibrium to occur when the rate of escape of molecules from the solid to the liquid state is equal to the rate of recapture of liquid phase molecules by the solid phase. At equilibrium, the average kinetic energy of molecules is the same whether in the liquid or solid phases, but the entropy per gram of the liquid is, of course, greater than that of the solid, and this difference could be calculated by using equation (6-2).

Under certain conditions of pressure and temperature, it is possible for a solid to reach equilibrium with its molecules in the vapor phase. The equilibrium vapor pressure under such circumstances is called the **equilibrium sublimation pressure.** Most pure substances have a unique pressure and temperature at which solid, liquid, and vapor can coexist. Such an equilibrium pressure and temperature is called the **triple point.** For water, the triple point is $0.01\,°C$ and 4.63 torr. Ice, liquid water, and steam are in equilibrium under these conditions, and all three phases will remain until external conditions are changed.

The melting point of a substance is only slightly affected by pressure, but boiling and sublimation are greatly affected by the external pressure. It is a matter

77

of experience to anyone who has lived at a high elevation that water boils more easily (at a lower temperature) on a mountain than at sea level.

A simple way to summarize the conditions under which equilibrium can be attained between the different phases of a pure substance is by means of a pressure-temperature phase diagram. Figure (6–8) is a generalized phase diagram with the curves labeled. The vaporization curve can be looked upon as giving the equilibrium vapor pressures at different temperatures or as giving the boiling points of the liquid at different pressures. This relationship has already been considered in Chapter 2. The fusion line gives melting points at different pressures, and the sublimation curve gives sublimation pressures at different temperatures. These three curves meet at the triple point. The critical point represents the pressure and temperature above which the liquid state of a substance cannot be realized.

The meaning of this diagram and the information that it provides is made clear by the following analysis: Moving in a horizontal direction beginning midway up the pressure axis (dotted line in Figure 6–8) tells us that at that fixed pressure, the substance is a solid. Continuing to move in the same direction means that the temperature is rising while the pressure is held constant. When the dotted line crosses the fusion line, liquid begins to form. All the material will now remain at that fixed temperature and pressure while heat is being applied until all of the solid has melted. Continuing to add heat then causes the temperature of the (now) liquid substance to rise. When the vaporization line is reached, again there is an equilibrium, but now it is between liquid and vapor, and the temperature and pressure are again fixed until all the liquid has been converted into vapor. Further heating beyond that point causes the temperature of the gaseous substance to resume its rise.

Figure (6–9) is the phase diagram for carbon dioxide. Notice that the triple point occurs at a pressure of 5 atm. At 1 atm, only sublimation is possible and no liquid phase is observed. Dry ice thus does not melt at ordinary pressures

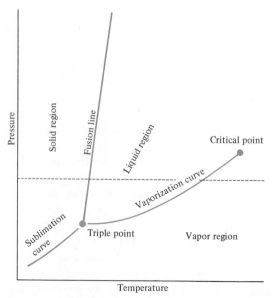

Figure 6-8. Generalized pressure-temperature phase diagram for a pure substance.

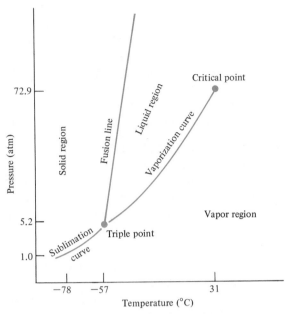

Figure 6-9. Pressure-temperature phase diagram for carbon dioxide (not to scale).

but passes directly to the vapor state. At pressures greater than the triple point pressure, liquid carbon dioxide can be realized.

The phase diagram for water (Figure 6-10) shows an interesting property of this substance. The solid phase is less dense than the liquid phase at the melting point. Consequently, it is possible to cause ice to liquify by the application of

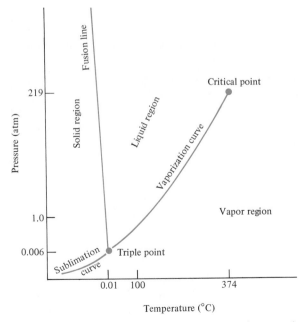

Figure 6-10. Pressure-temperature phase diagram for water (not to scale).

79

sufficient pressure. This property is reflected by the fact that the fusion line is tilted slightly to the left, showing that at constant temperature the application of pressure to ice can cause it to pass into the liquid phase. Most substances behave just the opposite and have their fusion lines tilted slightly to the right, for example, carbon dioxide (Figure 6–9). Thus, application of sufficient pressure to liquid carbon dioxide will cause it to solidify because the solid phase is denser than the liquid phase. This behavior is characteristic of most pure substances, that is their solid phases are denser than their liquid phases at their melting points.

6-5.   THE KINETIC MOLECULAR THEORY AS A SCIENTIFIC THEORY

We need to pause briefly to consider the impact of kinetic molecular theory on the physical sciences. At the beginning of Chapter 5, it was referred to as the "theory upon which much of modern physical science rests." What justification have we for this sweeping assertion? Let us consider the following points.

1. The basic kinetic molecular picture of matter is simple. It is simple to conceive and relatively few assumptions are necessary in its statement. Conceptually simple theories are superior to complex theories that involve many assumptions about the behavior of nature.
2. The kinetic molecular theory is a very general theory. It can be applied to explain very diverse phenomena. We have examined only a few so far, for example, pressure of gases, surface tension of liquids, melting of solids. We shall make use of it many times in our study to explain chemical phenomena. Its generality is so widely accepted among scientists that most "smaller theories" of today are built directly upon a kinetic molecular foundation without specific reference to the latter.
3. Quantitative agreement between observed and calculated properties of matter has been demonstrated time and time again. We looked at only a few quantitative applications with respect to gases, but many properties have been accurately calculated not only for gases but for condensed phases as well by using kinetic molecular theory.
4. Kinetic molecular theory has been a very fertile theory in suggesting new avenues of experimentation both for the purpose of testing its validity as well as uncovering new relationships. It has withstood the most active period of scientific activity in the history of man over the past hundred years. The basic postulates of kinetic molecular theory are accepted by all scientists.

To what degree has this theory affected our understanding of nature? Or, to put the question in a slightly different way—to what extent is the kinetic molecular model of matter true? A complete exploration of these questions would require a philosophical investigation far beyond the scope and purpose of this book. But a partial answer is that for many scientists and *most* chemists, the kinetic molecular view of matter is regarded as an accurate representation of the microscopic behavior of substances. In this sense, it is accepted as truth by many but not all scientists. However, all scientists, could probably agree that the kinetic molecular theory offers a convenient model for organizing our experience, and the general behavior of nature is as if the model were true.

PROBLEMS

1. Why did early workers detect no deviations from the ideal gas law? Under what conditions do significant deviations occur? What factors are principally responsible for such deviations?

2. The compressibility factor of a gas, z, is defined by the equation, $z = PV/nRT$, where P, V, and T represent actual, experimental values. Use the data of Table 6–1 to calculate the compressibility factors of H_2 and CO_2 at 40°C. What can be said about the magnitude of the compressibility factor and the relative degree of ideality of a gas? (Assume both gases exhibit ideal behavior at 1.0 atm in order to calculate n.)

3. What is the liquid-crystal phase? What types of molecules can assume this state?

4. Distinguish between crystalline and amorphous solids.

5. What physical interpretation is given to absolute zero by kinetic molecular theory? How does this interpretation differ from that which was developed in Chapter 4?

6. (a) How would the average kinetic energy of hydrogen molecules at 25°C. compare with the average kinetic energy of lead atoms at 25°C?
   (b) How would the average speed of hydrogen molecules compare with the average speed of lead atoms at 25°C?

7. If a block of gold and a block of silver are placed in contact and left for a long time so that diffusion occurs, will gold atoms diffuse farther into the silver or will silver atoms diffuse farther into the gold? Explain.

8. The Dutchman J. D. van der Waals proposed in 1873 an equation for gases of the form:

$$\left(P + \frac{n^2a}{V^2}\right)(V - nb) = nRT$$

where P, V, and T represent the pressure, volume, and absolute temperature of a real gas, n is the number of gram moles, R is the gas constant, and a and b are constants whose values must be determined experimentally for each gas considered. Compare van der Waals' equation with the ideal gas equation, $PV = nRT$, and explain qualitatively the purpose of the correction terms n^2a/V^2 and nb.

9. Explain qualitatively the following phenomena in terms of the kinetic molecular theory: (a) surface tension, (b) condensation of gases, (c) crystalline solids, (d) latent heat of vaporization, (e) equilibrium vapor pressure.

10. The molecular weights of ammonia, NH_3, and water, H_2O, are nearly the same. However, the melting and boiling points of water are much higher than those of ammonia. Which substance do you think should have the greater surface tension? Why?

11. Calculate the final temperature of 2.0 g of water if 500 cal of heat are absorbed, and if the 2.0 g of water originally consisted of 1.0 g of ice and 1.0 g of liquid water both at 0°C.

12. How much energy would be required to convert 1.0 g of solid benzene at its melting point to vapor at its boiling point. (See Tables 2–4, 6–2, and 6–3)

13. What is the melting point of a substance with a latent heat of fusion of 20 cal/g and an entropy of fusion of 0.15 cal/g °K?

14. (a) Calculate the entropy change when 1 g of solid benzene is changed to liquid benzene at 1 atm.
    (b) Calculate the entropy change when 1 gram-mole of benzene melts.

(c) According to kinetic molecular theory what is the difference between the solid and liquid benzene and how is this difference reflected by the entropy difference for the two phases?

15. Frederick Trouton in 1884 pointed out that many liquids have a *molar* entropy of vaporization at the normal boiling point of about 21 cal/mole °K.
    (a) Use the data of Table 6–2 and equation (6–2) to test the validity of this rule.
    (b) What is the physical significance of the fact that many liquids do have about the same molar entropy of vaporization at their normal boiling points?

16. What is the difference between the critical point and the triple point of a pure substance?

17. Ice skating is possible because a thin layer of liquid forms at the point where the blades of the skates come in contact with the ice and serve as lubrication. Use the phase diagram for water to explain this phenomenon. Do you think it would be possible to skate on solid carbon dioxide? Explain.

18. Consider the following pressure-temperature diagram for a pure substance:

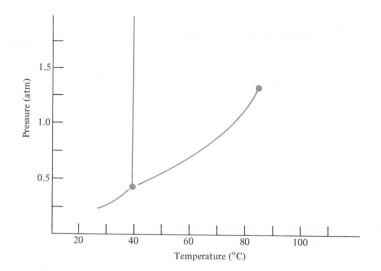

(a) Give the normal melting point.
(b) Give the normal boiling point.
(c) Give the pressure in atmospheres and the temperature in degrees Celsius below which sublimation is observed.
(d) Give the critical temperature in degrees Celsius and the critical pressure in atmospheres.
(e) At 1 atm and 60°C, in what phase is the substance?
(f) If the substance at 0.3 atm and 45°C were compressed while holding temperature constant, one would observe as the pressure increased that successive phase changes took place. Name the phases in order.

19. Sulfur dioxide has its triple point at −72.7°C and 16.3 torr. Solid SO_2 is more dense than liquid SO_2.
    (a) Sketch the pressure-temperature phase diagram for SO_2.
    (b) What would one observe if SO_2 at −75°C and 70 torr had its external pressure slowly reduced to 5 torr?

(c) What can be said about the average speed of SO_2 molecules in the solid, liquid, and vapor states at the triple point?

(d) What can be said about the entropy per gram of SO_2 in the three phases at the triple point?

20. What is the practical result if the triple point of a substance occurs at a pressure greater than 1 atm? less than 1 atm?

STOICHIOMETRY

7-1. CONSERVATION LAWS

Lavoisier's greatest contribution to chemistry was probably the insistence upon weight conservation in all chemical processes. The law of conservation of matter is undoubtedly one of the pillars upon which the science of chemistry is based. Lavoisier, however, was not able to find any generalizations that would explain the weights that combined in a given reaction. For example, in the two equations shown below, there is no obvious relation governing weights except for the equality of total weights on each side of the equations.

$$\text{sulfur} \quad + \quad \text{oxygen} \quad \longrightarrow \quad \text{sulfur dioxide}$$
$$(32.06 \text{ g}) \qquad (32.00 \text{ g}) \qquad\qquad (64.06 \text{ g})$$

$$\text{phosphorus} + \quad \text{oxygen} \quad \longrightarrow \quad \text{phosphorus pentoxide}$$
$$(123.88 \text{ g}) \qquad (160.00 \text{ g}) \qquad\qquad (283.88 \text{ g})$$

Dalton's atomic theory was a great step in understanding weight relations because his atoms had unique weights and they combined in definite proportions. Thus, the introduction of relative atomic weights made possible an understanding of combining weights. Moreover, another conservation law could be stated, which is really a restatement of conservation of matter: **atoms are conserved in a chemical reaction;** that is, the same number of each kind of atom is present before and after a chemical reaction.

Other conservation laws are employed, most notably the law of conservation of energy, which states that energy can neither be created nor destroyed. Einstein's theory of special relativity has altered these two conservation laws (matter and energy) by combining them into the statement that the sum of mass and energy must remain constant. This latter problem will concern us later and will not be pursued further here.

The law of combining volumes enunciated by Gay-Lussac and explained by Avogadro was a form of nonconservation law of molecules.

$$3\,H_2 + N_2 \longrightarrow 2\,NH_3 \tag{7-1}$$

$$Cl_2 + H_2 \longrightarrow 2\,HCl \tag{7-2}$$

$$2\,H_2 + O_2 \longrightarrow 2\,H_2O \tag{7-3}$$

The three equations shown are statements of the numbers of molecules of each reactant required to produce an equal weight of product. But note that, although the atoms are conserved in each process, the number of molecules is usually not. In equation (7-1) a total of four molecules of reactants (hydrogen and nitrogen) produce two molecules of product (ammonia). Although this is quite obvious and seldom emphasized, the student is urged to bear in mind that the numbers of molecules involved in chemical reactions are not additive; that is, **molecules are not conserved.** Of course, they should not be since a chemical reaction always means that new chemical substances (molecules) are formed.

7-2. CHEMICAL EQUATIONS AND MOLES

An equation such as (7-1), (7-2), or (7-3) above conveys much more information than is normally supposed by the beginning student. Equation (7-1) for example, states

1. The kinds of atoms that react to produce ammonia.
2. The molecular composition of each reactant and product.
3. The molecular ratios of reactants and products.
4. The direction in which reaction is expected to occur.

Thus in equation (7-1), we are told that three molecules of hydrogen are required to combine completely with one molecule of nitrogen to produce two molecules of ammonia. The combination of four molecules of hydrogen with one molecule of nitrogen can therefore only lead to two molecules of ammonia with one molecule of hydrogen being left over. Thus in determining the amount of ammonia produced from given quantities of hydrogen and nitrogen, one must first determine which reactant will limit the amount of product formed.

■ **Example 7-1**

Which reactant is in excess in each of the following mixtures? How much excess?
(a) $10\,g\,H_2 + 10\,g\,N_2$?
(b) $8.0\,g\,H_2 + 18\,g\,Cl_2$?
(c) $5.0\,g\,H_2 + 50\,g\,O_2$?

□ *Solution*

(a)
$$10\,g\,H_2 = \frac{10\,g}{2.0\,g/mole} = 5.0 \text{ moles } H_2$$

$$10\,g\,N_2 = \frac{10\,g}{28\,g/mole} = 0.36 \text{ mole } N_2$$

Required ratio of $H_2/N_2 = 3/1 < 5/0.36$. Therefore H_2 is in excess.

$$H_2 \text{ required} = 3 \times 0.36 = 1.08 \text{ moles}$$

$$\text{excess } H_2 = 5.00 \text{ moles} - 1.08 \text{ moles} = 3.92 \text{ moles}$$

$$3.92 \text{ moles} \times 2.0 \text{ g/mole} = 7.8 \text{ g } H_2$$

(b)
$$8.0 \text{ g } H_2 = \frac{8.0 \text{ g}}{2.0 \text{ g/mole}} = 4.0 \text{ moles } H_2$$

$$18 \text{ g } Cl_2 = \frac{18 \text{ g}}{70.92 \text{ g/mole}} = 0.25 \text{ mole } Cl_2$$

Required ratio of $H_2/Cl_2 = 1/1 = < 4.0/0.25$. Therefore H_2 is in excess.

$$H_2 \text{ required} = 0.25 \text{ mole} \times 1 = 0.25 \text{ mole}$$

$$\text{excess } H_2 = 4.0 \text{ moles} - 0.25 \text{ moles} = 3.8 \text{ moles}$$

(c)
$$5.0 \text{ g } H_2 = \frac{5.0 \text{ g}}{2.0 \text{ g/mole}} = 2.5 \text{ moles } H_2$$

$$50 \text{ g } O_2 = \frac{50 \text{ g}}{32 \text{ g/mole}} = 1.56 \text{ moles } O_2$$

Required ratio of $H_2/O_2 = 2/1 > 2.5/1.56$. Therefore O_2 is in excess.

$$O_2 \text{ required} = \tfrac{1}{2}(2.5 \text{ moles}) = 1.25 \text{ moles}$$

$$\text{excess } O_2 = 1.56 - 1.25 = 0.31 \text{ moles} \times 32 \text{ g/mole} = 9.9 \text{ g}$$

7-3. HEAT OF REACTION

All chemical reactions involve not only rearrangements in chemical composition but also concomitant changes in energy. Chemical substances like coal, TNT, or gasoline possess considerable amounts of chemical potential energy which can be released when the substance is caused to react. In these cases the reactions are

$$C + O_2 \longrightarrow CO_2 \tag{7-4}$$

$$4 C_7H_5N_3O_6 + 21 O_2 \longrightarrow 28 CO_2 + 10 H_2O + 6 N_2 \tag{7-5}$$

$$2 C_8H_{18} + 25 O_2 \longrightarrow 16 CO_2 + 18 H_2O \tag{7-6}$$

But although these equations are balanced in that the same numbers of each kind of atom appear on each side of the equations, there is one item left out of each equation—energy. All three of these reactions are known to produce great quantities of energy, mostly heat but also light and mechanical energy. Therefore energy must be included if the equations are to be completely meaningful. Energy can also be a reactant in chemical processes; that is, energy may be absorbed in some processes as well as released in others. As stated in Chapter 6, by convention, energy is considered to be a *reactant*. Thus, if energy is released, the amount of energy *required* is negative and an equation like (7–4) is written

$$C(\text{graphite}) + O_2(g) \longrightarrow CO_2(g) \qquad \Delta H = -17.9 \text{ kcal/mole}$$

The terms in parentheses refer to the physical state of the substance, that is, (g) = gas. To be completely unambiguous, conditions of temperature and pressure should also be specified.

■ Example 7–2

Calculate the amount of energy released during the combustion of one ton of coal.

□ *Solution*

$$C + O_2 \longrightarrow CO_2 \qquad \Delta H = -17.9 \text{ kcal/mole}$$

$$1 \text{ ton} = 2000 \text{ lb} \times \frac{454 \text{ g}}{\text{lb}} = 9.08 \times 10^5 \text{ g}$$

$$\text{gram-moles C} = \frac{9.08 \times 10^5 \text{ g}}{12.0 \text{ g/mole}} = 7.57 \times 10^4 \text{ moles C}$$

$$\text{heat liberated} = 17.9 \text{ kcal/mole} \times 7.56 \times 10^4 \text{ moles}$$

$$= 1.36 \times 10^6 \text{ kcal}$$

Example 7–2 demonstrates that the heat of reaction can be treated like a molar quantity.

■ Example 7–3

The complete combustion of an unknown amount of carbon was observed to evolve 25.0 kcal of heat energy. How much carbon was present in the sample?

□ *Solution*

The equation is

$$C + O_2 \longrightarrow CO_2 \quad \Delta H = -17.9 \text{ kcal/mole}$$

To solve for the amount of carbon

$$\text{moles C} = \frac{-25.0 \text{ kcal}}{-17.9 \text{ kcal/mole}} = 1.40 \text{ moles C}$$

7-4. PROBLEMS INVOLVING CHEMICAL REACTIONS

Of course, the most prominent use of chemical equations is in calculating the amount of product to be expected from given amounts of reactants. Knowledge of the meaning of a balanced equation makes this type of calculation very simple. Such calculations should always be carried out in terms of moles. An example will illustrate.

■ Example 7-4

Calculate the weight of carbon dioxide that would be produced by the complete combustion of 36 g of carbon.

☐ *Solution*

Step 1. Write and balance the equation.

$$C + O_2 \longrightarrow CO_2$$

Step 2. Convert the quantities given to moles.

$$\frac{36 \text{ g C}}{12 \text{ g/mole}} = 3.0 \text{ moles C}$$

Since the combustion is stated to be complete, an excess of O_2 can be assumed. Therefore the amount of carbon will determine the amount of product formed.

Step 3. Using the equation, determine how many moles of product are formed from 3.0 moles of carbon. The answer is 3.0.

Step 4. Convert moles of product to grams of product.

$$3.0 \text{ moles CO}_2 \times 44 \frac{g}{mole} = 132 \text{ g CO}_2$$

If the weights of all the reagents are stated, it is essential first to determine which one (the one not in excess) will determine the amount of product formed.

■ Example 7-5

How much silver chloride is formed by reacting 100 g each of sodium chloride and silver nitrate?

☐ *Solution*

$$NaCl + AgNO_3 \longrightarrow AgCl + NaNO_3$$

The mole ratio required by this equation is 1/1.

$$100 \text{ g NaCl} \div 58.44 \text{ g/mole} = 1.71 \text{ moles NaCl}$$

$$100 \text{ g AgNO}_3 \div 169.9 \text{ g/mole} = 0.589 \text{ mole AgNO}_3$$

Because $AgNO_3$ is in the smaller molar quantity, it determines the amount of AgCl formed.

$$\text{weight AgCl produced} = 0.589 \text{ mole} \times 143.4 \frac{\text{g}}{\text{mole}}$$

$$= 84.5 \text{ g AgCl}$$

■ **Example 7-6**

What volume of water vapor measured at STP can be obtained by the combination of 10.0 g of hydrogen with 10.0 g of oxygen.

☐ *Solution*

$$\text{moles of H}_2 = \frac{10.0 \text{ g}}{2.0 \text{ g/mole}} = 5.0 \text{ moles}$$

$$\text{moles of O}_2 = \frac{10.0 \text{ g}}{32.0 \text{ g/mole}} = 0.313 \text{ mole}$$

The equation

$$2 \text{ H}_2 + \text{O}_2 \longrightarrow 2 \text{ H}_2\text{O}$$

tells us that hydrogen is in excess and that $(5.00 - 0.626)$ moles of hydrogen will remain unreacted at the end of the reaction. The amount of water produced, then, is limited by the amount of O_2 present. Thus 0.626 mole of water will be produced.

$$\text{volume H}_2\text{O vapor at STP}^* = 0.626 \text{ mole} \times \frac{22.4 \text{ liters}}{\text{mole}} = 14.0 \text{ liters.}$$

■ **Example 7-7**

Calculate the weight and volume amount of carbon dioxide produced by the complete combustion of 1.00 g of carbon.

☐ *Solution*

The equation for the combustion of carbon is $C + O_2 \longrightarrow CO_2$, which tells us

---

*STP is an abbreviation for standard temperature and pressure, which is taken as 0°C (273°K) and 1 atm.

that for each mole of carbon burned, 1 mole of CO_2 is produced. To determine the number of moles of carbon, the use of proper units is helpful

$$\frac{1.00 \text{ g C}}{12.0 \text{ g/mole}} = 0.0833 \text{ mole C}$$

From the equation, we now know that 0.0833 mole of CO_2 is formed. The final solution is thus obtained by determining the amount of CO_2 in the units desired

$$\text{weight } CO_2 = 0.0833 \text{ mole} \times 44.0 \frac{\text{g}}{\text{mole}} = 3.67 \text{ grams}$$

$$\text{volume } CO_2 \text{ at STP} = 0.0833 \text{ mole} \times 22.4 \frac{\text{liters}}{\text{mole}} = 1.87 \text{ liters (STP)}$$

7-5. MOLARITY

The mole fraction concentration scale was discussed in Chapter 4. A frequently used method for expressing the concentration of a given solute in a solution is the molarity scale

$$\text{molarity} = \frac{\text{moles solute}}{\text{liter of solution}}$$

A capital M is used to designate molarity. A bottle labeled 0.50 M $C_{11}H_{22}O_{11}$ would indicate that each liter of the solution contains 0.50 mole of the sugar sucrose. The definition of molarity does not depend on the identity of the solvent.

Molarity is a convenient concentration unit because the moles of a particular solute can be calculated if the volume of solution used is known. In the above example, it can be seen that 10 ml (0.010 liter) of the 0.50 M $C_{11}H_{22}O_{11}$ contains 0.005 mole of $C_{11}H_{22}O_{11}$.

$$0.010 \text{ liter} \times 0.50 \frac{\text{mole}}{\text{liter}} = 0.0050 \text{ mole}$$

In order to prepare 250 ml (0.25 liter) of the 0.50 M $C_{11}H_{22}O_{11}$, the amount of sugar needed can be calculated.

$$0.50 \frac{\text{mole}}{\text{liter}} \times 0.25 \text{ liter} \times 329 \frac{\text{g}}{\text{mole}} = 41 \text{ g}$$

One would weigh out 41 g of sugar and add enough solvent to give a final volume of 250 ml. For very accurate work, carefully calibrated volumetric flasks are available in which solutions can be prepared. Note that you *do not* simply add 250 ml of solvent to the 41 g of sugar for, if you did, the final volume would be greater than 250 ml because the sugar itself occupies some volume. Only enough solvent is added to give a total volume of 250 ml.

PROBLEMS

1. On the basis of the law of conservation of matter, explain why atoms are conserved in a chemical reaction whereas molecules are not.

2. Explain what is meant by the term limiting factor in a chemical reaction.

3. Why is the heat of reaction negative when a reaction evolves heat?

4. Explain why the term molarity is not defined in terms of a specified volume of solvent?

5. Balance the following equations:

(a) $C_2H_4 + O_2 \longrightarrow CO_2 + H_2O$ (f) $Ca + O_2 \longrightarrow CaO$

(b) $SO_2 + O_2 \longrightarrow SO_3$ (g) $NaOH + HCl \longrightarrow NaCl + H_2O$

(c) $P_4 + O_2 \longrightarrow P_4O_{10}$ (h) $Ba(OH)_2 + HCl \longrightarrow BaCl_2 + H_2O$

(d) $N_2 + H_2 \longrightarrow NH_3$ (i) $C_6H_6 + O_2 \longrightarrow CO_2 + H_2O$

(e) $Ba + H_2O \longrightarrow Ba(OH)_2 + H_2$ (j) $Na + O_2 \longrightarrow Na_2O$

6. Consider the reaction

$$CH_4 + 2\,O_2 \longrightarrow CO_2 + 2\,H_2O$$

as the basis for the following problems.

Assume that the above reaction is carried out by mixing initially 12.0 g of methane and 24.0 g of oxygen and that reaction is as complete as possible.

(a) Calculate the limiting factor.

(b) Calculate the gram-moles of CO_2 formed.

(c) Calculate the gram-moles of H_2O formed.

(d) Calculate the grams of CO_2 and H_2O formed.

(e) Calculate the grams of reactant remaining.

7. Ethane burns in air according to the equation

$$2\,C_2H_6 + 7\,O_2 \longrightarrow 4\,CO_2 + 6\,H_2O$$

If 3.0 g of ethane and 12.8 g of oxygen react according to the above, calculate

(a) the total number of moles of reactants before reaction.

(b) the limiting factor.

(c) the number of moles of products formed.

(d) the weight of all substances present after reaction.

8. Consider the combustion of heptane.

$$C_7H_{16}(l) + 11\,O_2(g) \longrightarrow 7\,CO_2(g) + 8\,H_2O(l) \qquad \Delta H = -1150 \text{ kcal}$$

(a) If 2.80 g-moles of CO_2 are formed by this reaction, calculate the gram-moles each of heptane and oxygen used and the gram-moles of water found.

(b) If 1.00 g each of heptane and oxygen were mixed and burned according to the above equation, how much heat would be liberated under standard conditions?

(c) Calculate in grams the amount of H_2O that would be formed in part (b) above.

(d) Calculate the total volume of gases at STP after reaction in part (b) above.

9. The formula for ozone is O_3.

(a) How many gram-moles of ozone are there in 32 g?

(b) How many gram-moles of oxygen atoms are there in 32 g of ozone?

(c) If 32 g of ozone were to be converted completely into molecular oxygen, O_2, how many grams of O_2 would be formed?

(d) How many moles of O_2 would this be?

10. 4.45 g of the element, M, reacts with oxygen to form 10.0 g of the compound MO. Calculate the atomic weight of M.

11. How would you prepare 500 ml of 0.10 M NaCl?

12. Calculate the concentration of a solution that is prepared by adding 50 ml of 0.60 M HCl to sufficient pure solvent to make a total volume of 200 ml.

13. If two reactants A and B react completely according to the equation

$$2\,A + B \longrightarrow \text{products}$$

how many moles of A are required to react with 100 ml of 0.250 M B?

14. How many grams of HCl are left unreacted if 40 ml of 1.0 M HCl is reacted with 30 ml of 1.0 M NaOH?

15. For each of the following reactions determine which of the reactants is in excess and calculate how many grams will remain unreacted after the reaction has gone as far as it can.

(a) $2\,H_2 + O_2 \longrightarrow 2\,H_2O$ ($5.0\,g\,H_2 + 5.0\,g\,O_2$)

(b) $2\,Na + 2\,H_2O \longrightarrow 2\,NaOH + H_2$ ($5.0\,g\,Na + 5.0\,g\,H_2O$)

(c) $C + O_2 \longrightarrow CO_2$ ($5.0\,g\,C + 5.0\,g\,O_2$)

(d) $2\,Na + Cl_2 \longrightarrow 2\,NaCl$ ($5.0\,g\,Na + 5.0\,g\,Cl_2$)

(e) $S + O_2 \longrightarrow SO_2$ ($5.0\,g\,S + 5.0\,g\,O_2$)

(f) $4\,Fe + 3\,O_2 \longrightarrow 2\,Fe_2O_3$ ($5.0\,g\,Fe + 5.0\,g\,O_2$)

(g) $HCl + NaOH \longrightarrow H_2O + NaCl$ ($5.0\,g\,HCl + 5.0\,g\,NaOH$)

ATOMIC CONCEPTS

8-1. INADEQUACIES OF DALTON'S ATOMS

In spite of the great achievement of Dalton's theory, it suffered at least two inadequacies: (a) No mention was made of the mechanism by which atoms combine or of the forces that hold them together. (b) There was no explanation of the definite ratios in which atoms combine. For example, why do oxygen and hydrogen atoms combine in the ratios 1 to 2 and 2 to 2, but not any other ratio such as 1 to 1 or 2 to 1? Why does sodium not react with chlorine to form $NaCl_2$ or Na_2Cl? Certainly any theory of the atom must explain these glaring idiosyncrasies of nature.

8-2. EARLY ELECTROCHEMICAL OBSERVATIONS

By the end of the eighteenth century, Cavendish had employed the electrostatic generating machine to produce electric sparks that caused hydrogen and oxygen to combine. In 1800, Alessandro Volta in Italy put together his electric pile (battery). It consisted of alternating layers of silver and zinc plates, each separated by a layer of pasteboard or leather soaked with a conducting solution such as salt water (Figure 8-1). By the use of such a device, electricity could be produced that differed from the spark of the electrostatic generator in being capable of affording a continuous electric current.

This invention was followed almost immediately by a flurry of research activity on the uses of Volta's battery. In England (1800), Nicholson and Carlisle succeeded in decomposing water into hydrogen and oxygen by use of a battery. By 1807, Sir Humphry Davy in England had decomposed molten sodium oxide, thus preparing for the first time, the element sodium.

$$2\,Na_2O \xrightarrow{\text{electric current}} 4\,Na + O_2$$

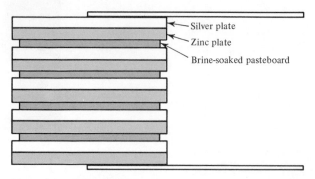

Figure 8-1. Volta's pile.

He similarly prepared potassium, magnesium, calcium, barium, and strontium. Most of these elements are very reactive toward air or moisture and are therefore not found free in nature.

Within the next 20 years, experimentation had led to a great number of facts about the nature of electricity, but it was Michael Faraday, a colleague of Davy's, who clarified its nature and interaction with matter.

Faraday was very interested in the effects of electric currents on matter. He measured the amount of current required to decompose a fixed amount of water. By a series of careful experiments he found that a fixed amount of current always liberates fixed amounts of each element from its compounds. Thus, for a given amount of electricity, 2 g of hydrogen and 16 g of oxygen are formed on the decomposition of water. This same amount of electricity also liberates 46 g of sodium and 78 g of potassium when passed through the respective pure molten salts (Figure 8–2). Farady thus arrived at the brilliant notion that electric current (like matter) consists of particles of electric charge and that the passage of a given amount of electric current through a pure molten sodium salt results in the production of the same amount of sodium, regardless of the compound in which it occurs. He proposed his **first law of electrolysis:** *the mass of an element produced by an electric current is proportional to the quantity of current used.*

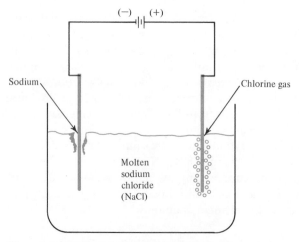

Figure 8-2. Electrolysis of NaCl (molten).

The quantity of electric current is given in terms of **coulombs,** which may be defined as *the quantity of electricity that will liberate 0.0011180 g of silver from a solution of silver nitrate.* The rate at which current flows in a circuit is measured in terms of the **ampere,** which is defined as a *rate of flow of one coulomb per second.*

■ **Example 8–1.**

An electric current flowing at a steady rate deposited 0.476 g of silver from a dilute solution of silver nitrate in 8.00 min. What was the amperage?

☐ *Solution*

$$\text{amperes} = \frac{\text{coulombs}}{\text{seconds}} = \frac{0.476 \text{ g Ag}}{\left(0.0011180 \dfrac{\text{g Ag}}{\text{coulomb}}\right)(8.00 \text{ min})\left(60 \dfrac{\text{sec}}{\text{min}}\right)}$$

$$= 0.887 \text{ amp}$$

Faraday was also able to see that, when an electric current was used to decompose a compound like sodium chloride, the amount of sodium and the amount of chlorine liberated were chemically equivalent. That is, they formed in the exact amounts required to reform sodium chloride. Faraday generalized this finding by the statement of the **second law of electrolysis:** *the same amount of electricity will produce all substances in exactly chemically equivalent amounts.*

■ **Example 8–2**

The electrolysis of a dilute solution of $CuCl_2$ produces copper at the negative electrode and chlorine gas at the positive electrode. How many grams of chlorine would be formed for each gram of copper produced?

☐ *Solution*

Since the elements are produced in chemically equivalent amounts, two atoms of chlorine would be produced per atom of copper.

$$1.00 \text{ g Cu} = \frac{1.00 \text{ g Cu}}{63.54 \text{ g Cu/mole Cu}} = 0.01574 \text{ mole Cu}$$

$$\text{g Cl} = 2(0.01574 \text{ mole}) \times \frac{(35.5 \text{ g Cl})}{\text{mole}} = 1.12 \text{ g Cl}$$

The amount of electric current required to deposit 1 mole of silver from a solution of silver nitrate was shown to be 9.65×10^4 coulombs. Exactly twice this amount was required to liberate 1 mole of copper. Farady found that all elements that are electrolytically prepared in this manner required multiples of this quantity. The quantity 9.65×10^4 coulombs, later came to be called the **Faraday, $\mathfrak{F}$,** which may be defined as 1 mole of electricity.

Faraday's laws of electrolysis proved to be quite useful to scientists working in the field of electrochemistry, but nearly 50 years were to elapse before a theory could emerge to explain these generalizations. Of one thing, however, one could be certain: there seemed to be little doubt that chemical combinations and electrical properties were intimately interrelated.

8-3. ELECTRIC DISCHARGES THROUGH GASES

By the late nineteenth century a wealth of information had accumulated on the phenomena of static electricity and electrical discharges. Shortly after Torricelli had demonstrated his barometer, the phenomenon of "barometric light" had been observed. This is a flash which is observed in the free space when the mercury in a barometer is made to run rapidly up and down the tube. Other related demonstrations were that a discharge of static electricity passes through a gas more readily when the pressure is reduced below atmospheric.

The study of these phenomena became immensely easier by the construction of an apparatus by Geissler in the middle of the nineteenth century. His apparatus consisted of two wires sealed into a glass tube. He also constructed an improved form of vacuum pump which could be used to evacuate the tube. By attaching metal electrodes to the wires inside the tube he obtained an apparatus similar to the one shown in Figure 8-3.

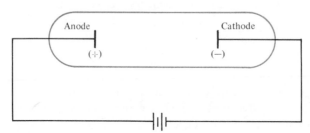

Figure 8-3. Cathode ray tube.

Using such tubes, experiments were performed on electrical discharges through gases under a variety of conditions: varying pressure, varying the gas, varying potential, as well as under varying external conditions such as magnetic and electric fields. By the late nineteenth century, these experiments had yielded the results summarized below:

1. Conduction of electricity through gases depends approximately directly on the **electrical potential*** applied and approximately inversely on the pressure of the gas. At sufficiently high potentials, conduction occurs even at

*Electrical potential may be defined as the amount of electrical work required to cause a specified amount of charge to flow from one point to another.

atmospheric pressure; that is, a potential of 50,000 volts produces a 5 cm spark in air.

2. Under reduced pressure, a different phenomenon is observed: 10,000 volts produces a discharge even when the electrodes are 50 cm apart.

3. Under even lower pressures—about 2 torr—a smooth discharge is observed whose color depends on the gas used. Each element gives a different color—air is pink, sodium vapor is yellow, mercury vapor is blue, neon is red, which, incidentally, is the origin of the modern neon lights. Each gas requires a different minimum critical voltage (electrical potential) for discharge to occur.

4. Reduction of the pressure inside the tube to below 0.1 torr caused the colored glow to disappear with a relatively dark space between the electrodes. This phenomenon was discovered by W. Crookes. Under these conditions, the walls of the tube fluoresce with a greenish glow. Even under these very low pressures, current continues to flow. The emanations were shown to originate from the cathode and were consequently called cathode rays.

J. J. Thompson, in the last decade of the nineteenth century had developed an interest in these phenomena. He and his students were particularly interested in elucidating the nature of these cathode rays. They devised a method for doing so by measuring the deflection of the rays by magnetic and electric fields. Thompson was thus able to determine that cathode rays consist of negatively charged particles, all of which have the same charge per unit of mass.

The calculation of the ratio of charge to mass of the cathode ray particles can be easily appreciated by the following reasoning:

The force exerted on a moving particle of mass m and charge e by the magnetic field, H is given by

$$\text{force} = Hev \tag{8-1}$$

in which v is the velocity of the particle. Because the magnetic field causes the particle to move in a circular path of radius r and because the acceleration of a particle moving in a circular path of radius r and velocity v is given by

$$\text{acceleration} = \frac{v^2}{r} \tag{8-2}$$

then the force on this particle is

$$\text{force} = \text{mass} \times \text{acceleration} = \frac{mv^2}{r} \tag{8-3}$$

Comparing equations (8–1) and (8–3)

$$\text{force} = Hev = \frac{mv^2}{r} \tag{8-4}$$

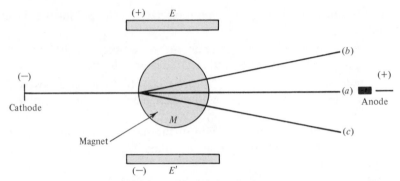

Figure 8-4. Schematic drawing of apparatus used in determining e/m of cathode rays. Particles deflected by the magnetic field M alone strike (c); those deflected by the electric field E alone strike (b) When the magnetic and electric fields are balanced, the particles strike (a).

which on rearrangement gives

$$\frac{e}{m} = \frac{v}{Hr} \tag{8–4a}$$

When an electric field is placed at right angles to the magnetic field, the deflection of the particle by the magnetic field can be cancelled by the electric field (Figure 8-4). The force exerted on the particle by the electric field E is

$$\text{force} = Ee \tag{8–5}$$

When the two forces are balanced so that the path of the particle is linear

$$Hev = Ee \tag{8–6}$$

and

$$v = \frac{E}{H} \tag{8–7}$$

substituting for v in equation (8–4) gives the final equation in which the quantity e/m of the cathode ray particles is given in terms of the measurable quantities E, H, and r.

$$\frac{e}{m} = \frac{E}{H^2 r} = -1.759 \times 10^8 \text{ coulombs/g} \tag{8–8}$$

Note that neither the charge, e, nor the mass, m, of these particles could be calculated independently. It was left to other scientists to make these determinations.

In these cathode ray experiments, Thompson also used a cathode in which a hole was drilled through the center (Figure 8-5). This experiment resulted, in addition to the cathode rays, in another kind of ray that emanated from the hole in the cathode. Thompson was able to demonstrate that these rays were positively

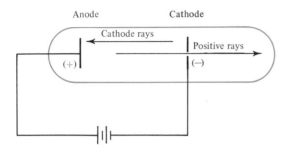

Figure 8-5. "Canal rays," or positive rays.

charged particles and that their mass was equal to that of the atoms of the gas used in the tube. Thompson concluded that these positive particles resulted when electrons were stripped from the atoms or molecules of gas. These charged atoms or molecules are called **ions.**

Although the mass of the positive particle depends upon the gas used, electrons are formed with all gases. Thus Thompson proposed that atoms consist of a positive mass into which electrons are embedded. This was the second step in the evolution of our concept of the atom. Dalton's atoms were simple spheres. Thompson's now had a slight complexity of structure.

8-4. MILLIKAN AND THE CHARGE ON THE ELECTRON

Thompson continued his experimentation on cathode rays in attempts to determine the exact charge on the electron. His method consisted of ionizing fog droplets and observing their rate of fall through the air and applying Stokes' law of falling bodies through fluid media. By counteracting the fall of the charged droplets by applying an electric field, Thompson hoped to determine the charge on such ionized water droplets.

Robert A. Millikan, in the second decade of the twentieth century, made much more refined measurements of the electronic charge. His method employed oil droplets, which have much lower volatility than water. The oil droplets could be viewed through a microscope placed horizontally (Figure 8-6). The size of the oil drop was determined by its rate of fall using Stokes' formula. Irradiation of the droplets with X rays produced negatively charged droplets that contained one or more electrons. By balancing the fall due to gravity with an electric field placed in opposition to gravity, the charge on each droplet could be obtained. After determinations of the charges on numerous oil droplets, the electronic charge

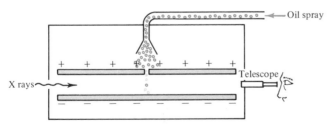

Figure 8-6. Schematic drawing of the essential parts of Millikan's oil drop apparatus for measuring the electronic charge.

was assumed to be the greatest common factor. The commonly accepted value of the electronic charge is -1.602×10^{-19} coulomb.

Employing this value of e in the previously given value of e/m gives a value of the electronic mass of 9.11×10^{-28} g which is about 1/1840 of the mass of a hydrogen atom.

8-5. ALPHA, BETA, AND GAMMA RAYS

A great series of experiments that delved deeply into the structure of the atom was performed by Ernest Rutherford and his associates in Canada and later in Cambridge, England, in the first 20 years of the twentieth century.

These experiments became possible as a result of the discovery by Becquerel in 1896 that a crystal of a uranium salt emitted rays that were able to penetrate black paper and affect a photographic plate even in the dark (Chapter 9). These rays were shown by Madame Curie and her husband to emanate from the uranium atom itself. They called this phenomenon **radioactivity.** Rutherford examined the nature of the rays by passing them through a magnetic field (Figure 8-7). Part of the rays were deflected in a direction, indicating that they were positively charged; part were deflected in the other direction, indicating that they were negatively charged; and part were undeflected. Rutherford named these three rays **alpha** (α), **beta** (β), and **gamma** (γ) rays, respectively. Employing both

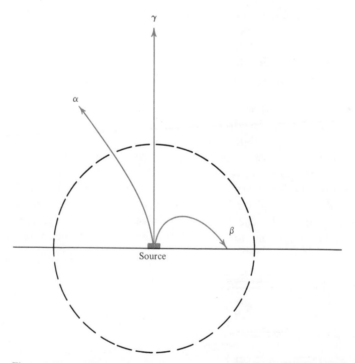

Figure 8-7. Deflection of radiation in a magnetic field. The magnetic poles face each other with the plane of the paper separating them and are represented by the broken circle. The α particles, being relatively heavy, are not deflected as much as the much lighter β particles. The γ rays are electromagnetic radiation and are therefore undeflected.

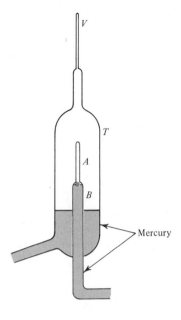

Figure 8-8. Schematic representation of Rutherford's apparatus for demonstrating that α particles are helium nuclei. The radium source was forced into the thin walled tube A by a column of mercury in B. α particles that passed through tube A could not pass through the thicker wall of T and were trapped. The presence of helium in the outer tube T was shown spectroscopically after forcing the gas into the vacuum tube V.

an electric and a magnetic field, as Thompson had done with cathode rays and positive rays (Section 8-3), and employing equation (8-8), Rutherford was able to determine the ratio of charge to mass (e/m) for the α and β rays. He thus determined that α rays are particles that have a positive charge twice as great in magnitude as that of the electron and a mass four times as great as the hydrogen atom. The β rays were shown to be identical to Thompson's cathode rays— electrons. The γ rays were found to be similar to X rays (Section 9-1) but of even greater energy.

Particle	Charge (coulomb)	Mass (g)
alpha	$+3.204 \times 10^{-19}$	6.64×10^{-24}
beta	-1.602×10^{-19}	9.11×10^{-28}

In an ingenious experiment, Rutherford was able to demonstrate that α particles are, in fact, the nuclei of helium atoms. He placed a sample of radioactive substance in a small thin-walled glass tube (Figure 8-8) through which the α-particles could pass. The outer tube was evacuated and, after several hours, the contents of the outer tube were analyzed spectroscopically and shown to contain helium. The presence of helium in the outer tube could only be explained by assuming that the highly penetrating α particles had acquired electrons from the glass or elsewhere and had become large, easily contained helium atoms.

8-6. DISCOVERY OF THE NUCLEUS

The experiments of interest at this point involved the bombardment of thin sheets of gold and other metals with α particles. The α particles were produced by

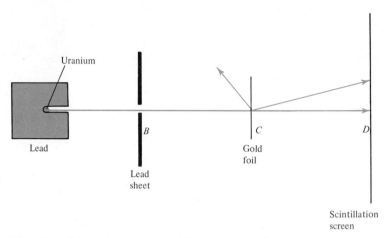

Figure 8-9. Schematic drawing of Rutherford's apparatus for the scattering of α particles by metal foils.

allowing uranium to decay naturally inside a lead block with a tiny hole in the side through which the α particles could be directed. Figure 8-9 is a schematic drawing of such an apparatus. The lead sheet, *B*, contained a tiny hole to direct the α particles in a thin stream. The scintillating screen consisted of zinc sulfide, which emits a visible flash or scintillation when hit by an α particle, and therefore allowed the detection of the position of each α particle as it passed from the foil.

The results of these experiments showed that most of the particles went through the foil completely unaffected. It was as though the foil were not there! However, a few were deflected by large angles, and some even bounced back. Rutherford explained these startling results by proposing that the atoms in the foil are mostly empty space because most of the α particles were undeflected. The rare deflections were caused by a tiny, positively charged nucleus within each atom. When a positively charged α particle comes near a positively charged atomic nucleus, the α particle, being much smaller than the nucleus, is repelled and its path is changed. A direct hit on the nucleus would cause it to bounce back. Collisions with electrons, on the other hand, do not deflect α particles because of their very great mass compared to that of the electron (7000 to 1).

Rutherford rationalized these results in terms of the following simple picture: The atom is composed of a central nucleus containing the positive charge and more than 99% of the mass. This nucleus is surrounded by electrons moving at relatively great distances so that the volume occupied by the nucleus is $1/10^{15}$ the total volume of the atom. Extending this picture to common sizes, if the nucleus were the size of an egg, the outermost electrons would revolve in a sphere with a radius of $1\frac{1}{2}$ miles!

Dalton's concept of hard little spheres had yielded to a rather complex system through the constant and devastating impact of new observations. Since hydrogen is the simplest element, with only one electron, Rutherford suggested that its nucleus is the fundamental particle of positive charge and named it the **proton.** The nuclei of the atoms were thus made up of protons, some of which, he postulated, contained embedded electrons. These latter combination particles contributed only mass to the nucleus. He later predicted a nuclear particle of mass nearly equal to that of the proton and bearing no electric charge, which he called the **neutron.** This particle was observed experimentally some 12 years after Rutherford's prediction. A summary of the properties of these particles in terms of atomic mass units (H = 1) and electronic charge units (electron = −1) is

	Mass	Charge
alpha	4	+2
electron	1/1840	−1
proton	1	+1
neutron	1	0

Rutherford's "solar system" model of the atom still left many questions unanswered as will become evident in the following discussion.

Our study of chemistry began with attempts to find the elementary substances in nature. Until the late nineteenth century, the search for the chemical elements had led to most of the ones that we know today. At that time, however, certain transitory particles appeared on the scene: Thompson's cathode rays, or electrons; his positive rays, which varied with the gas present in the evacuated tube; and, through the discovery of natural radioactivity, the α and β particles. This marked the beginning of a long series of discoveries of subatomic particles which has extended to the present. Chapter 9 will deal with the nature of the atomic nucleus.

One might question the use of the term **element** to describe a substance that is composed of simpler substances such as electrons, α particles, and so on. The usage is justified because these subatomic particles are not found free in nature as are the elements.

8-7. PERIODICITY AND THE PERIODIC TABLE

Before continuing our discussion of the role of electrical phenomena on atomic theory, an entirely different kind of chemical generalization must be examined. As will be seen in the pages that follow, building an atomic theory upon empirical foundations requires bringing together facts and observations from diverse sources. At the time when these discoveries were being made, little was imagined of the interrelationships of these facts and observations. It is our intention to draw them together in the following chapters.

The arrangement of the more than 100 elements into a simple rational scheme represents one of the great intellectual syntheses in the science of chemistry. The development of this scheme occurred over a period of nearly half a century beginning in the 1820's. One of the earliest attempts to coordinate the properties of elements was made by the German chemist Johann Wolfgang Dobereiner. In 1829, he published the observation that there are several triads of elements in which, when arranged in order of increasing atomic weights, the middle element has properties that are the average of the other two. Thus for the triad lithium (Li, at. wt. 6.9), sodium (Na, at. wt. 23), and potassium (K, at. wt. 39), 23 is the average of 6.9 and 39. Other properties of sodium that are roughly averages of those of lithium and potassium are shown in Table 8–1. See also Table 8–2 for the triad chlorine-bromine-iodine (Cl-Br-I).

Table 8-1. Some Properties of the Triad Lithium-Sodium-Potassium

	Li	Na	K	Average of Li and K
atomic number	3	11	19	11.5
atomic weight	6.94	22.99	39.10	23.02
atomic volume (cm^3)	13.1	23.7	45.3	29.2
melting point (°C)	180.5	97.8	63.2	121.9
heat of fusion (kcal/g-mole)	0.72	0.62	0.55	0.64
boiling point (°C)	1331	890	766	1049

Apparently little progress was made until 1864 when the Englishman John Neulands proposed that, if the elements were arranged in order of increasing atomic weight, the properties of the eighth resembled those of the first, the ninth resembled the second, and so on. He called this generalization the **law of octaves** because of the similarity to the musical scale. This is an example of the non-scientific, purely personal factors that influence the progress of science. Neuland's mother imparted to her son her great love for music. It was probably this interest in music that caused him to find an extra beauty in his law of octaves. His proposal before the scientific community was met with ridicule, however, and one of his colleagues in the audience sarcastically suggested arranging the elements in alphabetical order! The importance of Neuland's contribution was finally acknowledged in 1887 when he was awarded the Davy medal by the Royal Society.

The concept of chemical periodicity was developed independently by Lothar Meyer in Germany and Dimitri Mendeleev in Russia. Almost simultaneously, in 1869, these men published their versions of the periodic table. Meyer's work was based mainly on physical properties, such as atomic volume, fusibility, and brittleness, and included 56 elements. Mendeleev's table was based more on chemical properties. Probably the most convincing success of his system was the prediction of several hitherto undiscovered elements to fill gaps in his table. The most notable was the prediction of an element beneath silicon in the table. He called this element ekasilicon, and predicted its properties. Table 8-3 lists Mende-

Table 8-2. Some Properties of the Triad Chlorine-Bromine-Iodine

	Cl	Br	I	Average of Cl and I
atomic number	17	35	53	35
atomic weight	35.5	79.9	126.9	81.2
molecular volume of liquid at boiling point	45	54	68	56.5
melting point (°C)	− 102.4	− 7.2	113.6	5.6
heat of fusion (kcal/g-mole)	0.77	1.26	1.87	1.32
boiling point (°C)	− 34.0	58.2	184.5	75.2
heat of atomization (kcal/g-mole)	28.61	26.71	25.48	27.04

Table 8-3.

Property	Mendeleev's Prediction	Actual Value for Germanium
atomic weight	72	72.3
density, g/ml	5.5	5.36
oxide	EkO_2	GeO_2
density of the oxide	4.7	4.70
chloride:	$EkCl_2$	$GeCl_2$
boiling point, °C.	slightly below 100°C	83°C
density, g/ml	1.9	1.88

leev's predicted properties of ekasilicon along with the actual properties observed 15 years later when that element was discovered. It was later called germanium.

The insight of these men is one example of the great conceptual schemes that bring together many seemingly unrelated facts into a simple classification scheme. The periodic table combines the more than 100 elements into a small number of groups, each with its own set of properties. Mendeleev could not explain this amazing periodicity, but he offered that challenge to his fellow scientists to do so.

If all the elements are arranged in the order of their atomic weights, a periodic repetition of properties is obtained. This is expressed by the law of periodicity; the properties of the elements, as well as the forms and properties of their compounds are in periodic dependence or (expressing ourselves algebraically) form a periodic function of the atomic weights of the elements. . . . But just as without knowing the cause of gravitation, it is possible to make use of the law of gravity, so for the aims of chemistry, is is possible to take advantage of the laws discovered by chemistry without being able to explain their causes.

The challenge was accepted, but about 50 years were to elapse before a satisfactory explanation emerged (Chapter 10). Table 8–4 is a modern version of the periodic table. This table will be discussed in the chapters that follow.

PROBLEMS

1. List at least two inadequacies of Dalton's atomic theory.
2. It is known that sodium reacts spontaneously with oxygen to produce Na_2O. Explain how and why (in terms of energy) it is possible to convert Na_2O into elemental sodium and oxygen gas.
3. Extending the reasoning of problem 2, explain why elements such as sodium, potassium, calcium, and so on, were so late in being discovered.
4. State Faraday's first and second laws of electrolysis.
5. How many coulombs of current pass through a solution if the amperage is 3.5 amp for a period of 20 min?

Table 8-4. Periodic Table

Representative Elements		Transition Elements										Representative Elements					Noble Gases
																1 **H**	2 **He**
I	II	III	IV	V	VI	VII	VIII	VIII	VIII	I	II	III	IV	V	VI	VII	VIII
3 **Li**	4 **Be**											5 **B**	6 **C**	7 **N**	8 **O**	9 **F**	10 **Ne**
11 **Na**	12 **Mg**											13 **Al**	14 **Si**	15 **P**	16 **S**	17 **Cl**	18 **Ar**
19 **K**	20 **Ca**	21 **Sc**	22 **Ti**	23 **V**	24 **Cr**	25 **Mn**	26 **Fe**	27 **Co**	28 **Ni**	29 **Cu**	30 **Zn**	31 **Ga**	32 **Ge**	33 **As**	34 **Se**	35 **Br**	36 **Kr**
37 **Rb**	38 **Sr**	39 **Y**	40 **Zr**	41 **Nb**	42 **Mo**	43 **Tc**	44 **Ru**	45 **Rh**	46 **Pd**	47 **Ag**	48 **Cd**	49 **In**	50 **Sn**	51 **Sb**	52 **Te**	53 **I**	54 **Xe**
55 **Cs**	56 **Ba**	57 **La**	72 **Hf**	73 **Ta**	74 **W**	75 **Re**	76 **Os**	77 **Ir**	78 **Pt**	79 **Au**	80 **Hg**	81 **Tl**	82 **Pb**	83 **Bi**	84 **Po**	85 **At**	86 **Rn**
87 **Fr**	88 **Ra**	89 **Ac**	104	105													

Inner Transition Elements

58 **Ce**	59 **Pr**	60 **Nd**	61 **Pm**	62 **Sm**	63 **Eu**	64 **Gd**	65 **Tb**	66 **Dy**	67 **Ho**	68 **Er**	69 **Tm**	70 **Yb**	71 **Lu**
90 **Th**	91 **Pa**	92 **U**	93 **Np**	94 **Pu**	95 **Am**	96 **Cm**	97 **Bk**	98 **Cf**	99 **Es**	100 **Fm**	101 **Md**	102 **No**	103 **Lw**

6. What is the amperage of a current consisting of a total of 2000 coulombs flowing during the course of 1.0 hr?

7. What is the amperage of a current if it deposited 0.750 g of silver from a solution of silver nitrate in 10.00 min?

8. How much chlorine gas will be produced if the amount of current of problem 7 is allowed to flow through a solution of $CuCl_2$? (*Hint:* the formula of silver chloride is AgCl.)

9. Where do the cathode rays in a cathode ray tube originate?

10. Where do the positive rays originate?

11. How do you explain the fact that the voltage (electrical potential) must reach a minimum critical value before any gaseous discharge occurs?

12. Given the following values of electric charge on oil droplets, calculate the maximum unit charge of the electron. Why does none of these values coincide with the accepted value?

-4.8×10^{-19} $\qquad -16.8 \times 10^{-19}$

-9.6×10^{-19} $\qquad -28.8 \times 10^{-19}$

-14.4×10^{-19}

13. Describe α, β, and γ rays in terms of their material nature, mass, and electric charge.

14. In what ways do cathode rays and β particles differ?

15. In his experiments involving the bombardment of gold foil by α particles, why did Rutherford assume that the nucleus is positive and not negative? Why did he assume that the nucleus is so small compared with the size of the atom itself?

16. Contrast and compare the atom as described by Dalton, Thompson, and Rutherford.

17. Given the data in this chapter, calculate the charge to mass ratio (e/m) of each of the following: **(a)** proton, **(b)** α particle, **(c)** Li^+ ion.

18. Can John Dalton be criticized for not extending his atomic model beyond the simple notion of hard eternal spheres?

19. **(a)** Compare the properties that Dobereiner would have predicted for the element calcium from the following data for magnesium and strontium. Compare your predicted values with values obtained from a chemical handbook.

(b) Similarly predict the properties of germanium from the data below:

	Melting Point (°C)	Atomic Volume	Density	Atomic Weight
magnesium	650	14.0	1.74	24.32
calcium				
strontium	770	33.7	2.6	87.63

	Atomic Weight	Atomic Volume	Density	Melting Point (°C)
silicon	28.09	11.7	2.33	1410
germanium				
tin	118.70	16.2	7.28	232

(c) From a comparison of your predictions with the actual values, what can you conclude about Mendeleev's predictions?

9

THE ATOMIC NUCLEUS

The late eighteenth century witnessed one of the great intellectual upheavals of western civilization. The years 1775 to 1825 saw the development of Lavoisier's new chemistry and the beginnings of atomic theory. Thus was laid the foundation for our present chemical paradigm. But other ideas were also under attack. The American and French Revolutions brought with them the beginning of the end of the great monarchies and ushered in the age of democratic rule. The industrial revolution came into full swing with the growth of the iron and textile industries, the use of steam power for production and transportation, and the rise of the factory system with its method of mass production and mass living. During this period also, Mozart, Beethoven, and Schubert were creating the beautiful sounds made possible by instruments that had been developed over the preceding centuries. In short, every facet of man's life was profoundly affected by these years.

These changes brought to western man a period of growth and innovation that was to lead to enormous expansion in all areas. Thus science, music, government, and business were all in states of great activity as the world responded to the intellectual revolution.

Another great revolution has been in full swing during the middle of the twentieth century. It is more difficult to analyze because of its overpowering presence and our inability to become detached from it, but it certainly is closely linked to the discovery of nuclear energy. Nuclear energy has brought the unprecedented reality of unlimited power for the production of material goods. But nuclear energy has also brought an unprecedented ability for destruction that has raised the greatest problems with which man has yet had to grapple.

The events that led to the discovery of nuclear energy, besides being an integral part of chemistry, constitute one of the great dramas in the history of science, and will form the substance of this chapter.

9-1. THE DISCOVERY OF X RAYS

During the course of investigations of the effect of cathode rays on metal targets, Wilhelm Roentgen, in 1895, discovered that invisible rays were emitted from the metal target (anode) as the cathode rays struck it (Figure 9–1). This discovery was made because photographic plates wrapped in black paper and placed near the cathode ray tube were exposed by the rays. This meant that the rays had penetrated the glass of the cathode ray tube as well as the black paper. Roentgen was able to show that these rays, which he called X rays, were able to penetrate many materials, were not affected by a magnetic field, caused certain minerals to fluoresce, and were a form of electromagnetic radiation of extremely short wavelength, that is, high energy. Moreover, these rays were able to produce ionization in gases. Roentgen published his results including an X ray photograph of the bones of the hand. This caused a popular sensation since it apparently had a "peeping tom" appeal.

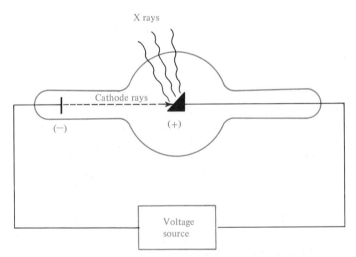

Figure 9-1. Production of X rays in a cathode ray tube.

9-2. DISCOVERY OF RADIOACTIVITY

Within the next few months, Henri Becquerel, who also was interested in this radiation and in the phosphorescence of certain minerals, was performing experiments with the mineral, potassium uranyl sulfate. This material is phosphorescent, that is, when irradiated with sunlight, it gives off a visible glow that continues even after it is removed from the sun. Becquerel observed that, after storing some protected photographic plates in the same desk drawer with a sample of potassium uranyl sulfate that had never been exposed to the sun, the plates became exposed. He concluded that the mineral was emitting radiation even without exposure to light, and that these rays were similar to X rays in that they could pass through thin glass or metal and could ionize gases. The actual nature of these rays has already been described in Section 8–5.

At this time a Polish student named Marie Sklodowska Curie, studying in the University of Paris, undertook to investigate the source of this radiation for her doctoral thesis. Along with her husband, Pierre Curie, a member of the physics faculty, she embarked on a monumental task. She had noticed that the ore of uranium was more radioactive than the amount of uranium salts called for. At that time, thorium was the only other element known that emitted these rays, and thorium was not present in the ore. She thus came to the realization that there must be some other element in the ore in small amounts, and that this highly radioactive element was the major source of radioactivity. She therefore set out to isolate this unknown substance.

Her method was to treat the ore with a chemical reagent that would preferentially dissolve part and leave the remainder undissolved, and then test the two parts (filtrate and precipitate) for radioactivity. The radioactive fraction was then subjected to another fractionation using a different chemical reagent, and the fractions tested again for radioactivity. After a series of many such fractionations, filtrations, and crystallizations, she was finally able to obtain a mixture of barium bromide ($BaBr_2$) and the bromide of a new element ($RaBr_2$) which had very similar properties. Because this new element was so extremely radioactive the Curies named it radium. Its similarity to barium suggested that it should fit into the same chemical family, that is, immediately below barium in the periodic table. Because of slight differences in the solubilities of the two salts, she was finally able to obtain a pure sample of this exciting new substance and study its properties.

Some of the properties of the new element were truly interesting and suggested that it was undergoing unique and deep-seated changes. For example, radium salts were observed to become colored yellow-pink on standing. They emitted a blue luminescence. Water solutions of radium salts were observed to evolve oxygen and hydrogen gas in the volume ratio 1 to 2. But probably most interesting of all was the observation that a sample of radium salt was always a few degrees warmer than the air around it. Measurements showed that a sample containing 1 g of radium emitted 100 cal/hr—enough to raise 1 g of water from a temperature of 0°C to its boiling point! Moreover, this heat emission apparently went on undiminished. This observation raised serious questions about the idea of conservation of energy and the indestructible atom. Finally, the atomic weight was found to be 226, which, along with its other chemical properties, placed it directly below barium in the periodic table.

The radioactivity of radium, as was to be observed with other radioactive elements, is completely unaffected by acids, bases, temperature or any other chemical treatment, and was eventually shown to be entirely a property of the atomic nucleus. The Curies also discovered another radioactive element in their analysis of pitchblende—polonium—and Madame Curie named it after her native Poland.

9-3. ATOMIC NUMBERS

Up until the early part of the twentieth century, the term atomic number had been used to designate the numerical position of an element in the periodic table. This number, moreover, with a very few exceptions follows the same sequence as that of the atomic mass.

In 1913, H. G. J. Moseley, a student of Rutherford, reported the results of a study of the characteristic X-ray spectra which gave a theoretical meaning to atomic number. When various metals are used as targets in a cathode ray tube, the X rays produced are of two types. There is a broad continuous spectrum of wavelengths and, superimposed upon this broad spectrum, are two very sharp and intense peaks of radiation (Figure 9-2). The exact wavelengths of the two peaks depend only on the nature of the metal and are independent of the energy of the cathode ray beam as long as it is above the required minimum.

Moseley's finding was that the wavelengths of the characteristic peaks in the X-ray spectra decrease in a very regular way with increasing atomic number of the metal used as target (Figure 9-3). He was able to interpret this regularity by concluding that the atomic number is a measure of the positive charge in the atomic nucleus. The atomic number, then, becomes a much more fundamental quantity than the atomic weight and the anomalies in the periodic table, when arranged according to atomic weight, for example, iodine (I) and tellurium (Te), disappear when arranged according to atomic number.

Thus the riddle of periodicity posed nearly 50 years earlier was finally solved. But, in addition, the concept of atomic number as the number of positive charges in the nucleus made possible for the first time an accurate theoretical definition of an element; namely, an **element** *is a substance all of whose atoms have the same nuclear positive charge*. Nuclear charge is usually stated in terms of the number of protons present. Section 9-4 will show the inadequacy of Dalton's original definition, which states that an element is a substance all of whose atoms are the same.

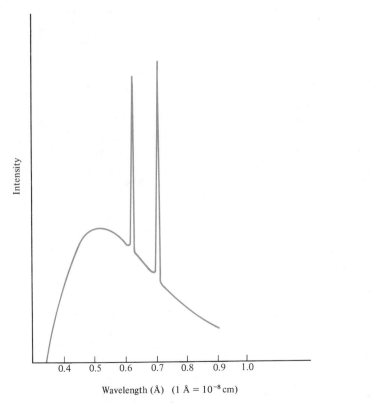

Wavelength (Å) (1 Å = 10^{-8} cm)

Figure 9-2. Characteristic X-ray spectrum of molybdenum.

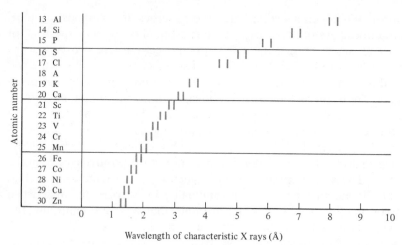

Figure 9-3. The regular variation of X rays emitted by metal targets as a function of atomic number of the metal.

9-4.   THE MASS SPECTROGRAPH—ISOTOPES

During his studies with positive rays, Thompson had developed a method for measuring their e/m ratio by passing the rays through a magnetic field. These positively charged particles would be deflected by this field, and the amount of deflection depended on the mass of the particle. He had observed in 1913 that positive rays produced from neon had particles of two different masses, 20 and 22. Thompson concluded that two kinds of neon atoms were present and that the observed atomic weight of neon (20.183) was an average value of the two as they occur naturally. The existence of different atomic weights of the same element had been predicted by Rutherford and Soddy not long before but this was the first experimental verification. Atoms of the same element that differ only in atomic weight are called **isotopes.**

F. W. Aston, a student of Thompson, later improved the apparatus by making it possible to vary the magnetic field and thereby obtain a spectrum of the masses of all the positive particles formed in the discharge (Figure 9-4). With his new mass spectrograph, Aston began an exhaustive study of all the elements to determine the presence of isotopes. At present, about 1400 isotopes have been observed, of which 332 occur naturally.

Of the first 80 elements only 19 are isotopically homogeneous. Table 9-1 lists some data on the number of isotopes of each element.

The existence of different atomic weights of the same element can be explained by assuming that nuclei are composed of protons and **neutrons.** Neutrons will be described more fully in Chapter 27. At this point they may be described simply as neutral particles of a mass nearly equal to that of the proton. Thus the atomic number (number of protons) determines an element, whereas the sum of the number of neutrons and protons determines atomic weights of isotopes.

The discovery of isotopes and that they had very nearly integral atomic masses helped solve, at least temporarily, a problem of long standing, namely, that of fractional atomic weights. It had much earlier been held that atoms are built of units of hydrogen and should therefore have masses that are integral multiples

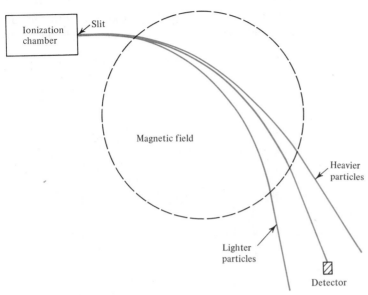

Figure 9–4. Mass spectrograph. After passage through the slit the positive particles pass through a magnetic field. Of the particles with the same charge, the lighter ones are deflected more. By varying the magnetic field strength, all particles can be focused on the detector. The e/m ratio is then determined from the field strength and other parameters.

of the atomic mass of hydrogen. Now one could explain that chlorine, with its atomic weight of 35.453, was in reality a mixture of isotopes of atomic weights 35 and 37.

The convention finally adopted for the designation of isotopes is to list the atomic number as a subscript and the atomic mass number as a superscript, both to the left of the chemical symbol. Thus the above chlorine isotopes are $^{35}_{17}\text{Cl}$ and $^{37}_{17}\text{Cl}$.

Table 9–1. Occurence of Isotopes

Elements	Number of Naturally Occurring Isotopes
Be, F, Na, Al, P, Sc, Mn, Co, As, Y, Nb, Rh, I, Cs, Pr, Tb, Ho, Tm, Ta, Au, At, Fr	1
H, He, Li, B, C, N, Cl, V, Cu, Ga, Br, Rb, Ag, In, Sb, La, Eu, Lu, Re, Ir, Ac	2
O, Ne, Mg, Si, Ac, K, Rn, Pa, U	3
S, Cr, Fe, Sr, Ce, Ra	4
Ti, Ni, Zn, Ge, Zr, W, Bi	5
Se, Kr, Pd, Ca, Er, Hf, Pt, Tl, Th	6
Ru, Ba, Nd, Sm, Hg, Mo, Gd, Dy, Yb, Os, Po	7
Ge, Te, Cd, Pb	8
Xe	9
Sn	10

Subsequent refinements of the mass spectrograph have shown that the actual masses of the isotopes are not exactly integers. These discrepancies will be taken up in Chapter 27.

PROBLEMS

1. Describe briefly the evidence that led Marie Curie to suspect that a yet undiscovered element was present in the ore pitchblende.

2. Correlate the descriptions in Column B with the items in Column A. (*Note:* There may be several entries from B for each item in A.)

Column A	Column B
α particle	(a) charge of +2
neutron	(b) high energy radiation produced by bombardment
electron	of a metal target by a stream of electrons
isotope	(c) helium nucleus
β particle	(d) charge of −1
γ ray	(e) charge of +1
X ray	(f) discovered radioactivity
proton	(g) high energy electromagnetic radiation produced
Marie Curie	in nuclear reactions
H. Becquerel	(h) neutral nuclear particle
W. Roentgen	(i) discovered the electron
	(j) negative particles ejected by atomic nuclei
	(k) discovered X rays
	(l) discovered radium
	(m) nuclei that differ only in their masses
	(n) nucleus of a hydrogen atom
	(o) mass about four times that of the hydrogen atom

3. Account for the constant emission of energy in an isolated sample of radium salt.

4. How do you account for the fact that the radioactivity of radium is not affected when it undergoes chemical reactions as with acids, bases, and so on.

5. Explain how X rays are produced.

6. What evidence can you cite to support the assumption that radioactivity originates in the nucleus of the atom?

7. Describe (a) α particles, (b) β particles, (c) γ rays.

8. The positions of the elements iodine and tellurium in the periodic table caused some confusion in the early announcement of the periodic tables. Why was this? How did Moseley's work on atomic numbers solve the problem?

9. In terms of our knowledge of nuclear compositions account for the existence of isotopes.

10. Tell whether each of the following pairs of nuclei are isotopes or not.

Nucleus X	Nucleus Y
(a) 6 protons + 7 neutrons	6 protons + 8 neutrons
(b) 8 protons + 8 neutrons	7 protons + 8 neutrons
(c) 6 protons + 8 neutrons	7 protons + 7 neutrons

11. The two isotopes of chlorine have atomic masses of 34.97 and 36.95. The experimentally determined atomic weight of ordinary chlorine is 35.46. Calculate the percentage of each isotope present in ordinary chlorine.

12. Consulting the periodic table or any other source, tell what element is represented by X in each of the following.

(a) $^{13}_{6}X$ **(b)** $^{26}_{12}X$ **(c)** $^{30}_{14}X$ **(d)** $^{18}_{8}X$ **(e)** $^{81}_{35}X$ **(f)** $^{34}_{16}X$

13. Describe the experimental method of producing an X-ray spectrum of a metal.

ELECTRONIC THEORY OF ATOMIC STRUCTURE

Our conception of the atom was extended considerably by the work of Rutherford and others in the decades around the turn of the century. However, the solar system model raised some serious questions. Classical physics holds that a charged body in nonlinear motion should radiate energy continuously. But, if the electron in an atom were emitting energy, its kinetic energy would decrease and it would spiral into the nucleus under the influence of the coulombic attraction between the positively charged nucleus and the negative electron. It is also well known that a stable atom does not emit radiation in this manner, and when gaseous atoms are excited by means of electrical or thermal energy, the energy that the atoms emit consists of only a few well-defined wavelengths—not a complete spectrum.

The solution to this problem of atomic stability occupied the efforts of many people during the early part of the twentieth century. We shall trace in broad outline the work that led to its solution. An understanding of atomic structure requires an understanding of the interactions between matter and energy—specifically energy in the form of light or electromagnetic radiation. Thus we must first review some early ideas concerning the nature of light.

10-1. EARLY WORK ON THE NATURE OF LIGHT

Isaac Newton in the early eighteenth century carried out experiments with light and established that ordinary sunlight, or "white" light, is actually a composite of several colors. He thought of light rays as composed of very small bodies emitted from shining substances. Indeed, a particle theory of light explains many observed properties; for example, reflected rays leave a mirror surface at the same angle as that between the incoming rays and the surface (Figure 10-1).

During this same period, Christian Huygens proposed a wave theory of light. The wave theory began to gain dominance over the corpuscular theory because

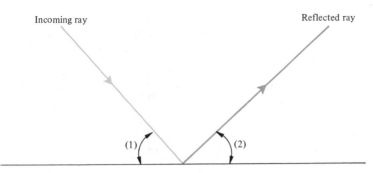

Figure 10-1. Angle of incidence (1) is equal to the angle of reflection (2) for light rays.

of experimental observations that seemed to be explainable only on the basis of wave phenomena. One of the most notable of such observations was the discovery of the interference of light by Thomas Young in 1803. He found that, when light from a common source is caused to pass through two tiny holes close together in a screen, the illumination of another screen some distance from the first does not result in two tiny spots. Rather, the spots of light are spread into patches that overlap and produce a number of alternating light and dark bands (Figures 10-2 and 10-3).

The interference phenomenon is readily explained if it is assumed that light is a form of wave motion. Light waves, which are originally in-phase (Figure 10-2) as they come from the same source, travel to the small holes in the screen where they cause waves to be emitted in all directions. This effect is called **diffraction** and is characteristic of all forms of wave motion. These waves then meet at the second screen and produce light or dark areas, depending upon whether the waves are still in-phase or whether they are out-of-phase (Figure 10-3). Because each color of light has its own wavelength (Figure 10-2a), the experiment is best carried out using light of a single wavelength—monochromatic light.

The observation of the interference phenomenon requires careful adjustment of the screens because the wavelength of the light, the distance between the holes

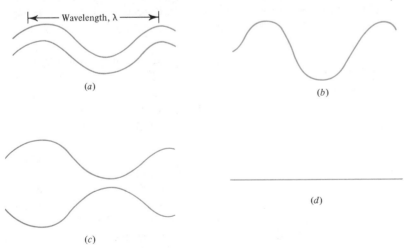

Figure 10-2. When in-phase waves (a) meet, reinforcement (b) occurs. When out-of-phase waves (c) meet, destruction (d) occurs.

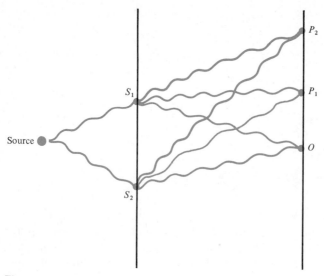

Figure 10-3. Illustration of the interference of light. Point O is equidistant from S_1 and S_2 so the light waves remain in phase and reinforce each other at point O. P_1 is $\frac{1}{2}$ wavelength nearer S_1 than S_2, so the waves meet at P_1 completely out-of-phase and cancel, producing a dark region. A bright region occurs at P_2, which is a whole wavelength farther from S_2 than S_1, thus causing the waves to be in-phase and reinforce.

S_1 and S_2 (Figure 10-3), and the distance between the screens are all related to the points at which the light and dark areas can be seen. As an example, if the holes are 1 mm apart and the screens are placed at a distance of 1 m from each other, then most wavelengths of visible light will cause the bright bands to appear a few tenths of a millimeter apart.

10-2. THEORY OF ELECTROMAGNETIC RADIATION

James Maxwell finally developed an elaborate mathematical theory in which light is identified with electric and magnetic waves. With the publication of his electromagnetic theory in 1865, Maxwell appeared to have settled once and for all the question of the nature of light. The electromagnetic theory (Figure 10-4) was quickly seen to embrace a variety of phenomena which includes not only visible light but "unseen" radiations as well, ranging from heat to X rays. All of these "rays" may be spoken of as electromagnetic radiations differing only in their wavelengths (Figure 10-5). In fact, the prediction of electromagnetic radiation with wavelengths much longer than those of light prompted Heinrich Hertz to attempt to produce such waves. The discovery of radio waves by Hertz in 1887 was the result. The discovery by Roentgen of electromagnetic radiation of very short wavelength, namely, X rays, followed soon after (Chapter 9).

Waves are characterized by their wavelength, λ. They also have a definite frequency, ν. Frequency can be physically viewed as the number of wavelengths passing a given point each second. The physical dimension of frequency is reciprocal time, normally expressed in seconds. The unit cycles per second (Hz) is sometimes used to emphasize the physical meaning of frequency, but the true

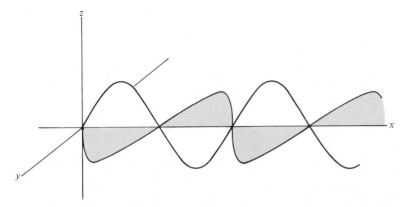

Figure 10-4. Electromagnetic waves according to Maxwell's theory. Electric field component is shown in the *xz* plane with the associated magnetic field in the *xy* plane. The wave is traveling along the *x* axis. Only one wave is pictured here; ordinarily, waves would be traveling along the *x* axis at all possible angles about the axis.

unit is simply, ν 1/sec. Because wavelength is a distance, usually centimeters, it is readily seen that the product of wavelength and frequency results in velocity units.

$$\lambda \text{ (cm)} \times \nu \left(\frac{1}{\text{sec}}\right) = c \left(\frac{\text{cm}}{\text{sec}}\right) \tag{10-1}$$

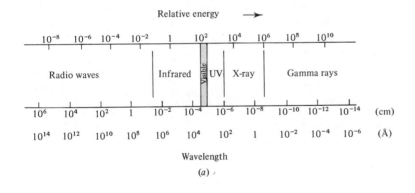

(a)

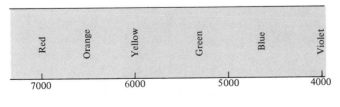

Wavelength (Å) (1 Å = 10^{-8} cm)

(b)

Figure 10-5. Electromagnetic radiation of various wavelengths.

For electromagnetic radiation, the velocity c in a vacuum has been found to have the value 3.00×10^{10} cm/sec. In a material medium, the velocity of electromagnetic radiation is less and depends on the particular medium and also on the wavelength of the radiation. In a vacuum, however, regardless of the wavelength of the radiation, the velocity is the same.

10-3. PLANCK'S QUANTUM THEORY

It is a fact of almost everyone's experience that solid objects that have been heated to incandescence change their color as the temperature is raised. An electric stove element will first appear dull red but will change to bright red and orange-red as it heats up. Very hot objects, such as the arc in a searchlight, are bluish white.

Even at lower temperatures, where the radiant energy no longer lies in the visible region, solid objects emit radiation of wavelengths that vary with the temperature of the object. A theoretically perfect absorber and radiator is referred to as a **black body.** Physicists refer to the distribution of energy emitted by solid objects as black body radiation (Figure 10-6). Analysis of black body radiation has shown that most of the energy is carried by radiation with wavelengths rather closely grouped. Little of the radiation has very high or very low wavelengths. Furthermore, as the temperature of the black body is increased, the radiation carrying most of the energy is shifted to shorter wavelengths.

In attempting to explain the energy distribution of black body radiation, Maxwell's theory met with failure. Max Planck, in 1900, developed an equation that satisfactorily represented the experimental data. However, in order to rationalize the equation, Planck was forced to assume that the radiation emitted by a black body is not given off continuously but rather in little bundles of radiant energy called **quanta.** This assumption was a complete break with classical electromagnetic theory, which viewed radiation as continuous waves. Planck imagined little oscillators in the black body that receive and emit radiant energy only in integral units of energy. Each unit or quantum of energy absorbed or emitted by a given oscillator is proportional to the frequency of the oscillator.

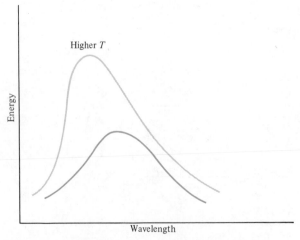

Figure 10-6. Energy distribution of black body radiation.

$$E \propto \nu$$

or
$$E = h\nu \qquad\qquad (10\text{--}2)$$

Where E is the energy of the quantum of radiant energy, ν is the frequency of the oscillator, and h is a proportionality constant, Planck's constant, with the value 6.63×10^{-27} erg sec. Planck explained the temperature effect by assuming that more oscillators of higher frequency were activated as temperature increases, thereby producing more radiant energy and shifting its frequency to higher values, or the wavelength to lower values.

10-4. PHOTOELECTRIC EFFECT

The phenomenon in which electromagnetic radiation causes the emission of electrons from the surface of a metal is known as the photoelectric effect (Figure 10-7). The electrons thus emitted are called photoelectrons. Notice that this effect is the converse of the manner in which X rays are produced. All metals exhibit the photoelectric effect if radiation of sufficiently high frequency is used, but the alkali metals—lithium, sodium, potassium, and so on—and some others are sensitive to visible light.

Hertz first observed this phenomenon in 1887, after which time it aroused the intense interest of scientists because it defied explanation in terms of classical electromagnetic theory. Specifically, the following facts were known by the early 1900s.

1. For each metal there is a certain minimum frequency for the incident radiation below which no photoelectrons are emitted. This minimum frequency is called the threshold frequency.
2. Radiation with a frequency above the threshold value produces electrons whose kinetic energy increases as the frequency increases.
3. The kinetic energy of a photoelectron is independent of the intensity of the radiation. That is, radiation below the threshold frequency will not

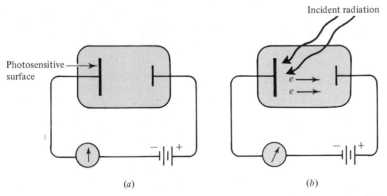

(a) (b)

Figure 10-7. The photoelectric effect. (a) No current flows in the absence of radiation. (b) Radiation causes the release of electrons from the photosensitive surface causing an electric current.

produce photoelectrons no matter how intense the radiation. Once the threshold value is reached, the number of electrons emitted per unit time is greater the more intense is the radiation.

According to Maxwell's theory, the intensity of the radiation should determine its energy. Thus even a feeble beam should steadily supply energy so that after a sufficient interval of time enough energy would be delivered to begin to detach electrons from the metal. No such time interval has ever been observed, and classical electromagnetic radiation theory can no more explain the photoelectric effect that it can the energy distribution of black body radiation.

In 1905, Albert Einstein offered an explanation for the photoelectric effect using Planck's quantum idea. Einstein boldly suggested that all electromagnetic radiation is quantized and travels through space in the form of small wave-bundles, which he called **photons** (synonymous with quanta). He calculated the energy of each photon using Planck's equation, $E = h\nu$. In addition, Einstein assumed that a photoelectron receives its energy from a single photon. The energy received by an electron is used to overcome the attraction of the electron for the metal, and then any excess energy gives the electron kinetic energy. The equation for this process is

$$h\nu = W + \tfrac{1}{2} m v^2 \qquad (10\text{--}3)$$

where $h\nu$ represents the energy of the photon (Planck's constant times the frequency of the radiation), W is the work necessary to detach an electron from the metal surface, and $\tfrac{1}{2} m v^2$ is the kinetic energy of the photoelectron.

Equation (10–3) and the assumptions behind its derivation offer simple explanations for the three observations cited above.

1. The threshold frequency is the frequency just necessary to remove electrons and impart no kinetic energy to them, namely

$$h\nu_0 = W \qquad (10\text{--}3a)$$

Where ν_0 represents the threshold frequency. Below this frequency, a photon would have insufficient energy to remove an electron.
2. The kinetic energy given the electrons represents the excess energy of the photons over that required to remove the electrons from the metal, that is

$$\tfrac{1}{2} m v^2 = h\nu - W \qquad (10\text{--}3b)$$

3. Intensity of radiation is simply the number of photons impinging on each square centimeter of surface per second. If more photons of sufficient energy strike the surface, of course, more electrons will be released. However, regardless of the number of photons, no photoelectrons will be produced unless the photons individually possess at least the energy given by equation (10–3a).

Whereas Planck's assumption of the quantization of radiation had been concerned only with the black body phenomenon, the extension of this idea to *all* electromagnetic radiation embodied in Einstein's explanation of the photo-electric effect suggested that radiation is indeed corpuscular in nature. To be sure, the properties of frequency and wavelength that are associated with waves still apply, but the wave motion of electromagnetic radiation is not continuous; rather it is transmitted in the form of a wave particle. The energy of each wave particle can be calculated by using Planck's relation (equation 10-2). The old particle versus wave controversy about the nature of light appeared to have come full circle by the early 1900s with partial truth to be found in both explanations.

10-5. ATOMIC BRIGHT-LINE SPECTRA

Not long after the discovery of sodium and potassium, it was noticed that sodium compounds always produce an intense yellow color when heated in a flame, and potassium compounds always give a lavender color. However, it remained for two Germans, a chemist, Robert Wilhelm Bunsen, and a physicist, Gustav Robert Kirchhoff, to bring quantitative accuracy to the identification of elements by their characteristic colors in a flame. In 1859, these two men invented the spectro-scope—an instrument for the accurate determination of each wavelength present in radiation consisting of a complex mixture of wavelengths (Figure 10-8).

Unlike the continuous spectrum produced by a heated solid object (black body radiation), elements in the gaseous state give a unique combination of specific wavelengths, which is characteristic of the particular element. Such a combination is called the bright-line spectrum of an element (Figure 10-9). Because no two elements give the same pattern, Bunsen and Kirchhoff recognized the power of this instrument as an analytical tool. Spectroscopic analysis gives both qualitative identification and quantitative determination of the elements present. From the specific wavelengths, the identity of each element can be established. Measurements of the amount of the element present can be made from the intensities of the lines. Spectroscopy has become a highly sophisticated instrumental method for chemical analysis. Today, instruments are commercially available that will operate with wavelengths in the electromagnetic spectrum far beyond those of the visible region.

With the advent of spectroscopy, some scientists turned their attentions to a search for order among the wavelengths characteristic of a given bright-line spectrum. Johann Jakob Balmer found such an order in the bright-line spectrum of hydrogen. Using data that were available for the visible region, Balmer reported

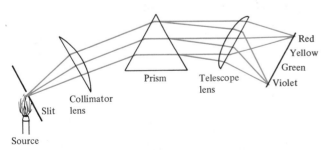

Figure 10-8. Components of a simple spectroscope. A narrow beam of light emerges from the slit and is focused to parallel rays by the collimator lens. The prism separates the wavelengths of the radiation and these rays are focused by the telescope lens for observation.

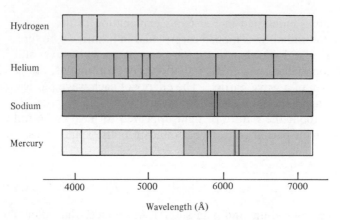

Figure 10-9. Bright-line spectra in the visible region for several elements.

in 1885 that the wavelengths at which the lines occur can be calculated using the empirical equation.

$$\lambda = C\frac{m^2}{m^2 - n^2} \qquad (10\text{-}4)$$

Where C is a constant and m and n are integers, but with the restrictions that the fraction never be negative nor $m = n$. Balmer simply "played around" with numbers in formulating his equation. He had no idea why equation (10-4) reproduced the bright-line spectrum for hydrogen. With $n = 2$, the value of C is 3.646×10^{-5} cm and equation (10-4) becomes

$$\lambda = 3.646 \times 10^{-5}\frac{m^2}{m^2 - 4} \qquad (10\text{-}4\text{a})$$

The values of $m = 3, 4, 5 \ldots 16$ reproduced the wavelengths of 14 known lines in the visible region for hydrogen (Figure 10-9). The divergence between the calculated wavelengths and the observed values was no greater than 0.1% and was much less in most cases. Clearly, this correspondence was no accident. Some fundamental property of the hydrogen atom was responsible. But what? The answer was to remain a mystery for nearly a quarter of a century.

Balmer predicted that series of lines would be found corresponding to the values for $n = 1, 3, 4 \ldots$ This prediction was borne out as these series corresponding to different values of n in equation (10-4) were discovered. The development of better spectroscopes and the prediction of the exact wavelengths to be expected hastened the investigations of bright-line spectra following Balmer's work.

10-6. BOHR'S THEORY OF THE ATOM

By the early part of the twentieth century, much was known concerning the behavior and composition of atoms. For example, we have seen how the work of Thomson and Rutherford led to a planetary model for the atom with the positive charges concentrated in a small but massive nucleus and the negative electrons arranged around this nucleus. The fact that the atomic number of an element is equal to the positive charge on the nucleus and hence the number of electrons in the neutral atom was also known. Yet the key to an understanding of the details of atomic structure was missing. If radiation is quantized, then the absorption and emission of radiation by atoms must occur by some mechanism that is quantized. The Danish theoretical physicist, Niels Bohr, finally supplied the key in 1913.

Bohr realized that, since classical mechanics is unable to account for the existence of a stable atom much less any of its properties, it would be necessary to make some assumptions completely outside the realm of classical mechanics. Planck had proceeded in this manner in treating black body radiation, and, as a result, a totally different picture of electromagnetic radiation had emerged. A similar, revolutionary approach was called for in dealing with atomic structure. Starting with a basic planetary model (Figure 10–10), Bohr made the following assumptions:

1. Within an atom there are a number of **stationary states** or stable, circular orbits in which the electron can move without emitting radiant energy. The electron must occupy some stationary state at all times except when moving from one state to another.

2. When the electron moves from a stationary state of higher energy E_2 to one of lower energy E_1, radiation is emitted whose frequency is determined by the Planck relation

$$E_2 - E_1 = h\nu \qquad (10\text{–}5)$$

3. An election in a stationary state must obey the condition

$$m\upsilon(2\,\pi r) = nh \qquad (10\text{–}6)$$

where m is the mass of the electron, υ is its velocity, and r the radius of its orbit. Planck's constant is h, and n is an integer 1, 2, 3, 4, ...

It is clear that Bohr leaned heavily on the Planck relation in formulating his assumptions. Assumption 3 was necessary because if just *any* values of υ and r

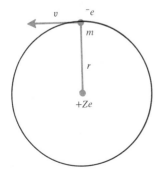

Figure 10-10. Planetary model of a one-electron atom that served as Bohr's basic model. An electron of mass m and charge $-e$ moves in a circular orbit of radius r around a fixed nucleus of charge $+Ze$.

are allowed, no meaningful results can be obtained. By limiting the values of v and r in terms of whole multiples of Planck's constant, as given by equation (10–6), Bohr found that he was able to account for a number of atomic properties. It must be understood that equation (10–6) cannot be derived. It is a basic assumption concerning the behavior of electrons in atoms. Perhaps some insight into equation (10–6) can be gained by noting that the units on h are energy multiplied by time. In terms of the properties of the electron, units of energy times time are obtained if mass, velocity, and distance the electron travels in one revolution ($2\pi r$) are all multiplied together. Equation (10–6) is often written

$$mvr = nh/2\pi \tag{10–6a}$$

because the quantity mvr has a special meaning to physicists. It represents the **angular momentum** of an object of mass m traveling with a velocity v in a circular orbit of radius r. Equation (10–6a) says that the angular momentum of an electron in an orbit is limited to whole multiples of the quantity $h/2\pi$. Equations (10–6) and (10–6a) are, of course, equivalent, and neither one is more fundamental than the other because they represent equally well a basic assumption of Bohr's theory.

A particle traveling in a circular orbit requires a centripetal (center seeking) force to keep it from flying off along a tangent. Bohr assumed that the electrical force of attraction between the negative electron and the positive nucleus was this centripetal force. He then proceeded to use ordinary classical laws of physics plus his nonclassical assumption given by equation (10–6) to derive an expression for the energy of the electron.

$$E = \frac{-2\pi^2 m Z^2 e^4}{n^2 h^2} \tag{10–7}$$

The energy is seen to depend on the value of n. This value is called the **principal quantum number** of the electron. Physically, the value of n determines which orbit the electron is in. The other quantities are constant for a given atom, namely, the mass of the electron m, the atomic number Z, the electronic charge e, and Planck's constant h. The negative sign appears in front of the expression because the point of zero energy corresponds to complete separation of the electron from the nucleus ($n = \infty$). Because the electron is attracted by the nucleus, work is done by the system as the electron falls into some orbit. The negative sign shows that work is done by the atomic system (energy given off) as the electron moves to orbits closer to the nucleus.

It is not possible to measure the energy of an electron in any given orbit, but a transition from one orbit to another should absorb or emit radiation. According to equation (10–7) the energy of the electron in two different stationary states should be

$$E_2 = \frac{-2\pi^2 m Z^2 e^4}{n_2^2 h^2} \tag{10–7a}$$

$$E_1 = \frac{-2\pi^2 m Z^2 e^4}{n_1^2 h^2} \tag{10–7b}$$

Expressing the difference $E_2 - E_1$ [using equations (10–7a) and (10–7b)] in combination with equation (10–5) and removing the common factor from the two terms gives

$$h\nu = \left(\frac{2\pi^2 mZ^2e^4}{h^2}\right)\left(\frac{1}{n_1^2} - \frac{1}{n_2^2}\right)$$

or

$$\nu = \left(\frac{2\pi^2 mZ^2e^4}{h^3}\right)\left(\frac{1}{n_1^2} - \frac{1}{n_2^2}\right) \tag{10-8}$$

An alternate form of equation (10–8) can be derived using (10–1) so that wavelength rather than frequency is calculated

$$\lambda = \left(\frac{ch^3}{2\pi^2 mZ^2e^4}\right)\left(\frac{n_1^2 n_2^2}{n_2^2 - n_1^2}\right) \tag{10-8a}$$

If now, the constants are introduced and the hydrogen atom specifically considered ($Z = 1$), when $n_1 = 2$, an interesting relationship results

$$\lambda = 3.646 \times 10^{-5}\left(\frac{n_2^2}{n_2^2 - 4}\right) \tag{10-8b}$$

Equation (10–8b), derived theoretically, is essentially identical with equation (10–4a), Balmer's empirical equation, for calculating the wavelengths of the visible bright-line spectrum of hydrogen! Bohr had solved the problem of accounting for the existence of atomic line spectra. When an atom is excited by thermal or electrical means, the electron is raised to a higher orbit. As the electron returns to a lower orbit, energy is emitted with the specific frequency corresponding to the difference in energy between the two orbits. Bohr could identify the electronic transition giving rise to a given line in the spectrum because the values of n_1 and n_2 correspond to the principal quantum numbers (orbit numbers) for the electron in the two energy states. These orbits or energy states defined by the various values of n are often referred to as quantum levels.

Other one-electron species such as He^+ and Li^{2+} can be treated using equation (10–8) or (10–8a). The bright-line spectra are quantitatively predicted by these equations. However, Bohr's equations cannot be used for any species with more than one electron because no account is taken of electronic repulsions. Attempts to allow quantitatively for such repulsions met with failure, but the Bohr model at least provided a qualitative insight into the bright-line spectra of all atoms. Because of its success in explaining experimental observations, Bohr's theory was accepted immediately. Dalton's rather simple atoms had become considerably more complex.

10-7. EXTENSIONS OF BOHR'S THEORY

The triumph of Bohr's theory in explaining quantitatively the bright-line spectra of one-electron atoms and ions was short lived. Better spectroscopes were soon constructed with greater abilities to distinguish the wavelengths of a pair or more of closely spaced lines. This ability is called the resolving power of a spectroscope. As instruments of higher resolving power were developed it became apparent that the bright-line spectra of hydrogen and other atoms were far more complex than had previously been thought. The more lines, the more electronic transitions there are. Also, closely spaced lines indicate that there are a number of transitions that are not very different in energy. Bohr's model did not allow for any such similar energy states because the energy of the electron was defined by the value of the principal quantum number alone, and different n values corresponded to relatively large energy differences. Bohr's atom was clearly too simple. Either the theory would have to be modified in some way or discarded. A modification that solved the problem for a time was suggested by Arnold Sommerfeld in 1916.

As a result of his special theory of relativity, Einstein, in 1905, had developed an equation that shows that the mass of a moving object increases with increasing velocity. Only at velocities approaching the speed of light would this relativistic mass increase become appreciable, but an electron in a Bohr orbit could have a fairly high velocity. Sommerfeld assumed that the electron might travel in elliptical orbits as well as circular orbits (Figure 10–11). Because an electron in such a "flattened" orbit would sometimes be closer to the nucleus and at other times farther away, its velocity would vary during each revolution. Taking into account the relativistic mass change that this variation in velocity would produce, Sommerfeld was able to account for the small energy differences between certain electronic states as demanded by the line spectra.

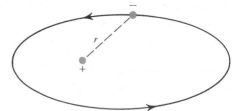

Figure 10–11. An elliptical electronic orbit. The radius r is not a constant as in the case of circular orbits.

As a result of this theoretical treatment, an additional quantum number was found to be necessary to define the energy state of an electron. This quantum number is called the **azimuthal quantum number** and is designated by the letter l. The values that l may have depend upon the particular value of the principal quantum number, n, according to the rule: $l = 0, 1, 2, \ldots n - 1$. Thus if $n = 1$, $l = 0$ only, whereas if $n = 2$, l may have the values 0 and 1. Physically, the particular value of l determines the relative "flatness" of the electronic orbit. For a given n value, the higher the l value, the more nearly circular the orbit (Figure 10–12).

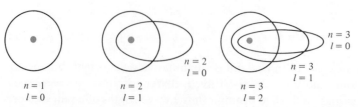

Figure 10–12. Orbits of various combinations of the quantum numbers n and l.

We now see that for $n = 1$, there is only one circular electronic orbit with $l = 0$. However, for $n = 2$, the electron may travel a somewhat elliptical orbit ($l = 0$) or it may travel in an essentially circular orbit ($l = 1$) of only slightly higher energy. The value of $n = 3$ gives rise to three possible suborbits, $l = 0$, which is fairly elliptical; $l = 1$, somewhat less elliptical; or $l = 2$, a circular orbit, and so on for higher values of n.

Spectroscopic analysis soon led to yet another complication of the picture. In 1896, Pieter Zeeman had reported that, while examining the line spectrum produced by a sodium flame, the "lines were seen to widen" when the flame was placed between the poles of a strong magnet. Later experiments using spectroscopes of high resolution revealed that the widening was due to the splitting of each single line into three or more closely spaced lines by the action of a magnetic field. This phenomenon is called the **Zeeman effect.** In order to account for the Zeeman effect in terms of the Bohr-Sommerfeld model, it became necessary to find additional energy states for the electron. These states must have the property that, although ordinarily they are energetically indistinguishable, they assume slightly different energies under the influence of an externally applied magnetic field. That some sort of interaction of this nature should occur was not too surprising. A negative particle in motion constitutes an electric current, and electric currents had long been known to have an associated magnetic field that interacts with an external magnetic field. However, the details of this interaction had to be worked out for electrons in atoms if the basic theory were to survive.

An additional quantum number, m, called, appropriately, the **magnetic quantum number,** was assigned in order to fix more definitely the energy of an electron in an atom. This number has values which depend upon the particular l quantum number according to the rule: $m = -l, -l + 1, -l + 2, \ldots 0, 1, 2, \ldots l$, or, more compactly, $m = 0, \pm 1, \pm 2, \ldots \pm l$. Thus it is that for each value of l, there are, in general, a number of different energy states available to the electron. If $l = 0$, only one state, $m = 0$, is available, but if $l = 1$, m can have values of $-1, 0$, or $+1$ indicating that three different states are available. In the absence of a magnetic field, each of the states corresponding to the various m values (for a given l value) have the same energy. But in the presence of a magnetic field, small differences in energy arise between the different states corresponding to each value for m. The nature of this energy difference can be imagined in a crude way by thinking of several different possible orientations in space for the electronic orbits. The presence of a magnetic field causes small differences in energy to appear among these orientations (Figure 10–13). For the case where $l = 1$, we can imagine the orbital motion of the electron for one m value to be aligned with the magnetic field, for another m value to be perpendicular to the field, and still a third orbit to be aligned against the field. In the case where $l = 2$, m can take on five values, $-2, -1, 0, +1, +2$, corresponding to one orbit perpendicular to a magnetic field, whereas two orbits each are aligned with the field and against the field.

Spectroscopic evidence showed that still more electronic energy transitions can occur. The electron in its revolutions around the nucleus might also rotate about an axis. Rotation could occur either in a clockwise or counterclockwise direction and would give rise to a magnetic field that would either augment or diminish somewhat the magnetic field associated with the orbital motion of the electron. Wolfgang Pauli suggested assignment of a fourth number, the **spin quantum number,** which can have only two possible values corresponding to the two

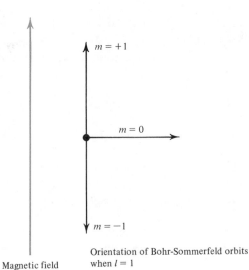

Figure 10-13. Differences in the orientation of electronic orbits in the presence of a magnetic field.

Magnetic field

Orientation of Bohr-Sommerfeld orbits when $l = 1$

possible directions of rotation. In order to distinguish the magnetic effect associated with the magnetic quantum number from that of the spin phenomenon, the magnetic and spin quantum numbers are designated m_l and m_s, respectively. The possible values of m_s are $+\frac{1}{2}$ or $-\frac{1}{2}$.

The four quantum numbers, n, l, m_l, m_s, serve to define completely the energy state of an electron in an atom. We have seen that the first two found theoretical justification in the Bohr-Sommerfeld treatment. The last two were added out of the practical necessity for accounting for all of the energy transitions that are observed in the bright-line spectra of atoms obtained with high resolution spectroscopes. In terms of the Bohr-Sommerfeld model, n defines the *size* of an electronic orbit, l its *shape,* m_l the *orientation,* and m_s the *spin* of the electron.

The Bohr-Sommerfeld theory was highly successful in explaining quantitatively the bright-line spectra of one-electron systems and providing a qualitative explanation for the line spectra of many-electron atoms. We shall see in the following chapter that it clarified the fundamental basis of the periodic nature of elemental properties. The greatest failure of the theory from a chemical point of view is that it offers no explanation for covalent bonding. The Bohr-Sommerfeld atom is a dynamic mechanical system. It is very difficult, if not impossible, to use such a system to explain the formation of stable bonds between atoms in a molecule. But, a static picture, obviously, will not explain other atomic properties so successfully treated by the dynamic model. Scientists in the 1920s faced just this dilemma. Several answers to the problem were advanced, and in a later section we shall consider the one best suited to the solution of chemical problems. But first we need to investigate in more detail the meaning of the quantum numbers.

10-8.   ELECTRON DISTRIBUTIONS IN ATOMS

The four quantum numbers completely assign an electron to a specific energy state. In a sense, the quantum numbers serve the electron as a complete address serves to locate a person. There are four necessary parts to a complete address for an individual—the state, city, street, and house number. A statement only of his home state would make location of a person rather difficult, but supply of his city would narrow the search considerably. His street would locate him much more fully, and the house number would complete the identification.

If we assume that electrons in atoms ordinarily occupy the lowest energy states available, then all electrons in all atoms would have a principal quantum number, $n = 1$ and little if any differentiation among the elements would result. Clearly, a "ground rule" for the assignment of electrons to energy states is necessary. Such a "ground rule" was suggested by Pauli in 1925 and is now known as the **Pauli exclusion principle:** *No two electrons in the same atom can have the same set of four quantum numbers.*

Application of this rule clears up the problem, and we can now assign electrons to their ground state (lowest energy state) arrangements in atoms. If $n = 1$, then $l = 0$ and $m_l = 0$. However, m_s can be either $+\frac{1}{2}$ or $-\frac{1}{2}$ so that there are two different combinations of the four numbers, namely, 1, 0, 0, $+\frac{1}{2}$ and 1, 0, 0, $-\frac{1}{2}$, respectively. Each of these two combinations corresponds to an energy state available to an electron. Since no other unique combinations are possible for $n = 1$, there can only be two electrons in the first quantum level. One of the above combinations would designate the ground state configuration of hydrogen, whereas both combinations would be used to designate the ground state configuration of helium (one combination for each electron). The next element, lithium, with three electrons would have two electrons in the first quantum level with the same quantum number combinations as given above. The third electron, however, would have $n = 2$, $l = 0$, $m_l = 0$, $m_s = -\frac{1}{2}$ (the negative value is conventionally taken to indicate a slightly lower energy than the positive). Immediately, it becomes evident why lithium so readily forms the Li^+ ion: Lithium has an electron much farther from its nucleus, and consequently attracted less strongly, than the electrons in either helium or hydrogen. As we shall see in the following chapter, many other periodic properties are explained by such simple considerations of electronic states.

Table 10–1 gives all possible unique combinations of the four quantum numbers for values of n through 3. Remember that each electron in an atom must have its own unique combination of the four numbers, so that the total number of such combinations for a given value of n represents the maximum of electrons which a given quantum level may contain.

10-9. WAVE PROPERTIES OF MATTER

If electromagnetic radiation has some properties best explained by assuming a particlelike or quantized nature for radiation, could not particles have some properties that could only be explained by assuming a wavelike nature for the particles? In 1924, Louis de Broglie suggested just such a wave nature for particles, and derived a theoretical equation for calculation of the wavelengths of the so-called matter waves.

$$\lambda = \frac{h}{mv} \tag{10-9}$$

Table 10-1. Unique Combinations of the Four Quantum Numbers

Principal, n	Azimuthal, l	Magnetic, m_l	Spin, m_s	Total Number of Unique Combinations of All Four Numbers
1	0	0	$+\frac{1}{2}, -\frac{1}{2}$	2
2	0	0	$+\frac{1}{2}, -\frac{1}{2}$	8
	1	1	$+\frac{1}{2}, -\frac{1}{2}$	
		0	$+\frac{1}{2}, -\frac{1}{2}$	
		-1	$+\frac{1}{2}, -\frac{1}{2}$	
3	0	0	$+\frac{1}{2}, -\frac{1}{2}$	18
	1	1	$+\frac{1}{2}, -\frac{1}{2}$	
		0	$+\frac{1}{2}, -\frac{1}{2}$	
		-1	$+\frac{1}{2}, -\frac{1}{2}$	
	2	2	$+\frac{1}{2}, -\frac{1}{2}$	
		1	$+\frac{1}{2}, -\frac{1}{2}$	
		0	$+\frac{1}{2}, -\frac{1}{2}$	
		-1	$+\frac{1}{2}, -\frac{1}{2}$	
		-2	$+\frac{1}{2}, -\frac{1}{2}$	

The m and v refer to the mass and velocity, respectively of the particle and h is Planck's constant. Where the mass is very small, as in the case of electrons, the wavelengths might be large enough to observe experimentally, reasoned de Broglie. Confirmation of this theoretical prediction was not long in coming. Clinton Davisson and Lester Germer found that a beam of electrons can be made to undergo diffraction and interference phenomena characteristic of wave motion.

The relation between the idea of matter waves and electronic structure in atoms still presented a problem, however. In an attempt to solve this problem, de Broglie imagined electrons in atoms to have the characteristics of standing waves. A standing wave results when two wave trains of the same wavelength and traveling at the same speed, but in opposite directions, superimpose to give the appearance of stationary waves. Such waves are easily demonstrated with a rope if one end is tied or held securely and the other end is agitated to produce a wave train in the rope. As the waves are reflected from the fixed end, there will be found certain wave velocities (depending on the length of the rope) for which one, two, three, (or more) loops "stand" along the rope (Figure 10–14).

de Broglie assumed that the electron in an atom could be treated as if it were a standing wave which is closed upon itself (Figure 10–15). He further assumed that these standing waves must have a circumference that corresponds to an integral multiple of the electronic wavelength.

$$n\lambda = 2\pi r \tag{10-10}$$

If a substitution is made for λ in equation (10–10) using equation (10–9), there results

$$\frac{nh}{mv} = 2\pi r \tag{10-10a}$$

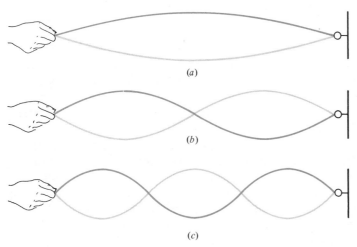

Figure 10-14. Demonstration of standing waves. Velocities for which the standing waves are established depend on the length of the rope: (*a*) fundamental vibration, (*b*) first overtone, (*c*) second overtone.

which upon rearrangement yields

$$\frac{nh}{2\pi} = mvr \qquad (10\text{-}10b)$$

Notice that equation (10-10b) is identical with equation (10-6a), one of Bohr's basic assumptions. Apparently de Broglie's approach, although physically quite different, is in some respects mathematically equivalent to Bohr's treatment.

10-10. SCHRÖDINGER WAVE MECHANICAL THEORY

Using de Broglie's theory of the wave characteristics of electrons, Erwin Schrödinger in 1926 developed his famous wave equation. Solution of this equation

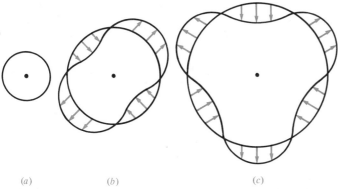

(*a*)   (*b*)   (*c*)

Figure 10-15. Representation of standing, electronic matter waves in an atom. For electronic orbits of radius r and electronic wavelength λ, $r = n\lambda/2\pi$. (*a*) $n = 1$, (*b*) $n = 2$, (*c*) $n = 3$.

for the hydrogen atom gives an expression for the electronic energies identical with that derived by Bohr. However, an additional and quite important feature arises from Schrödinger's treatment. Wave functions (mathematical expressions) describing the distributions around the nucleus of the various energy states available to electrons are part of the mathematical results. Each wave function or energy state is found to depend mathematically upon the assignment of three integral numbers—numbers whose values are arrived at in a manner identical with that for the quantum numbers n, l, and m_l!

Theoreticians prefer a theory that gives the greatest number of valid results for the fewest number of assumptions. Since the wave mechanical treatment immediately gives rise to three of the four quantum numbers, it is preferred to the Bohr treatment. The Schrödinger theory can do everything the old quantum theory (as the Bohr-Sommerfeld treatment is now called) can do plus much more. The way in which the wave approach lends itself to a consideration of covalent bonding is of chemical importance.

The exact mathematical forms of the wave functions depend on assignment of the n, l, and m_l quantum numbers. The resulting function can be looked upon as a mathematical description of the energy state corresponding to the particular values of the quantum numbers. Such an energy state is called an **orbital**. Notice that this term has a definite meaning of its own which is only vaguely related to the idea of a Bohr *orbit*. Because an orbital is an energy state for which only three of the four quantum numbers have been defined, there are always two unique combinations of the four quantum numbers corresponding to each orbital. For example, if an orbital with $n = 3$, $l = 1$, $m_l = 1$ is considered, m_s may have a value of $+\frac{1}{2}$ or $-\frac{1}{2}$ so that two unique combinations for this orbital are possible, namely, $n = 3$, $l = 1$, $m_l = 1$, $m_s = \frac{1}{2}$ and $n = 3$, $l = 1$, $m_l = 1$, $m_s = -\frac{1}{2}$. Thus any orbital can accommodate two, but no more than two, electrons. Two electrons whose quantum numbers differ only in the value of m_s are said to be **paired**. Two electrons in any given orbital are, therefore, always paired. It should be noted that any two electrons repel because they bear the same charge, but paired electrons do not repel as strongly as unpaired electrons because paired electrons have opposite magnetic orientations that are attractive.

In describing the orbitals, a shorthand notation has been developed to simplify matters. Usually, only the quantum numbers n and l need be specified because the energy of an electronic state is largely determined by these two quantum numbers. The n quantum number is designated by the number corresponding to its value. The l quantum number is identified by a letter—one for each l value. Table 10–2 gives these designations.

Writing the electronic distributions of atoms in their ground states is fairly simple. The ground state for hydrogen would be $1s^1$, and for helium $1s^2$. The superscript is used to show the number of electrons in each type of orbital. The lithium notation would be $1s^2 2s^1$. Appendix D gives the ground state electronic distributions for all the elements through atomic number 105.

Table 10–2. Designations for the l Quantum Number

l value	0	1	2	3	4	5
Letter designation	s	p	d	f	g	h

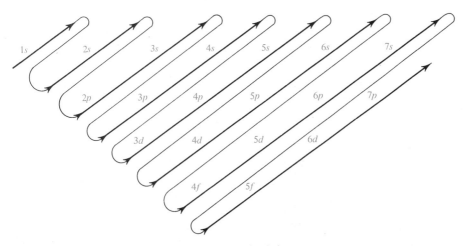

Figure 10-16. Order of filling of electronic energy levels in atoms.

A detailed consideration of orbital filling shows that one electron will fill each one of a given set of equal energy orbitals (the three p, five d, or seven f orbitals) before pairing occurs. This process is reasonable because electrons repel each other. This rule of orbital occupancy is often referred to as **Hund's rule.** Notice too, that the orbitals do not always fill in order of increasing values of n. Thus, the $3d$ orbitals in potassium are empty while the $4s$ orbital has one electron. It is a general rule that the next higher s orbital will fill before a given d level begins to fill. Likewise, the p orbital of the next higher level and the s orbital two levels higher both fill before an electron goes into an f orbital (note elements 47 through 58, Appendix D.) A useful chart that demonstrates the order of filling is shown in Figure 10-16. The energies of the orbitals increase strictly as n increases only for one-electron systems. The repulsions among electrons in many-electron atoms are primarily responsible for the orbital energy "inversions" that appear to occur. The removal of electrons, however, usually occurs from the orbital of highest n value. Thus, even though the ground state configuration of vanadium (V) is $1s^2 2s^2 2p^6 3s^2 3p^6 3d^3 4s^2$ (where the $4s$ is filled even though the $3d$ is not), for V^{2+} we have $1s^2 2s^2 2p^6 3s^2 3p^6 3d^3$. That is, the $4s$ electrons are ionized before the $3d$ electrons.

10-11. PHYSICAL INTERPRETATION OF WAVE FUNCTIONS

We have said that an orbital represents an energy state available to one or two (paired) electrons in an atom. The mathematical equation describing an orbital is called a wave function, and the specific form of the equation depends on the values of the quantum numbers n, l, and m_l. The wave functions corresponding to the various orbitals are found to have certain spatial characteristics. For one-electron systems such as H, He^+, Li^{2+}, the spatial characteristics of the various orbitals are known with certainty. Figure 10-17 gives these shapes for the s and the three p orbitals for one-electron systems.

Two alternate physical interpretations of the meaning of this spatial dependence, both equally valid, can be given. One view is that the electronic standing

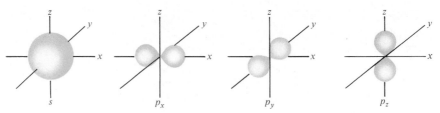

Figure 10-17. Spatial characteristics of s and p orbitals for one-electron systems.

wave is arranged in space as indicated. The negative charge of the electron is thus spread out in space around the nucleus in a definite way. An alternate view treats the electron more as a particle of negative charge that has a high probability of being found at any given time in certain definite regions of space around the nucleus and low or zero probabilities of being found in other regions. Thus an electron in a p_x orbital has a high probability of being found in a region of space concentrated along the x axis on either side of the nucleus and zero probability of being found either in the nucleus or along the y or z axes. Either point of view can be used in treating electronic phenomena.

Students often ask: How can an electron in a p orbital get from one side of the nucleus to the other without passing through the nucleus? In terms of wave mechanics, this question amounts to asking how an ordinary standing wave "gets" from one region of high amplitude to another without "passing through" the point of zero amplitude in between (note Figure 10-14). Standing waves simply have certain properties characteristic of them, and if the electron is being treated *as if it were* a standing wave, questions about how a *particle* might behave really have no meaning.

Note that the orbital directions illustrated in Figure 10-17 are known only for one-electron systems. Atoms with many electrons will not necessarily have orbitals with the same directional properties because of the effect of electronic repulsions.

PROBLEMS

1. Distinguish between wavelength and frequency.
2. What is meant by the bright-line spectrum of an atom?
3. What did Bohr mean by a stationary state of an atom?
4. How did the Zeeman effect play a role in the elucidation of the electronic structure of atoms?
5. What is the Pauli exclusion principle?
6. In terms of the assignment of quantum numbers, what is meant by paired electrons?
7. State Hund's rule and give an example of its application.
8. If the index and third finger are held in front of one eye, with the other eye closed while looking at a light, a series of dark lines will appear as the fingers are brought together until they almost touch. Explain the origin of the dark lines.

9. List the essential differences between Planck's theory of black body radiation and Maxwell's theory of electromagnetic radiation.

10. If a radio station operates at a frequency of 900 kilocycles (kc), calculate the wavelength of the radiowaves in centimeters, in feet [1 kc = $1 \times 10^3(1/sec)$].

11. (a) In order to produce photoelectrons from potassium, the wavelength of light must not be greater than $7.0 = 10^{-5}$ cm. Calculate the threshold frequency for potassium.

(b) Calculate the velocity of photoelectrons from potassium produced by radiation with a wavelength of 4.0×10^{-5} cm (mass of the electron = 9.1×10^{-28} g).

12. Calculate the wavelengths of the first five lines of the hydrogen spectrum using Balmer's equation.

13. (a) Use a diagram to describe the Bohr model of the atom.

(b) State the assumptions which Bohr made in his quantitative treatment of the atom.

(c) What quantum number(s) did Bohr utilize? Give symbol, name, possible values and their physical interpretation.

14. Calculate the frequency and wavelength of the radiation emitted when the electron undergoes a Bohr transition from $n = 2$ to $n = 1$ in the hydrogen atom.

15. Compare the relative wavelengths of radiation emitted when the electron undergoes Bohr transitions from $n = 2$ to $n = 1$ and $n = 3$ to $n = 2$ and $n = 3$ to $n = 1$.

16. (a) How did Sommerfeld alter Bohr's model of the atom?

(b) What quantum numbers did Sommerfeld utilize, and what physical meaning did they convey?

17. (a) Calculate the de Broglie wavelength of an electron traveling at a speed of 1×10^8 cm/sec.

(b) Calculate the de Broglie wavelength of a 1×10^3 kg automobile traveling at a speed of 3×10^3 cm/sec.

(c) What experimental problem would be encountered in attempting to study the wave nature of large objects?

18. (a) In the broadest conceptual terms, how does the Schrödinger wave mechanics treatment of the atom differ from the Bohr-Sommerfeld approach?

(b) What quantum numbers arise from the Schrödinger treatment?

(c) Why is the wave mechanical approach more useful than the Bohr-Sommerfeld treatment of the atom for chemical applications?

19. Write out all possible combinations of the four quantum numbers when $n = 2$. Indicate which combinations correspond to s states and which to p states. How many electrons can be accommodated in the second quantum level?

20. (a) Assign quantum numbers for a $2s$ orbital.

(b) Assign quantum numbers for an electron in a $2s$ orbital.

(c) Assign quantum numbers for the valence electron of the ground state sodium atom.

21. Write ground state electronic configurations ($1s^2, 2s^2, 2p^6 \ldots$) for the following: (a) Si (b) Sc (c) N^{3-} (d) K (e) O^+ (f) V^{2+} (g) Ar (h) Co^{2+} (i) S^{2-} (j) Ca

22. Examine Appendix D carefully. Notice that the filling of d and f orbitals does not always follow a consecutive pattern (see elements number 24, 29, 58 and many others).

(a) How do you think the actual electron configurations are determined?

(b) Can you find any regularity for the nonconsecutive filling of the d and f orbitals in certain elements?

23. Extend Appendix D for the hypothetical elements beyond number 105 until the next noble gas element is reached. What is its atomic number?

24. How would an electron in a hydrogen 2s orbital differ spatially from an electron in a 2p orbital?

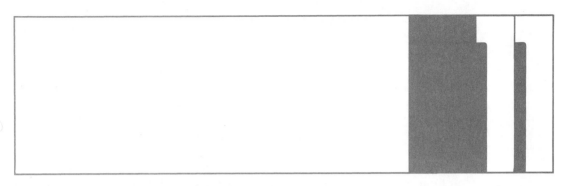

PERIODIC PROPERTIES OF THE ELEMENTS

Following the work of Moseley on atomic numbers, it became evident that the fundamental property which determines the position of an element in the periodic table is its atomic number, that is, its nuclear charge. In an electrically neutral atom, the number of electrons is, of course, equal to the atomic number, and it is the electronic configurations of atoms that determine their periodic relationships. Of particular importance is the number of electrons that are in the highest quantum level. This level is commonly referred to as the valence shell of an atom, and the electrons in it are called the valence electrons. The valence electrons primarily determine the bonding characteristics and most of the other properties of an element as well. Reference to the periodic table in Section 8-7 reveals four basic types of elements: the representative elements, the transition metals, the inner transition metals, and the noble gases.

The representative elements are those atoms with a valence shell containing only electrons in s and p orbitals. All underlying sets of orbitals are either completely filled or completely empty. The noble gas atoms have filled s and p orbitals and show an exterior of eight electrons, with the exception of helium which has only the $1s^2$ electrons. As with the representative elements, the underlying sets of orbitals in the noble gas atoms are either completely filled or empty. The transition metal atoms, on the other hand, have underlying sets of orbitals that are partially filled. The ordinary or outer transition elements (atoms of atomic numbers 21–30, 39–48, 57 and 72–80) have an underlying set of d orbitals which fill in passing from one element to the next along the period. The inner transition elements (atoms of atomic numbers 58–71 and 90–103) have an underlying set of f orbitals which fill in passing from one element to the next along the period.

11-2.   IONIZATION POTENTIALS

If a gas is placed in an apparatus similar to the cathode ray tube described previously, the voltage can be increased until a current begins to flow between the electrodes. This minimum voltage required to cause ionization in the gas is called the ionization potential and is a measure of the difficulty in removing the most weakly held electron from an atom of the gas to form a positively charged ion. In similar fashion more than one electron can be removed from each atom to form multiply charged ions, and the ionization potential of each successive electron can be measured.

The reactions associated with these potentials are

$$M \longrightarrow M^+ + e^- \qquad \text{first ionization potential}$$

$$M^+ \longrightarrow M^{2+} + e^- \qquad \text{second ionization potential}$$

$$M^{2+} \longrightarrow M^{3+} + e^- \qquad \text{third ionization potential}$$

$$\vdots \qquad\qquad\qquad \vdots$$

A low ionization potential corresponds to an easily removed electron, whereas a high ionization potential is associated with a strongly bound electron. For any

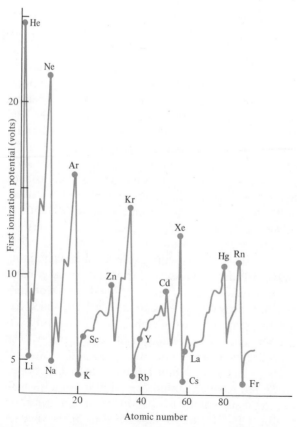

Figure 11-1.   First ionization potential versus atomic number.

given atom, the successive ionization potentials always increase because the positive charge against which the electron is withdrawn is greater.

Figure 11-1 is a plot of first ionization potentials as a function of atomic number. Notice that the noble gases are elements of maximum ionization potential with respect to the other elements near them in the periodic table. This fact is consistent with the electronic configurations in which the stable octet is realized in each case. The drastic decrease in ionization potential for the elements immediately following each noble gas is consistent with the outermost electron residing in a higher quantum shell. The valence electrons are farther from the nucleus and hence less strongly held. Metallic behavior is associated with relatively low ionization potentials. Proceeding across a given row or period, each additional electron is added to the same quantum level. At the same time, the nuclear charge increases and thus attracts the electrons more strongly. This effect leads to an increase in ionization potential in moving left to right in the periodic table. Nonmetal behavior is associated with relatively high ionization potentials; hence the elements progressively become more nonmetallic in going from left to right across a period. Finally, the particular valence quantum level achieves the s^2p^6 configuration of a noble gas and the process of filling the next higher shell begins.

Nearly any property of the elements that one cares to consider will show a periodic variation when plotted as a function of atomic number. Ionization potential has been considered in some detail because it is a direct measure of the relative ease of removal of electrons from an atom. We shall explore the consequences of variations of this property as it pertains to chemical bonding in the next chapter. Before considering the details of bonding, however, some of the properties and reactions of the elements will be investigated.

11-3. THE ALKALI METALS

Table 11-1 lists some pertinent data for the representative elements of group I, known as the alkali metals. These reactive metals are good electrical and thermal conductors and are soft enough to cut with a knife.

Francium is a highly radioactive element that exists in nature only in minute amounts. Its properties have therefore never been fully determined although they may be estimated from its position in the periodic table. Compounds of the other

Table 11-1. Electronic Structures and Ionization Potentials of Alkali Metals

Element	Atomic Number	Electronic Configuration*	Ionization Potential, volts	
			First	Second
lithium (Li)	3	[He]$2s^1$	5.39	75.6
sodium (Na)	11	[Ne]$3s^1$	5.14	47.3
potassium (K)	19	[Ar]$4s^1$	4.34	31.8
rubidium (Rb)	37	[Kr]$5s^1$	4.18	27.4
cesium (Cs)	55	[Xe]$6s^1$	3.89	23.4
francium (Fr)	87	[Rn]$7s^1$	3.83	22.5

*The inner electronic configuration of each atom is the same as that of the noble gas in brackets.

alkali metals are readily found in nature and the elements are usually prepared in the same manner that they were discovered, namely, electrolysis of their fused salts (Section 8–2).

The rather large difference between the first and second ionization potentials for these elements is explained by the electronic configurations of the atoms. Removal of the first electron is not too difficult, but removal of a second electron involves disruption of a noble gas type configuration. Furthermore, the outermost electron is seen to lie in a quantum level more remote from the nucleus than the remaining electrons. It is therefore not surprising to find that removal of more than one electron from these atoms is not observed in ordinary chemical reactions. Some physical properties of the alkali metals are given in Table 11–2. Note the fairly regular changes in these properties.

The alkali metals are so reactive that they must be kept out of contact with the air or moisture. This is usually accomplished by storing them under mineral oil or kerosene. In contact with water, all the alkali metals react to produce hydrogen and a solution of the hydroxide of the metal. A typical reaction is that between potassium and water.

$$2\,K + 2\,H_2O \longrightarrow 2\,KOH + H_2$$

The reactions of the metals with oxygen, however, lead to different types of products, depending on the metal.

$$4\,Li + O_2 \longrightarrow 2\,Li_2O$$
(lithium monoxide)

$$2\,Na + O_2 \longrightarrow Na_2O_2$$
(sodium peroxide)

$$K + O_2 \longrightarrow KO_2$$
(potassium superoxide)

Although each of these reactions yields different products, the metal always loses only one electron to form the +1 ion.

Rubidium and cesium react in the same way as potassium. The monoxides of potassium, rubidium, and cesium can be prepared by reaction between the metal and its nitrate, for example, in the case of cesium

$$10\,Cs + 2\,CsNO_3 \longrightarrow 6\,Cs_2O + N_2$$

Table 11–2. Physical Properties of the Alkali Metals

Element	Atomic Radius (Å)	Density at 25°C (g/cm³)	Melting Point (°C)
lithium	1.58	0.53	186
sodium	1.86	0.97	98
potassium	2.38	0.86	62
rubidium	2.53	1.53	39
cesium	2.72	1.87	29

The reactions of these oxides with water are summarized in the following equations.

$$Li_2O + H_2O \longrightarrow 2\,LiOH$$

$$Na_2O_2 + 2\,H_2O \longrightarrow 2\,NaOH + H_2O_2$$

$$2\,KO_2 + 2\,H_2O \longrightarrow 2\,KOH + H_2O_2 + O_2$$

The compound, H_2O_2, hydrogen peroxide, is unstable and readily decomposes to give water and oxygen.

The alkali metals all react directly with hydrogen to give what are called saline hydrides. As an example, sodium reacts

$$2\,Na + H_2 \longrightarrow 2\,NaH$$

These hydrides react with water to produce the same products as in the case of the metals themselves. The stoichiometry is different, however.

$$RbH + H_2O \longrightarrow RbOH + H_2$$

All of the metals react readily with the halogens (the group VII elements, fluorine, chlorine, bromine, iodine), for example

$$2\,Na + Br_2 \longrightarrow 2\,NaBr$$

Only lithium reacts with nitrogen to form the nitride.

$$6\,Li + N_2 \longrightarrow 2\,Li_3N$$

This substance reacts with water to form ammonia.

$$Li_3N + 3\,H_2O \longrightarrow 3\,LiOH + NH_3$$

11-4. ALKALINE EARTH METALS

The representative metals of group II are often called the alkaline earth metals. Like the alkali metals, they are good thermal and electrical conductors. They are harder than the alkali metals and are somewhat less reactive, although they rank among the more reactive metals. Table 11-3 shows the similarities between the electronic structures of the atoms of the two groups. The removal of two electrons results in the noble gas electronic configuration. The stability of this arrangement is reflected once again by the large jump in the third ionization potential for the alkaline earth metals. Some physical properties are given in Table 11-4.

Note that the radius of each alkaline earth metal is less than the radius of the alkali metal (Table 11-2) immediately to its left in the periodic table. In general, we find that atomic size decreases in going from left to right across a period because each element gains an additional proton and the additional electron is placed in the same quantum level. The overall result is a drawing in

Table 11-3. Electronic Structure and Ionization Potentials of Alkaline Earth Metals

Element	Atomic Number	Electronic Configuration*	Ionization Potential, volts		
			First	Second	Third
beryllium (Be)	4	[He]$2s^2$	9.3	18.2	154
magnesium (Mg)	12	[Ne]$3s^2$	7.6	15.0	80.1
calcium (Ca)	20	[Ar]$4s^2$	6.1	11.9	51.2
strontium (Sr)	38	[Kr]$5s^2$	5.7	11.0	43
barium (Ba)	56	[Xe]$6s^2$	5.2	10.0	36
radium (Ra)	88	[Rn]$7s^2$	5.3	10.1	36

*The inner electronic configuration of each atom is the same as that of the noble gas in brackets.

of the electron cloud by the increased nuclear positive charge. Of course, in going down a group in the periodic table, the atoms are expected to increase in size since the outer electrons reside in a higher quantum level as you go from element to element.

These elements are produced primarily in the same way as alkali metals—electrolysis of a suitable fused salt. The alkaline earth metals also react with water to give the hydroxide of the metal and hydrogen, although beryllium and magnesium do so only at high temperatures. As an example, the reaction of calcium with water can be expressed

$$Ca + 2H_2O \longrightarrow Ca(OH)_2 + H_2$$

All of the alkaline earth metals react readily with the halogens, for example

$$Mg + Cl_2 \longrightarrow MgCl_2$$

The reaction of these metals with oxygen is simpler than in the case of the alkali metals in that the alkaline earth metals give only the simple oxides. The reaction of strontium with oxygen is typical.

$$2\,Sr + O_2 \longrightarrow 2\,SrO$$

Only barium forms a stable peroxide at high temperatures by direct reaction of oxygen with the oxide.

$$2\,BaO + O_2 \longrightarrow 2\,BaO_2$$

Table 11-4. Physical Properties of the Alkaline Earth Metals

Element	Atomic Radius (Å)	Density at 25°C (g/cm³)	Melting Point (°C)
beryllium	0.89	1.86	1283
magnesium	1.36	1.75	650
calcium	1.74	1.55	850
strontium	1.91	2.60	770
barium	1.98	3.60	704

Except for BeO, all the alkaline earth oxides react with water to form slightly soluble hydroxides, for example

$$CaO + H_2O \longrightarrow Ca(OH)_2$$

Calcium and the heavier alkaline earth metals react directly with hydrogen at elevated temperatures to form hydrides.

$$Ca + H_2 \longrightarrow CaH_2$$

These compounds react with water in a manner analogous to the alkali metal hydrides.

$$CaH_2 + 2 H_2O \longrightarrow Ca(OH)_2 + 2 H_2$$

Except for beryllium, the alkaline earth metals react directly with nitrogen at high temperature to form nitrides.

$$3 Mg + N_2 \longrightarrow Mg_3N_2$$

Like their alkali metal analogs, these nitrides react with water

$$Mg_3N_2 + 6 H_2O \longrightarrow 3 Mg(OH)_2 + 2 NH_3$$

Large quantities of $CaCO_3$ (limestone) are found in nature. This material is mined to produce calcium oxide (commercial "lime") by thermal decomposition at about 1000°C.

$$CaCO_3 \longrightarrow CaO + CO_2$$

Calcium oxide finds uses in the manufacture of other chemicals, in plaster, in the paper industry, and in certain metallurgical processes (Section 29–5).

11-5. GROUP III REPRESENTATIVE ELEMENTS

An examination of Table 11–5 shows a large increase in ionization potential between the third and fourth potentials as should be expected from the electronic configuration of the group III elements.

Table 11-5. Electronic Structures and Ionization Potentials of Group III Representative Elements

Element	Atomic Number	Electronic Configuration*	Ionization Potential, volts			
			First	Second	Third	Fourth
boron (B)	5	$[He]2s^22p^1$	8.3	25.1	37.9	259
aluminum (Al)	13	$[Ne]3s^23p^1$	6.0	18.8	28.4	120
gallium (Ga)	31	$[Ar]3d^{10}4s^24p^1$	6.0	20.6	30.8	64
indium (In)	49	$[Kr]4d^{10}5s^25p^1$	5.8	18.9	28.0	54
thallium (Tl)	81	$[Xe]4f^{14}5d^{10}6s^26p^1$	6.1	20.4	29.8	51

*The inner electronic configuration of each atom is the same as that of the noble gas in brackets.

Table 11-6. Physical Properties of Group III Representative Elements

Element	Atomic Radius (Å)	Density at 25°C (g/cm³)	Melting Point (°C)
boron	0.7	2.5	2300
aluminum	1.48	2.70	660
gallium	1.33	5.91	30
indium	1.45	7.3	156
thallium	1.80	11.85	304

Consideration of the extremely high melting point of boron in comparison with the melting points of the heavier elements in this family (Table 11-6) suggests that boron is different from the other elements. Even more striking differences are found in that boron is an extremely poor thermal and electrical conductor whereas the heavier elements are good conductors—especially aluminum. Boron is brittle whereas the other elements are metallic in character. In fact, boron is classified as a nonmetal, and with this group the mixed metal and nonmetal trend which extends over the next three groups is established. Thus the lighter elements are more nonmetallic and the heavier are metallic until group VII, the halogens, is reached. In group VII all the elements are nonmetals.

Aluminum is an important metal, widely distributed in nature largely in the form of its hydrated oxide, bauxite ($Al_2O_3 \cdot xH_2O$), or in conjunction with silicon in complex silicates. Aluminum metal is produced by electrolysis of its oxide (Section 29-6).

Table 11-7. Electronic Structures and Ionization Potentials of Group IV Representative Elements

Element	Atomic Number	Electronic Configuration*	Ionization Potential, volts				
			First	Second	Third	Fourth	Fifth
carbon (C)	6	[He]$2s^2 2p^2$	11.3	24.4	47.9	64.5	392.0
silicon (Si)	14	[Ne]$3s^2 3p^2$	8.1	16.3	33.4	45.1	167
germanium (Ge)	32	[Ar]$3d^{10} 4s^2 4p^2$	8.1	15.9	34.3	45.8	94
tin (Sn)	50	[Kr]$4d^{10} 5s^2 5p^2$	7.3	14.5	30.5	40.6	81
lead (Pb)	82	[Xe]$4f^{14} 5d^{10} 6s^2 6p^2$	7.4	15.0	31.9	43.9	69

*The inner electronic configuration of each atom is the same as that of the noble gas in brackets.

Table 11-8. Physical Properties of Group IV Representative Elements

Element	Atomic Radius (Å)	Density at 25°C (g/cm³)	Melting Point (°C)
carbon	0.77	3.51 (diamond)	3500
		2.26 (graphite)	
silicon	1.17	2.4	1420
germanium	1.22	5.36	959
tin	1.51	7.36	232
lead	1.37	11.34	327

Table 11-9. Electronic Structures and Ionization Potentials of Group V Elements

Element	Atomic Number	Electronic Configuration*	Ionization Potential, volts			
			First	Second	Third	Fourth
nitrogen (N)	7	[He]$2s^2 2p^3$	14.5	29.6	47.4	77.5
phosphorus (P)	15	[Ne]$3s^2 3p^3$	11.0	19.7	30.2	51.4
arsenic (As)	33	[Ar]$3d^{10} 4s^2 4p^3$	10.5	20.3	27.4	50.2
antimony (Sb)	51	[Kr]$4d^{10} 5s^2 5p^3$	8.6	18.8	24.7	43.9
bismuth (Bi)	83	[Xe]$4f^{14} 5d^{10} 6s^2 6p^3$	7.3	16.7	25.6	45.3

*The inner electronic configuration of each atom is the same as that of the noble gas in brackets.

Table 11-10. Physical Properties of Group V Representative Elements

Element	Atomic Radius (Å)	Density at 25°C (g/cm^3)	Melting Point (°C)
nitrogen	0.53	0.0012*	−210
phosphorus	0.93	1.8	44 (white)
arsenic	1.25	5.7	815 (3 atm)
antimony	1.44	6.7	631
bismuth	1.46	9.8	271

*Gaseous.

11-6. GROUP IV REPRESENTATIVE ELEMENTS

The five elements of this group are about evenly divided in terms of metallic behavior. Tin and lead are metals, germanium is a borderline case, and silicon and carbon are nonmetals. However, although carbon in the form of diamond is a nonconductor, in its other crystalline modification, graphite, carbon is a good electrical conductor. Likewise tin has two crystalline forms. Above 13°C, tin is metallic, but below that temperature it exists in the form of a brittle, nonmetallic material.

11-7. GROUP V REPRESENTATIVE ELEMENTS

Only bismuth displays unequivocal metallic properties. Arsenic and antimony are borderline cases because each posseses a crystalline form that is metallic in appearance.

Table 11-11. Electronic Structures and Ionization Potentials of Group VI Representative Elements

Element	Atomic Number	Electronic Configuration*	Ionization Potential, volts			
			First	Second	Third	Fourth
oxygen (O)	8	[He]$2s^2 2p^4$	13.6	35.1	54.9	77.4
sulfur (S)	16	[Ne]$3s^2 3p^4$	10.4	23.4	35.0	47.3
selenium (Se)	34	[Ar]$3d^{10} 4s^2 4p^4$	9.8	21.5	32	43
tellurium (Te)	52	[Kr]$4d^{10} 5s^2 5p^4$	9.0	18.6	31	38
polonium (Po)	84	[Xe]$4f^{14} 5d^{10} 6s^2 6p^4$	8.4	19.4	27	?

*The inner electronic configuration of each atom is the same as that of the noble gas in brackets.

Table 11-12. Physical Properties of Group VI Representative Elements

Element	Atomic Radius (Å)	Density at 25°C (g/cm^3)	Melting Point (°C)
oxygen	0.60	0.0014*	−219
sulfur	1.06	2.1	119
selenium	1.16	4.8	220
tellurium	1.44	6.2	450

*Gaseous.

11-8. GROUP VI REPRESENTATIVE ELEMENTS

Only polonium of these elements is typically metallic although tellurium is somewhat metallic and selenium has a crystalline form with a metallic appearance. Polonium is highly radioactive, and precautions must be taken when working with it to shield the intense α radiation given off by the metal and its compounds.

Table 11-13. Electronic Structures and Ionization Potentials of the Halogens

Element	Atomic Number	Electronic Configuration*	Ionization Potential, volts			
			First	Second	Third	Fourth
fluorine (F)	9	[He]$2s^22p^5$	17.4	35.0	62.6	87.2
chlorine (Cl)	17	[Ne]$3s^23p^5$	13.0	23.8	39.9	53.5
bromine (Br)	35	[Ar]$3d^{10}4s^24p^5$	11.9	22.8	36.0	50.2
iodine (I)	53	[Kr]$4d^{10}5s^25p^5$	10.5	19.1	33	?
astatine (At)	85	[Xe]$4f^{14}5d^{10}6s^26p^5$	9.5	20.1	29	?

*The inner electronic configuration of each atom is the same as that of the noble gas in brackets.

Table 11-14. Physical Properties of the Halogens

Element	Atomic Radius (Å)	Density at 25°C (g/cm^3)	Melting Point (°C)
flourine	0.64	0.0017*	−220
chlorine	0.99	0.0032*	−102
bromine	1.14	3.1†	−7
iodine	1.33	4.9	114

*Gaseous.
†Liquid.

11-9. GROUP VII REPRESENTATIVE ELEMENTS—THE HALOGENS

All of the halogens are typical nonmetals. They have no physical characteristics of metals and their high ionization potentials assure nonmetallic chemical behavior. Astatine is a very radioactive element and exists in nature only in minute amounts. Its properties have never been fully determined.

Summary of the Representative Elements □ The general conclusions from this brief consideration of the representative elements may be summarized.

1. Increasing ionization potentials (nonmetallic characteristics) are associated with passing from left to right along a period of the periodic table.
2. Decreasing ionization potentials (metallic characteristics) are associated with moving from top to bottom within a given group or chemical family of the periodic table.
3. Atomic size decreases from left to right along a period and increases from top to bottom in a group of the periodic table.

11-10. NOBLE GASES

These elements have variously been known as the inert gases, rare gases, and most recently as the noble gases. Their inertness delayed discovery until the late nineteenth century. The yellow spectral line of the lightest of these gases had been noted in the spectrum of the sun in 1868, but little note was taken of this discovery. Later it was named helium after helios, the Greek word for sun.

It was not until 1894 that Lord Rayleigh and Sir William Ramsey isolated argon, the most abundant of these elements, from the atmosphere. Their report of the discovery of argon (lazy one) included a lengthy discussion of all the fruitless attempts to make it react. Another problem that arose from the failure of argon to react was in assigning a molecular weight. Finally it was decided on the basis of physical properties that the molecules are monatomic. Still another problem was the location in the periodic table. Within the next 5 years all the other known noble gases were discovered and these problems were solved. It was seen that the noble gases form a logical "buffer" zone of relatively unreactive elements separating the very reactive alkali metals and the very reactive halogen nonmetals. Table 11–15 lists some of the properties and abundances of the noble gases.

In 1962, it was discovered that xenon reacts with fluorine to give the fluorides, XeF_2, XeF_4, and XeF_6. Since then, fluorides of krypton have also been prepared.

Table 11-15. Properties of the Noble Gases

Element	Atmospheric Concentration (ppm)*	Melting Point (°C)	Boiling Point (°C)
helium (He)	5.2	−272.1 (25 atm)	−269.0
neon (Ne)	18	−248.6	−246.4
argon (Ar)	9340	−189.4	−185.9
krypton (Kr)	1.1	−156.6	−152.9
xenon (Xe)	0.09	−111.5	−107.1
radon (Rn)	6×10^{-14}	−71	−65

*ppm = parts per million.

11-11. TRANSITION METALS

An examination of the periodic table (Section 8–7) shows that, in the horizontal row beginning with potassium (K), there are ten elements that interrupt the regular sequence of the groups of representative elements. That is, the element scandium (Sc) does not have the properties expected of an element that falls under aluminum (Al) in the table. Similarly, the properties of titanium (Ti) do not qualify it for a position under silicon (Si). Instead, the ten elements scandium through zinc (Zn) seem to have a set of properties all their own, and periodicity among the representative groups is resumed with gallium (Ga), germanium (Ge), arsenic (As), and so on.

These so-called **transition** elements are similar in that they are all metals; and most all exhibit variable ionic charges. Most of the compounds of these metals are highly colored in contrast with compounds of the representative elements which are usually colorless. The transition metals all have one or two valence electrons (chromium and copper have one valence electron). The difference in electron configuration within this series of metals is in the electron shell *next* to the outermost shell. Thus, these elements of the fourth period have the electron distributions shown in Table 11–16. Because the outermost electronic configurations are similar, these elements are quite similar in their properties. In the fifth and sixth periods, transition series exist whose electronic structures are similar to those in the fourth period.

Although a group or vertical chemical family similarity exists among the transition elements, a strong horizontal or period similarity is also found for these elements. This horizontal relationship for the transition metals in the fourth period can be seen in the similarity of their ionization potentials (Table 11–16). Note particularly the constancy of the first ionization potentials. This behavior is in marked contrast with the representative elements where, it has been noted, a marked increase in ionization potentials occurs in going across a period. The horizontal similarity can also be seen in other properties of the transition elements (Table 11–17).

Table 11-16. Electronic Structures and Ionization Potentials of Transition Elements of the Fourth Period

Element	Atomic Number	Electronic Configurations	Ionization Potential, volts			
			First	Second	Third	Fourth
scandium (Sc)	21	$[Ar]3d^14s^2$	6.5	12.8	24.8	74
titanium (Ti)	22	$[Ar]3d^24s^2$	6.8	13.6	27.5	43
vanadium (V)	23	$[Ar]3d^34s^2$	6.7	14.7	29.3	48
chromium (Cr)	24	$[Ar]3d^54s^1$	6.8	16.5	31.0	50
manganese (Mn)	25	$[Ar]3d^54s^2$	7.4	15.6	33.7	52
iron (Fe)	26	$[Ar]3d^64s^2$	7.9	16.2	30.6	57
cobalt (Co)	27	$[Ar]3d^74s^2$	7.9	17.1	33.5	?
nickel (Ni)	28	$[Ar]3d^84s^2$	7.6	18.2	35.2	?
copper (Cu)	29	$[Ar]3d^{10}4s^1$	7.7	20.3	36.8	?
zinc (Zn)	30	$[Ar]3d^{10}4s^2$	9.4	18.0	39.7	?

Table 11-17. Physical Properties of Transition Elements of the Fourth Period

Element	Atomic Radius (Å)	Density at 25°C (g/cm³)	Melting Point (°C)
scandium	1.44	2.99	1200
titanium	1.32	4.54	1800
vanadium	1.22	6.11	1710
chromium	1.17	7.19	1615
manganese	1.17	7.44	1260
iron	1.16	7.87	1535
cobalt	1.15	8.90	1492
nickel	1.15	8.9	1453
copper	1.17	8.94	1063
zinc	1.25	7.13	420

11-12. INNER TRANSITION METALS

Period six contains, in addition to the expected transition series of metals, an **inner transition series,** commonly called the **rare earth metals,** which vary in the electron configuration of the shell twice removed from the outermost shell. There 14 elements lie between lanthanum (La, at. no. = 57) and hafnium (Hf, at. no. 72). In this series the energy level that previously had filled to 18 electrons now can accept electrons to a total of 32. This is another case in which the lowest energy level available is not the valence shell. Table 11-18 shows the electron

Table 11-18. Electronic Structures and Ionization Potentials of the Rare Earth Elements

Element	Atomic Number	Electronic Configuration*	Ionization Potential, volts First	Second
cerium (Ce)	58	$[Xe]4f^26s^2$	6.5	12.3
praseodymium (Pr)	59	$[Xe]4f^36s^2$	5.7	?
neodynium (Nd)	60	$[Xe]4f^46s^2$	5.7	?
promethium (Pm)	61	$[Xe]4f^56s^2$	?	?
samarium (Sm)	62	$[Xe]4f^66s^2$	5.6	11.2
europium (Eu)	63	$[Xe]4f^76s^2$	5.7	11.2
gadolinium (Gd)	64	$[Xe]4f^75d^16s^2$	6.2	12
terbium (Tb)	65	$[Xe]4f^96s^2$	6.7	?
dysprosium (Dy)	66	$[Xe]4f^{10}6s^2$	6.8	?
holmium (Ho)	67	$[Xe]4f^{11}6s^2$	?	?
erbium (Er)	68	$[Xe]4f^{12}6s^2$	?	?
thulium (Tm)	69	$[Xe]4f^{13}6s^2$	?	?
ytterbium (Yb)	70	$[Xe]4f^{14}6s^2$	6.2	12.1
leutetium (Lu)	71	$[Xe]4f^{14}5d^16s^2$	6.2	14.7

*The inner electronic configuration of each atom is the same as that of the noble gas in brackets.

Table 11-19. Physical Properties of the Rare Earth Elements

Element	Atomic Radius (Å)	Density at 25°C (g/cm³)	Melting Point (°C)
cerium	1.65	6.66	795
praseodymium	1.65	6.78	935
neodymium	1.64	7.00	1024
promethium	?	?	?
samarium	1.66	7.54	1072
europium	1.85	5.26	826
gadolinium	1.61	7.90	1312
terbium	1.59	8.27	1356
dysprosium	1.59	8.54	1407
holmium	1.58	8.80	1461
erbium	1.57	9.05	1497
thulium	1.56	9.32	1545
ytterbium	1.70	6.98	824
leutitium	1.56	9.5	1652

configurations of this inner transition series. There is another inner transition series following actinium (Ac, at. no. = 89), all of whose elements are radioactive. Detailed chemistry is known only for the first few elements of this series. In interpreting the electronic configurations in Table 11–18, it must be remembered that the xenon [Xe] configuration has a $5s^2 5p^6$ exterior, so that in all cases the $4f$ levels are two quantum levels below the levels containing valence electrons.

Since chemical and physical properties are determined primarily by the outer electrons, it might be expected that the inner transition elements should all be even more similar to one another than the normal transition series since they all have the same electron configurations in the outer *two* energy levels. This expectation is confirmed by the fact that these elements created tremendous problems of separation and identification especially because they occur together in nature. The most common ionic charge in this series is $+3$ although other charges are observed. Promethium (Pm) is radioactively unstable and is not found in nature. It has been prepared by artifical means however. Some physical properties of the rare earth elements are given in Table 11–19.

The rare earth metals are used in alloys with other metals such as iron to increase hardness and strength. They are used extensively in cigaret lighter flints.

PROBLEMS

1. What fundamental property of an atom determines its position in the periodic table? What property was used by Mendeleev to construct his table?

2. Name the four basic types of elements and differentiate among them according to

(a) their general physical properties; (b) their metallic or nonmetallic characteristics; (c) their electronic structure.

3. Why do the successive ionization potentials for a given atom all increase? Why do the first ionization potentials of the elements from lithium (Li) to neon (Ne) increase across the period? Why do the first ionization potentials of the alkali metals decrease in going from lithium to francium (Fr)?

4. The first ionization potentials of the transition metals of the fourth period (Table 11–16) show little variation. How can you rationalize this fact with the higher nuclear charge as one moves across the period? (*Hint:* remember that a 4s electron is removed in forming the ion in each case).

5. Plot the density as a function of atomic number for the elements of the second period (lithium through neon); also plot these variables for one of the representative element groups, for example, carbon through lead. Try to rationalize the trends. Can you explain any abrupt change in the trend of this property?

6. Write equations for the reaction of the following metals with oxygen.
   (a) Li   (b) Na   (c) K   (d) Rb   (e) Mg   (f) Ca

7. For each of the following, fill in the expected products and balance the equation.
   (a) $Li + N_2 \longrightarrow$ (d) $K + KNO_3 \longrightarrow$ (g) $NaH + H_2O \longrightarrow$
   (b) $Na + H_2O \longrightarrow$ (e) $Rb + H_2 \longrightarrow$ (h) $RbO_2 + H_2O \longrightarrow$
   (c) $K + F_2 \longrightarrow$ (f) $Cs + Cl_2 \longrightarrow$ (i) $Na_2O_2 + H_2O \longrightarrow$
                                                   (j) $K_2O + H_2O \longrightarrow$

8. For each of the following, fill in the expected products and balance the equation:
   (a) $Be + Cl_2 \longrightarrow$ (e) $CaH_2 + H_2O \longrightarrow$ (h) $Mg + Br_2 \longrightarrow$
   (b) $BaO + O_2 \longrightarrow$ (f) $Ba + H_2 \longrightarrow$ (i) $Ca + H_2O \longrightarrow$
   (c) $MgO + H_2O \longrightarrow$ (g) $Mg_3N_2 + H_2O \longrightarrow$ (j) $Ba + H_2O \longrightarrow$
   (d) $Sr + N_2 \longrightarrow$

9. Write balanced equations to show how one might prepare
   (a) NH_3 from Li_3N (d) KBr
   (b) CO_2 from $CaCO_3$ (e) CaH_2
   (c) H_2 from Li

10. Why are the metals of the representative elements of the later groups not as reactive as those of groups I and II?

11. Name four physical properties of boron that set it apart from the other representative elements of group III. How do these properties differ?

12. Nonmetal behavior is associated with electrical nonconductors. Graphite is an electrical conductor yet carbon is classified as a nonmetal. Explain.

13. Tin roofs are common in warm climates but not in cold climates. Why?

14. What property of polonium causes problems in its study or use?

15. What element should be chemically the most metallic? the most nonmetallic?

16. Most gases and vapors are diatomic (H_2, O_2, Cl_2, and so on). The noble gases are unusual in that they are monatomic. Suggest a reason why the noble gases have this property.

17. On the basis of electronic configuration why are there 10 elements in each row of transition elements and 14 elements in each row of inner transition elements?

18. List the atomic radii for representative elements across several periods. Compare the trends with the atomic radii across the transition metals of the fourth period (Table 11–17). Try to rationalize the differences.

19. Compare the trends in atomic radii across periods for representative and transition elements with the atomic radii of the rare earth metals (Table 11–19). Try to rationalize any differences among the three types of elements.

20. The chemistries of the rare earth elements are very similar. How did this fact delay knowledge of these chemistries?

21. Prepare an overall summary of properties of the elements (take 105 as the number of known elements) by calculating (a) the percentage of elements which are metals, and (b) the percentage of elements in each category (metal and nonmetal), which are solid, liquid, and gaseous at 25°C and 1 atm [mercury (Hg) is the only metal that is a liquid at 25°C.]

22. Why should lack of reactivity of the noble gases present a problem in determining their molecular weights?

23. Should the elements from number 104 through 112 constitute an ordinary or an inner transition series? Explain.

12

CHEMICAL BONDING

The previous two chapters have dealt with atomic theory and its application to explaining the periodic properties of the elements. The present chapter deals with chemical bonding. Like periodicity, chemical bonding is explained logically and ultimately only through atomic theory. But before taking up the theoretical explanation of bonding, it will be instructive to introduce some of the practical ways in which the various types of chemical bonds affect the nature of substances. We shall begin our discussion with a simple property of gross matter—electrical conductivity.

12-1. ELECTRICAL CONDUCTIVITY AND PHYSICAL PROPERTIES

We all know that certain substances, such as copper wire, conduct an electric current. Others, like rubber and glass, are very poor conductors and are frequently used as insulators to prevent the flow of electric current. Water is an interesting substance in this regard: although water is generally believed to be a good electrical conductor, it is actually a poor conductor if pure. It is the dissolved impurities that account for the conductivity of water. Water solutions of salts are excellent conductors. Figure 12–1 shows an apparatus used to demonstrate conductivity. Interestingly enough, a sodium chloride crystal is not a conductor, although the pure molten salt is.

Table 12–1 lists some common substances with their electrical conductivities and some other physical properties. A fact that emerges from Table 12–1 is that the arbitrary separation of substances into conductors and nonconductors automatically results in a separation based on other physical properties: nonconductors generally are substances with low melting and boiling points, low densities and high vapor pressures, whereas conductors have high melting and boiling points, high densities, and low vapor pressures. To be sure, not all of the substances in

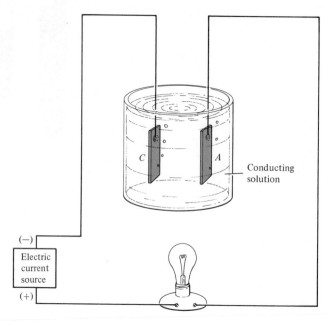

Figure 12-1. Apparatus for demonstrating the conductivity of solutions. Current flows from the negative end of the source through the wire to the plate C. The current flows from C to A through the solution. From A to the positive terminal of the current source, the current flows through the wire. Conduction through the solution is signalled by the light bulb in the path.

Table 12-1 obey this generalization completely. For example carbon in the form of diamond is a nonconductor, but also has high melting and boiling points and density as well as a zero vapor pressure. As we shall see later, quite satisfactory explanations exist for the exceptions.

The problem at this point is to demonstrate that sets of physical properties "go together," so to speak. Why, for example, is a high electrical conductivity associated with high melting point, high density and low vapor pressure? We can find the answer to this question, as well as to more complex ones, in the minute detailed structure of matter.

12-2. NATURE OF CONDUCTORS

The transport of an electric current requires that electrically charged particles move through the conducting medium. In the case of a copper wire or other metallic conductor, electrons move through the action of "electromotive force," which is analogous to the motion of water through a pipe under the influence of a difference in water level or "hydrostatic pressure."

An electric current does not flow readily through pure water, therefore, we can assume that there is no appreciable number of charged particles present. Dissolving a small amount of salt in the water produces a solution that is a good conductor. The sodium chloride solution must therefore contain electrically charged particles. Molten sodium chloride similarly allows the passage of electric current and thus must contain charged particles. Solid sodium chloride does not

Table 12-1. Physical Properties of Some Common Substances
Group A: Nonconductors

Substance	Melting Point (°C)	Boiling Point[a] (°C)	Density at 25°C[a]	Vapor Pressure at 25°C (torr)
water	0	100	1.00[b]	24
methane	− 184	− 162	0.72[c]	very high
ethylene	− 169	− 104	0.57[c]	750,000
acetylene	− 81.8	− 83.6	0.62[c]	40,000
ethyl alcohol	− 114	78.5	0.79[b]	50
isopropyl alcohol	− 89	82.3	0.79[b]	40
chloroform	− 69	61.3	1.50[b]	250
carbon tetrachloride	− 23	76.8	1.6[b]	100
sulfur (solid)	113	445	2[b]	1
sulfur (molten)	–	445	2[b]	1
carbon (diamond)	3500	4200	3.51[b]	0
benzene	5.5	80.0	0.879[b]	100
isooctane (gasoline)	− 107	99.3	0.69[b]	50
ethyl ether	− 116	34.6	0.71[b]	600
naphthalene (moth balls)	80.2	217.9	1.15[b]	1
sucrose (sugar)	186 (decompose)	–	1.59[b]	0

[a] Measured at atmospheric pressure (760 torr).
[b] Density given in grams per milliliter.
[c] Density given in grams per liter.

Group B

Substances	Electrical Conductivity	Melting Point (°C)	Boiling Point (°C)	Density at 25°C (g/cm³)	Vapor Pressure at 25°C (torr)
sodium chloride (solid)	nonconductor	801	1413	2.165	0
(molten)	good conductor	801	1413	—	–
carbon (graphite)	good conductor	3600	4200	2.25	0
sodium hydroxide (solid)	nonconductor	318	1390	2.13	0
(molten)	good conductor	—	—	—	–
copper (solid or molten	good conductor	1083	2300	8.94	0

conduct electricity, but we must not be too hasty in concluding that it does not contain charged particles. Instead, the charged particles in solid sodium chloride are held in a rigid, immobile framework.

Any reasonable arrangement of these oppositely charged ions in a crystal must involve maximum separation of like charges and minimum separation of unlike charges. In sodium chloride crystals, the sodium and chloride ions are arranged as shown in Figure 12–2. Of course, other similar arrangements are also possible for ionic crystals and, in fact, do occur. The exact geometry will depend on the relative sizes of the two ions and the number of electric charges on each.

Regardless of the exact structure, the important features of substances whose atoms are bound to each other in this way are

157

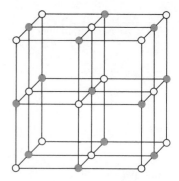

Figure 12-2. The sodium chloride crystal structure. The diagram is intended to show relative positions only. Actually the ions will be arranged in the closest packing possible. Dark spheres represent sodium ions and light spheres represent chloride ions.

1. The forces that hold the ions together are electrostatic* and these electrostatic forces extend in all directions equally.
2. Each ion is equally strongly bonded to each of its closest neighbors.

The consequences of this type of chemical bonding are

1. Breakage of the crystal requires the rupture of many bonds between oppositely charged ions and therefore a considerable amount of energy. This predicts high melting and boiling points and low vapor pressures for ionic crystals.
2. Because opposite charges attract each other and each ion is closer to its oppositely charged neighbors than it is to ions of the same charge, the ions will coalesce to the tightest arrangement possible. This predicts a high density for ionic crystals.
3. A fixed crystal structure does not allow any mobility of ions with respect to one another, hence crystals should not conduct an electric current. Melting of the crystal, however, would allow the mobility required for conduction.
4. Solution of an ionic crystal occurs primarily if the bonds formed between each ion and the solvent molecules are stronger than the bonds between ions in the crystal. This is true only of a relatively few solvents, the principal one of which is water (Figure 12-3).

12-3.   NATURE OF NONCONDUCTORS

Many of the properties of conducting substances can thus be explained by their ionic structure. But what of nonconductors? The physical properties of these

---

*Electrostatic forces are governed by Coulomb's law, which states that the force of attraction of unlike charges, or of repulsion of like charges, is directly proportional to the product of the charges and inversely proportional to the square of the distance separating them.

$$F = \frac{q_1 q_2}{\varepsilon \, d^2}$$

where F = force, q_1 and q_2 are the charges of particles 1 and 2 respectively, d = the distance between the particles, and ε is a proportionality constant called the dielectric constant.

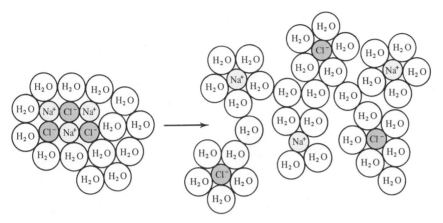

Figure 12-3. Model of the solution of sodium chloride (NaCl) in water.

substance (Group A, Table 12–1) indicate that the units of matter of which they are composed are not held together nearly as firmly as those of ionic materials. Thus, they are low melting and low boiling materials of rather low densities and high vapor pressures. These substances, though, do not exhibit the properties of their constituent atoms any more than ionic substances. For example, water is entirely different from hydrogen or oxygen, the elements that combine chemically in its formation. In nonionic substances, atoms combine to form molecules that are more or less independent of each other, but which exhibit some degree of intermolecular attraction.

The nature of the bonding that holds atoms together in the form of molecules will be discussed in subsequent sections. At this point, only the necessary gross features of this kind of bond will be summarized. These gross features result from the bulk properties of the bonds.

1. Discrete molecules are formed with relatively low intermolecular attraction (attraction between one molecule and another).
2. The law of constant composition implies that the number of bonds is fixed for each element.
3. Molecules are electrically neutral.

Thus, whereas ionic substances are extremely large three-dimensional networks of tightly held ions, molecular substances consist of discrete neutral molecules that are attracted to each other relatively weakly. Thus the properties outlined above and in Table 12–1 can be easily rationalized.

The bonding in molecular substances is called **covalent** and will be discussed in Section 12–6. Covalent substances make up many if not most of the substances of everyday experience. Whereas ionic compounds are virtually always saltlike in physical appearance, covalent substances have a wide range of physical states. All the gases including our atmosphere, water, all living things, and even minerals such as diamond, graphite (coal), and silica (SiO_2) are covalent.

12-4. NATURE OF METALLIC CONDUCTORS

The metals must be considered in a third group because of their unique properties. Although ionic substances can conduct an electric current only when molten or

Table 12-2. Physical Data on Lithium

Internuclear distance in the metal	3.03 Å[a]
Ionic radius of Li^+ (as in LiCl)	0.60 Å
Internuclear distance in (covalent) Li_2	2.67 Å
Binding energy[b] in (covalent) Li_2	1.9 kcal/g
Binding energy in the metal	5.6 kcal/g
Number of atoms surrounding each Li atom in the metal	14
Number of fluorine atoms closely surrounding each Li atom in (ionic) LiF	4

[a] The Ångstrom unit (Å) is equal to 10^{-8} cm.
[b] Binding energy is the energy released when the separated atoms are united to form the compound.

dissolved, metals are conductors in the solid state. They also have the unusual properties of malleability (capability of being deformed by beating) and ductility (capability of being drawn). Moreover, X-ray diffraction analysis (Chapter 13) demonstrates that, in the crystalline state, the atoms in metals are at unusually large distances from one another compared with the atoms of ionic or covalent substances. Table 12–2 lists relevant data for lithium, which is a typical metal.

As can be seen, there are great differences between the parameters involving atoms in metallic lithium and those for either ionic lithium (in LiCl) or in covalent lithium (in Li_2). The most notable feature of metallic lithium is the interatomic distance of 3.03 Å. Even in covalent Li_2 the distance is only 2.67 Å. If the metal consisted of Li^+ ions, then the distance would be twice the ionic radius, or 1.20 Å. Therefore, the bonding in the metal is neither ionic nor covalent. That the bonding in the metal is quite stable is borne out by the much higher binding energy compared to that for covalent Li_2. The physical properties of the metal—malleability and ductility—can be explained by the large interatomic distances that allow considerable room for atomic movement when the metal is stressed.

12-5. IONIC BONDING

Theories of bonding were relatively slow in coming. Although electrically conducting salts had been known for many years, the first comprehensive theory explaining ionic bonding was presented as late as 1916. In that year Walter Kossel in Germany observed that metals and nonmetals that combine to form ionic substances occur, respectively, just after and just before noble gases. In terms of ionization potentials, ionic substances are formed by the combination of metals of low ionization potentials with nonmetals of high ionization potentials. Kossel suggested that, by a simple electron transfer from metal to nonmetal, both elements could achieve the noble gas electronic structure, and in the process, form ions.

In the formation of sodium chloride, the ions result from the transfer of one electron from each sodium atom to each chlorine atom. The loss of one electron by a sodium atom leaves an outer shell of eight electrons. Similarly, chlorine possesses seven electrons, and the addition of one more yields a stable octet.

$$\text{Na } 2 \text{ } 8 \text{ } 1 + \text{Cl } 2 \text{ } 8 \text{ } 7 \longrightarrow [\text{Na } 2 \text{ } 8]^+ + [\text{Cl } 2 \text{ } 8 \text{ } 8]^-$$

The reaction of magnesium with fluorine similarly results in the transfer of electrons from magnesium to fluorine. From the ionization potentials, magnesium possesses two relatively loosely held electrons and fluorine, like chlorine, possesses seven outer electrons. Stable octets result if one magnesium atom transfers two electrons (one each) to two fluorine atoms.

$$\text{Mg } 2 \text{ } 8 \text{ } 2 + 2 \text{ F } 2 \text{ } 7 \longrightarrow [\text{Mg } 2 \text{ } 8]^{2+} + 2 [\text{F } 2 \text{ } 8]^-$$

But what of the nature of the ionic crystals that form? The separation of oppositely charged ions from one another requires a considerable expense of energy as is evidenced by the physical properties given in Table 12–1—high melting and boiling points, low vapor pressure, and so on. A measure of the stability of such crystals is offered by the crystal **lattice energy,** which is defined as the energy evolved when separate ions in the gaseous phase are allowed to form 1 mole of the crystalline compound. Such lattice energies are given in units of kilocalories per gram-mole at 25°C and 1 atm pressure. Table 12–3 lists the lattice energies for some common salts. Note that lattice energies are **exothermic,** that is, energy is released in the process of forming the crystal.

Table 12–3. Lattice Energies of the Alkali Halides (kcal/mole)

	F	Cl	Br	I
Li	246.8	202.0	190.7	176.8
Na	218.7	185.9	176.7	165.4
K	194.4	169.4	162.4	153.0
Rb	185.9	164.0	157.5	148.7
Cs	178.7	155.9	151.1	143.7

Because all of the ions involved in Table 12–3 have the same charge magnitude, a simple correlation can be observed between lattice energy and ionic size. Ionic size increases on going from lithium to cesium and from fluorine to iodine. A simple Coulomb's law treatment would predict that CsI should have the smallest lattice energy because, in this crystal structure, the opposite charges are the most widely separated. Similarly, LiF should exhibit the highest lattice energy. This type of generalization must be made with caution, however, because other factors, such as differences in the crystal lattice geometry, will often play a major role.

12-6. COVALENT BONDING

Also in 1916, Gilbert N. Lewis in California proposed a theory to explain the nature of covalent bonding. He suggested that the outer shell of an atom be considered to contain its electrons at the corners of a cube. He argued that a solar system model in which electrons are in constant motion cannot be reconciled with the formation of bonds that have fixed geometries.

As early as 1874, van't Hoff and Le Bel had proposed that the four bonds of carbon atoms are oriented towards the corners of a regular tetrahedron with the carbon atom at the center.

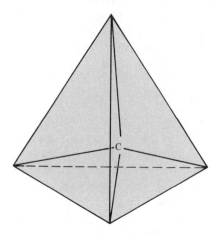

Within the next 50 years the tetrahedral geometry of carbon and many other atoms as well became completely accepted. It was this apparently fixed geometry of covalent bonding that drove Lewis to find an explanation in the electrical nature of atoms.

Lewis' concept of the covalent bond involved the sharing of a pair of electrons by the bonded atoms. This aspect of his theory with slight modifications is the commonly held model of covalent bonding today. His complete theory, however, was somewhat fanciful and pictured the valence electrons in an atom as occupying the corners of a cube.

A modernized form of Lewis' postulates for covalent bonding follow.

1. The **kernel,** containing the nucleus and inner shell electrons, and denoted by the chemical symbol, remains unaltered during chemical reactions.
2. Atoms tend to hold even numbers of electrons in the outer shell, and especially to hold eight electrons, which are normally arranged in pairs with one pair at each corner of a tetrahedron.
3. Two atomic shells are mutually interpenetrable.
4. Electric forces between particles that are very close together do not obey the simple law of inverse squares which holds at greater distances (Coulomb's law).

Although it is difficult to imagine "static" electrons in fixed positions about an atom, Lewis' objection to the solar system atomic model was valid. His cubical atom theory in which electrons are localized in space also was quite prophetic of wave mechanics, which was to be introduced 10 years later (Sections 10–10 and 10–11).

The most important features for present purposes are postulates 2 and 3. According to these postulates, a chemical bond is formed when two electrons are shared. Lewis suggested that electrons be represented by dots and that covalent bonding could be represented by the pairing of these dots.

$$:\overset{..}{\underset{..}{Cl}}\cdot \quad + \quad \cdot \overset{..}{\underset{..}{Cl}}: \quad \longrightarrow \quad :\overset{..}{\underset{..}{Cl}}:\overset{..}{\underset{..}{Cl}}:$$

(7 electrons)   (7 electrons)   (14 electrons)

Thus, the combination of two chlorine atoms, each with a deficiency of one electron, results in a molecule in which both atoms have completed shells. Similarly, double and triple bonds can form by sharing two or three pairs of electrons, respectively:

$$:\overset{.}{\underset{.}{O}}: \quad + \quad :\overset{.}{C}: \quad + \quad :\overset{.}{\underset{.}{O}}: \quad \longrightarrow \quad :\overset{..}{O}::C::\overset{..}{O}:$$

(6 electrons)   (4 electrons)   (6 electrons)   (16 electrons)

$$:\overset{.}{N}\cdot \quad + \quad \cdot \overset{.}{N}: \quad \longrightarrow \quad :N:::N:$$

(5 electrons)   (5 electrons)   (10 electrons)

Some of the salient features of the covalent bond are

1. In general, each atom contributes one electron to form an **electron pair bond.**
2. Bonds may contain one, two, or three pairs of electrons, in which case they are called **single, double,** or **triple** bonds, respectively.
3. Electron pair bonds will arrange themselves so as to minimize repulsion. See Table 12–4 for examples of the resulting possible geometries.
4. Physically, the bond may be pictured as the mutual attraction of each

Table 12-4. Some Possible Geometries in Covalent Compounds

Example	Geometry*	
CH_4		tetrahedral, all angles = $109\frac{1}{2}°$
NH_3		pyramidal (tetrahedral if the unshared electron pair is included)
H_2O		angular (tetrahedral if the unshared electron pairs are included)
CH_2O		planar trigonal
CO_2	$:\overset{..}{O}=C=\overset{..}{O}:$	linear
C_2H_2	$H—C\equiv C—H$	linear

*For clarity in writing formulas a straight line is often used to represent an electron pair bond.

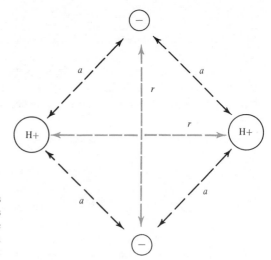

Figure 12–4. Representations of the attractions and repulsions of the hydrogen molecule (H₂). H⁺ represents the nucleus (proton), ⊖ represents the electron. The positions need not be fixed as shown but only serve to show average positions in which attractive forces (a) are greater than repulsive forces (r).

positive nucleus and the negative electron pair of the bond (Figure 12–4) along with the mutual repulsion of the two nuclei and of the two electrons. The balance of these repulsive and attractive forces results in an optimum distance between the nuclei and therefore average bond lengths that can be measured.

The consequences of this model of the covalent bond are

1. Covalent bonding results in discrete molecules that are only relatively weakly attached to other molecules in the sample. This results in a relatively small amount of energy required to separate the molecules from one another, hence the low melting and boiling points, low densities, and high vapor pressures of most covalent substances.
2. Each kind of atom will, in general, form only a definite number of bonds, determined by the number of electrons that it requires to complete its octet. This number is called the **valence** of the atom.
3. Minimization of the mutual repulsion of electrons normally results in a tetrahedral arrangement of electron pairs as shown in Figure 12–5 for methane (CH₄). Molecules that have fewer than four bonds still attain the tetrahedral arrangement in which unshared (unbonded) electron pairs occupy the corners of the tetrahedron.

Figure 12–5. Tetrahedral arrangement of the electron pairs in the methane and water molecules.

Methane (CH₄)   Water (H₂O)

12-7. BONDING IN METALS

The facts presented in Section 12–4 are best accomodated by a model for metals in which the atoms, including all but the valence electrons, are relatively far apart and in which these positive ions are held together by a loosely held "sea" of electrons. In this model, the high conductivity in the solid is accounted for by the high mobility of the electrons (low ionization potentials) under the influence of an applied electric potential. The great atomic separations account for the properties of flexibility, malleability, and ductility observed in metals.

A more detailed treatment of the nature of metallic bonding is postponed until Chapter 17. At this point, the essential feature is the extreme electronic mobility, which accounts for the conductivity of metals in the solid phase.

12-8. COVALENT BONDING AND ORBITAL OVERLAP

The results of wave mechanics were quickly applied to the problem of the covalent bond. In 1927, Wolfgang Heitler and Fritz London showed theoretically that a stable bond can result from sharing two electrons by two atoms provided the shared electrons are **paired,** that is, that their spins are opposed. Writing Lewis dot structures takes on a deeper meaning in the light of this interpretation. The representation of the hydrogen molecule as, H:H, now implies that the shared electrons are *paired.* The electrons are not thought of as "frozen" between the hydrogen nuclei; rather the two dots represent the most probable position of the electrons or, alternatively, they represent the region of space where the negative charge density is greatest. Sometimes emphasis is given to this latter interpretation by means of the diagram

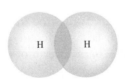

where the shading is used to represent charge density—the heavier the shading the greater the negative charge. However, chemists commonly employ the simpler dot structures but with the wave mechanical understanding attached to them.

A covalent bond is still primarily electrical in nature—the force holding the bonded atoms together arising from the mutual attraction of the nuclei for the shared electrons. The wave mechanical view of the nature of the electron, however, gives a better understanding of how the electrons are shared.

Linus Pauling and John Slater soon extended the work of Heitler and London in an effort to explain more fully the directional characteristics of covalent bonds. They suggested that covalent bonding occurs when two orbitals on two different atoms *overlap* and share the paired electrons in the overlapped region of space. In general, they said the more extensive the degree of overlap the stronger the bond. This general theory of the manner in which covalent bonding occurs is called the **valence bond theory.**

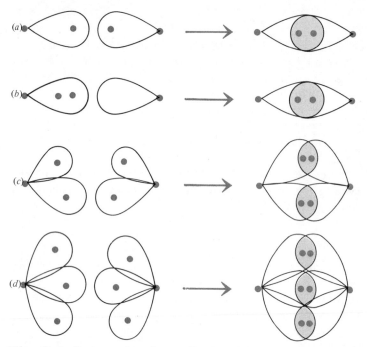

Figure 12-6. Representation of some different covalent bond types. (*a*) Ordinary covalent single bond formation, (*b*) coordinate covalent single bond formation, (*c*) double covalent bond formation, (*d*) triple covalent bond formation.

Notice that there is no requirement that *each* atom supply an electron for the formation of the electron pair bond. Often times, one atom will provide an empty orbital that can overlap an orbital on another atom which contains a pair of electrons. This type of situation is referred to as a **coordinate** covalent bond, but the bond, once established, is no different from one in which each atom provides an electron. Two atoms sharing a pair of electrons in overlapped orbitals are said to have established a single covalent bond. If sufficient, suitable orbitals are available, atoms may share more than two electrons and form double or triple bonds. Figure 12–6 gives a pictoral representation of these different bond types.

A relationship exists between the number of shared electrons and the length and strength of a covalent bond. The strength of a covalent bond is the energy necessary to break the bond so as to divide the electrons between the atoms. Table 12–5 illustrates this relationship for some N-N bonds. An increase in bond strength and decrease in bond distance as the number of shared electrons increases is typical for any two atoms.

12-9. ELECTRON REPULSION AND THE SHAPES OF MOLECULES

Only valence shell electrons are involved in bonding because the underlying electrons are too tightly bound by the nuclear attraction to engage in the sharing process. We can thus confine our attention to the valence shell in treating bonding in compounds. A useful but not infallible guide in writing electronic structures

Table 12-5. Properties of some N-N Bonds

Molecule	Electrons Shared N-N	N-N Distance (Å)	N-N Strength (kcal/mole)
hydrazine H:N̈:N̈:H Ḧ Ḧ	2	1.47	39

dinitrogen
  difluoride
 	
F:N::N:F
  F̈   F̈
	4	1.24	100

Nitrogen :N:::N:	6	1.10	225

is the octet rule. Atoms of elements in the third and higher periods of the periodic table have *d* orbitals available in the valence shell in addition to the *s* and *p* orbitals and may, therefore, have more than eight electrons in their valence shells. We shall have occasion later on to consider a number of examples where this "expansion of the octet" occurs. Atoms of elements in the second period, however, have only *s* and *p* orbitals available in their valence shells (there are no *d* orbitals in the second quantum level) and can thus never have more than eight electrons in their valence shells. The octet rule is quite useful for these elements.

When four separate pairs of electrons are arranged around an atom in its valence shell, the electronic repulsions should cause the pairs to avoid each other as much as possible. Avoidance would best be accomplished if each pair, on the average, were situated at the corner of a tetrahedron around the atom.

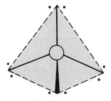

If only three separate pairs were involved, we would expect a planar, triangular arrangement

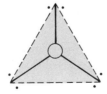

whereas two pairs would lead to a linear situation.

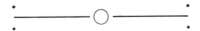

167

Let us apply this rather simple idea to a consideration of all the compounds of the second row elements with chlorine, namely,

$$LiCl \quad BeCl_2 \quad BCl_3 \quad CCl_4 \quad NCl_3 \quad Cl_2O \quad ClF$$

The first, LiCl, is a typically ionic compound and will not be treated, but the remaining compounds of this series display predominantly covalent characteristics.

The first step in attempting to write an electronic structure is to consider the electron distribution in the valence shells of the atoms involved.

$$Be:1s^2 \mid 2s^2 \qquad\qquad Cl:1s^22s^22p^6 \mid 3s^23p^5$$

valence shell   valence shell

It is obvious that although Be will not be able to obtain eight electrons in its valence shell, Cl needs but one to complete the octet. Both Cl atoms may thus complete their octets by gaining a share of one Be electron each. The two electrons originally paired in the Be valence shell can become unpaired in order to engage in the sharing process with the unpaired electron on each of the Cl atoms.

$$:\overset{..}{\underset{..}{Cl}}:Be:\overset{..}{\underset{..}{Cl}}:$$

Because there are only two electron pairs around Be, a linear molecule is predicted which is experimentally found to be the correct structure. Always check to make sure *all* valence electrons are accounted for by a dot structure. Originally, the Be atom had two valence shell electrons and each Cl atom had seven for a total of 16 $(2 + (2 \times 7))$. The structure of the BeCl$_2$ molecule must show 16 electrons as, indeed, it does. Notice that the Be in this molecule represents a violation of the octet rule which, as mentioned above, is only a guide to help predict electronic structure.

For BCl$_3$, we have

$$B:1s^2 \mid 2s^22p^1 \qquad Cl:1s^22s^22p^6 \mid 3s^23p^5$$

Here too we see that the B atom cannot get eight electrons through sharing, although each Cl atom can.

$$:\overset{..}{\underset{..}{Cl}}:$$
$$\overset{..}{\underset{..}{B}}$$
$$:\overset{..}{\underset{..}{Cl}} \qquad \overset{..}{\underset{..}{Cl}}:$$

The predicted planar triangular structure has been observed; and the Cl-B-Cl angles are 120°. The necessary 24 electrons are shown.

Now consider CCl$_4$.

$$C:1s^2 \mid 2s^2sp^2 \qquad Cl:1s^22s^22p^6 \mid 3s^23p^5$$

The C atom with four electrons in its valence shell can complete the octet by gaining a share in one electron from each of the four Cl atoms. Notice that a total of 32 electrons must be shown in the dot structure.

$$\begin{array}{c} \text{:} \ddot{C}\text{l:} \\ \text{:} \ddot{C}\text{l} \text{:} C \text{:} \ddot{C}\text{l:} \\ \text{:} \ddot{C}\text{l:} \end{array}$$

A tetrahedral structure is predicted and has been confirmed. Each Cl-C-Cl bond angle is found to have the value expected for a regular tetrahedron, 109°.

With NCl_3, we have

$$N:1s^2 \ \big| \ 2s^2 2p^3 \qquad Cl:1s^2 2s^2 2p^6 \ \big| \ 3s^2 3p^5$$

Because nitrogen has five valence electrons, it can easily complete the octet.

$$\begin{array}{c} \text{:} \ddot{C}\text{l:} \ddot{N} \text{:} \ddot{C}\text{l:} \\ \text{:} \ddot{C}\text{l:} \end{array}$$

Each atom has eight electrons around it, but one pair on the N atom is not involved in bonding. Such a pair of electrons is called a **nonbonding** or **lone pair** in contrast with the bonding pairs. We would predict an overall tetrahedral structure based on the electron pairs around nitrogen. However, since a pair of electrons cannot be located experimentally, and only the positions of the *atoms* can be determined, the shape of NCl_3 would be described as pyramidal.

The tremendous effect of the lone pair on the molecular geometry is evident by comparison with the planar structure of BCl_3 where there is no lone pair. The Cl-N-Cl bond angles would be expected to be close to regular tetrahedral angle of 109°. In fact, the bond angles in NCl_3 are about 107°, which fits well.

For the molecule Cl_2O

$$O:1s^2 \ \big| \ 2s^2 2p^4 \qquad Cl:1s^2 2s^2 2p^6 \ \big| \ 3s^2 3p^5$$

we obtain

$$\begin{array}{c} \text{:} \ddot{O} \text{:} \\ \text{:} \ddot{C}\text{l} \quad \ddot{C}\text{l:} \end{array}$$

which is described as an *angular* molecule. The Cl-O-Cl bond angle has been found to be 111°, a fairly close approximation to the predicted 109° angle.

The last molecule, ClF

$$F:1s^2 \ \big| \ 2s^2 2p^5 \qquad Cl:1s^2 2s^2 2p^6 \ \big| \ 3s^2 3p^5$$

leads to the electronic arrangement

$$\text{:} \ddot{C}\text{l} \text{:} \ddot{F} \text{:}$$

which must be linear.

169

12-10.   HYBRIDIZATION OF ORBITALS

The shapes of molecules have been rationalized by a consideration of electronic repulsions. If a covalent bond occurs by means of the overlap of two orbitals, then there would have to be orbitals suitably oriented to give the observed structures. It could be argued that under the influence of electronic repulsions, the orbitals in many-electron atoms would be oriented as discussed in Section 12-9, but is there any *theoretical justification for the assumption of such orientations?*

Linus Pauling showed that it is possible to combine mathematically the wave functions for the orbitals and produce new wave functions corresponding to orbitals with different spatial characteristics. These new orbitals are called **hybrid orbitals.** The orientations of hybrid orbitals using the *s* and *p* orbitals are shown in Figure 12-7. The same number of hybrid orbitals result as the number of ordinary orbitals that are combined mathematically.

Hybrid orbitals using the *d* orbitals have also been constructed. For example, the combination of one *d* orbital and one *s* and two *p* orbitals gives a combination referred to as *dsp²* hybrid orbitals. These four orbitals are directed toward the corners of a square.

The use of hybrid orbitals in describing the bonding orbitals used by atoms has an additional attractive feature. The energies of the hybrid orbitals are all the same, whereas the energies of ordinary *s, p,* and *d* orbitals differ slightly from each other. Since each covalent bond formed by an atom with several other

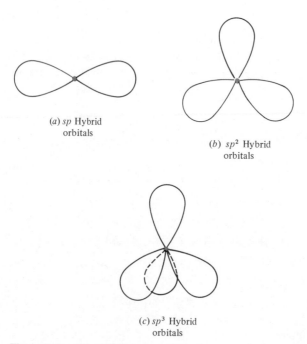

(a) *sp* Hybrid
orbitals

(b) *sp²* Hybrid
orbitals

(c) *sp³* Hybrid
orbitals

Figure 12-7.   Orientations of hybrid orbitals using the *s* and *p* orbitals. (*a*) Combination of an *s* and *p* orbital to form two *sp* hybrid orbitals. (*b*) Combination of an *s* and two *p* orbitals to form three *sp²* hybrid orbitals. (*c*) Combination of an *s* and all three *p* orbitals to form four *sp³* hybrid orbitals.

identical atoms is equally strong (for example, the chlorides of Section 12–9), equal energy hybrid orbitals give a better description of the bonding picture than do the use of orbitals with different energies.

We can effectively combine the ideas of electron repulsions and hybridization by considering the electron pair repulsions to be physically responsible for the basic shapes of many molecules, whereas a consideration of the orbitals employed involves the use of the appropriate hybrid orbitals.

12-11. PROBLEM OF ELECTRON DELOCALIZATION

It has been found that the electronic structures of some species cannot be represented satisfactorily by a simple Lewis dot structure. Consider, for example, sulfur dioxide, SO_2. There are eighteen valence electrons on the three atoms of this molecule. To avoid violation of the octet rule, it is necessary to have a double bond between the S atom and one O atom.

The overall angular geometry is correctly predicted by this structure. Notice that the four electrons of the double bond must be counted as only one group because they are restricted to the same region of space. Thus, there are essentially three groups around the S atom, and an angular shape is expected with a bond angle of about $120°$, which fairly well approximates the actual O-S-O bond angle of $117°$.

However, this structure predicts that one S-O bond is shorter than the other. Remember that double bonds are always shorter than single bonds for any two given atoms. In fact, the S-O bonds are equal in length and are intermediate with respect to the expected lengths for S-O single and double bonds. The conclusion is that each of the bonds must be a little more than a single bond. Pauling suggested that the term **resonance** be applied to such situations and the electronic structure indicated as

The actual structure cannot be represented by means of a single simple dot diagram and is rather a sort of average of the indicated structures. This representation does *not* mean that there is any jumping of the double bond from one place to another. The actual structure is a definite electronic pattern which simply cannot be illustrated by a single diagram. The S-O bonds may be considered as $1\frac{1}{2}$ bonds.

The + and − signs indicate the **formal charges** on the atoms. Formal charge is an important concept in the consideration of electronic structures because it helps to assess the overall charge distribution for a given dot diagram. The formal charge of an atom is arrived at by the following rules:

1. Count the average number of electrons, S, residing on an atom by adding all of the nonbonding electrons on the atom and one half of its bonding electrons.
2. The difference $(N - S)$, where N is the number of valence shell electrons in the neutral atom, is the formal charge on the atom.

The algebraic sum of the formal charges on the atoms in a species equals the net charge on the species. Naturally, the particular electron dot structure one has drawn will determine the formal charge distribution, and because of this the distribution will give some indication of the probability that a given dot structure is a reasonable representation. For example, the dot structure

$$\ddot{S}^{2+}$$
$$_{-}\!:\!\overset{..}{\underset{..}{O}}\!:\quad:\!\overset{..}{\underset{..}{O}}\!:_{-}$$

is not reasonable because it places a $+2$ charge on one atom. Electrons would be strongly attracted by the high positive charge. It is unlikely to find more than a $+1$ or -1 charge on one atom.

Wherever identical structures can be drawn, a resonance situation exists. Sometimes more than two identical structures must be considered, as for example with NO_3^-. This ion must have 24 electrons in the valence shells of the atoms involved. (Do not forget to add an "extra" electron for the negative charge.) In placing these electrons so as not to violate the octet rule for any atom, a N-O double bond becomes necessary. This double bond may be represented between any one of the three N-O pairs, so three identical structures arise

$$\left[\quad\begin{array}{ccc}\overset{..}{O} & \overset{..}{:O:} & \overset{..}{:O:} \\ \| & & \\ N^+ & N^+ & N^+ \\ _{-}\!:\!\overset{..}{O}\!:\;:\!\overset{..}{O}\!:_{-} & :\!\overset{..}{O}\!:\;\;\overset{..}{O} & \overset{..}{O}\;\;:\!\overset{..}{O}\!:_{-}\end{array}\quad\right]^{-}$$

The actual N-O bond lengths are equal and correspond to the expectations for $1\frac{1}{3}$ bonds. The ion is a planar species as a consideration of electron pair repulsions predicts.

The problem that resonance theory attempts to solve is found whenever electrons cannot be written as localized either on a given atom or between two atoms. When a pair of electrons is engaged in bonding more than two atoms, it is said to be **delocalized.** Simple dot structures deal in terms of reasonable fixed average positions for electrons. When the positions of electrons are averaged over several atoms, it becomes impossible to represent the electronic arrangement by means of a single conventional dot structure. The resonance approach is an attempt to enable one to represent the delocalized situation in terms of the simple dot diagrams.

An alternate answer to the problem raised by electron delocalization is found in **molecular orbital theory.** This theory was developed first by Robert Mulliken and F. Hund at roughly the same time that valence bond theory was emerging. In principle, whereas valence bond theory considers bonding from the standpoint of *atomic orbitals* from different atoms overlapping to create a suitable energy level for the shared electrons, molecular orbital theory simultaneously considers all the nuclei in a molecule and attempts to construct (mathematically) entirely new energy levels (molecular orbitals) which are associated with the molecule as a whole. The valence electrons of all the atoms in the molecule are then thought to occupy these molecular orbitals two at a time, analogous to the occupation of the atomic orbitals by the electrons in atoms.

In practice, as long as the electrons in a molecule are fairly localized on individual atoms or between two atoms, the valence bond and molecular orbital pictures are not too different. Molecular orbital theory has found its greatest utility in treating electrons that are delocalized. Indeed, the theory is ready made for such situations, with its emphasis on energy levels associated with the entire molecule.

The examples of SO_2 and NO_3^- can be represented

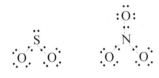

where the dotted lines represent the pair of delocalized electrons residing in a molecular orbital which extends over all the atoms.

Resonance theory, which is an outgrowth of the valence bond approach to bonding, and molecular orbital theory are both successful in dealing with the problem of delocalization of electrons. Some systems are dealt with more easily using one approach than the other. Some chemists out of a personal or educational bias will defend one as "better" than the other. Both theories are useful and both of them will no doubt continue to be used by chemists to rationalize and explain the properties of molecules in terms of their electronic structures.

12-12. BONDING AS A PERIODIC PROPERTY

The periodic arrangement of elements was discussed in Chapter 11. As was discussed earlier, many properties, both chemical and physical, are found to vary periodically with atomic number. Table 12-6 gives the common valences of the first three periods excluding the transition metals.

From a knowledge of the periodicity of ionization potentials, a rather simple explanation for the tendency of some elements to form ionic compounds and others to form covalent compounds can be rationalized. Elements found to the left of the broken line in Table 12-6 are the metals, those to the right are nonmetals. Further, the metals nearly all possess one, two, or three electrons that are available for bonding and, at least in principle, space for eight. Loss of these valence electrons leaves an ion with one to three units of positive charge. The other alternative—formation of covalent bonds—would result in the acquisition of a share in five to seven electrons. Apparently the former alternative leads to greater stability, especially when the recipient atom is a nonmetal and has six or seven valence electrons. In this case, complete transfer of electrons results in ions with small electric charges. The second alternative—covalent bonding—on the other hand, is much more likely when either of two conditions exist.

Table 12-6. Common Valences of the First 36 Elements

Number of Valence Electrons $\longrightarrow$	1	2	3	4	5	6	7	8
element	H	He						
atomic number	1	2						
valence	1	0						
element	Li	Be	B	C	N	O	F	Ne
atomic number	3	4	5	6	7	8	9	10
valence	1	2	3	4	3	2	1	0
element	Na	Mg	Al	Si	P	S	Cl	A
atomic number	11	12	13	14	15	16	17	18
valence	1	2	3	4	3	2	1	0
element	K	Ca	Ga	Ge	As	Se	Br	Kr
atomic number	19	20	31	32	33	34	35	36
valence	1	2	3	4	3	2	1	0

1. When both atoms forming the bond have a large number of valence electrons. In this case, complete transfer of electrons would result in one ion with less than eight electrons in its valence shell (or with an exceedingly large resultant charge)

$$:\ddot{C}l\cdot + \cdot\ddot{C}l: \xrightarrow{\ \ \times\ \ } :\ddot{C}l:^- + \ddot{C}l:^+$$

2. When one or both atoms have an intermediate number of valence electrons. Electron transfer would result in a high amount of charge in one of the atoms.

$$\cdot\dot{\underset{\cdot}{C}}\cdot + 4\ \cdot\ddot{C}l: \xrightarrow{\ \ \times\ \ } C^{4+} + 4\ :\ddot{C}l:^-$$

$$\cdot\dot{\underset{\cdot}{C}}\cdot + 4\ \cdot H \xrightarrow{\ \ \times\ \ } :\ddot{\underset{\cdot\cdot}{C}}:^{4-} + 4\ H^+$$

The above can be generalized by the statement: *Ionic substances result from the combination of atoms from opposite sides of the periodic table and covalent substances result from the combination of atoms in the right hand region of the periodic table.*

The melting point behavior of the chlorides of the first two dozen elements (Table 12-7) affords still another periodic property and simultaneously under-

Table 12-7. Melting Points of the Chlorides of the Elements (°C)

LiCl	BeCl$_2$	BCl$_3$	CCl$_4$	NCl$_3$	OCl$_2$	FCl
613	440	−107	−23	−40	−20	−154
NaCl	MgCl$_2$	AlCl$_3$	SiCl$_4$	PCl$_3$	SCl$_2$	Cl$_2$
801	708	190	−70	−91	−78	−101
KCl	CaCl$_2$	ScCl$_3$	GeCl$_4$	AsCl$_3$	SeCl$_2$	BrCl
776	772	939	−49	−18	−80	−66

scores this generalization. The transition from ionic to covalent bonding on moving from left to right is signaled by the abrupt drop in melting point. The discontinuity in melting points on going from left to right in Table 12–7 is a result of a discontinuity in the lattice energies. Crystals composed of neutral molecules will have considerably lower lattice energies than those composed of ions.

PROBLEMS

1. State whether each of the proposed structures of the substances below is reasonable solely on the basis of the properties given.
   (a) Lithium fluoride (LiF): melting point 870°C, boiling point 1670°C, density 2.6 g/cm^3, conductor of electric current in molten state only. Proposed structure: Li^+F^- (crystal with alternating Li^+ and F^- ions in three dimensions).
   (b) Silicon carbide (SiC): melting point 2600°C, boiling point very high, density 3.2 g/cm^3, nonconductor of electric current either in solid or molten state. Proposed structure: $Si^{4+}C^{4-}$ (crystal as in (a) above).
   (c) Boron trichloride (BCl_3): melting point $-107°C$, boiling point 12.5°C, density 1.4 g/cm^3. Proposed structure:

$$\underset{Cl \quad\quad Cl}{\overset{\overset{\displaystyle Cl}{|}}{B}} \quad \text{(planar)}$$

2. Give one example of each of the following:
   (a) A substance whose water solution should conduct an electric current.
   (b) A nonconductor that has high melting and boiling points.
   (c) A molecule containing four or more atoms all of which lie in the same plane.
   (d) A molecule containing three or more atoms all of which lie in a straight line.

3. Employing Coulomb's law explain why crystals of sodium chloride are stable even though the negative chloride ions are very close together. Use Figure 12–2 in which $\bullet = Na^+$.

4. The conduction of an electric current by a medium depends on the following factors:
   (1) the presence of charged species, for examples, ions, electrons.
   (2) mobility of the charged species.
   (3) An applied electromotive force.
   Which of these factors accounts for (a) The nonconductivity of solid NaCl. (b) The nonconductivity of diamond. (c) The conductivity of graphite. (d) The conductivity of solid copper.

5. What problems were answered by Lewis's theory of electron pair covalent bonding?

6. What problems were answered by Kossel's theory of ionic bonding?

7. What experimental evidence suggests that methane (CH_4) consists of discrete molecules whereas carbon consists of an infinite covalent network.

8. List as many ways as you can in which ionic bonding differs from covalent bonding (cite both theoretical and experimental differences).

9. How can you account for the malleability and ductility of metals on the basis of their structure?

10. Why is the bonding in metals considered apart from ionic and covalent bonding?

11. Cite experimental evidence for the octet rule; i.e., that atoms tend to acquire a total of eight valence electrons in chemical reactions.

12. What is the difference between lone pair electrons and bonding electrons?

13. Predict the charge on the ions that form from electron transfer among each of the following atoms:

   (a) Na + Br $\longrightarrow$ (e) K + Cl $\longrightarrow$ (i) K + N $\longrightarrow$
   (b) Mg + I $\longrightarrow$ (f) Cs + F $\longrightarrow$ (j) Ca + O $\longrightarrow$
   (c) Al + F $\longrightarrow$ (g) Li + O $\longrightarrow$ (k) Ba + F $\longrightarrow$
   (d) Li + Cl $\longrightarrow$ (h) Na + S $\longrightarrow$

14. Predict the electron dot structures and numbers of atoms required when atoms combine to produce a covalent compound:

$$\left(\text{Example: } \cdot \overset{\cdot}{\text{C}} \cdot \; + \; \cdot \text{H} \longrightarrow \text{H} \overset{\overset{\text{H}}{\cdot\cdot}}{\underset{\overset{\cdot\cdot}{\text{H}}}{\text{C}}} \text{H} \right)$$

   (a) $\cdot \overset{\cdot}{\text{C}} \cdot \; + \; \cdot \overset{\cdot\cdot}{\underset{\cdot\cdot}{\text{Cl}}} : \; \longrightarrow$ (e) $: \overset{\cdot\cdot}{\underset{\cdot\cdot}{\text{Cl}}} \cdot \; + \; : \overset{\cdot\cdot}{\underset{}{\text{O}}} : \; \longrightarrow$

   (b) $\cdot \overset{\cdot}{\text{C}} \cdot \; + \; : \overset{\cdot}{\underset{\cdot}{\text{O}}} : \; \longrightarrow$ (f) $\cdot \overset{\cdot}{\text{B}} \; + \; \cdot \overset{\cdot\cdot}{\underset{\cdot\cdot}{\text{Cl}}} : \; \longrightarrow$

   (c) $: \overset{\cdot\cdot}{\underset{\cdot\cdot}{\text{Cl}}} \cdot \; + \; \cdot \overset{\cdot\cdot}{\underset{\cdot\cdot}{\text{Br}}} : \; \longrightarrow$ (g) $\cdot \text{Be} \cdot \; + \; \cdot \overset{\cdot\cdot}{\underset{\cdot\cdot}{\text{Cl}}} : \; \longrightarrow$

   (d) $\text{H} \cdot \; + \; \cdot \overset{\cdot\cdot}{\underset{\cdot\cdot}{\text{F}}} : \; \longrightarrow$

15. Predict the geometry of each of the molecules in problem 14.

16. What factors determine the geometry of (a) an ionic crystal lattice, (b) a covalent molecule.

17. Draw Lewis dot structures for the following species:
   (a) H_2S (b) SCl_2 (c) PH_3 (d) HCl (e) O_3 (f) CO_3^{2-} (g) BF_3 (h) SiF_4 (i) CO_2 (j) CO

18. Predict the geometry of each of the species of problem 17.

19. In very general terms, explain the difference between the valence bond and molecular orbital approaches to bonding.

20. Give the hybrid orbitals employed for covalent bonding by the following species and the configurations you would expect for the resulting molecules. (a) divalent Be (b) trivalent N (c) divalent O (d) tetravalent Zn (e) trivalent Al (f) tetravalent Al

21. Label each of the structures below with the correct formal charges.

   (a) $: \overset{\cdot\cdot}{\underset{\cdot\cdot}{\text{O}}} - \text{C} \equiv \text{C} - \text{C} \equiv \text{O} :$ (c) $: \overset{\cdot}{\underset{\cdot\cdot}{\text{O}}} \quad \overset{\text{N}}{\diagup \diagdown} \quad \overset{\cdot}{\underset{\cdot\cdot}{\text{O}}} .$

   (b) $CH_3 - C \overset{\diagup \overset{\cdot\cdot}{\underset{\cdot}{\text{O}}} :}{\diagdown \underset{\cdot\cdot}{\text{O}} : \text{H}}$ (d) $\text{H} : \overset{\cdot\cdot}{\underset{\cdot\cdot}{\text{O}}} - \overset{\overset{\displaystyle : \overset{\cdot\cdot}{\text{O}} :}{|}}{\underset{\underset{\displaystyle : \overset{\cdot\cdot}{\text{O}} :}{|}}{\text{S}}} - \overset{\cdot\cdot}{\underset{\cdot\cdot}{\text{O}}} : \text{H}$

22. What is meant by electron delocalization in molecules? How is it dealt with theoretically?

23. Draw all the Lewis dot structures including formal charges that contribute importantly to the resonance hybrid of each of the following:
    (a) carbon monoxide (CO)
    (b) Nitrous oxide (NNO)

    (c) Carbonate ion $\left(\begin{array}{c} O \\ \| \\ C \\ O \diagup \diagdown O \end{array} \right)^{2-}$

    (d) Nitromethane $\left(CH_3 - N \diagup^{\textstyle O}_{\diagdown O} \right)$

24. Describe and explain the trends in lattice energy observed in Table 12–3.

CHEMICAL STRUCTURE

STEREOCHEMISTRY—THE SHAPES OF MOLECULES

We have already seen that atoms that have four other atoms bonded to them usually have the four bonds directed to the corners of a tetrahedron. The reasons for this have been given in Chapter 12, and are based primarily on mutual repulsion among the electron pairs of each bond.

It is interesting and scientifically important to realize that this tetrahedral arrangement was discovered quite some time before the discovery of electrons or bonding! The tetrahedral arrangement of the four bonds of a carbon atom (Figure 13–1) was conceived independently and almost simultaneously in 1874 by J. H. van't Hoff in Holland and J. A. le Bel in France. Their bold idea was met with considerable hostility at first, but within 15 years it was nearly universally accepted.

Before going into their arguments, it will be necessary to present several ideas. We shall return to van't Hoff and le Bel in Section 13–7.

13-1. CONSTITUTIONAL ISOMERS

Isomers (iso = same, mer = unit) is a term that describes compounds that have the same molecular formulas and compositions but in some way are different. To say that two pure substances are different is to say that they have different sets of properties. An example of isomers is ethyl alcohol and dimethyl ether. It can be seen that both of these compounds have the same molecular formula—C_2H_6O—and thus have the same molecular weights. They are, however, put together differently and

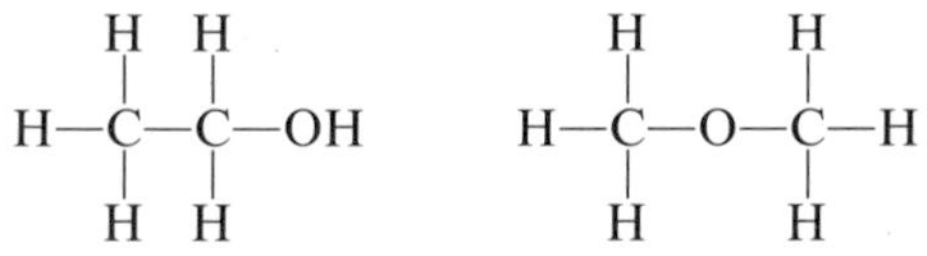

ethyl alcohol   dimethyl ether

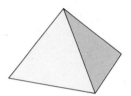

Regular tetrahedron

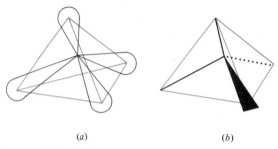

(a)   (b)

Figure 13-1. (a) Tetrahedral arrangement of electron orbitals in a regular tetrahedron. (b) The convention that solid lines are in the plane of the paper, the wedge extends out toward the reader, and the dashed line backs away from the reader behind the paper.

are therefore different substances. These two compounds differ in their **constitutions;** that is, they differ in the exact sequence in which their atoms are bonded: ethyl alcohol contains the sequence C-C-OH, for example, whereas dimethyl ether contains the sequence C-O-C. Such isomers are called **constitutional isomers.**

Normal and isobutane provide another example of constitutional isomers. In normal butane, the two inner carbon atoms are each bonded to two carbon atoms and two hydrogen atoms; the two outer carbon atoms are each bonded to one carbon atom and three hydrogen atoms. In isobutane, three carbon atoms are each bonded to the same carbon atom and to three

$$CH_3-CH_2 \atop \qquad CH_2-CH_3$$

n-butane

$$\overset{CH_3}{\underset{CH_3}{CH_3-C-H}}$$

isobutane

hydrogen atoms. The remaining carbon atom is bonded to three other carbon atoms and one hydrogen atom.

Some other examples of constitutional isomers are given in Table 13–1.

13-2. STEREOISOMERS I—CIS-TRANS ISOMERS

Another kind of isomerism exists of which two of the butenes (C_4H_8) are examples. Four of the isomers of C_4H_8 all possess a double bond and can thus be made to react with hydrogen to form butanes (C_4H_{10}). Three of the butenes (the normal or straight chain butenes), on reaction with hydrogen yield normal butane, the fourth (isobutylene) yields isobutane. The latter butene must have the structure

Table 13-1. Examples of Constitutional Isomers

$$[H-\overset{\overset{\displaystyle H}{|}}{\underset{\underset{\displaystyle H}{|}}{N}}-H]^+[O{=}C{=}N] \qquad\qquad H-\overset{\overset{\displaystyle H}{|}}{N}-\overset{\overset{\displaystyle O}{\|}}{C}-\overset{\overset{\displaystyle H}{|}}{N}-H$$

ammonium cyanate   urea

$$H-\overset{\overset{\displaystyle H}{|}}{\underset{\underset{\displaystyle H}{|}}{C}}-N\overset{\nearrow O}{\underset{\searrow O}{}} \qquad\qquad H-\overset{\overset{\displaystyle H}{|}}{\underset{\underset{\displaystyle H}{|}}{C}}-O-N{=}O$$

nitromethane   nitrosomethane

$$CH_3-CH_2-CH_2-CH_3 \qquad\qquad CH_3-\overset{\overset{\displaystyle CH_3}{|}}{CH}-CH_3$$

n-butane   isobutane

$$CH_3-\overset{\overset{\displaystyle CH_3}{|}}{\underset{\underset{\displaystyle CH_3}{|}}{C}}-CH_2-\overset{\overset{\displaystyle CH_3}{|}}{CH}-CH_3 \qquad\qquad CH_3-CH_2-CH_2-CH_2-CH_2-CH_2-CH_2-CH_3$$

isooctane ("100 octane")   n-octane

shown in equation (13–1) because it is known that reaction with hydrogen occurs to transform the double bond into a single bond and that no change takes place in the carbon skeleton.

$$CH_3-C\overset{\nearrow CH_2}{\underset{\searrow CH_3}{}} \xrightarrow{\;H_2\;} CH_3-\overset{\overset{\displaystyle CH_3}{|}}{\underset{\underset{\displaystyle CH_3}{|}}{C}}-H \qquad\qquad (13\text{–}1)$$

isobutylene   isobutane

The other three butenes may have structures in which the double bond joins the first and second carbon atoms or the second and third carbon atoms. The existence of three isomers can only be explained if rotation

$$CH_3-CH_2-CH{=}CH_2 \xrightarrow[\text{Pt catalyst}]{\;H_2\;} CH_3-CH_2-CH_2-CH_3$$

n-butane

$$CH_3-CH{=}CH-CH_3 \xrightarrow[\text{Pt catalyst}]{\;H_2\;} CH_3-CH_2-CH_2-CH_2$$

n-butane

about a C-C double bond is restricted, which would account for two structures with the double bond in the middle position.

$$\overset{\displaystyle H}{\underset{\displaystyle CH_3}{}}{\Large\diagdown}C{=}C{\Large\diagup}\overset{\displaystyle H}{\underset{\displaystyle CH_3}{}} \qquad\qquad \overset{\displaystyle H}{\underset{\displaystyle CH_3}{}}{\Large\diagdown}C{=}C{\Large\diagup}\overset{\displaystyle CH_3}{\underset{\displaystyle H}{}}$$

cis   trans

183

Table 13-2. Properties of *cis-* and *trans*-Butene

	cis	*trans*
melting point (°C)	− 139	− 106
boiling point (°C)	4	1
heat of combustion (kcal/g-mole)	648.1	647.1

That these two isomers do exist is demonstrated by the isolation of two substances with different sets of properties (Table 13-2). The isomers are designated cis and trans as shown above. Cis means 'on the same side' and trans means 'on the other side'. Since these isomers are not constitutionally different they are an example of **stereoisomers** (Greek: stereo = solid). All compounds that possess a double bond may not exist as cis and trans isomers. For example there is only one isomer of isobutylene. In order for cis-trans isomers to exist, both carbon atoms forming the double bond must hold two different groups attached to them. Thus any

$$\begin{matrix} A \\ \diagdown \\ B \end{matrix} C = C \begin{matrix} A \\ \diagup \\ B \end{matrix} \qquad \begin{matrix} A \\ \diagdown \\ B \end{matrix} C = C \begin{matrix} A \\ \diagup \\ D \end{matrix} \qquad \begin{matrix} A \\ \diagdown \\ B \end{matrix} C = C \begin{matrix} D \\ \diagup \\ E \end{matrix}$$

$$(I) \qquad\qquad (II) \qquad\qquad (III)$$

of the structures (I, II, III) is capable of existing as cis-trans isomers, but the

$$\begin{matrix} A \\ \diagdown \\ A \end{matrix} C = C \begin{matrix} B \\ \diagup \\ B \end{matrix} \qquad \text{or} \qquad \begin{matrix} A \\ \diagdown \\ A \end{matrix} C = C \begin{matrix} B \\ \diagup \\ D \end{matrix}$$

$$(IV) \qquad\qquad\qquad (V)$$

structures (IV, V) are incapable of cis-trans isomerism.

Stereoisomers may be defined as isomers that have the same constitutions and differ only in the spatial arrangement of their atoms. The study of the shapes of molecules is called **stereochemistry.**

The inability of rotation about a C-C double bond is further justification for the orbital overlap model of covalent bonding, for if overlapping of orbitals leads to stability, reduction of overlapping, required in rotation, should require considerable energy.

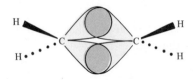

Stereoisomers of this type are frequently also called **cis-trans isomers** or **geometrical isomers.** As we shall see in subsequent chapters, there are many examples in chemistry of stereoisomers of this kind. But at this time we will discuss a different kind of stereoisomer.

13-3. STEREOISOMERS II—CHIRAL MOLECULES

Objects may be classified into two groups on the basis of a very interesting criterion which is called **chirality** (Greek: chiro = hand). A chiral object is one that, like a human hand, can exist in a right-handed and a left-handed form. Examples of common chiral objects are ears, feet, screws, golf clubs, and some sea shells (Figure 13–2). Objects that can exist in only one form are called **achiral** (Greek: prefix *a-* means *not*). Examples of achiral objects are bricks, screwdrivers, hammers, nails, and coffee cups (Figure 13–3).

Molecules can also be classified as chiral or achiral. In order to do so a simple rule is available that allows us to distinguish chiral from achiral objects: *A chiral object and its mirror image are not superposable.* Thus, if an object and its mirror image are superposable, then the object is achiral. *Superposable objects are objects that can be placed upon one another so that all parts of one coincide with the corresponding parts of the other.*

A simple example will illustrate this point: The mirror image of your right hand is identical to your left hand. Now attempts to superpose your right and left hands will always fail. One way to show this failure is by trying to fit a left handed glove on your right hand.

This difference in chirality between the two hands and between all chiral pairs of objects is a rather interesting phenomenon. It is often difficult to describe the

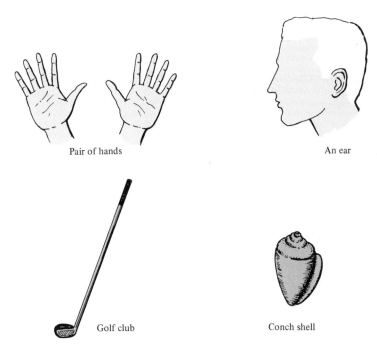

Pair of hands

An ear

Golf club

Conch shell

Figure 13–2. Some chiral objects.

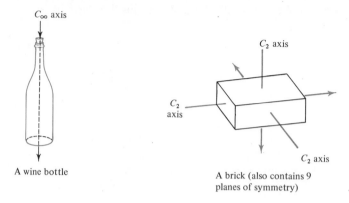

A wine bottle

A brick (also contains 9
planes of symmetry)

Coffee cup
(a plane of symmetry
cuts the cup and handle)

Figure 13-3.   Some achiral objects.

difference without access to the above rule. The description is made especially difficult because both hands are put together in the same way and therefore have identical constitutions. A pair of isomers that differ from each other in the ways described above are called **enantiomers.**

13-4.   SYMMETRY—A CRITERION OF CHIRALITY

Our discussion of the chirality of molecules can be made much more quantitative through a consideration of symmetry. Discussions of objects using the rather simple concepts of symmetry can be very helpful in visualizing complex three-dimensional shapes in the two-dimensional world of our written language.

An object can be classified according to the symmetry elements that it possesses. Symmetry elements can be defined in terms of symmetry operations:

Symmetry Element		Symmetry Operation
axis	(C_n)	rotation
plane	(σ)	reflection
center	(i)	inversion

If an object possesses an axis of symmetry, rotation about that axis by an angle, θ, results in a new orientation that is indistinguishable from the original. The axis is referred to as a C_n axis where $n = 360/\theta$. The plane (σ) is defined in terms of the reflection operation. Thus, for every point on the object a corresponding point is found by moving from the original point along a line perpendicular to the plane to an equal distance on the other side of the plane. The center of symmetry (i) is a point in the object through which is applied the inversion operation. In this operation, an imaginary line is drawn from any point on the object through the center (i) and extended beyond it for an equal distance. Inversion means that this latter point will correspond to the original point. These elements and operations are shown schematically in Figure 13-4. The best way for the student to familiarize himself with these concepts is to apply them to some common objects.

Using the above concepts, objects may be divided into two classes:

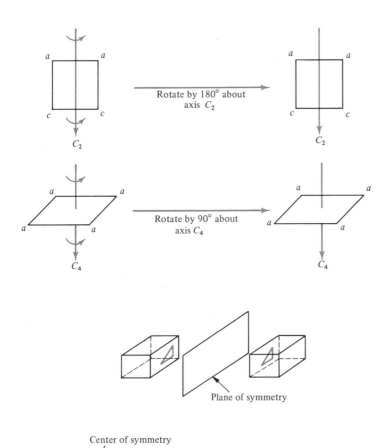

Figure 13-4. Examples of symmetry elements and symmetry operations.

Class I: Objects that possess neither a plane (σ) nor a center (i) of symmetry. These objects are not superposable on their mirror images. Class I objects are capable of existing in "right-handed" and "left-handed" forms and are therefore *chiral*. Objects typical of this group have been shown in Figure 13–2.

Class II: Objects that possess either a plane (σ) or a center (i) of symmetry or both. These objects are superposable on their mirror images and are said to be *achiral*. Examples are hammers, coffee cups, ice cream cones, barrels, balls, simple chairs, and tables (Figure 13–3).

The rule for chirality given in section 13–3 is therefore seen to be a functional means of stating that a chiral object cannot have a plane nor a center of symmetry. The student is encouraged to examine some common objects, both chiral and achiral, to satisfy himself of the equivalence of these two means of describing the same idea. Table 13–3 gives some common molecular structures along with their symmetry elements and their chirality.

By far the most common and important type of chiral molecules are those that possess an **asymmetric carbon atom**—a carbon atom that is attached to four *different* atoms or groups. It can readily be seen that molecules such as the last two in Table 13–3 can exist as two enantiomers by attempting mentally to superpose them.

(mirror)   (mirror)

It should be pointed out, however, that **asymmetry,** that is, the absence of symmetry elements, is not necessary for chirality because hydrogen peroxide as shown in Table 13–3 possesses a C_2 axis and is chiral. (This axis bisects both the O-O bond and the angle of the two intersecting planes and is perpendicular to the O-O bond.)

The obvious question to ask at this point is whether an actual difference can be detected between such enantiomers. Enantiomeric pairs always have identical melting points, boiling points, densities; and in fact nearly all their physical properties are identical. The only difference in the behavior of a pair of enantiomers is in their interaction with another chiral object. An analogy will serve to illustrate this. A right hand and a left hand can equally well use a hammer, screwdriver, or saw because each of these objects is achiral. In contrast, a chiral object will not fit the two hands equally well. For example, a right handed glove fits the right hand but not the left; shaking hands is equally easy whether both parties use right hands or left hands, but difficult if one uses his right hand and the other uses his left hand.

The most important case of such an interaction of an enantiomeric pair with something chiral is discussed in Section 13–5. That phenomenon also serves to detect physically the presence of chiral molecules.

Table 13-3. Symmetry Properties of Some Molecular Structures

Molecular Structure	Symmetry Elements	Chirality	Common Object Having the Same Symmetry Elements
H—Cl	$\begin{Bmatrix} 1\ C_\infty \\ \infty\ \sigma \end{Bmatrix}$	achiral	ice cream cone
H—C≡C—H	$\begin{Bmatrix} 1\ C_\infty \quad \infty\ C_2 \\ 1\ \sigma\ (\perp \text{ to } C_\infty) \\ \infty\ \sigma\ (\text{containing } C_\infty) \end{Bmatrix}$	achiral	cylinder
$\begin{array}{c} H \quad\quad H \\ \diagdown \quad \diagup \\ C{=}C \\ \diagup \quad \diagdown \\ H \quad\quad H \end{array}$	$\begin{Bmatrix} 3\ \sigma \\ 3\ C_2 \\ i \end{Bmatrix}$	achiral	rectangle
$\begin{array}{c} H \quad\quad Cl \\ \diagdown \quad \diagup \\ C{=}C \\ \diagup \quad \diagdown \\ H \quad\quad H \end{array}$	$\{1\ \sigma\}$	achiral	coffee cup
$\begin{array}{c} \ddot{N}{-}H \\ H \quad H \end{array}$	$\begin{Bmatrix} 3\ \sigma \\ 1\ C_3 \end{Bmatrix}$	achiral	tripod
$\begin{array}{c} Cl \\ \mid \\ Cl{-}P{-}Cl \\ \mid \\ Cl \end{array}$ (with extra Cl)	$\begin{Bmatrix} 4\ \sigma \\ 1\ C_3 \\ 3\ C_2 \end{Bmatrix}$	achiral	trigonal bipyramid
$\begin{array}{c} O \\ O \quad\ O \end{array}$ (ozone)	$\begin{Bmatrix} 2\ \sigma \\ 1\ C_2 \end{Bmatrix}$	achiral	boomerang
$\begin{array}{c} H \\ \diagdown \\ O \\ \mid \\ O \\ \diagdown \\ H \end{array}$ (H_2O_2)	$\{1\ C_2\}$	chiral	scissors
$\begin{array}{c} H \\ \mid \\ C{-}{-}Br \\ \diagup \quad \diagdown \\ Cl \quad\quad CH_3 \end{array}$	$\begin{Bmatrix} \text{no elements} \\ \text{(asymmetric)} \end{Bmatrix}$	chiral	hand

13-5. OPTICAL ACTIVITY AND CHIRALITY

Ordinary visible light may be "filtered" so that the light that emerges from the filter consists only of electromagnetic vibrations in a single plane. Such a beam of **plane polarized** light may be visualized as a planar ribbon (Figure 13–5). When a beam of such plane polarized light is passed through a chiral substance (either pure or in solution), the plane of polarization is twisted and the plane of the emergent beam is at an angle to that of the original beam.

Louis Pasteur attributed this ability to rotate the plane of polarized light to chirality (he called it dissymmetry) of the sample interacting with the plane

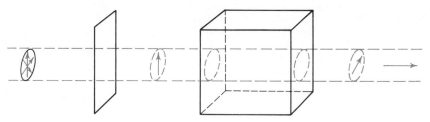

Figure 13-5. Schematic drawing of the production of polarized light and its rotation by an optically active sample.

polarized light beam which is also chiral. The phenomenon is known as **optical activity.** Pasteur found that some solids exhibit optical activity only in the crystalline state whereas others retain their optical activity on dissolving. He deduced that in the former substances, optical activity is due only to chirality in the arrangement of the atoms in the crystal lattice. When the lattice is destroyed, the chirality is lost with it. The latter substances, on the other hand, retain their optical activity because the molecules themselves are chiral, and dissolving does not destroy molecular chirality. van't Hoff and le Bel were to use this later as evidence for the tetrahedral nature of carbon atoms.

As might be expected, molecules that differ only in their chirality affect plane polarized light to exactly the same extent but in opposite directions. If the plane of polarization is rotated to the right (clockwise), the sample substance is said to be **dextrorotatory** (dextro = right). If it is rotated to the left, the sample substance is said to be **levorotatory.** Such substances are designated $(+)$ and $(-)$, respectively, and a pair of substances that are so related may be called **enantiomers.** An exactly equal mixture of enantiomers does not exhibit optical activity and is called a **racemic mixture.**

As stated above, the most common chiral molecules are those that have a carbon atom in which all four atoms or groups attached are different. An important example is the amino acid alanine, $CH_3CH(NH_2)COOH$. This molecule is chiral and therefore exists in two enantiomeric forms.

The two enantiomers of alanine have identical properties as long as they are observed in an *achiral* medium. These two isomers have the same density, boiling and melting points, and vapor pressures. In the interaction with a different molecule which is also chiral, however, the enantiomers will behave quite differently. For example, reaction with one enantiomer of the alcohol, $CH_3CH_2CH(CH_3)OH$, results in two forms of the ester (a) and (b)

(a)

(b)

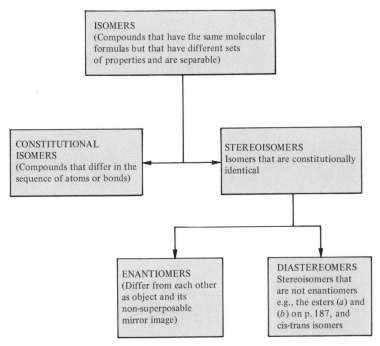

Figure 13-6. Relationship among the kinds of isomers.

The use of molecular models reveals that these two isomers are not mirror images and, moreover, are quite dissimilar in geometry. They have entirely different optical activities and different sets of physical and chemical properties, and, in the above reaction, they are produced at different rates. Molecules that are related in this way, that is, that are constitutionally identical but are not enantiomers, are called diastereomers along with the cis-trans isomers of Section 13-2. Thus we see that stereoisomers are divided into two groups—enantiomers and non-enantiomers. Figure 13-6 is a diagram relating the different kinds of isomerism. We shall see later that there are other examples of diastereomers.

13-6. HYDROGEN BONDING

Before continuing our discussion of the effects of molecular shapes, a digression is in order. The physical properties of substances can frequently be rationalized on the basis of molecular features. In fact, this rationalization is one of the major objectives of the science of chemistry. The influence of the weight of a given molecule on its boiling point was discussed in Chapter 6. In general, barring complications, the boiling points of substances should increase regularly with increasing molecular weight. Figure 13-7 shows this relationship for hydrides of group IV, V, VI, and VII elements. The expected behavior is observed only with the hydrides of the carbon group. The other hydrides show a sharp irregularity with the lightest member of each group: water, for example, has a much higher than expected boiling point compared with H_2S, H_2Se, and H_2Te. Hydrogen fluoride (HF) and ammonia (NH_3) also behave anomalously.

These observations can be explained nicely by assuming that, with the lighter members of each group, the molecules are strongly attracted to each other (asso-

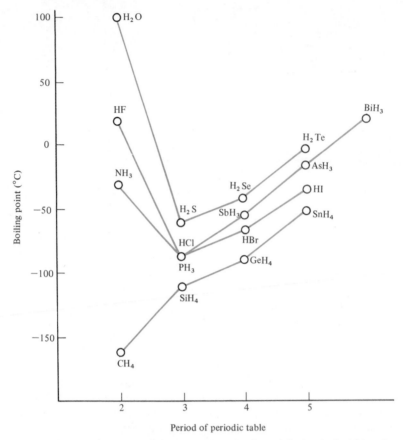

Figure 13-7. Variation of boiling point with molecular weight for the hydrides of group IV, V, VI, and VII elements.

ciated) in the liquid phase giving the effect of a much higher molecular weight. In the liquid phase, HF has been shown to exist as conglomerates $(HF)_5$ and $(HF)_6$. This phenomenon is observed to a significant extent only with compounds that contain hydrogen atoms covalently bonded to the elements in the top right corner of the periodic table (fluorine, oxygen, nitrogen). A large body of experimental evidence of this type suggested to W. M. Latimer and W. H. Rodebush in 1920 that this intermolecular association occurs through a "hydrogen bond" as shown for $(HF)_5$ in which five HF molecules are held together in a ring. Water molecules also exhibit this form of association but the pattern is less regular and undoubtedly exists in three dimensions. (Note that the "hydrogen bond" is shown as a dotted line to distinguish it from the normal covalent bond.)

Ordinary ice contains a high degree of hydrogen bonding. In ice, each water molecule is attached tetrahedrally to four other water molecules by four hydrogen bonds. There are two O-H bond distances, one is 1.00 Å, and the other is 1.76 Å. The O-O distance is 2.76 Å. This indicates that each hydrogen atom is strongly bonded to one oxygen atom and weakly bonded to another.

No other element besides hydrogen exhibits this kind of bonding because hydrogen is the only atom whose nucleus is not shielded by inner shell electrons. The bare positive nucleus of a bonded hydrogen atom can implant itself into the unshared electron cloud of an adjacent atom. This hydrogen nucleus, however, will be most "bare" of its bonding electrons when the atom to which it is covalently bonded has a very great affinity for electrons; namely, the elements flourine, oxygen, and nitrogen.

Another example of a hydrogen bonded substance is acetic acid (CH_3COOH). Determination of the molecular weight of acetic acid indicates a value about double the expected value. Hydrogen bonding accounts very well for the nature of this "double molecule."

$$
\underset{\underset{\uparrow}{1.00\ \text{Å}}}{CH_3-C} \begin{matrix} O\text{----}H-O \\ \diagup \qquad \diagdown \\ O-H\text{----}O \end{matrix} \underset{\underset{\uparrow}{1.60\ \text{Å}}}{C-CH_3}
$$

Hydrogen bonds usually require 5–10 kcal/mole to be broken. The corresponding dissociation energies of covalent bonds are in the range 50–100 kcal/mole. Hydrogen bonds are therefore about one tenth as strong as typical covalent bonds.

13-7. TETRAHEDRAL GEOMETRY

We began this chapter with the assertion that van't Hoff and le Bel had proposed that the geometry of carbon compounds and those of similar elements is tetrahedral. We shall now return to a brief discussion of their evidence using the principles of stereochemistry developed in this chapter.

Their hypothesis was based upon the number of isomers that exist for certain compounds. For example, the molecule of methylene chloride (CH_2Cl_2) has a central carbon atom to which are attached two hydrogen atoms and two chlorine atoms. Several spatial arrangements are possible as shown in Figure 13–8. Each of the geometries in Figure 13–8 suggests a different number of isomers. Since no more than one isomer of methylene chloride or, in fact, of any compound of the formula CX_2Y_2, has ever been observed, the tetrahedral geometry, which predicts only one form, is suggested.

Symmetry considerations are useful here also. We have seen that optical activity can be explained by the presence of chiral molecules. Molecules in which four different groups are attached to the central carbon atom (CWXYZ) are

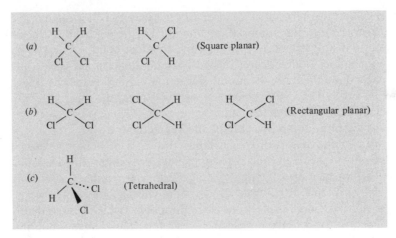

Figure 13-8. Some possible geometries of methylene chloride (CH_2Cl_2): (a) square planar predicts two isomers, (b) rectangular planar predicts three isomers, and (c) tetrahedral predicts only one isomer.

asymmetric and can have the geometries shown in Figure 13-9. Both planar forms (a) and (b) possess a plane of symmetry and therefore are not capable of existence as enantiomers. Structure (c), the tetrahedral structure, is chiral and predicts that molecules of formula CWXYZ should be optically active, as is observed in practice.

The kind of evidence offered above is negative in that it is based upon the failure to isolate isomers. Nonetheless, it induced van't Hoff and le Bel to put forth their bold hypothesis. This is an example of one of the many kinds of evidence upon which scientific theories are built.

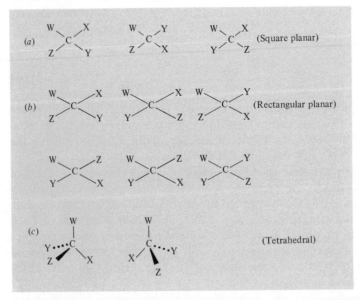

Figure 13-9. Some possible geometries of the hypothetical molecule CWXYZ: (a) square planar predicts three isomers (not enantiomeric because all possess a plane of symmetry, σ); (b) rectangular planar predicts six isomers (not enantiomeric); (c) tetrahedral predicts two enantiomers (asymmetric).

13–8. DIFFRACTION ANALYSIS TECHNIQUES

Before closing this chapter, it should be made clear that modern chemists have far better tools with which to work than were available to van't Hoff and le Bel. Only one of these—diffraction techniques—will be discussed briefly here. With these techniques and many others currently available, it has been possible unequivocally to establish all the preceeding assertations made about the stereochemistry of molecules as well as the assertions that will be made in the following chapters. Nearly everyone has seen a cut-glass chandelier that reflects light in small spots all around a room. Consider the problem of reconstructing the chandelier, if all one could observe were the spots that it reflected. A simpler problem and perhaps more illustrative is that of trying to reconstruct the shape of a diamond by analyzing the spots of light that are reflected off its surfaces from a single light source.

Diffraction techniques for the study of chemical structures employ an analogous idea. By far the most widely used diffraction technique employs X rays. X rays are electromagnetic radiation of frequencies approximately 10^{18} cycles per second (Hz) (wavelengths near 1 Å). Because their wavelengths are of about the same magnitude as atomic spacings in crystals, X rays may be used to determine the atomic or ionic patterns in crystals. It is thus found that, if a beam of X rays is focused onto the surface of a crystal, the reflection and diffraction form a pattern that may be recorded on a photographic plate. The interpretation of these patterns in terms of the spacings of their lines and spots along with a knowledge of the wavelength of the X rays used allows an exact reconstruction of the relative positions of the atoms in the crystal (Figure 13–10).

X ray diffraction technique is applicable only to crystalline solids and has been extremely valuable in the determination of the structures of many complex substances (Figure 13–11). In fact, it is the only direct and unequivocal method available for the determination of structures. The detailed methods employed in the analysis and interpretation of X ray diffraction patterns are very elaborate and complex and require special training.

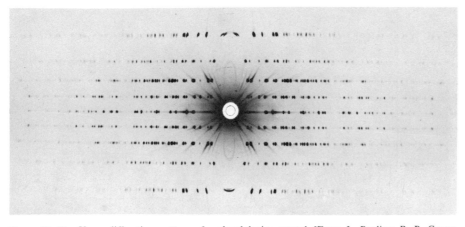

Figure 13–10. X ray diffraction pattern of a glycylglycine crystal. [From L. Pauling, R. B. Corey, and R. Hayward: The structure of protein molecules. *Scientific American.* July, 1954, p. 53. Reprinted through the courtesy of Dr. R. B. Corey.]

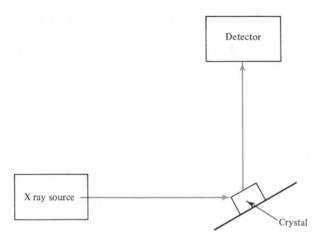

Figure 13-11.   Diagram of X ray crystallography apparatus.

The following simple analysis, however, is easy to follow and demonstrates the basic principle involved. Consider the parallel X ray beams *a*, *b*, and *c*, issuing from a common X ray source of a single wavelength (Figure 13–12). The beam *a* is reflected at point *A* to give beam *a′*, beam *c* is reflected at point *C* to give beam *c′*. The two beams *a′* and *c′* can reinforce each other and thus produce an intense spot at *X* only if the distance *y* is an integral number of wavelengths, λ. Otherwise the two beams *a′* and *c′* would cancel.

Such a bright spot (or series of bright spots) occurs only when the angle of incidence θ is adjusted so that $y = n\lambda$. Thus, the method consists in varying the angle θ to obtain the best reinforcement patterns, from which the distance *d* between planes in the crystal can be determined.

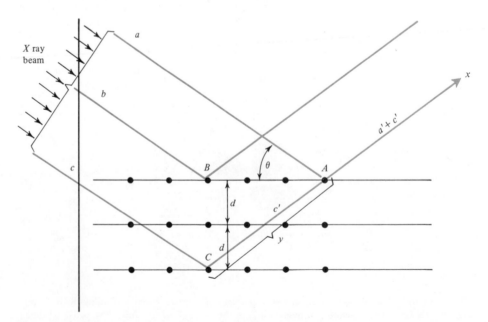

Figure 13-12.   The reinforcement of X rays in diffraction by crystal lattice.

By the method of X ray diffraction analysis bond distances reliable to within ± 0.002 Å can be obtained. One of the main problems with the method, however, arises in structure determinations involving hydrogen. The reason is that X rays are scattered by the electrons in atoms and not by their nuclei. Hydrogen exhibits very little scattering ability because its electrons are involved in bonds and are not centered on the nucleus, and it has no inner electrons as do all other atoms. Therefore, while the position of hydrogen atoms can be determined in some cases, the accuracy is usually less than with other atoms.

Neutron diffraction is similar to X ray diffraction. If one considers the de Broglie wave nature of nuclear size particles, one can see that neutrons and even other particles should be capable of diffraction in the same way as electromagnetic radiation. In the case of neutron diffraction, however, it is the atomic nuclei that produce the scattering. Hydrogen nuclei, which are comparable in size with neutrons, exhibit a high scattering efficiency. Because the experimental difficulties are somewhat greater than in X ray diffraction analysis, neutron diffraction analysis is used principally to locate hydrogen atoms.

Electron diffraction analysis has also been employed to illucidate chemical structures. In this case, because of the nature of the electron beams, analysis is carried out on gas samples in which intermolecular forces are absent. Here again, scattering is by nuclear charges so that hydrogen atoms can be accurately located.

PROBLEMS

1. Describe each of the following pairs of structures as identical, constitutional isomers, enantiomers, diastereomers, or not isomers (The structures are tetrahedral unless otherwise stated).

(a) CH_2ClBr and CH_2BrCl

(b) H—C—C—O—C—H and H—C—O—C—C—H

(c) [structure] and [structure]

(d) CH_3—C—C—H and CH_3—C—C—H

(e) [Pt structure] and [Pt structure] (square planar)

(f) CH_3—C—H and CH_3—C—Cl

(g)   [structure: C bonded to Cl, H, Br, CH₃]   and   [structure: C bonded to CH₃, Cl, Br, H]

(h)   [structure: H, Cl — C=C=C — Cl, H]   and   [structure: Cl, H — C=C=C — H, Cl]

2. List five common objects not mentioned in the text that are achiral.

3. List five common objects not mentioned in the text that are chiral.

4. Give the symmetry elements for each of the objects given in questions 2 and 3.

5. What is the distinction between the terms *chiral* and *asymmetric*.

6. Under what conditions do enantiomers exhibit different properties? Under what conditions do they exhibit the same properties?

7. Using the definition of kinetic energy ($KE = 1/2\ mv^2$) and the fact that two substances at the same temperature have the same average kinetic energy, explain why the substance that has the higher molecular weight should be expected to have the higher boiling point.

8. Which three elements exhibit the most pronounced hydrogen bonding when attached to hydrogen?

9. Give several examples of substances in which hydrogen bonding accounts for unusual properties.

10. Using Figure 13–7, estimate the boiling points of H_2O, HF, and NH_3 if there were no hydrogen bonding.

11. Account for the fact that methane does not hydrogen bond.

12. Account for the fact that H_2Se does not hydrogen bond.

13. Consult Figure 13–7. Would you say that there is any hydrogen bonding in HCl? Explain.

14. If the compound CHFClBr were square planar, how many isomers would there be? Draw their structures.

15. If CHFClBr were rectangular planar, how many isomers would there be? Draw their structures.

16. Give the symmetry elements for each of the structures in Figures 13–8 and 13–9.

17. Why is the diffraction analysis technique for determining the structures of pure substances so useful when X rays are employed?

18. Describe what is meant by *reinforcement* in X ray diffraction analysis.

19. Why cannot hydrogen atoms be located using X ray diffraction analyses?

20. What are the advantages of neutron and electron diffraction analysis over X ray diffraction analysis?

COVALENT ARCHITECTURE I. GEOMETRICAL VARIATIONS

A. TETRAHEDRAL GEOMETRY

In the preceding chapter we discussed some of the principles of stereochemistry. Using those ideas and concepts, the present chapter will explore some of the chemical consequences of tetrahedral geometry, as well as two other molecular geometries of importance—octahedral and square planar.

14-1. HYDROCARBONS

Of the hundred-odd elements that form the million or more diverse substances on earth, carbon is the only one that has the ability to form large stable networks of covalent bonds to itself. Silicon is the nearest competitor to carbon in this regard; however, silicon is able to form "chains" of no more than eight silicon atoms covalently bonded to each other. All attempts to form longer chains of the same element have failed. It would therefore be illuminating to examine the nature and the results of this unique ability of carbon.

One of the simplest hydrocarbons is ethane, C_2H_6. The two forms of its structure are

"eclipsed" form   "staggered" form

Viewed through the carbon-carbon bond, the two forms may be drawn as

"eclipsed" form   "staggered" form

The C-C bond distance in ethane is 1.54 Å, the same as in diamond, and the bond angles are very close to 109.5° as expected for the tetrahedral structure. If one considers the mutual repulsions between electron pairs in orbitals on adjacent carbon atoms, the more stable arrangement should be the "staggered" form. This was shown to be the case by an analysis of the spectrum of ethane. Although under ambient conditions, rotation about the C-C bond occurs quite freely, the stability of the eclipsed form is less than that of the staggered form by about 3 kcal/mole due to electronic repulsions among the C-H bonds. Most of the molecules exist in or near the staggered form.

A slightly more complex situation is found in the isomers of the butanes, C_4H_{10}, which are found in petroleum and are gases under ordinary conditions. These are

normal-butane   isobutane
(*n*-butane)

Some of the properties of the two isomeric butanes are listed in Table 14–1. In both isomers, the bond angles and distances are essentially the same as those in ethane.

Compounds are known with much longer carbon chains than butane. In fact, there is probably no limit to the number of carbon atoms that can be bonded together to form stable molecules. The common plastic polyethylene contains molecules made up of thousands of carbon atoms bonded together.

Table 14–1. Physical Properties of the Isomers of Butane

	n-Butane	*Isobutane*
boiling point (°C)	0	− 12
melting point (°C)	− 138	− 159
density at −20° (g/ml)	0.622	0.604

Another arrangement of atoms commonly found in nature is the ring. Many substances have been found in nature or synthesized in the laboratory that contain carbon atoms bonded to each other in the form of rings containing from three to scores of atoms per ring. Four examples are shown in Figure 14–1, which illustrates some of the interesting properties of ring compounds.

Table 14–2 lists some physical properties of these cyclic compounds. The first two are gases under ambient conditions; the last two are liquids. Cyclopropane is the most potent of the inhalation anesthetics producing unconsciousness and loss of feeling within seconds. If the bond between two carbon atoms is taken to be a straight line between the two atoms, then the C-C-C angles in cyclopropane are 60°. This is a rather large deviation from the most stable tetrahedral angle of 109.5°. The bonds are bent, however, so that overlap between orbitals is not only reduced, but also strained in the same way that a steel spring would be if distorted (Figure 14–2). The cyclopropane molecule suffers still another kind of strain. Figure 14–2*b* shows that the C-H bonds are all eclipsed; and therefore are somewhat strained because of the mutual repulsion of electron pair bonds on adjacent carbon atoms.

Cyclobutane also suffers from the strain due to eclipsing of C-H bonds. The strain due to deviation from the tetrahedral angle, however, is less than that in cyclopropane. Planar cyclopentane should have the least angle distortion strain because its C-C-C angles should be 108° (the internal angles of a regular pentagon). The strain due to eclipsing of C-H bonds, however, causes a slight puckering

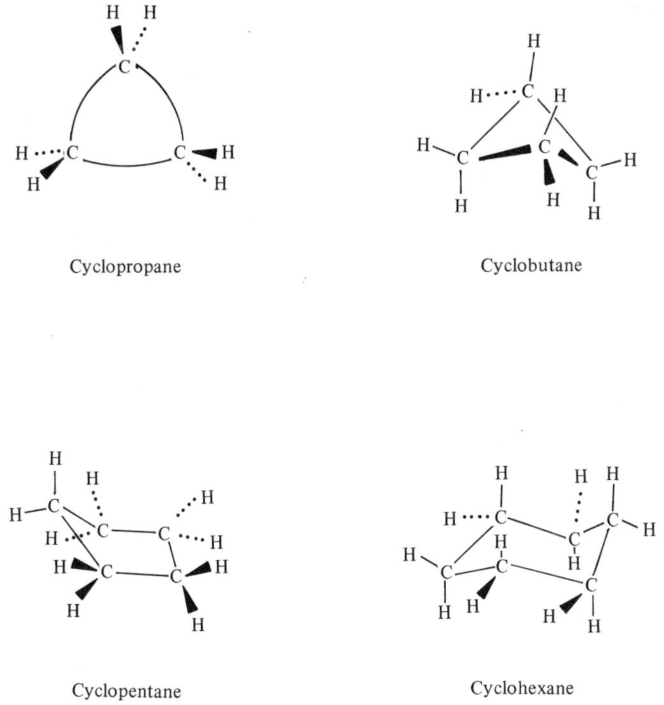

Cyclopropane

Cyclobutane

Cyclopentane

Cyclohexane

Figure 14–1. Some cyclic hydrocarbons.

Table 14-2.   Physical Properties of Some Cyclic Hydrocarbons

Compound	Melting Point (°C)	Boiling Point (°C)	Density of Liquid (g/ml)	C—C Bond Distance (Å)
cyclopropane	− 127	−33		1.51
cyclobutane	− 80	13		
cyclopentane	− 94	49	0.746	1.54
cyclohexane	6.5	81	0.778	1.54

in the rings so that one of the C atoms is out of the plane of the other four (Figure 14-1). The C-H bonds on this carbon are now staggered with its neighbors. This staggering relieves the strain enough to compensate for the decreased C-C-C angle that results from puckering. The ring is somewhat flexible so that the pucker moves around the ring.

Cyclohexane is the least strained of the cyclic compounds shown. Its six-membered ring (Figure 14-1) allows the necessary puckering to satisfy both the angular demand of 109.5° for the C-C-C angles and the complete staggering of all C-H bonds.

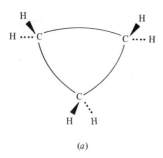

(a)

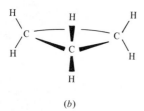

(b)   Figure 14-2.   Cyclopropane.

14-3.   MOLECULAR STRAIN AND CHEMICAL ENERGY

One might ask at this point what this type of strain has to do with the actual properties of these substances. If it is understood that molecular strain is a form of potential energy, then strain can be associated with reactivity or instability. That is, the more strain in a molecule, the more energy will be released when that molecule reacts to form a more stable molecule. One rather easy way to measure strain energy is to burn a sample of the compound and measure the quantity of heat energy evolved. Because all hydrocarbons burn in oxygen to form carbon dioxide and water, they may be mutually compared in energy content. This comparison is made by considering the amount of heat energy released per mole of carbon dioxide formed.

It must be emphasized that the greatest portion of the energy content in these molecules resides in the covalent bonds themselves. For this reason the combustion of a hydrocarbon releases an amount of energy approximately equal to the difference between the sum of the energies of the C=O bonds and H—O bonds formed (CO_2 and H_2O) and the sum of the energies of the C—H and O=O bonds broken (hydrocarbon and O_2). Thus in the reaction

$$CH_4 + 2\,O_2 \longrightarrow CO_2 + 2\,H_2O$$

the heat effect is due primarily to the difference

$$\Sigma(\Delta H \text{ of 4 O—H bonds} + \Delta H \text{ of 2 C=O bonds})$$
$$- \Sigma(\Delta H \text{ of 4 C—H bonds} + \Delta H \text{ of 2 O=O bonds})$$

Table 14-3 lists some hydrocarbons along with the amount of angle strain (measured by the angular deviation from 109.5°) and the heat of combustion given per mole of hydrocarbon and per mole of carbon dioxide. As expected, the heat released (heat of combustion) per carbon atom is greater the smaller the ring because, in addition to the bond energy described above, there is also additional energy due to angular strains and to C—H bond eclipsing. The relationship does not continue with the larger rings, however. In fact, in rings greater than about 12 carbon atoms, there is apparently no more strain energy than there is in the open chain compounds.

This discussion has provided examples of the many forms of stored energy in molecules. All fuels contain stored molecular strain energy, bond energy, and other forms of potential energy.

Table 14-3. Strain and Heats of Combustion of Hydrocarbons

Ring Size	Angular Distortion from 109.5°	Heat of Combustion (kcal/mole hydrocarbon)	Heat of Combustion (kcal/mole CO_2)
3	49.5	499.8	166.6
4	19.5	655.9	164.0
5	1.5	793.5	158.7
6	0	944.5	157.4
8	0	1269	158.6
12	0	1891	157.6
14	0	2204	157.4
n-butane	0	629.6	157.4

14-4.   UNSATURATION

Another form of molecular potential energy exists in molecules that contain double or triple covalent bonds. Ethylene and acetylene are examples of such substances. Ethylene gas is obtained from petroleum and is used commercially to make polyethylene. It has the interesting property of causing fruit to ripen. Acetylene is obtained quite readily from the action of water on calcium carbide (CaC_2) which in turn is made by heating calcium oxide (CaO) and coke at high temperatures.

$$2\,CaO + 5\,C \longrightarrow 2\,CaC_2 + CO_2$$

$$CaC_2 + 2\,H_2O \longrightarrow C_2H_2 + Ca(OH)_2$$

Acetylene is used commercially in oxyacetylene welding.

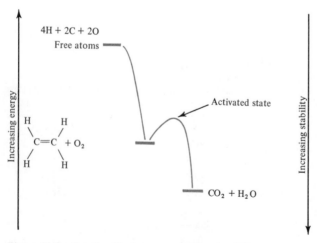

ethylene   acetylene

The formation of a double bond is best pictured as the simultaneous overlapping of two half filled orbitals of each of two atoms. The triple bond can then be pictured as a similar overlap of three orbitals of each of two atoms. This model (Section 12–9) requires that ethylene be planar and that acetylene be linear, and these geometries have been verified experimentally. Bond angle strain results from the distortion required to bring about overlap. You should bear in mind that, although we consider that bond angle distortion represents instability in molecules, the molecules are stable enough to exist indefinitely unless caused to undergo reaction. That is, they are stable with respect to the separated atoms, but unstable relative to other, less strained molecules. The relationship is demonstrated in Figure 14–3.

Figure 14-3. Relationship between energy and stability.

The problem immediately arises: If ethylene contains strain energy and is therefore unstable relative to carbon dioxide, why does it exist indefinitely? That is, why does it not spontaneously combine with oxygen in the air and immediately form carbon dioxide? This problem is analogous to the following physical situation: A block of wood (Figure 14-4) can be placed on the table in two ways: standing on end (*a*), and lying on its length (*c*). Either of these positions is stable because the block will not move spontaneously unless it is disturbed. Position (*a*), however, is immediately recognized as less stable than position (*c*). In order for the block in position (*a*) to fall, it must be disturbed, that is, energy must be furnished to get it into the "activated" position (*b*). It may then either topple on to (*c*), or fall back to (*a*). The block in position (*b*) is analogous to the ethylene and oxygen molecules in the "activated state."

The important point here is that the stability of ethylene, and any other molecule that contains strain energy in whatever form, is due to the fact that the molecule must be activated before it will rearrange itself into a more stable molecule or set of molecules. This, of course, is a very happy situation because without it, all substances in the universe would immediately react to form the most stable substances possible and no further events could occur.

In order to make this concept clearer let us look at a typical chemical change in which strain energy is released. The reaction of ethylene with hydrogen forms ethane.

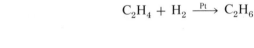

$$C_2H_4 + H_2 \xrightarrow{\text{Pt}} C_2H_6$$

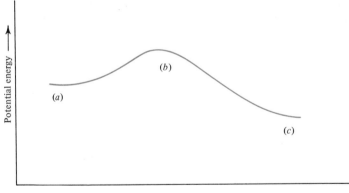

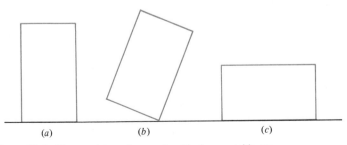

Figure 14-4. Energy states of a wooden block on a table top.

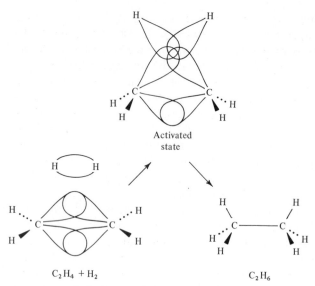

Figure 14-5. Activation of ethylene and hydrogen to form ethane.

Using our molecular model, the reaction is shown schematically in Figure 14–5. In the activated state the overlap in the H—H and C═C bonds is reduced and the overlap between the incipient C—H bonds has begun. This system represents a highly unstable arrangement and must either form products or "topple" back into reactants. To complete our analogy with the block of wood, we must compare the "ground zero" and "driving force" for the two processes. Ground zero for the wooden block was taken as the table top. The driving force was the force of gravity that caused the block to fall in either direction. In the chemical analogy, ground zero is the maximum overlap between orbitals with the least angle distortion from the tetrahedral angle. The driving force is the electrical attraction between the electrons and the nuclei of the atoms being bonded.

14-5. THE BENZENE MOLECULE

Benzene, C_6H_6, is an interesting substance whose molecular structure was a theoretician's headache for many years. Benzene is a common solvent and is used as a cleaning agent. It is produced from the distillation of coal tar and from petroleum. Its molecular structure has been shown by diffraction techniques to possess a planar ring of six carbon atoms each joined to one hydrogen atom. Structure (a) satisfied these requirements but did not satisfy the valency of four for carbon. Structure (b) accounted for the required tetravalency by the use of alternating single and double bonds. X-ray analysis has demonstrated, however, that all the C-C bonds in benzene are identical in length and equal to 1.39 Å. This value is intermediate between a single (1.54 Å) and a double bond (1.33 Å). Moreover, benzene possesses an unusual lack of reactivity for a molecule containing three double bonds. Its chemical properties are more like ethane than like ethylene.

(a)   (b)   (c)   (d)

The explanation of the benzene structure that accounts for these apparently anomalous properties is based on a resonance hybrid of the two structures

and

The benzene ring is very common in nature, probably because of its unique stability. It is frequently designated by structure (d), which abbreviates the resonance forms and implies total electron delocalization as in (c), in which the circles represent orbitals perpendicular to the plane of the paper and the shaded areas represent overlap of these orbitals. The inscribed circle in (d) thus represents the equal distribution of electrons around the entire ring.

B. OCTAHEDRAL GEOMETRY

14-6. STRUCTURE ELUCIDATION

Thus far, we have seen that many atoms form four stable bonds with other atoms and that these bonds are arranged so that they are directed toward the corners of a tetrahedron. Atoms in the third and higher periods can form more than four bonds because they possess d orbitals in their valence shells. The most common numbers of bonds found greater than four are six and eight. Examples of hexavalent atoms are sulfur in SF_6 and platinum in $Pt(NH_3)_2Cl_4$. In these and many other hexavalent atoms the bonds are directed toward the corners of a regular octahedron (Figure 14-6). Two isomers of $Pt(NH_3)_2Cl_4$ exist. One is an orange crystalline solid, the other a yellow crystalline solid. In 1900, a Swiss chemist named Alfred Werner explained the different properties of these two isomers by assuming that they possess the octahedral configurations shown in Figure 14-6. In such a configuration, two isomers are expected—one with the NH_3 groups adjacent to each other (cis) and the other with the NH_3 groups opposite each other (trans). Werner's assignments were confirmed many years later by X ray diffraction studies of these substances.

Typical of the problems that Werner faced in this interesting but confusing group of compounds was the following. The four compounds of cobalt, chlorine,

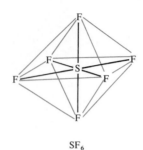

SF$_6$

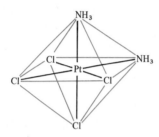

cis-Pt(NH$_3$)$_2$Cl$_4$

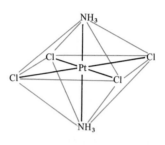

trans-Pt(NH$_3$)$_2$Cl$_4$

Figure 14-6. Octahedral configuration of SF$_6$ and *cis*- and *trans*-Pt(NH$_3$)$_2$Cl$_4$.

and ammonia are shown below with their reactivities toward silver nitrate solution* and their electrical conductivities.

	Silver Nitrate Solution	*Electrical Conductivity*
CoCl$_3$(NH$_3$)$_6$	white precipitate (3 moles AgCl formed)	conducts in water solution
CoCl$_3$(NH$_3$)$_5$	white precipitate (2 moles AgCl formed)	conducts in water solution
CoCl$_3$(NH$_3$)$_4$	white precipitate (1 mole AgCl formed)	conducts in water solution
CoCl$_3$(NH$_3$)$_3$	no reaction	nonconductor

Werner brilliantly conceived the notion that the cobalt atom could be covalently bonded to six other atoms, ions, or molecules in an octahedral "inner coordination sphere" while being bonded ionically to other ions in an "outer sphere." He formulated the above compounds with the inner sphere within brackets and the outer sphere outside the brackets as follows (Figure 14-7)

$$[Co(NH_3)_6]^{3+} \ 3 \ Cl^- \qquad \text{ionic}$$

$$[Co(NH_3)_5Cl]^{2+} \ 2 \ Cl^- \qquad \text{ionic}$$

---

*Silver nitrate solution reacts instantly with ionic chlorides to form the insoluble white silver chloride. Since reaction with covalent chloride is much slower, this solution is used as a test for ionic chlorides.

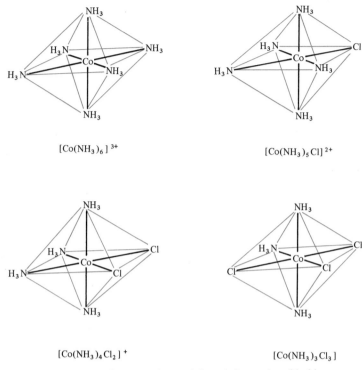

$[Co(NH_3)_6]^{3+}$

$[Co(NH_3)_5Cl]^{2+}$

$[Co(NH_3)_4Cl_2]^{+}$

$[Co(NH_3)_3Cl_3]$

Figure 14-7.   Inner sphere complexes of the cobalt ammine chlorides.

$[Co(NH_3)_4Cl_2]^{+} Cl^{-}$ ionic

$[Co(NH_3)_3Cl_3]$ neutral, totally covalent molecule

Werner was, of course, not able to explain such bonding in terms of any theoretical model because the theoretical foundation for bonding had not yet been laid. In fact, had it not been for the dogma of the tetrahedral carbon atom proposed earlier, the structural theory of bonding in octahedral molecules might have come years earlier.

In terms of the atomic theory, the formation of more than four bonds by an element is accounted for by the involvement of d orbitals in the valence shell. Coordinate covalent bonds in the cobalt ammine chlorides are formed by the overlap of a filled orbital on a chloride ion or an ammonia molecule with an empty d orbital of the cobalt ion. The use of the inner shell plus the valence shell of cobalt allows up to six bonds to be formed. Although in principle more than six bonds are possible, the practical maximum of six is maintained by two factors: (1) the impossibility of fitting more than six atoms or small molecules around a relatively small ion like Co^{3+}, and (2) the instability arising from the accumulation of too much negative charge on the central ion through donated electrons. Recall that the bonds are coordinate covalent and that the Cl^- and NH_3 furnish all the electrons.

Another interesting aspect of octahedral complexes is their stereochemistry. We have seen that cis-trans isomers are possible with octahedral complexes (Figure

209

14-6). Optical activity is also possible. One of the interesting ligands* that has been used to form complexes is ethylene diamine, $H_2N—CH_2CH_2—NH_2$. This molecule has two reactive ends and therefore forms two bonds with transition metal ions. The resulting octahedral complex ion is chiral and therefore can exist in enantiomeric forms.

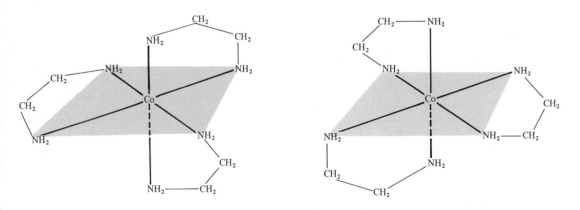

The pure enantiomeric forms of this and many other such complexes have been separated.

14-7. FERROCENE AND METALLOCENES

Ferrocene was first prepared in 1951. This substance is a crystalline orange compound (melting point 173°C) and can be prepared by heating cyclopentadiene

$$Fe + 2\,C_5H_6 \longrightarrow Fe(C_5H_5)_2 + H_2$$

(C_5H_6) with iron. Ferrocene is stable even when heated to 400°C. It is not decomposed by bases nor, in the absence of oxygen, by acids, and is reasonably unreactive. This substance has been shown to have the structure

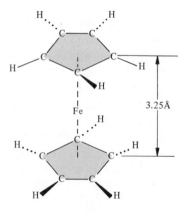

*Ligand: an atom, group, or ion that can form a covalent bond with a metal to form a complex species.

an atom of iron being sandwiched in between two cyclic hydrocarbon rings of formula C_5H_5.

Compounds of this "sandwich" structure, containing other metal atoms in the center and containing rings of five or six carbon atoms, have also been prepared. The measured C-Fe bond length in ferrocene is 2.04 Å and the perpendicular distance between the rings is 3.25 Å. These data suggest that the equivalent of nearly a full single bond exists between the iron atom and the carbon atoms.

The C-C bond distances in both rings is 1.44 Å, which is intermediate between the single bond length, 1.54 Å, and the double bond length, 1.33 Å. Lastly, the distortion in bond angles from the normal tetrahedral value of 109.5° is very great. Nonetheless, the ferrocene molecule is quite stable and capable of undergoing a great many reactions throughout which the sandwich structure remains intact. Bonding probably involves overlap of the inner vacant electron shells of iron with the unused valence orbitals on carbon, and may be represented by resonance structures like the following:

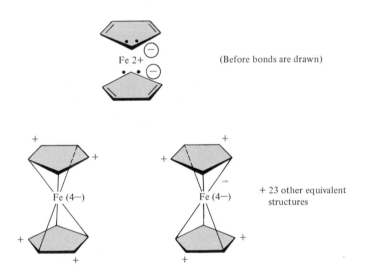

Fe 2+   (Before bonds are drawn)

Fe (4—)   Fe (4—)   + 23 other equivalent
                              structures

C. SQUARE PLANAR GEOMETRY

14-8. NONMETAL COMPOUNDS

Some species are structurally square planar in terms of the positions of their atoms but are really octahedral if the unshared electron pairs around the central atom are included. For example, the tetrachloroiodate ion consists of four chlorine atoms situated at the corners of a square centered upon an iodine atom.

$$\begin{bmatrix} Cl & Cl \\ & I & \\ Cl & Cl \end{bmatrix}^-$$

However, the total number of electrons in the valence shell of the iodine atom is twelve, that is, six pairs.

electrons in valence shell of neutral I atom	$= 7$
"extra" electron to give negative charge	$= 1$
unpaired electron provided by each of 4 Cl atoms for sharing	$= \underline{4}$
total electrons around the I atom in ICl_4^- ion	$= 12$

Iodine, of course, has ample room for these electrons with its empty valence shell d orbitals. The six pairs of electrons are arranged in an octahedral configuration to reduce electron repulsions. The structure may be represented

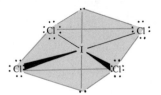

Another example of this type of square planar geometry is found with the xenon tetrafluoride molecule, XeF_4.

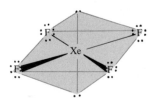

14-9. TRANSITION METAL COMPLEXES

A relatively small number of transition metal ions form complexes in which the ligands are in a square planar arrangement. Figure 14-8 shows the metals which commonly display this geometry.

The Pt^{2+} ion is the most extensively studied ion of this group. Almost all of the complexes of this ion are quite stable and have the square planar arrangement.

Figure 14-8. Metal ions that commonly form square planar complexes.

Notice the contrast with complexes of Pt^{4+}, which are typically six-coordinate and octahedral. Numerous examples of cis-trans isomerism have been observed with the Pt^{2+} ion. A classic example is the molecule $[Pt(NH_3)_2Cl_2]$.

$$
\begin{array}{cc}
\text{Cl} \quad NH_3 & \text{Cl} \quad NH_3 \\
\diagdown \diagup & \diagdown \diagup \\
Pt & Pt \\
\diagup \diagdown & \diagup \diagdown \\
\text{Cl} \quad NH_3 & H_3N \quad \text{Cl} \\
(cis) & (trans)
\end{array}
$$

In general, there are only two possible isomers of this type in square planar compounds. However, in the special case where all four ligands are different, three such isomers are possible. The experimental difficulties involved in attaching four different ligands to the same metal ion largely preclude the actual preparation of such species. An example in which the three diastereomers have been separated is

$$[Pt(NH_3)(NH_2OH)(py)(NO_2)]NO_2$$

where

$$
py = :N \diagup\!\!\!\!\diagdown \; \begin{array}{c} H \quad\quad H \\ C\!=\!C \\ C\!-\!H \\ C\!=\!C \\ H \quad\quad H \end{array}
$$

The three isomers have the structures

$$
\left[\begin{array}{cc} O_2N & NH_3 \\ \diagdown & \diagup \\ & Pt \\ \diagup & \diagdown \\ py & NH_2OH \end{array} \right]^+
\quad
\left[\begin{array}{cc} py & NH_3 \\ \diagdown & \diagup \\ & Pt \\ \diagup & \diagdown \\ O_2N & NH_2OH \end{array} \right]^+
\quad
\left[\begin{array}{cc} py & NH_3 \\ \diagdown & \diagup \\ & Pt \\ \diagup & \diagdown \\ HOH_2N & NO_2 \end{array} \right]^+
$$

The case of the Ni^{2+} ion is interesting in that it forms four-coordinate complexes, some of which are tetrahedral and some of which are square planar. Furthermore, the nickel ions in square planar complexes have no unpaired electrons, whereas those in the tetrahedral species have two unpaired electrons. The ground-state electron configuration of Ni^{2+} predicts two unpaired electrons:

$$1s^2 2s^2 2p^6 3s^2 3p^6 3d^8$$

where the $3d$ orbitals would have the arrangement

$$3d$$

$$\underline{\uparrow\downarrow} \;\; \underline{\uparrow\downarrow} \;\; \underline{\uparrow\downarrow} \;\; \underline{\uparrow} \;\; \underline{\uparrow}$$

213

The difference in these geometries can be related to the number of unpaired electrons in the following way: *Empty* hybrid orbitals on the nickel ion are required to receive the electron pairs from the ligands.

In the tetrahedral case:

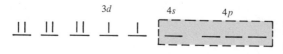

sp^3 hybrid orbitals on Ni^{2+} into which electron pairs on the ligands are donated to form coordinate covalent bonds.

In the square planar case:

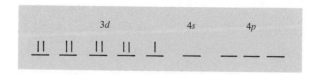

dsp^2 hybrid orbitals on Ni^{2+} into which electron pairs on the ligands are donated to form coordinate covalent bonds.

Recall that sp^3 hybrid orbitals are directed toward the corners of a tetrahedron while dsp^2 hybrid orbitals point to the corners of a square. The Pd^{2+}, Pt^{2+}, and Au^{3+} ions all have no unpaired electrons in their square planar complexes, which allows dsp^2 hybridization.

It is easier to pair electrons the higher their quantum level. Thus, the pairing energy is less in the heavier metal ions than it is for Ni^{2+}. Apparently, only very strong electron donor ligands such as CN^- can form bonds strong enough to release sufficient energy to cause the d electrons to pair up in Ni^{2+}. (The dsp^2 orbitals form stronger covalent bonds than the sp^3 orbitals.) For this reason, only the strong electron donors form square planar complexes with Ni^{2+}, whereas the weaker electron donors such as Cl^- or Br^- form tetrahedral complexes.

The Cu^{2+} ion has only one unpaired electron, so there is no way for it to pair. Because the d electrons in this ion would have the ground-state configuration

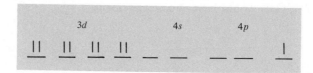

one would assume that the empty s and p orbitals would result in a tetrahedral, sp^3 configuration. However, square planar geometry is realized in practically all four-coordinate complexes of Cu^{2+}, even with ligands of weak electron donating ability. To achieve the empty d orbital required for the dsp^2 hybridization, it is necessary to assume that the unpaired $3d$ electron is promoted into a $4p$ orbital.

The resulting empty dsp^2 hybrid orbitals can now accept electron pairs from the four ligands. With an electron in a higher energy orbital than usual, it would normally be expected that removal of this electron to form Cu^{3+} would be facilitated. Thus a square planar complex of Cu^{2+} should be easier to convert to Cu^{3+} than uncomplexed Cu^{2+}. In fact, this is *not* found to be the case, and the presumed promotion of the unpaired $3d$ electron has no effect on the ease of ionization. The theory is weak at this point, and the explanation of the square planar geometry of Cu^{2+} complexes is obviously not as simple as suggested.

Other theoretical approaches have been used to treat the square planar problem. Some success has been achieved in explaining why square planar rather than tetrahedral geometry is to be expected with Cu^{2+}. Molecular orbital approaches in particular have given fairly satisfactory descriptions of the square planar complex ions.

However, none of these theories has fully answered the basic question of why only four ligands bind to these metal ions. There is ample room to group six ligands around these metal ions in most cases, and just why the highly symmetrical, six-coordinate octahedral structure is not attained is difficult to explain. It would seem that the extra bond energy released by the formation of two additional bonds would be sufficient in many cases to overcome any unfavorable effects of additional ligands. The interest of chemists in these square planar structures is heightened because of unanswered questions like the ones above.

PROBLEMS

1. In what important way is the element carbon unique?
2. Which of the following configurations of ethane is more stable? Explain. Which is the staggered and which is the eclipsed form?

3. Demonstrate whether or not propane is capable of isomerism.
4. Why cannot the isomers of butane interconvert as do the staggered and eclipsed forms of ethane?
5. Would you expect butane or cyclobutane to have the higher entropy? Explain. (See Chapter 6)
6. Name and describe the sources of strain energy (potential chemical energy) in (a) cyclopropane, (b) cyclopentane, (c) cyclohexane.
7. How do you account for the much greater reactivity of cyclopropane compared with cyclopentane?
8. What is the major form of chemical potential energy in butane relative to combustion to CO_2 and H_2O. What is it in cyclobutane?
9. Account for the source of energy in the combustion of (a) methane, (b) cyclopropane, (c) hydrogen sulfide.

10. Judging from the data in Table 14–3 would you expect cyclopropane or ethylene to have the higher heat of combustion per carbon atom? Explain.

11. Referring to the structures of ethylene and acetylene in Section 14–4, explain why acetylene has a shorter C-C bond distance than ethylene.

12. Most processes can be characterized as having an activation energy, an activated state, and a driving force. Describe the activated state and driving force for each of the following:
    (a) A large boulder placed on the top of a cliff.
    (b) A water reservoir held by a dam.
    (c) The combustion of methane (CH_4).

13. What experimental results conflict with the simple structure,

for benzene?

14. What is implied by the structure

15. Write all the resonable structures that you can for the substance naphthalene, $C_{10}H_8$, which has the skeletal structure

16. How is the violation of the octet rule allowed in compounds like K_2SF_6 (the SF_6^{2-} ion)? Which elements do not necessarily obey this rule? Which elements *must* obey the octet rule?

17. Draw the cis and trans forms for the hypothetical ion $SF_4Cl_2^{2-}$. List all the symmetry elements of each.

18. Judging from the reactivity of the cobalt ammine chlorides described in Section 14–6, what can be said about the Co-Cl bond in the inner coordination sphere? Explain.

19. How does the presence of nonbonded electron pairs on the iodine atom explain the square planar geometry of ICl_4^-?

20. Can the existence of cis-trans isomers of $Pt(NH_3)_2Cl_2$ be used as evidence for square planar geometry? Explain.

21. What is meant by the term electron pairing when referring to the formation of complex ions?

22. Is pairing energy greater or less in heavier metal ions than it is for the lighter ones?

23. Explain in general terms why $NiCl_4^{2-}$ is tetrahedral while $Ni(CN)_4^{2-}$ is square planar.

24. For each pair below, indicate the one that represents the more reasonable geometry for that molecule. Explain your choice.

(a) :Ö=C=Ö: or   :Ö=C⟨:Ö:

(b) H—T̈e—H   or   H—T̈e⟨H

(c)
F—C—F (with F above and F below)   or   C with F above, F—F to the right and F to lower left

(d) H—C̈l: or   H=C̈l:

15

COVALENT ARCHITECTURE II.
ELEMENTS OF THE FIRST
THREE PERIODS

The present chapter will explore the structure and stereochemistry of the elements in the first three periods of the periodic table, their hydrides, halides, oxides, and hydroxides.

15-1. HYDROGEN

Many of the properties of hydrogen have been discussed elsewhere in this book. It is the simplest element, whose atoms have only one nuclear proton and one planetary electron. It is the lightest element and is very reactive. Having only one valence electron, it superficially resembles the alkali metals (lithium, sodium, and so on, which have only one valence shell electron) in forming the positive ion, H^+. It is also, however, capable of gaining one electron (as do the halogens) to form the hydride ion, H^-. Hydride ions only occur when hydrogen is made to react with metals of low ionization potential. It is thus seen that hydrogen does not really belong with the metals of group I nor the representative nonmetals of group VII, and probably is best classified in a catagory all its own.

H

							He
Li	Be	B	C	N	O	F	Ne
Na	Mg	Al	Si	P	S	Cl	Ar

Since hydrogen forms compounds with nearly all of the elements, its chemistry will not be discussed separately, but instead will be taken up as it arises in the following discussions.

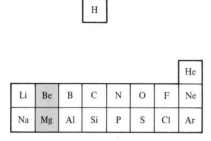

Elemental magnesium is typically metallic and its compounds are typically ionic. They will therefore be treated in subsequent chapters.

One of the most dramatic sources of beryllium in nature is emerald which is a beryllium aluminum silicate, $Be_3Al_2(SiO_3)_6$. The discovery of beryllium has been attributed to the geometrical similarity between the mineral beryl and emerald, which have the same elemental compositions. According to the French chemist Fourcroy, "It is to geometry that we owe . . . the source of this discovery; it is that (science) that furnished the first idea of it, . . . since according to the analysis of the emerald by M. Klaproth and that of beryl by M. Bindheim one would not have thought it possible to recommence this work without the strong analogies or almost perfect identity that Citizen Hauy found for the geometrical properties between these two stony fossils."

Pure beryllium has a dark metallic luster and forms hexagonal crystals. It is less reactive than magnesium and most of its compounds are highly toxic.

Beryllium Hydride □ As expected for hydrides of the elements of groups I and II, berylluim hydride, BeH_2, is saltlike in its properties and therefore is ionic. The hydrogen is present as the hydride ion, H^-. Berylluim hydride is extremely reactive with water or air and, in fact, has never been prepared in the pure state. It has only been obtained in ether solution.

Beryllium Chloride □ In contrast to the ionic nature of the hydride, beryllium chloride is covalent and, in the solid state, forms extended chains of indefinite length.

$$\underset{Cl}{\overset{Cl}{>}}Be\underset{Cl}{\overset{Cl}{<}}Be\underset{Cl}{\overset{Cl}{<}}Be\underset{Cl}{\overset{Cl}{<}}Be\overset{}{<}$$

The Be-Cl-Be angle has been determined by X-ray diffraction analysis to be 82°, and the Be-Cl bond length to be 2.05 Å.

The electronic structure of beryllium, $1s^2,2s^2$, suggests that it can form the ion, Be^{2+}. In the chloride, however, four covalent bonds can form by the use of the vacant $2p$ orbitals. In two of the four bonds of beryllium, both beryllium and chlorine furnish one electron to form an ordinary covalent bond. In the other two bonds, the Cl atom furnishes both electrons to form a coordinate covalent bond. The Cl atoms are tetrahedrally arranged around the Be atom, which uses sp^3 hybrid orbitals. The Lewis representation is

In the vapor state $BeCl_2$ exists as individual $BeCl_2$ molecules. As expected from the electron pair repulsion theory, the molecule is linear. The beryllium atom in this case uses sp hybrid orbitals.

$$:\overset{..}{Cl}:Be:\overset{..}{Cl}:$$

Beryllium Oxide □ The oxide, BeO, is actually a three-dimensional infinite network in which each Be atom is surrounded tetrahedrally by four O atoms and each O atom is surrounded tetrahedrally by four Be atoms. The structure is the same as the diamond structure. Two Be-O bonds are ordinary covalent bonds, and two are coordinate covalent bonds.

Beryllium Hydroxide □ As shall become evident in this chapter, hydroxides of the elements may be acidic or basic. Moreover, the acidic or basic nature of the hydroxide is a periodic property and may be used as one means of distinguishing chemically between metals and nonmetals: metals form alkaline hydroxides and nonmetals form acidic hydroxides.

Acids and bases will be discussed later, however, they may be defined here as follows: Acids in water solutions release hydrogen ions, H^+, to the water to form hydrated protons, H_3O^+, whereas bases release hydroxide ions, OH^-, the complementary relationship between acids and bases is that they neutralize each other to form the very stable water molecule.

$$H_3O^+ + OH^- \longrightarrow 2\,H_2O$$

Metal hydroxides are generally ionic. Nonmetals are nearer to oxygen in the periodic table and hence have similar electron affinities. Therefore nonmetals tend to form convalent bonds with oxygen. The covalent bond depletes the O atom of electron density and thus renders the O-H bond relatively more ionic in character.

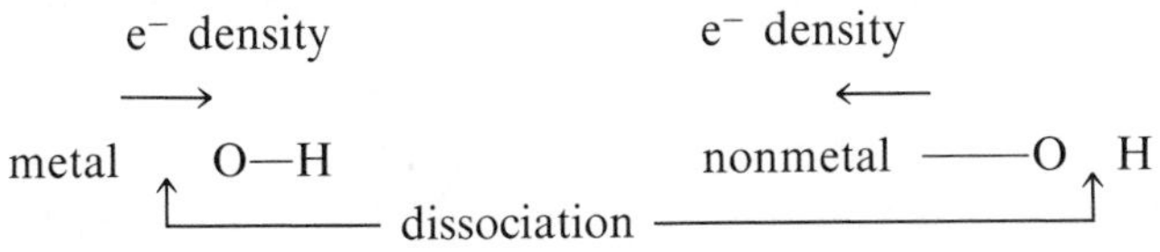

e^- density $\longrightarrow$

e^- density $\longleftarrow$

metal O—H   dissociation   nonmetal —O H

Dissociation of a H^+ ion is therefore easier in nonmetal hydroxides than in metal hydroxides.

Metals with small ions and high charges—+2 or +3—can withdraw enough charge from the O atom of OH^- groups or hydrated water molecules to confer acidic character to the hydroxide. Beryllium is a small atom and forms a +2 ion, thus it exhibits this behavior. The beryllium ion in water solution is tetra-hydrated

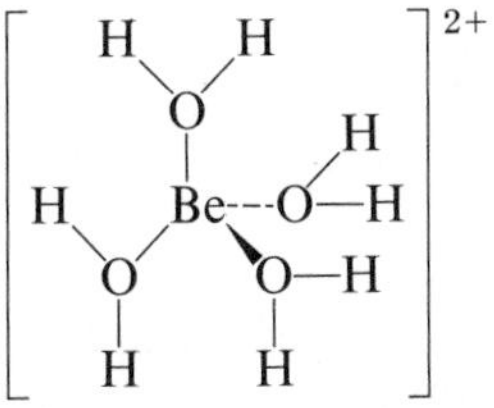

The hydroxide exists in water solution as $Be(H_2O)_2(OH)_2$ and is uncharged and insoluble. It reacts with both acids and bases to form soluble ions and is therefore said to be **amphoteric;** that is, *neutralizes both acids and bases.*

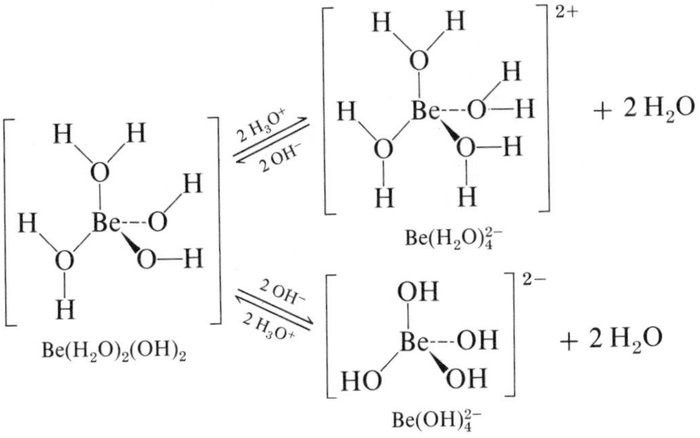

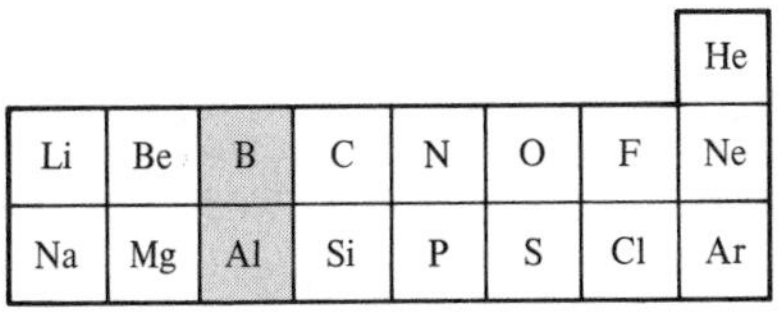

$Be(H_2O)_4^{2-}$

$Be(OH)_4^{2-}$

$Be(H_2O)_2(OH)_2$

							He
Li	Be	B	C	N	O	F	Ne
Na	Mg	Al	Si	P	S	Cl	Ar

15-3. BORON AND ALUMINUM

Boron is the only nonmetal in group III. In the pure form it is a dark gray brittle solid and a semiconductor. It is nearly as hard as diamond and, in the pure element, forms up to six bonds. In this respect it resembles the metals (Chapter 17). Its melting point is about 2300°C.

Boron Hydrides □ Boron is the only member of the group III elements to form an extended series of hydrides. These hydrides, in contrast to the ionic hydrides of the metals, are volatile and extremely reactive with air or water. Because the electronic structure of boron is $1s^2 2s^2 2p^1$, the simplest member of the series should be BH_3. All attempts to prepare BH_3 yield instead the dimer, diborane, B_2H_6. Because of the electronic structure of boron, the structure of diborane has presented the problem of explaining an insufficient number of electrons for the number of bonds needed. The structure involves two "bridge" hydrogens, which are bonded fractionally to the two boron atoms.

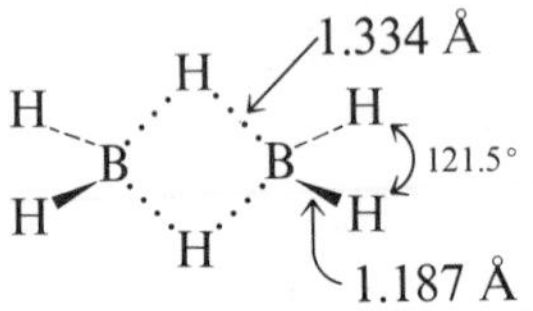

(B-B distance = 1.770 Å)

(A)

In this representation, dotted lines represent partial bonds. The structure A, in terms of resonance, may be described simply as a hydrid of the structures B through E, in which boron has a tetrahedral configuration.

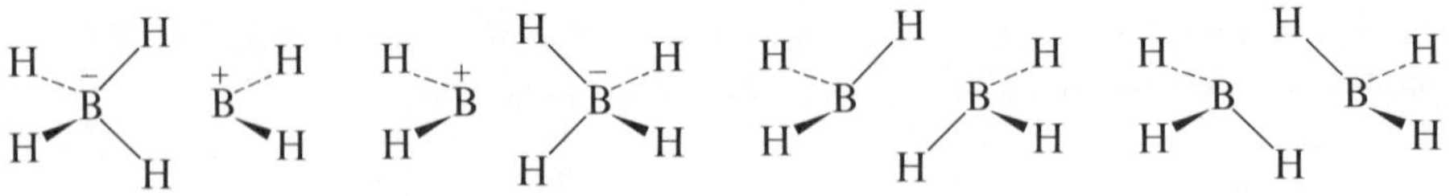

(B)   (C)   (D)   (E)

This representation is justified by the bond lengths observed for the molecule.

Other boranes are known, all of which are electron deficient also. Their structures similarly require drawing resonance structures involving bridged hydrogen. Decaborane, $B_{10}H_{14}$, is an example.

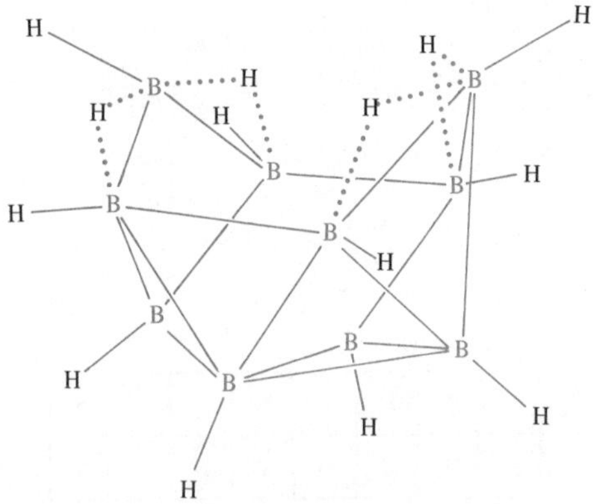

The absence of a B-B bond accounts also for the reactivity of diborane with molecules that contain unshared electron pairs, for example

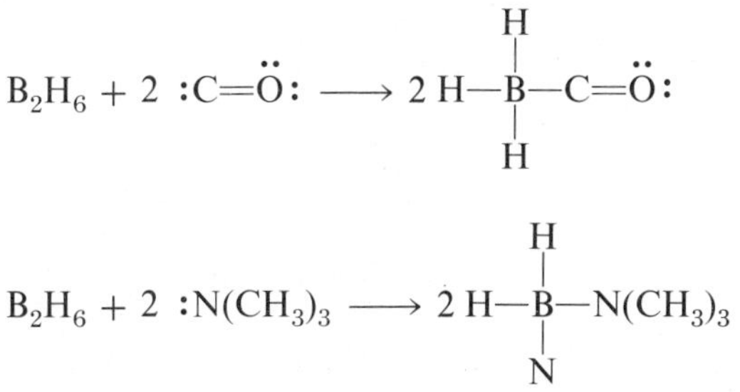

$$B_2H_6 + 2 \ :C{=}\overset{..}{\underset{..}{O}}: \longrightarrow 2 \ H{-}\underset{\underset{H}{|}}{\overset{\overset{H}{|}}{B}}{-}C{=}\overset{..}{\underset{..}{O}}:$$

$$B_2H_6 + 2 \ :N(CH_3)_3 \longrightarrow 2 \ H{-}\underset{\underset{N}{|}}{\overset{\overset{H}{|}}{B}}{-}N(CH_3)_3$$

In these cases the new bond formed is a coordinate covalent bond.

Aluminum Hydride □ Aluminum hydride has not been obtained free of the ether solvent in which it is prepared. It is believed to be an extended chain of AlH_3 units, and is probably saltlike in containing hydride ions, H^-.

Boron Halides □ In contrast to the hydrides, boron fluoride and boron chloride exist as the simple symmetrical molecules.

The difference is undoubtedly due to the greater electron affinity of fluorine and chlorine compared to hydrogen, which should leave the boron atoms relatively positively charged and therefore mutually repulsive. In these cases, the molecules are planar trigonal since the boron atom has no unshared electrons in its fourth orbital.

Aluminum Halides □ Aluminum chloride is a white solid that reacts vigorously with water. In the vapor phase it is dimeric, Al_2Cl_6, with the bridged structure shown.

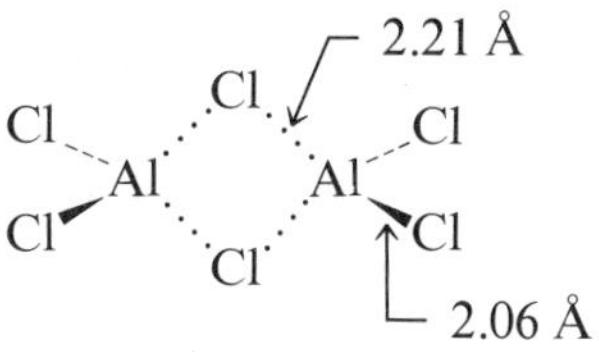

Again, this structure is a hybrid of the contributing tetrahedral structures

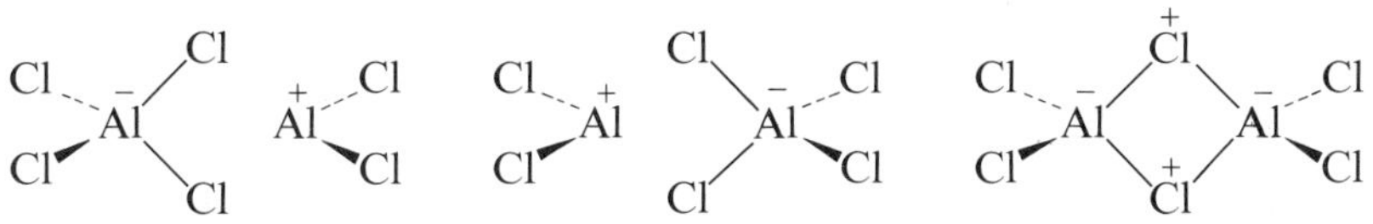

223

and so on, as well as the nonionic structures

Above 400°C, the dimer decomposes to the simple $AlCl_3$.

$$Al_2Cl_6 \xrightarrow{400°C} 2 \, AlCl_3$$

Boron Oxide □ Boron occurs in nature exclusively as oxygen compounds. It is found chiefly as calcium borate, CaB_4O_7, found in South America and California, and as borax $Na_2B_4O_7$ found in North America and India. The oxide of boron is a white solid that readily forms glasses. Its formula is $(B_2O_3)_n$. The crystalline form probably contains tetrahedral BO_4 units linked in a structure similar to that of BeO. Gaseous boron oxide consists of simple angular molecules of B_2O_3.

Aluminum Oxide □ Aluminum oxide $(Al_2O_3)_n$ exists in many physical forms. Because of its stability at high temperatures and its high melting point (~2000°C) aluminum oxides are good refractory materials.

Boron Hydroxide □ Boron is typically nonmetallic in that its hydroxides are acidic. The common commercial boric acid, more precisely orthoboric acid, H_3BO_3, is only weakly acidic but is not amphoteric like aluminum hydroxide (see p. 225).

As would be anticipated, H_3BO_3 is a trigonal planar molecule because the

fourth orbital on boron is vacant. Another form of boric acid, metaboric acid, exists. Its formula is usually written as HBO_2 but it does not consist of units of HO—B=O. Instead, it is a trimer (contains three units of formula

in which the molecule is a planar hexagon). Meta acids differ from ortho acids by the elimination of a molecule of water.

$$H_3BO_3 \longrightarrow HBO_2 + H_2O$$

A third form of boric acid, pyroboric acid, has the formula $H_4B_2O_5$. Its structure is

$$\begin{array}{ccc} HO & & OH \\ | & & | \\ B & & B \\ \diagup \quad \diagdown & \diagup \quad \diagdown \\ HO & O & OH \end{array}$$

$$2\,H_3BO_3 \longrightarrow H_4B_2O_5 + H_2O$$

and consists also of a dehydrated form of orthoboric acid.

$$2\,H_3BO_3 \longrightarrow H_4B_2O_5 + H_2O$$

Pyroacids are related to ortho acids by the elimination of a water molecule between two molecules of the ortho acid. In all three of the boric acids, the geometry of the boron atoms is the same—trigonal planar.

Aluminum Hydroxide □ Aluminum hydroxide occurs naturally as the minerals bayerite and hydrargillite. Like beryllium hydroxide, aluminum hydroxide is amphoteric. Thus, although it is insoluble in water, it is soluble in and reacts with both acidic and basic solutions. Aluminum ions, like beryllium ions in water solution, do not exist as free Al^{3+} ions, but instead as hexahydrated ions. The Al^{3+} ion, being larger than the Be^{2+} ion is able to accommodate six water molecules instead of four. The reactions of aluminum hydroxide [$Al(H_2O)_3(OH)_3$] with acids and bases are discribed by the following equations.

15-4. CARBON AND SILICON

Carbon is one of the many elements that exist in more than one physical form. Pure carbon is found in nature in the form of graphite and diamond, each having its own set of properties. Diamond is extremely hard (in fact, the hardest substance

							H	

								He
Li	Be	B	C	N	O	F	Ne	
Na	Mg	Al	Si	P	S	Cl	Ar	

known), and is a nonconductor of electricity. Graphite is soft, unctuous, useful as a lubricant, and an excellent conductor of electricity. The properties of diamond and graphite can be quite satisfactorily rationalized by their molecular structures.

Both diamond and graphite contain vast networks of covalently bonded atoms with no easily described ends. A portion of the structure of diamond is shown.

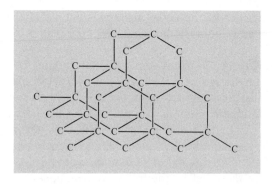

This tetrahedral structure was determined by W. H. Bragg and W. L. Bragg in 1914 using the method of X-ray diffraction analysis which they developed. In this network each atom is bonded to four other atoms by bonds of equal lengths (1.54 Å) and all the available electrons are used in bonding. The result is an extremely rigid, completely interconnected framework and a relatively high density. A perfect diamond, therefore, is an example of a single molecule. Breakage of such a diamond requires the rupture of countless covalent bonds, hence its hardness. The localization of all the electrons accounts for the inability to conduct an electric current.

Graphite has a quite different structure which consists of planar sheets of atoms. X-ray diffraction analysis has shown that each sheet contains atoms that

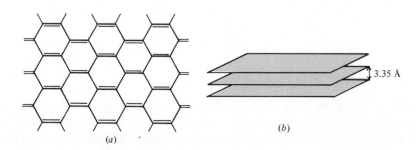

(a)

(b)

are each bonded to three other atoms. The C-C bond distances within a sheet are all equal (1.42 Å) and the bond angles are 120°. The distance between sheets is 3.35 Å. The normal double bond length between carbon atoms, such as in ethylene, is known to be 1.33 Å.

The only way to explain the equivalence of all the bond lengths and bond angles is to assume that all the bonds are identical. If they are all single bonds, then there are extra unused electrons (one per carbon atom). Since there is no evidence that these electrons are unpaired, we must assume that they are either paired and accumulated on alternate carbon atoms or that they are employed in double bonds as shown above in (a). Neither of these alternatives is satisfactory because there is no reason why an electron pair would be located on one carbon atom giving it a net negative charge whereas on an adjacent carbon atom there would be a deficiency of electrons with the consequent unit positive charge. The accepted explanation is that the orbitals that contain these "extra" electrons are able to overlap with orbitals on all three adjacent atoms.

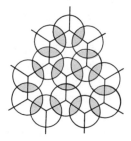

Circles represent the orbitals which contain the "extra" electrons

Such an arrangement accounts for all the observed parameters: Equal orbital overlap in several directions results in weaker than normal C-C double bonds, the bond is intermediate in length between a single and a double bond. An alternate way of expressing the structure of graphite is as a resonance hybrid of the large number of structures that can be written. Some of these contributing structures for a small portion of a graphite molecule are as follows.

Regardless of the method of representation, the result is symmetrical overlap over all bonds and therefore equal bond lengths. Moreover, the extended overlap results in an extended delocalization of the "extra" electrons over the entire structure. This high mobility of electrons results in good electrical conductivity. The large separation between layers does not allow any significant amount of bonding between them. Hence the layers are able to slide past one another with little resistance which accounts for the lubricating properties of graphite. The graphite structure is unique among the elements.

Silicon, having the same valence shell electron configuration as carbon, might be expected to behave like carbon. In fact, it bears only a superficial resemblance. The pure element has the same crystal structure as diamond. Although the simple compounds of silicon have structures similar to simple carbon compounds, as will be shown below, their properties are quite different. Unlike carbon, silicon does not complete its electron octet by forming a small number of multiple bonds; instead, only single bonds are formed. Carborundum, one of the hardest substances known, is a compound of silicon with carbon, SiC, also called silicon carbide. It also has the diamond structure and exists in several physical forms.

Carbon Hydrides ☐ The hydrides of carbon, called **hydrocarbons,** can exist in literally an infinite variety as has been discussed in Chapter 14. Carbon, in this sense, is unique among the elements. The nearly infinite variation in structure that results has been discussed in the preceeding chapter.

Silicon Hydrides ☐ The hydrides of silicon are known from SiH_4 to Si_6H_{14}. Attempts to prepare the higher members have always met with decomposition to the lower ones. Heating the higher hydrides, for example, Si_6H_{14}, leads to decomposition to the lower members. The silicon hydrides are gases that are spontaneously flammable in air. Like their carbon analogs, the silicon hydrides possess tetrahedral geometry about the silicon atoms.

Carbon Halides ☐ The halides of carbon offer no surprises. They are all tetrahedral. In addition to the simple members—CF_4, CCl_4, CBr_4, CI_4—longer chain compounds are also known.

Halocarbons are generally toxic and should always be handled accordingly. Of the simple members, all the combinations of halogens and hydrogens with carbon are known

$$CH_3X \qquad CH_2X_2 \qquad CHX_3 \qquad CX_4$$

in which X = F, Cl, Br, or I.

Silicon Halides ☐ The silicon halides are also known to have tetrahedral structures exemplified by $SiCl_4$.

$$\begin{array}{c} Cl \\ | \\ Cl\diagup \overset{\displaystyle Si}{} \text{---} Cl \\ Cl \qquad Cl \end{array}$$

In addition, because of empty d orbitals, silicon can form additional bonds. This only happens with fluorine and oxygen, elements of the greatest electron-attracting power. The resulting hexafluoride, for example, potassium fluorosilicate (K_2SiF_6), is octahedral in geometry (see Chapter 14).

$$\left[\begin{array}{c} F \\ F \diagdown \; | \; \diagup F \\ Si \\ F \diagup \; | \; \diagdown F \\ F \end{array} \right]^{2-}$$

The Si-F bonds in SiF_6^{2-} are quite ionic in character.

Carbon Oxides ☐ Carbon forms three oxides: carbon monoxide, CO; carbon dioxide, CO_2; and carbon suboxide, C_3O_2. The least reactive of the three is the dioxide and the most reactive is the suboxide. These substances are interesting in their structures.

Carbon monoxide presents an interesting structural problem as its formula cannot be expressed simply by one structure. Several resonance structures are required to describe the molecule, and carbon monoxide has a structure that is a hybrid of the following

$$:C{=}\overset{\cdot\cdot}{\underset{}{O}}: \qquad :\overset{\cdot\cdot}{\underset{-}{C}}{-}\overset{\cdot\cdot}{\underset{+}{O}}: \qquad \overset{-}{:C}{\equiv}\overset{+}{O}: \qquad \overset{+}{:C}{-}\overset{\cdot\cdot}{\underset{\cdot\cdot}{O}}\overset{-}{:}$$
$$\quad\text{I}\qquad\qquad\quad\text{II}\qquad\qquad\quad\text{III}\qquad\qquad\quad\text{IV}$$

The sum of the contributions of structures II and III must be nearly equal to that of structure IV because the dipole moment* of the molecule is found to be nearly zero.

Carbon monoxide, unlike the dioxide, is very toxic. It reacts with the iron atoms in the blood to form a stable Fe-CO bond that prevents the iron atoms from becoming attached to oxygen molecules. Iron in the blood functions by transporting breathed oxygen to all the cells in the body. Thus death from carbon monoxide poisoning results from lack of oxygen.

Carbon dioxide is linear, as predicted on the basis of electron pair repulsion theory. The electronic structure can be represented as

$$\overset{\frown\; 1.163\ \text{Å}}{\overset{\cdot\cdot}{:}O{=}C{=}\overset{\cdot\cdot}{O}\overset{\cdot\cdot}{:}}$$

although the bond length, which is intermediate between a double bond (1.22 Å

in $\begin{array}{c} H \\ \diagdown \\ \; C{=}\overset{\cdot\cdot}{O}: \\ \diagup \\ H \end{array}$) and a triple bond (1.13 Å in carbon monoxide), suggests that polar

structures like $\overset{+}{:}O{\equiv}C{-}\overset{\cdot\cdot}{\underset{\cdot\cdot}{O}}\overset{-}{:}$ contribute to the resonance hybrid.

Carbon dioxide dissolves to a slight extent in water to form carbonic acid, H_2CO_3, of structure

---

*__Dipole moment__ (μ) is defined by the equation $\mu = el$ where e = the charge due to an unsymmetrical displacement of electrons in a bond and l = the distance of separation of the charges. Dipole moment thus is one measure of the ionic character of a covalent bond.

$$\overset{\displaystyle O}{\underset{\displaystyle H-O \qquad O-H}{\overset{\parallel}{C}}}$$

This is a very weak acid and accounts for the taste of carbonated water. It dissociates to a slight extent to give bicarbonate ions. HCO_3^- which, to an even slighter extent, dissociate to form carbonate ions, CO_3^{2-}.

$$H_2O + H_2CO_3 \rightleftharpoons H_3O^+ + HCO_3^-$$

$$H_2O + HCO_3^- \rightleftharpoons H_3O^+ + CO_3^{2-}$$

The carbonate ion must be written as a resonance hybrid of three exactly equivalent forms.

Carbon suboxide, C_3O_2, is more reactive than the other two oxides. It is also a gas at room temperature (b.p. 6°C). Electron diffraction studies indicate that carbon suboxide molecules are linear

$$O=C=C=C=O$$

with bond lengths 1.28 Å and 1.16 Å.

Again shorter than expected bond lengths for C=C and C=O suggest contributions from resonance structures involving triple bonds.

$$:\ddot{O}=C=C=C=\ddot{O}: \qquad {}^+:O\equiv C-C\equiv C-\ddot{O}:^-$$

Silicon Oxide ☐ Only one stable oxide of silicon is known, $(SiO_2)_n$, of which several crystalline modifications exist. In all of them four oxygen atoms are arranged tetrahedrally around each silicon atom. Unlike carbon, silicon does not form double bonds. Quartz is a commonly occurring form of silicon dioxide. Of course, ordinary sand is an impure form of silicon dioxide.

Carbon Hydroxide ☐ Generally, two or more hydroxy groups do not exist attached to the same carbon atom unless a strongly electron-attracting atom is also attached. The usual case results in the loss of water.

$$\overset{\displaystyle O-H}{\underset{\displaystyle H}{\overset{\mid}{\underset{\mid}{H-C-O-H}}}} \longrightarrow \overset{\displaystyle O}{\underset{\displaystyle H}{\overset{\diagup\parallel}{\underset{\mid}{H-C}}}} + H_2O$$

Carbonic acid is an example of a compound in which two hydroxy groups are attached to the same carbon atom. In this case, the double bonded oxygen atom stabilizes the structure.

Silicon Hydroxide ☐ Hydroxides of silicon are readily formed by the reaction of silicon halides with water. The products are usually complex molecules consisting of long chains of $(Si-O)_n$ units.

15-5. NITROGEN AND PHOSPHORUS

Nitrogen gas, which forms about 80% of the air, is a very stable, almost inert, colorless, odorless gas. It is composed of diatomic molecules, N_2. Consistent with the electronic configuration, $1s^2 2s^2 2p^3$, two nitrogen atoms can readily form a triple covalent bond, $:N{\equiv}N:$, of length 1.097 Å.

H

							He
Li	Be	B	C	N	O	F	Ne
Na	Mg	Al	Si	P	S	Cl	Ar

Because of the unshared pair of electrons in tricovalent nitrogen compounds, some interesting geometries arise, as shown in Table 15-1. Note that, as expected, the geometries are identical to those of the corresponding carbon compounds showing that the identity of the atom is of little or no importance in determining geometry when the outer electronic configurations are the same. It might be noted that examples of carbon compounds in which a carbon atom has an unshared

Table 15-1. Stereochemistry of Nitrogen Compounds

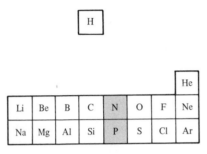

Atomic Unit	Geometry	Examples	Example of Carbon Analog
N̈---	trigonal pyramidal	$:NH_3, :NF_3$	—
N⁺---	tetrahedral	$\overset{+}{N}H_4$	CH_4
N⁺	trigonal planar	$HO-N\overset{\displaystyle O}{\underset{\displaystyle O}{}}$	$H-C\overset{\displaystyle O}{\underset{\displaystyle H}{}}$
—N̈	angular	$HO-\overset{..}{N}\diagdown O$	—
=N⁺= or —N⁺≡	colinear	$:\overset{-}{N}=\overset{+}{N}=\overset{..}{\underset{..}{O}}:$	$:\overset{..}{O}=C=\overset{..}{O}:$ $H-C{\equiv}C-H$

electron pair are less common than in the nitrogen analogs. This is due to the greater electron affinity of nitrogen compared to carbon; that is, such a carbon atom would be extremely reactive.

Phosphorus, immediately beneath nitrogen in the periodic table, has, of course, the same valence electronic configuration as nitrogen and might therefore be expected to exhibit similar bonding. Phosphorus, however, has the same structure as nitrogen gas, P_2, only at temperatures above 2000°C. Below 800°C, P_2 is converted to P_4, white (or yellow) phosphorus. Molecules of white phosphorus are tetrahedral with bond angles of 60°:

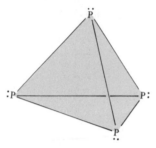

White phosphorus has a boiling point of 280°C, a melting point of 44.1°C, and a density of 1.82 g/cm³. It is very reactive—ignites spontaneously in air—and is therefore stored under water.

At a pressure of approximately 1.12×10^5 atm phosphorus is converted into "black phosphorus" (P_∞), a much more compact structure whose density is 2.69 g/cm³. The structure of black phosphorus contains parallel sheets of atoms in a zig-zag arrangement. Each sheet is joined to the one above and below in such a way that each phosphorus atom is bonded singly to three other atoms.

Still another modification is known which forms by treating phosphorus with ultraviolet light or by heating to 260°C. This form is known as "red phosphorus" and has a density of 2.35 g/cm³. Its structure is not known.

The failure of phosphorus to form a stable diatomic molecule at ordinary temperatures might be explained by its larger atomic size* and the consequent distortion that would result from the formation of a triple bond with one other atom.

Nitrogen Hydrides □ The principle hydride of nitrogen is ammonia, NH_3. Its structure is trigonal pyramidal (Table 15-1) as evidenced by its dipole moment of 1.44 debye. (The debye unit $= 10^{-18}$ esu cm.) The presence of the unshared

electron pair confers on ammonia the ability to react with electron-deficient species to form coordinate covalent bonds. Its reaction with a hydrogen ion is typical.

$$H^+ + :NH_3 \longrightarrow NH_4^+$$

Ammonia is a pungent gas at ordinary temperatures (boiling point $-30.9°C$). This boiling point is high compared with that of PH_3, a fact that is explained by the existence of hydrogen bonds in ammonia (Chapter 13).

Another hydride of nitrogen is the liquid hydrazine, N_2H_4, which has been used as a rocket fuel. The structure is again pyramidal at each nitrogen atom, and because of repulsions between unshared electron pairs, orientation of the NH_2 groups about the N-N bond is hindered. Hydrazine, too, is highly asso-

ciated by hydrogen bonding as evidenced by its unusually high boiling point, 114°C, and its high viscosity.

The compound dimethyl hydrazine has been used extensively as a rocket fuel.

Phosphorus Hydrides □ In addition to phosphine, PH_3, there are a number of higher hydrides. P_2H_4 decomposes in the presence of water to even higher complex hydrides, for example, $P_{12}H_6$. The structure of phosphine is similar to that of ammonia except for smaller H-P-H angles.

*Recall that on moving down a column in the periodic table, atoms have the same number of bonding electrons but larger numbers of inner shell electrons. The consequence of this is that these heavier atoms cannot approach each other as closely and therefore the amount of orbital overlap required in multiple bonding is reduced.

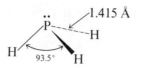

Nitrogen Halides ☐ Of the nitrogen trihalides only the bromide is unknown. All have the pyramidal structure of ammonia. The trifluoride, NF_3, is quite stable when pure. The low dipole moment (0.2 debye) of NF_3 suggests that the molecule might be planar. Instead it is found to be pyramidal with the N-F dipoles cancelling the dipole caused by the orbital containing the unshared electron pair.

NXYZ, should be chiral and therefore resolvable into optically active isomers. Such is not the case because of the ease of inversion.

Difluorodiazine, N_2F_2, is a stable colorless gas which may be separated into cis-trans isomers.

trans   cis

Tetrafluorohydrazine, N_2F_4, has also been prepared.

Phosphorus Halides ☐ Like the nitrogen trihalides, the phosphorus trihalides are all pyramidal and present no stereochemical surprises. All of the trihalides are known. Table 15-2 contains some data on bond lengths and bond angles of the phosphorus halides.

Phosphorus has the ability to form more than four bonds and thus go beyond an octet of electrons due to the availability of $3d$ orbitals which can accept electrons to form additional bonds. Although as many as nine bonds could be formed in this way, an upper limit of six is never exceeded because of the crowding that would result.

Table 15-2. Bond Lengths and Bond Angles of the Phosphorus Trihalides and Pentahalides

Compound	P-X Distance (Å)	X-P-X Angle
PF_3	1.535	100°
PCl_3	2.043	100°6′
PBr_3	2.18	$101\frac{1}{2}$°
PI_3	2.43	102°
PF_5	1.57 (all equal)	same as PCl_5
PCl_5	$\begin{Bmatrix} 2.04 \\ 2.19 \end{Bmatrix}$	(see structure below)

Phosphorus pentafluoride is a colorless gas (b.p. $-84.5°C$). It is a symmetrical molecule, all of whose bond lengths and bond angles are the same. Phosphorus pentachloride, however, is unusual. It is a yellowish solid that fumes in moist air. It does not melt but instead sublimes at 162°C. It is a very reactive substance that decomposes when heated and yields phosphorus trichloride and chlorine gas.

$$PCl_5 \xrightarrow{\text{heat}} PCl_3 + Cl_2$$

It is also very vigorously attacked by water. In the vapor state phosphorus pentachloride has the structure shown, in which the vertical P-Cl bonds are longer and therefore weaker than the horizontal ones. The three horizontal Cl atoms and the P atoms lie in one plane. The solid form of phosphorus pentachloride probably does not contain covalent molecules as shown below for the vapor.

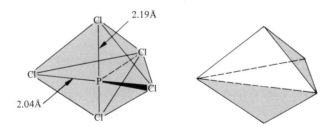

Conductivity measurements in solvents that are inert toward phosphorus pentachloride indicate an ionic structure. The ions are believed to be tetrahedral PCl_4^+ and octahedral PCl_6^- which result from

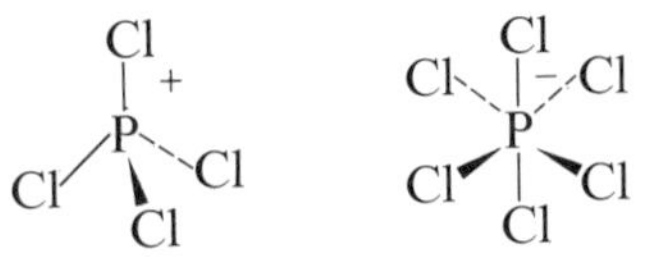

Cl
| +
P
Cl   Cl
Cl

PCl_4^+

Cl
|
Cl   P   Cl
Cl   Cl
Cl

PCl_6^-

the transfer of a chloride ion from one phosphorus pentachloride molecule to another.

The molecules, P_2Cl_4 and P_2I_4 have also been prepared. The chloride is a very reactive liquid that fumes in the air. The iodide is a reactive solid.

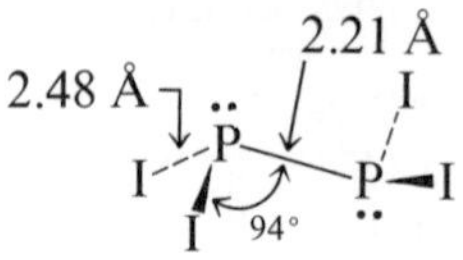

$$2.48 \text{ Å} \qquad 2.21 \text{ Å}$$

$$I\text{---}P\overset{}{\underset{I}{\diagup}}\!\!P\text{---}I \quad 94°$$

Nitrogen Oxides ☐ There are five oxides of nitrogen, all of which are well known. All but one are gases at ordinary temperatures. Nitrogen pentoxide is a white solid that melts at 30°C and decomposes at 47°C.

Nitrous oxide, N_2O, is a colorless gas that is used as a general anesthetic and is called laughing gas because of this interesting side effect. Electron diffraction studies have shown it to have a linear structure.

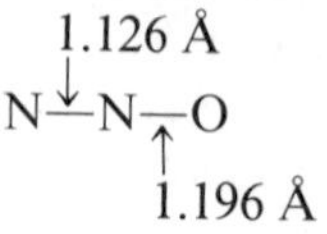

$$1.126 \text{ Å}$$
$$N\text{---}N\text{---}O$$
$$1.196 \text{ Å}$$

Equal contributions from the two electron structures

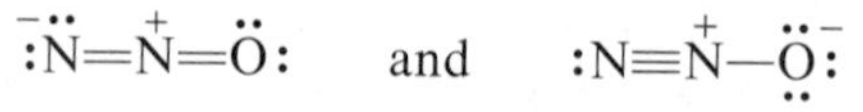

$$:\overset{..}{\overset{-}{N}}=\overset{+}{N}=\overset{..}{O}: \qquad \text{and} \qquad :N\equiv\overset{+}{N}-\overset{..}{\overset{-}{O}}:$$

should produce a hybrid structure with zero dipole moment because the dipole moments of the two structures above are opposed. This is borne out by the actual dipole moment of 0.17 debye.

Nitric oxide, NO, is a colorless gas that becomes blue in the liquid and solid phases. As predicted by its odd number of electrons it is paramagnetic* in the gas phase in which it exists as simple NO molecules that are resonance hybrids of the two structures

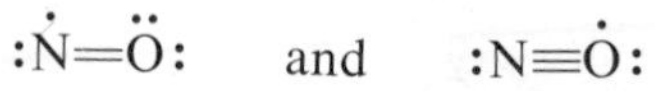

$$:\overset{.}{N}=\overset{..}{O}: \qquad \text{and} \qquad :N\equiv\overset{.}{O}:$$

In the liquid and solid phases nitric oxide is diamagnetic,† which suggests that it is associated to form clusters like the following

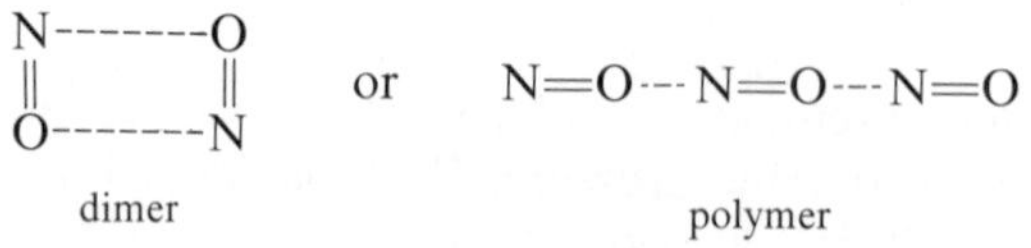

$$\begin{array}{ccc} N\text{-------}O \\ \| \qquad\qquad \| \\ O\text{-------}N \end{array} \qquad \text{or} \qquad N{=}O\text{---}N{=}O\text{---}N{=}O$$

dimer   polymer

Nitrogen dioxide, NO_2, is also an odd-electron molecule. It is a brown gas that readily associates at lower temperatures to form dinitrogen tetroxide, N_2O_4. The tetroxide-dioxide equilibrium favors the dioxide form at temperatures above 150°C.

$$N_2O_4 \rightleftharpoons 2\ NO_2$$

---

*Paramagnetic: magnetic; capable of being attracted by a magnet.
†Diamagnetic: nonmagnetic; incapable of being attracted by a magnet.

The dioxide molecule is angular and may be thought of as a hybrid of the following structures.

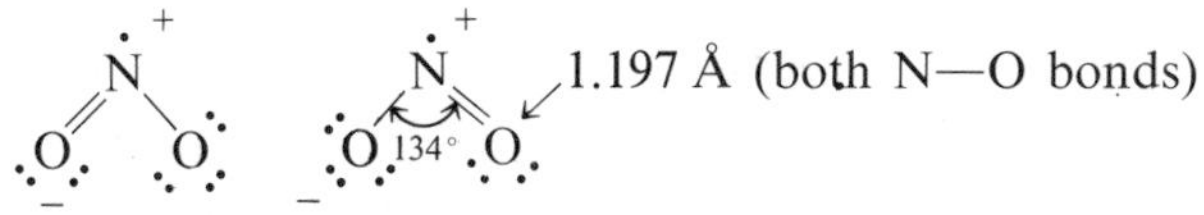

1.197 Å (both N—O bonds)

The tetroxide is a colorless gas whose structure is given by

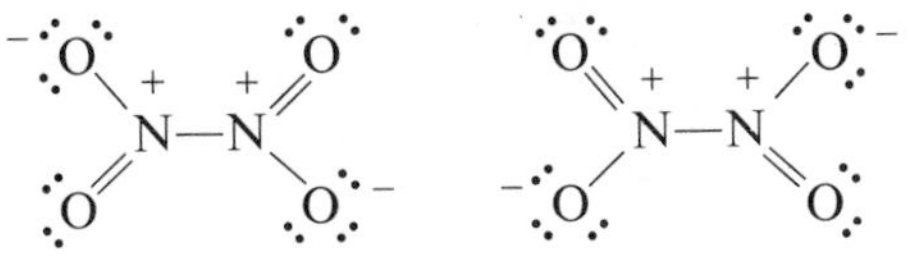

N-O distance = 1.17 Å
N-N distance = 1.64 Å
N∠O O angle = 126°

The N-N bond is extraordinarily long compared with that in hydrazine (1.45 Å), partially explaining the ease with which N_2O_4 decomposes.

Dinitrogen trioxide, N_2O_3, probably exists only in the liquid and solid states. In the vapor state it is almost completely dissociated into NO and NO_2

$$N_2O_3 \rightleftharpoons NO + NO_2$$

as seen from the brown color produced. In the condensed phases N_2O_3 is blue. The structure of N_2O_3 in the crystalline form has proved too complex to be analyzed.

The structure of gaseous dinitrogen pentoxide, N_2O_5, has not been accurately determined. In the crystalline form, it consists of equal numbers of planar NO_3^- ions and linear NO_2^+ ions.

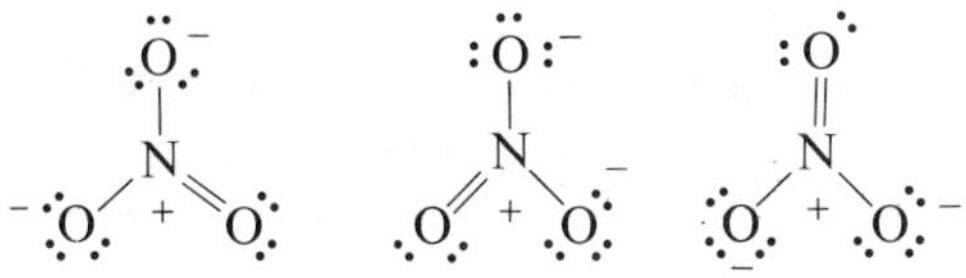

N-O distance = 1.24 Å

Dinitrogen pentoxide is the anhydride of nitric acid, that is, its reaction with water produces nitric acid.

$$N_2O_5 + H_2O \longrightarrow 2\ HONO_2$$

Phosphorus Oxides ☐ Two stable oxides of phosphorus are known, the tri-oxide, P_4O_6, and the pentoxide, P_4O_{10}. The names are a carryover from the time when these compounds were believed to be P_2O_3 and P_2O_5, respectively.

The structures of these oxides in the vapor phase can be visualized by placing oxygen atoms beyond the midpoints of the edges of the P_4 tetrahedron. Similarly, the pentoxide contains in addition, one oxygen atom attached to each phosphorus atom.

237

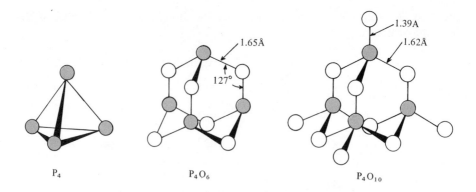

The P-O bond length of 1.65 Å in P_4O_6 corresponds to a considerable amount of double bond character since a normal P-O single bond is 1.71 Å in length. Also the angle is closer to that of a double bonded phosphorus atom

$$-\overset{|}{P} \underset{\diagdown}{\diagdown} \ \ (125°)$$

than to the tetrahedral single bond angle of 109.5°. The structure of crystalline P_4O_6 is not yet known.

The pentoxide exists in several solid forms. In one form, rhombohedral, there are discrete P_4O_{10} molecules as in the vapor. The most stable form, orthorhombic, consists of an infinite array of PO_4 tetrahedra each of which shares three oxygen atoms with nearest neighbor PO_4 tetrahedra. The P-O bond distances in molecular P_4O_{10} also suggest considerable double bond character.

Nitrogen Hydroxides ☐ There are no hydroxides as such of nitrogen. The most important substances that can be considered as hydroxides or hydrated oxides are nitric acid, $HONO_2$, and nitrous acid, HONO. These are the hydrates of N_2O_5 and N_2O_3, respectively.

$$N_2O_3 + H_2O \longrightarrow 2 \ HONO$$

Nitric acid is a strong acid because it dissociates completely in water solution to form nitrate ions and hydronium ions.

$$HONO_2 + H_2O \longrightarrow H_3O^+ + NO_3^-$$

The molecule of nitric acid has the electron structure

As contrasted with the hydroxides of the metals, nitrogen is so strongly electron withdrawing that it is the H-O bond that dissociates rather than the O-N bond.

Nitrous acid is also an acid, but a weaker one; that is, dissociation is incomplete in water solution.

$$HONO + H_2O \longrightarrow H_3O^+ + NO_2^-$$

The molecule of HONO is angular as is the nitrite ion, NO_2^-, because of repulsion due to the unshared electron pair on nitrogen.

nitrous acid   nitrite ion

Phosphorus Hydroxides ☐ The reaction of the trioxide and the pentoxide separately with water produces respectively, phosphorous and phosphoric acids.

$$P_4O_6 + 6\,H_2O \longrightarrow 4\,H_3PO_3$$
orthophosphorous acid

$$P_4O_{10} + 6\,H_2O \longrightarrow 4\,H_3PO_4$$
orthophosphoric acid

Orthophosphorous acid contains only two H atoms that can dissociate in water solution. The third is attached directly to the P atom.

In orthophosphoric acid, H_3PO_4, all three H atoms are attached to O atoms, and all are dissociable in water solution. The phosphorus bonds are arranged nearly tetrahedrally.

This acid also exists in two other forms—the pyro and meta acids.

$$2\,H_3PO_4 \longrightarrow H_4P_2O_7 + H_2O$$
pyrophosphoric acid

The pyrophosphoric acid molecule has the structure

$$\text{O} \overset{1.63\,\text{Å}}{\underset{\text{HO}}{\text{P}}} \overset{\text{O}}{\underset{134°}{\text{P}}} \overset{1.47\,\text{Å}}{\underset{\text{O—H}}{\text{P}}}$$

Metaphosphoric acid is a more dehydrated form of phosphoric acid.

$$H_3PO_4 \longrightarrow HPO_3 + H_2O$$

It is a thick syrup or glassy substance that does not exist as molecules of HPO_3 but, instead, as a complex mixture of highly associated molecules of varying lengths such as

There is yet another oxyacid of phosphorus called hypophosphorous acid, H_3PO_2. Its structure is not known with certainty, but it has one hydrogen atom than can dissociate in water solution and probably has the structure

15-6. OXYGEN AND SULFUR

The structure of oxygen gas might be expected to be similar to that of nitrogen except for having a double instead of a triple bond. The bond between the two oxygen atoms in O_2 might be expected to be

$$\ddot{\text{O}}::\ddot{\text{O}}$$

The properties of oxygen do not conform to this structure, however, because, unlike nitrogen which is almost completely inert to chemical combination, oxygen

Li	Be	B	C	N	O	F	Ne
Na	Mg	Al	Si	P	S	Cl	Ar

is fairly reactive. Moreover, oxygen has a distinct blue color which is noticeable in the pure liquid. Most small molecules with double bonds are colorless. Hence, it seems unlikely that oxygen would have the structure shown. An alternative structure has a single bond in which two electrons remain unpaired.

$$:\overset{\cdot\cdot}{\underset{\cdot\cdot}{O}}{-}\overset{\cdot}{\underset{\cdot\cdot}{O}}:$$

Substances whose molecules contain unpaired electrons usually exhibit color, and, at least for simple molecules, the presence of color is usually taken as evidence that there are unpaired electrons. Oxygen also exhibits the property of para-magnetism, further confirming the presence of unpaired electrons.

The problem of structure is not so easily disposed of, however, because oxygen has a bond length of 1.208 Å, which corresponds closely to that of a double bond. Two theoretical explanations are in current use. The molecular orbital method predicts that two electrons remain unpaired because they are in orbitals of equal energy (Hund's rule); at the same time, a double bond is predicted. The other theory is due to Pauling and employs two "three-electron" bonds and an electron pair bond to account for the observations. A three-electron bond between two atoms of element A is considered to be a resonance hybrid of the two structures

$$A:.A \quad \text{and} \quad A.:A$$

in which one orbital of each atom is involved. Neither of these structures can be expected to form a stable bond. Resonance between the two structures, how-ever, results in considerable stabilization. In the case of oxygen, the resonance structures would all have one ordinary electron-pair covalent bond and two three-electron bonds.

$$:\overset{\cdot\cdot}{\underset{\cdot}{O}}{-}\overset{}{\underset{\cdot\cdot}{O}}: \qquad :\overset{}{\underset{\cdot\cdot}{O}}{-}\overset{\cdot\cdot}{\underset{\cdot}{O}}:$$

Pauling has introduced the following notation to signify the two three-electron bonds in oxygen.

$$:O{:\!:\!:}O:$$

Although these theoretical models are quite different, they account equally well for all the observed properties of oxygen.

Ozone ☐ There is another modification of elemental oxygen that is produced when O_2 is treated with ultraviolet radiation or with an electrical discharge. Under these conditions ozone, O_3 is formed.

$$3\,O_2 \longrightarrow 2\,O_3$$

The above conversion in the outer layers of the atmosphere results in the absorp-tion of much of the sun's ultraviolet radiation, which would otherwise be very destructive to life on the earth's surface. The ozone molecule is bent with an angle of 117°.

$$\overset{\displaystyle O}{\underset{\underset{117°}{O\,\diagdown\,O}}{\diagup\ \diagdown \leftarrow 1.28\ \text{Å}}}$$

Its bond lengths suggest less double bond character than O_2 and may be represented as a resonance hybrid of the two structures

Oxygen reacts with all of the elements except, of course, the noble gases. With the very reactive alkali metal cesium, the **superoxide** ion is formed.

$$Cs + O_2 \longrightarrow CsO_2$$

The superoxide ion is an oxygen molecule that has acquired one electron. Its bond length is 1.30 Å suggesting considerably more single bond character than O_2.

$$:\overset{..}{O}-\overset{..}{O}: \underset{1e^-}{\longrightarrow} [:\overset{..}{O}-\overset{..}{O}:]^-$$

According to Pauling's notation it may be described as

$$[:\overset{..}{O}\cdots\overset{..}{O}:]^-$$

In any event, the superoxide ion is paramagnetic since it has one unpaired electron.

Reaction of oxygen with sodium, on the other hand produces sodium peroxide.

$$2\,Na + O_2 \longrightarrow Na_2O_2$$

In the formation of the **peroxide** ion, two electrons have been added to an oxygen molecule. Thus this ion is not paramagnetic. Its bond length of 1.49 Å indicates a single O-O bond as found in H_2O_2.

$$:\overset{..}{O}-\overset{..}{O}: + 2\,e^- \longrightarrow [:\overset{..}{O}-\overset{..}{O}:]^{2-}$$

In contrast to sodium and cesium, lithium forms neither a peroxide nor a superoxide. Instead, the oxide ion, O^{2-}, is formed.

$$4\,Li + O_2 \longrightarrow 2\,Li_2O$$

In the formation of the oxide ion, the oxygen molecule acquires four electrons, thus rupturing the bond, entirely.

$$:\overset{..}{O}-\overset{..}{O}: + 4\,e^- \longrightarrow 2\,[:\overset{..}{O}:]^{2-}$$

The oxide ion is diamagnetic. The oxides of potassium, rubidium, and cesium can be formed by heating oxygen with an excess of the metal. With the exception of barium, which reacts with oxygen to form the peroxide, BaO_2, all the other metals react with oxygen to form the oxides.

In many cases, especially the more reactive metals of groups I and II, the oxides react with water spontaneously to form the hydroxides which are quite basic.

$$Na_2O + H_2O \longrightarrow 2\,NaOH$$

$$CaO + H_2O \longrightarrow Ca(OH)_2$$

Oxygen Hydrides □ Besides water, oxygen forms one other hydride—hydrogen peroxide, H_2O_2. Hydrogen peroxide is a very explosive liquid when nearly pure, and has a boiling point of 150°C. Its melting point is -0.4°C. Commercial hydrogen peroxide is normally sold as 3%, 10%, and 30% solutions, although for industrial purposes, 90% solutions are prepared.

Decomposition of dilute hydrogen peroxide solutions occurs gradually with the evolution of oxygen gas

$$2\,H_2O_2 \longrightarrow 2\,H_2O + O_2$$

The molecule has an unusual geometry brought about by mutual repulsion of nonbonded electron pairs. Such a configuration accounts for the relatively

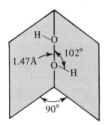

high dipole moment of 2.13 debye, which would be difficult to explain on the basis of a planar structure such as

$$\text{H—O}\diagdown_{\text{O—H}}$$

Sulfur □ Sulfur exists in several solid modifications, the most common of which is the yellow orthorhombic crystalline form, density 2.06 g/cm³. A monoclinic modification also exists which is stable above 95°C. It has a density of 1.96 g/cm³. At room temperature, monoclinic sulfur spontaneously changes into the orthorhombic modification (Figure 15–1). Both crystalline forms consist of molecules of eight atoms, S_8. The molecular configuration of sulfur is different from all the preceeding examples in that it forms eight-membered rings. Structure determinations have shown that the rings are not planar, but instead are puckered as shown in structure B below with bond angles of 108° and bond lengths of 2.04 Å. The unpuckered planar configuration A requires bond angles of 135°, considerably in excess of the stable tetrahedral angle of 109.5°. The situation is relieved by puckering.

(A)   (B)

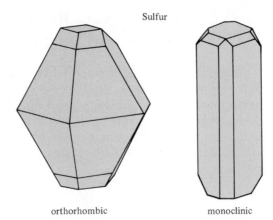

Sulfur

orthorhombic   monoclinic

Figure 15-1. Crystalline modifications of sulfur.

Sulfur has another unusual property: On heating, orthorhombic sulfur melts at 112°C. Further heating results in an increased viscosity rather than a decreased viscosity as is observed with other liquids. The explanation of this unusual viscosity behavior lies in the conversion of the S_8 rings into very long chains, S_∞, at elevated temperatures. Such long chains become entangled and result in the observed viscosity increase. At higher temperatures the viscosity decreases as the sulfur chains are cleaved. When the temperature of the vapor exceeds 1200°C, decomposition to single atoms occurs.

Of the element pairs nitrogen-phosphorus and oxygen-sulfur, the first of each pair is gaseous and the second of each pair is solid. Can we explain the differences in physical state in terms of their structures? P_4, P_∞, and S_8 molecules are much heavier than the others, hence the energy required to move them is greater than for the lighter diatomic and triatomic molecules. The process of boiling must involve imparting sufficient kinetic energy to the molecules so that their random motion becomes so violent that they shoot out into the air space above the liquid. This motion must be violent enough not only to propel the molecules into space, but also to overcome the mutual attractive forces among the molecules. Clearly the greater this attraction, the more energy is required, hence the higher the boiling point. In summary, boiling point is determined by two factors: Energy must be imparted to the molecules (1) to cause them to move very rapidly and (2) to cause them to become separated from each other so that they may leave the body of liquid to become vapor. The physical state of phosphorus and sulfur can best be explained by the greater molecular weights.

Sulfur Hydride □ Sulfur forms only one hydrogen sulfide, the colorless gas, H_2S. It boils at −62°C and it melts at −83°C. It has the characteristic odor of rotten eggs. Not commonly known is the fact that hydrogen sulfide is toxic to an extent comparable with hydrogen cyanide, HCN. Hydrogen sulfide dissolves to a moderate extent in water to produce a weakly acidic solution.

$$H_2S + H_2O \rightleftharpoons H_3O^+ + HS^-$$

Sulfur Oxides □ Two oxides of sulfur are common—sulfur dioxide, SO_2, and sulfur trioxide, SO_3. The dioxide is a pungent colorless gas formed directly on

244

burning sulfur in air. Sulfur dioxide liquifies on cooling to $-10°C$. Its relatively high boiling point makes it an interesting solvent for the study of many chemical reactions. The sulfur dioxide molecule is angular as suggested by its dipole moment of 1.6 debye. The angularity is also predicted by the unshared electron pair on sulfur. The S-O bond lengths are identical at 1.43 Å. Resonance structures are

Sulfur trioxide is highly reactive and is formed through the reaction of sulfur dioxide with oxygen.

$$2 SO_2 + O_2 \longrightarrow 2 SO_3$$

The reaction is very slow at ordinary temperatures and is therefore conducted at between 400 and 700°C commercially. A catalyst such as V_2O_5 or Pt is required even at these temperatures in order for the reaction to proceed at a reasonable speed (see Chapter 29). Under ordinary conditions, SO_3 exists as a white solid that sublimes readily. Molecules in the gas phase are planar trigonal, and can be described as a resonance hybrid of three structures.

Sulfur Hydroxides □ Hydroxides of the nonmetals were stated above to be acidic. This is quite true for sulfur. The hydroxides are formed by the reaction of the respective oxides with water.

$$SO_2 + H_2O \longrightarrow H_2SO_3$$
$$SO_3 + H_2O \longrightarrow H_2SO_4$$

Sulfurous acid, H_2SO_3, is not stable in the pure state. It exists only in solution in which it dissociates.

$$H_2SO_3 + H_2O \longrightarrow H_3O^+ + HSO_3^-$$

It is a relatively weak acid compared to sulfuric acid, H_2SO_4. The bisulfite ion, HSO_3^-, is also capable of dissociation, but to a lesser degree, to give sulfite ion, SO_3^{2-}.

$$HSO_3^- + H_2O \longrightarrow H_3O^+ + SO_3^{2-}$$

Sulfuric acid, H_2SO_4 is a strong acid and dissociates completely in dilute aqueous solution.

$$H_2SO_4 + H_2O \longrightarrow H_3O^+ + HSO_4^-$$

245

The hydrogen sulfate (bisulfate) ion, HSO_4^-, also dissociates but to a much smaller extent to give the sulfate ion

$$HSO_4^- + H_2O \longrightarrow H_3\overset{+}{O} + SO_4^{2-}$$

Pyrosulfuric acid is readily formed by the reaction of SO_3 and H_2SO_4.

$$H_2SO_4 + SO_3 \longrightarrow H_2S_2O_7$$

The geometries of the two acids are as follows:

sulfurous acid   sulfite ion

sulfuric acid   sulfate ion

In sulfuric acid and the sulfate ion, the sulfur is tetrahedral. More will be said about the acidic properties of these species in Part III.

15-7. THE HALOGENS

The elements of the seventh vertical group of the periodic table, called the halogens, are all quite reactive. Fluorine, the first member of the group is the most reactive nonmetal. Fluorine and chlorine are gases under ordinary conditions. Bromine is a dark red liquid and iodine is a purple solid. All of these elements are toxic, fluorine being the most dangerous. Some physical properties of the halogens are given in Table 15-3.

All the halogen atoms contain seven electrons in their valence shells. Thus the formation of diatomic molecules,

$$:\overset{..}{\underset{..}{X}}:\overset{..}{\underset{..}{X}}:,$$

is not surprising.

							H	

								He
Li	Be	B	C	N	O	F	Ne
Na	Mg	Al	Si	P	S	Cl	Ar

Table 15-3. Physical Properties of the Halogens

	Fluorine	Chlorine	Bromine	Iodine
Density (g/cm³)	1.108 (liq. at b.p.)	1.57 (liq. at b.p.)	3.14	4.94
Crystal structure	gas (F_2)	tetragonal (Cl_2)	orthorhombic layer lattice (Br_2)	orthorhombic layer lattice (I_2)
Color of gas	pale yellow	yellow green	brown red	violet
Melting point (°C)	−219.6	−102	−7.2	113.6
Boiling point (°C)	−188	−34	58	184.5

Halogen Hydrides ☐ All the halogens form stable hydrides, HX, which are acids in water solutions (X = any halogen).

$$HX + H_2O \longrightarrow H_3O^+ + X^-$$

The dissociation is complete in dilute aqueous solutions except for HF. Of interest is the great degree of hydrogen bonding observed, especially with HF. As discussed in Chapter 13, hydrogen fluoride forms several hydrogen bonded species including the cyclic $(HF)_5$ and $(HF)_6$, as well as linear chains. These hydrogen bonded species result in unexpectedly high boiling and melting points.

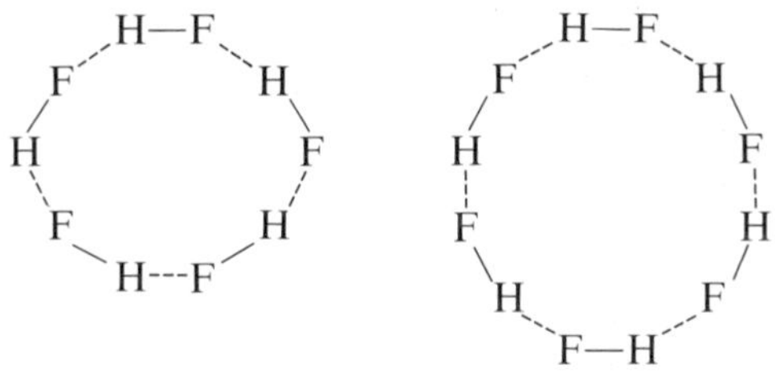

$(HF)_5$ $(HF)_6$

Halogen Oxides ☐ Fluorine forms two oxides, F_2O, and F_2O_2. The former is relatively stable and has been studied. It is angular as expected.

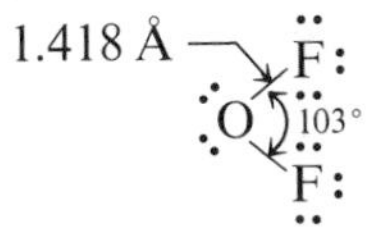

The oxide F_2O_2 is unstable above −40°C and its structure has not been determined.

Bromine forms no oxides that are stable at room temperature, and structural studies have therefore not been carried out.

Chlorine forms an extensive series of oxides: Cl_2O, ClO_2, ClO_3, Cl_2O_6, ClO_4, Cl_2O_7, although little is known about the latter four. The first, Cl_2O has a structure

247

similar to that of F_2O. Chlorine dioxide, ClO_2, is similar to SO_2 in geometry but has an odd electron and is therefore paramagnetic. Its structure can be given as a resonance hybrid involving an electron pair bond and a three-electron bond.

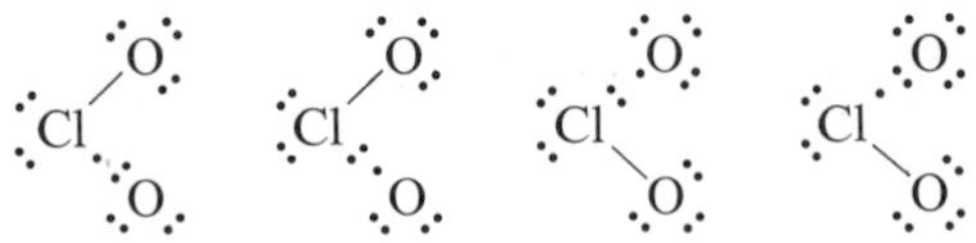

Liquid Cl_2O_6, which is not paramagnetic, dissociates on vaporization to paramagentic ClO_3. The heptoxide, Cl_2O_7, is the anhydride of perchloric acid, $HClO_4$

$$Cl_2O_7 + H_2O \longrightarrow 2\ HClO_4$$

This oxide has an unusually large Cl-O-Cl angle which can be explained by the bulkiness of the ClO_3 groups.

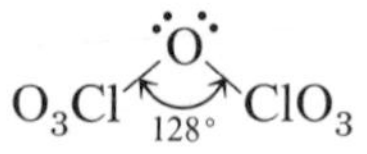

Iodine forms several oxides whose structures are not known with certainty. A pentoxide, I_2O_5, which is a white powder, is stable up to 300°C. Above 300°C, it decomposes to give iodine and oxygen.

$$2\ I_2O_5 \longrightarrow 2\ I_2 + 5\ O_2$$

Both I_2O_4 and I_4O_9 are yellow powders which probably exist in the ionic forms

$$IO^+\ IO_3^- \quad \text{and} \quad I^{3+}\ 3(IO_3^-)$$

Halogen Hydroxides □ Like the sulfur compounds, these should better be called oxyacids since they are all acidic. Fluorine forms no oxyacids.

The simplest oxyacid of the halogens is HOX, hypohalous acid. These acids are of low stability decreasing in the order hypochlorous > hypobromous > hypoiodous, or $HOCl > HOBr > HOI$.

Only chlorine forms the next member, chlorous acid—HOClO—although it has never been isolated. Only its salts, the chlorites, have been prepared. In the crystalline state at −35°C, ammonium chlorite has the chlorite ion structure shown.

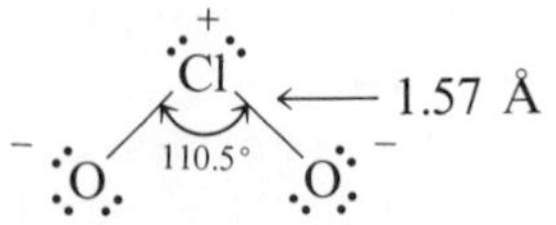

Halic acids, $HOXO_2$, are known for chlorine, bromine, and iodine. In this case their stability decreases in the order: $HOIO_2 > HOClO_2 > HOBrO_2$. Only iodic acid, $HOIO_2$, has been obtained as pure crystals. In their crystalline salts, the ions, XO_3^-, exist as rather flat pyramids because of the unshared electrons on the central halogen atom.

$$\left[\begin{array}{c} :\ddot{O}:^- \\ \overset{\cdot|}{\underset{\cdot\cdot}{Cl^{2+}}} \longleftarrow 1.48 \text{ Å} \\ :\ddot{O}. \quad .\ddot{O}: \end{array}\right]^- \quad \left[\begin{array}{c} :\ddot{O}:^- \\ \overset{\cdot|}{\underset{\cdot\cdot}{Br^{2+}}} \longleftarrow 1.64 \text{ Å} \\ :\ddot{O}. \quad .\ddot{O}: \end{array}\right]^-$$

The structure of the pure crystalline iodic acid, is known to be

$$\begin{array}{c} :\ddot{O}:^- \; \underleftarrow{\quad} 1.81 \text{ Å} \\ \overset{\cdot|}{I^{2+}} \; \underleftarrow{\quad} 1.89 \text{ Å} \quad \text{(Mean O—I—O angle = 98°)} \\ :\ddot{O}. \quad .\ddot{O}\text{—H} \end{array}$$

The final acid of the series occurs with chlorine, bromine, and iodine— perchloric acid, $HOClO_3$, perbromic acid, $HOBrO_3$, and periodic acid, $HOIO_3$. These compounds form the tetrahedral ions.

$$\left[\begin{array}{c} :\ddot{O}:^- \\ | \\ \overset{\cdot}{\underset{\cdot\cdot}{Cl^{3+}}} \text{--} \ddot{O}:^- \\ :\ddot{O}. \quad .\ddot{O}: \end{array}\right]^-$$

An ion of interesting geometry is formed by the reaction of hydrogen fluoride with potassium iodate.

$$2 \text{ HF} + \text{KIO}_3 \longrightarrow \text{KIO}_2\text{F}_2 + \text{H}_2\text{O}$$

The structure of the $IO_2F_2^-$ ion is

$$\begin{array}{c} 2.00 \text{ Å} \\ \downarrow \\ \text{F—I}\overset{\leftarrow}{\;}\text{F} \\ \swarrow \quad \underset{\longleftarrow}{\big\|} 1.93 \text{ Å} \\ \text{O} \qquad \text{O} \\ \underset{100°}{\smile} \end{array}$$

The strange shape is made clear by considering the unshared electron pair. Counting these, the bonds are directed toward the five corners of a trigonal bipyramid

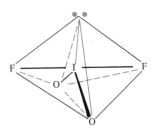

15-8.   THE NOBLE GASES

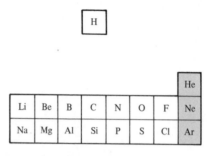

The noble gases were discussed in Section 11–10. Only a relatively few molecules are known that contain noble gas atoms. The structure of one of these, XeF_4, is discussed in Section 14–8.

15-9.   PERIODIC TRENDS

Many of the properties and structural features of the elements of the first three periods have been presented in this chapter. It will now be of interest to examine some of the trends within the entire collection of elements studied.

It was noted earlier in this chapter that the gradation from metals to nonmetals on moving from left to right in the table was attended by other concurrent changes. Thus the oxides and hydroxides of the elements go from basic to acidic on moving from left to right. Table 12–7 demonstrated the abrupt changes from ionic to covalent bonding in the chlorides of these elements.

Compounds in which a central atom has four pairs of electrons should be tetrahedral according to the electron pair repulsion theory. Many molecules, however, show deviations from this angle. Table 15–4 shows the structural features of the hydrides of the elements in the upper right section of the periodic table.

Table 15-4.   Bond Angles and Lengths of the Hydrogen Compounds of Some Nonmetals

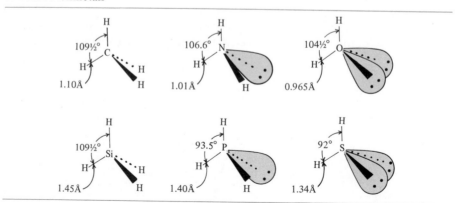

Note that the angle H-X-H decreases on moving from left to right and from top to bottom in this part of the periodic table. The deviations from tetrahedral can best be explained by the mutual repulsions of the unbonded electron pairs. The volume occupied by the orbital of an electron pair is reduced when the pair forms a bond between two atoms. This follows if one considers that in a nonbonded electron pair mutual repulsion between electrons is not reduced by the electron-nucleus attraction that exists when the electron pair is involved in a covalent bond. Hence, these nonbonding pairs force the bonding electron pairs into a configuration of smaller bond angles. This effect is more important the more nonbonded pairs there are (moving from left to right) and more important the larger the orbitals (moving down the table to the larger atoms).

Another correlation that can be domonstrated is the decrease in ionization potential on moving down any vertical group of the periodic table. Because metallic character of an element is associated with low ionization potential, one can say that metallic character increases on moving down a vertical group. Thus magnesium is considerably more metallic than beryllium—the chloride and hydride of magnesium are more ionic and magnesium hydroxide is strictly basic, whereas that of beryllium is amphoteric. A similar trend is even more evident with boron and aluminum. In this case boron is by all definitions nonmetallic whereas aluminum is quite metallic. Although this vertical trend is less evident with the lighter members of the groups on the right of the periodic table, it is still present and can be shown with the heavier members.

Because metallic character decreases on moving to the right in the table and increases on moving down the table, one might expect to find a diagonal trend: Thus, boron is more metallic than carbon, its nearest neighbor to the right; silicon is more metallic than carbon, its nearest neighbor down. Therefore one might expect some similarity between boron and silicon. This trend is found to be quite general throughout the table, especially with the lighter elements. For example, as was stated earlier, beryllium and aluminum are both amphoteric. Table 12-7 shows that the change from ionic to covalent chlorides occurs in a diagonal fashion in the periodic table.

The periodic table has probably been the single most important unifying principle in the development of chemistry. Adding to the drama of its discovery is the fact that it was purely an empirical endeavor. Nor for nearly 40 years was chemical theory to develop to the stage where an explanation of periodicity could be possible. It might be said, moreover, that the development of theoretical chemistry was given great impetus by the discovery of periodicity.

PROBLEMS

1. What problems arise in attempting to place hydrogen in a group or chemical family in the periodic table?

2. Discuss differences among the following hydrides: LiH, BeH_2, B_2H_6, CH_4, NH_3, H_2O, and HF with respect to (a) bond character and (b) molecular geometry.

3. Describe the structure of BeO. How does this structure differ from that of the oxides of the other alkaline earth metals? What causes this difference?

4. In terms of electron densities on the atoms involved, explain why the hydroxides of elements in a given period of the periodic table become more acidic in character as you go from left to right along the period.

5. In what way does boron differ from the other elements of its group and period in terms of its compounds with hydrogen?

6. Compare and contrast the chlorides of B and Al with respect to molecular structure.

7. The hydroxides of boron and aluminum both react readily with NaOH solution. However, although aluminum hydroxide reacts with HCl solutions, boron hydroxide does not. Explain these similarities and differences in behavior.

8. Pure diamond and pure graphite differ greatly in their properties although they are chemically identical. Make a list of the properties of these two substances and explain the differences in terms of bonding and structure.

9. Why are there so many different carbon containing compounds?

10. The substances CF_4 and SiF_4 are both known. The ion SiF_6^{2-} also forms readily whereas CF_6^{2-} is unknown. Why?

11. Explain differences in the properties of CO_2 and SiO_2 in terms of bonding and structure.

12. At room temperature, nitrogen is a fairly inert gas whereas phosphorus is a reactive solid. Why?

13. The molecules NH_3 and NF_3 have approximately the same structure. Describe this structure and explain why NH_3 has an appreciable dipole moment whereas NF_3 has a very small dipole moment.

14. How does PCl_5 differ in the vapor and solid states?

15. Draw electron dot structures for the following: (a) CO; (b) PH_3; (c) PCl_5; (d) NO; (e) NO_2; (f) N_2O_4; (g) CO_3^{2-}; (h) NO_2^+; (i) NO_2^-.

16. The two oxides of phosphorus are often written P_2O_3 and P_2O_5. In what respect are these formulations incorrect?

17. Speculate on the most probable structure of (a) N_2O_3 and (b) F_2O_2.

18. Write formulas for the following oxyacids: (a) orthophosphoric acid; (b) metaphosphoric acid; (c) pyrophosphoric acid; (d) orthophosphorus acid; (e) orthoboric acid; (f) pyroboric acid; (g) pyrosulfuric acid.

19. What problem arises in writing II. electron dot structure of O_2?

20. Draw electron dot structures that adequately explain bond lengths and the magnetic properties of the species (a) O_3; (b) O_2^-; (c) O_2^{2-}.

21. Unlike most substances, the viscosity of liquid sulfur increases at first as it is heated before it finally decreases at higher temperatures. Explain this behavior.

22. In terms of molecular structure, explain why the first elements in the representative element groups V and VI are gaseous while the other elements in those groups are solids at 25°C. Why are fluorine and chlorine of group VII both gaseous at 25°C?

23. In what ways do the following compounds of fluorine differ from the analogous compounds of the other halogen? (a) hydrides and (b) oxyacids.

24. Name the following compounds: (a) $HOClO_3$; (b) $HOBrO_3$; (c) $HOCl$; (d) $HOIO_2$; (e) $NaClO_2$; (f) $KClO_4$; (g) $Ca(OCl)_2$.

25. Explain what is meant by the diagonal trend in the periodic table and give some specific examples of this trend.

26. The bond angles in CH_4 and SiH_4 are all precisely 109.5°, but the bond angles in NH_3 and PH_3 differ somewhat (Table 15-4). Explain.

27. What is the shape of the $IO_2F_2^-$ ion? Rationalize the structure in terms of electron pair repulsion theory.

28. Use electron pair repulsion theory to explain why the H—O—H angle in water is less than the H—N—H angle in ammonia.

29. Use electron pair repulsion theory to explain the difference in structure between SO_3 and SO_3^{2-}.

30. Give the gross geometry of the following species: **(a)** H_2S; **(b)** SO_2; **(c)** N_2O; **(d)** NO_3^-; **(e)** NH_3; **(f)** $SiCl_4$; **(g)** PCl_3; **(h)** PCl_4^+; **(i)** PCl_6^-; **(j)** C_3O_2.

31. Assess the relative importance of the following resonance forms to the actual electronic structure of nitric oxide.
    (a) $:\overset{+}{N}:\overset{..}{O}:$ **(b)** $:N::O:$ **(c)** $:N:\overset{..}{O}:$ **(d)** $:\overset{-}{N}::\overset{+}{O}:$ **(e)** $:\overset{..}{N}:\overset{.}{O}:^{2+}$

32. Explain why one compound exists with the formula N_2F_4 but two compounds are known with the formula N_2F_2.

33. Give the anhydrides of the following acids: **(a)** HNO_2; **(b)** HNO_3; **(c)** H_3PO_3; **(d)** H_2SO_3; **(e)** $HOCl$.

34. Explain how structural differences account for the fact that in a solution H_3PO_4 can lose three H^+ ions, H_3PO_3 can lose two H^+ ions, and H_3PO_2 can lose only one H^+ ion.

COVALENT ARCHITECTURE III.
POLYMERS

16-1. CHARACTERISTICS OF POLYMERIC SUBSTANCES

Most of the substances that have been discussed are characterized by definite sets of properties. Table 12–1 showed many common substances whose melting and boiling points, densities, and vapor pressures are fixed within very narrow limits. Mixtures of substances, however, do not possess fixed properties; in fact, the properties of mixtures frequently vary widely as the composition changes. For example, dilute solutions of sugar in water are visibly quite similar to water. More concentrated sugar solutions are, however, very viscous.

There are some substances that, even when chemically pure, have properties that either vary greatly from sample to sample or are quite diffuse and difficult to establish. For example, a pure sample of sodium chloride, heated gradually, remains solid until a temperature of 801°C is reached where it rapidly melts over a 1° or 2° range of temperature. On the other hand, a pure sample of gelatin, when treated similarly, will gradually soften and become liquid over a wide temperature range.

Another specific property of substances like gelatin is the viscosity of their solutions: a dilute solution of sugar dissolved in water on visual examination is indistinguishable from pure water. The same amount of gelatin dissolved in water produces a very viscous solution.

16-2. PROBLEM OF MOLECULAR WEIGHT

The molecules of the substances that exhibit these unusual properties have been found to have a common structural characteristic that sets them apart from molecules of other substances: they have extremely high molecular weights— occasionally in the millions. Careful examinations of the molecular structures of these high molecular weight substances have shown that they invariably contain repeating units of low molecular weight, and they are thus called **polymers** (many units). A simple example is polyethylene, which is made by the catalyzed addition of ethylene molecules (monomer) to each other.

$$\begin{array}{ccccccc}
H & & H & H & & H & \\
\diagdown & & \diagup & \diagdown & & \diagup & \\
& C=C & + & & C=C & + \cdots \longrightarrow \\
\diagup & & \diagdown & \diagup & & \diagdown & \\
H & & H & H & & H &
\end{array} \qquad \text{(16-1)}$$

Eventually, the process is terminated by one of several freak reactions, but not in a *precisely* predictable way. The result is that the polyethylene molecules in a sample will all differ somewhat in molecular weight. The intellectual problem that results is twofold. If a pure sample of polyethylene contains molecules of varying molecular weight, then our previously defined concept of chemical purity must change. Secondly, with polymers, one must speak of *average* molecular weights.

Examples of polymers abound in today's world, both in nature and as products of modern technology. Some examples of naturally occurring polymeric substances are hair, silk, cellulose, starch, wool, and rubber. Examples of commercial polymers are nylon, dacron, orlon, plexiglas, synthetic rubber, and bakelite.

16-3. CLASSIFICATION ACCORDING TO PHYSICAL BEHAVIOR

Polymeric materials may be classified according to their physical behavior into three classes:

1. **Elastomers**—possess the property of elasticity.
2. **Thermoplastic polymers**—soften and become fluid on heating and may therefore be molded.
3. **Thermosetting polymers**—hard, infusible (do not melt or soften) solids that are insoluble. Often a solvent will dissolve *in* the thermosetting polymer causing swelling.

The infusibility of polymers is dependent mainly on the presence and number of **crosslinks** which join adjacent chains (Figure 16-1). These crosslinks tie the polymer mass into a framework whose rigidity depends upon the number of crosslinks. Polymers whose individual molecular chains are not joined by crosslinks are called linear polymers.

16-4. PREPARATION OF POLYMERS

Because the properties of polymers depend so much on chain length, small amounts of impurities can often lead to chain termination and consequent poor physical characteristics. Therefore, the preparation of polymers requires extreme purity of starting materials and very clean conditions. Moreover, due to their nature, polymers are frequently difficult or impossible to purify.

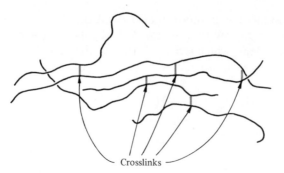

Figure 16-1. Schematic drawing of crosslinks in a polymer.

Within these limitations the preparation of polymers is carried out employing otherwise ordinary chemical reactions. For example, an amide can be made by heating an amine with a carboxylic acid in the presence of a suitable catalyst. The result is the loss of a molecule of water, hence this reaction is termed a **condensation reaction.** If both molecules have two such groupings, a polymer can result.

$$CH_3CH_2NH_2 + HO-\overset{O}{\underset{\|}{C}}-CH_3 \longrightarrow CH_3CH_2\overset{H}{\underset{|}{N}}-\overset{O}{\underset{\|}{C}}-CH_3 + H_2O$$

amine   carboxylic acid   an amide

$$H_2N-CH_2CH_2-NH_2 + HO-\overset{O}{\underset{\|}{C}}-CH_2CH_2-\overset{O}{\underset{\|}{C}}-OH \longrightarrow \sim N-CH_2CH_2\overset{O}{\underset{|}{N}}\overset{\|}{C}-CH_2CH_2-\overset{O}{\underset{\|}{C}}\sim$$

diamine   diacid   polyamide

or

$$HO-\overset{O}{\underset{\|}{C}}-CH_2NH_2 + HO-\overset{O}{\underset{\|}{C}}-CH_2NH_2 \longrightarrow \sim\overset{O}{\underset{\|}{C}}-CH_2-N-\overset{O}{\underset{\|}{C}}-CH_2N\sim$$

amino acid   amino acid   polyamide

Note that a polymer can result from either (1) the combination of a diamine with a diacid, or (2) the combination of an amino acid with another amino acid. Polymers formed by condensation reactions are termed **condensation polymers.**

Polymers may also be prepared by the mutual addition of molecules containing double bonds. Such addition reactions are brought about by a variety of catalysts. An example has been given in equation (16–1) using ethylene. These polymers are called **addition polymers.** Some examples of each type are shown in Table 16–1.

256

Table 16-1. Types of Polymers According to Synthetic Method Used

A. Condensation Polymers

1. Dacron, a polyester

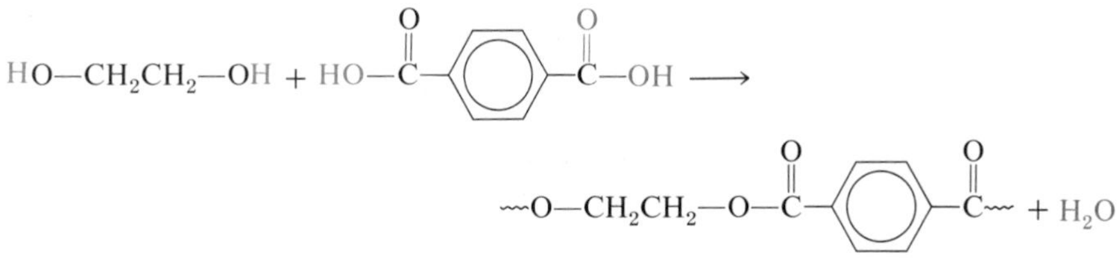

2. Nylon, a polyamide

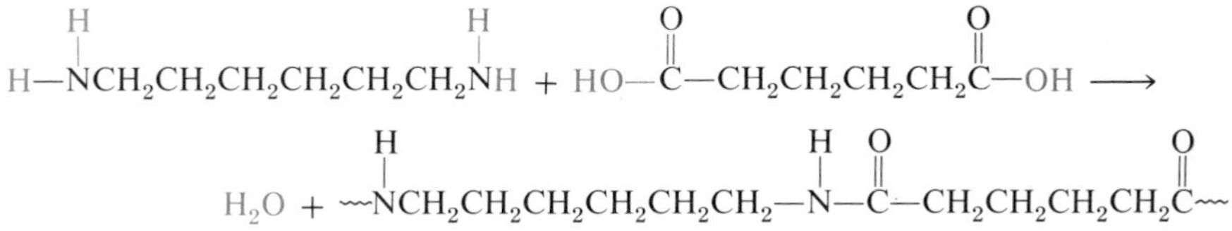

3. Glyptal from glycerine and phthalic acid anhydride, a polyester

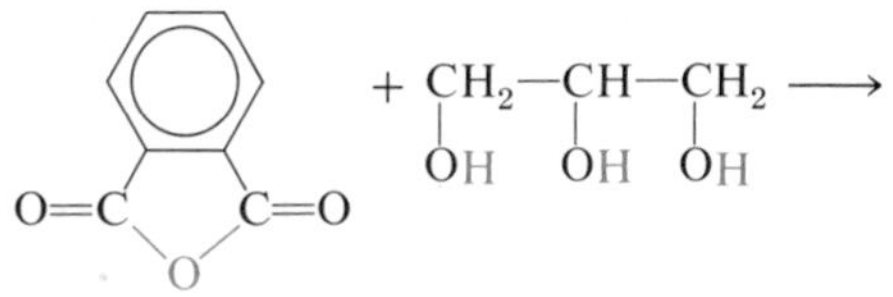

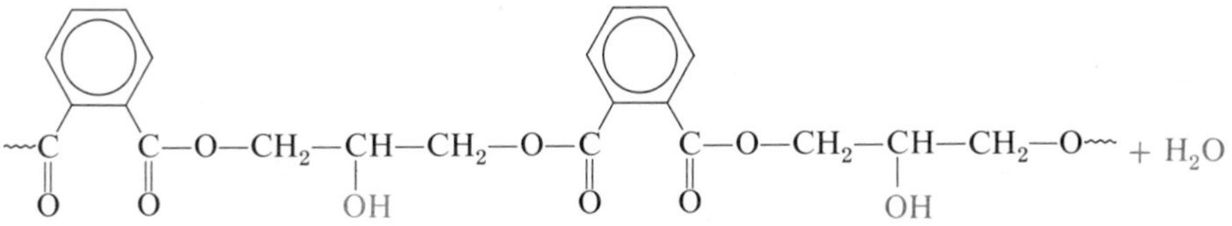

(thermoplastic)

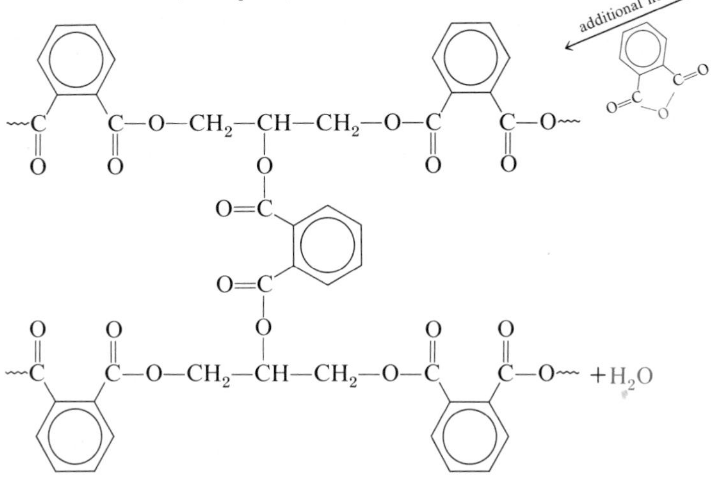

additional heat

(thermosetting)

257

Table 16-1   (continued)

B. Addition Polymers

1. Polyethylene

$$n \; CH_2{=}CH_2 + \text{acid catalyst} \xrightarrow[\text{and temperature}]{\text{high pressure}} \{CH_2{-}CH_2\}_n$$

2. Polystyrene (styrofoam)

$$n \; CH{=}CH_2 + \text{acid catalyst} \xrightarrow[\text{and temperature}]{\text{high pressure}} \{CH{-}CH_2\}_n$$

3. Polyacrylonitrile (orlon)

$$n \; CH{=}CH_2 \xrightarrow[\text{and temperature}]{\text{high pressure}} \{CH{-}CH_2\}_n$$

16-5.   EFFECT OF MOLECULAR STRUCTURE ON PHYSICAL PROPERTIES—CRYSTALLINITY

The most important use of polymeric substances is in structural materials. Examples are textiles, machine parts, surface coatings, films, and elastic objects. The reason why very large molecules are useful in making such objects is that there are many opportunities for such long molecules to become "entangled" through intermolecular forces. As will be seen in Section 16–11 and Figure 16–5, the usefulness of silk is due to its fiber strength, which can be explained by the high degree of hydrogen bonding between parallel molecular chains. These intermolecular forces confer upon the entire mass a degree of crystallinity. Because there are many areas of the polymer where random orientation prevents crystallinity, the mass does not appear crystalline. In fact, an entirely crystalline polymer would be brittle and unsuitable for most uses. The degree of orientation of molecular chains, that is, the degree of crystallinity, determines most of the mechanical properties of the polymer. The degree of intermolecular forces depends mainly on the nature of the monomer units in the chain. For example, the attractive forces between molecular chains of polyethylene are weak compared to the hydrogen bonding between molecular chains of nylon.

In general, polymeric substances can be placed into four groups according to their degree of crystallinity:

1. **Amorphous**—no crystallinity, weak forces between chains, motion not restricted. Low tensile strength, suffers plastic flow. Example: gelatine.
2. **Unoriented crystalline polymer**—considerable crystallinity, but crystalline regions are randomly oriented. Polymer can be molded above its melting point. Higher tensile strength than amorphous polymer. Example: undrawn nylon.

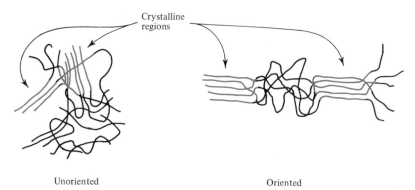

Figure 16-2. Diagram showing the effect of stretching on the orientation of a partially crystalline polymer.

3. **Oriented crystalline polymer**—considerable crystallinity with crystalline regions oriented (usually by a "cold-drawing" process). Very high tensile strength. Example: drawn nylon. When nylon fiber is produced, it is quite weak and may be easily snapped by quick stretching (unoriented crystalline polymer). By stretching the fiber slowly to two or three times its original length, the tensile strength is tremendously increased so that it becomes very difficult to snap (Figure 16-2).

4. **Elastomer**—intermediate between 1 and 2. The requirements for an elastomer are relatively short chains and a small amount of crystallinity (or crosslinking) to restore amorphous form after stress is removed. No plastic flow. Example: rubber.

16-6. ENERGY TRANSFORMATIONS IN POLYMER ORIENTATION

An interesting experiment can be performed that illustrates the interconversion of heat energy and entropy (energy lost due to randomization—Chapter 6). Touch a relaxed rubber band to the lips (a very sensitive temperature sensing area). Quickly stretch the rubber band to its limit and once more touch it with the lips. The marked rise in temperature is partly explained by the loss of entropy as the unoriented polymer is oriented on stretching. The entropy lost is converted into kinetic energy of the molecules which manifests itself as a rise in temperature. Part of the temperature rise is due to the conversion of mechanical work (stretching) into heat energy.

16-7. FACTORS THAT AFFECT POLYMER STRENGTH

In general, as might be expected, mechanical strength increases with the length of the polymer molecule. This is reasonable because of cohesive forces among overlapped chains, especially in a highly oriented polymer. Figure 16-3 bears this out for two types of polymers: polyamides (curve I) and polyhydrocarbons (curve II). The difference between curves I and II is explained by hydrogen

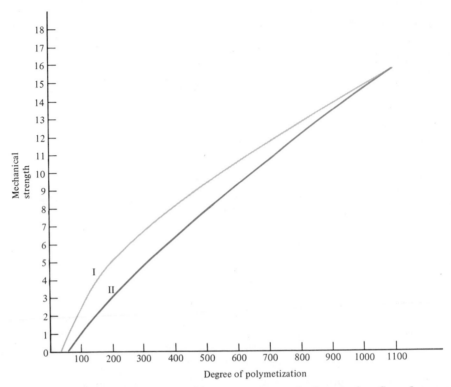

Figure 16–3. Strength of polymers as a function of degree of polymerization. Curve I represents polyamides and curve II represents polyhydrocarbons.

bonding in the polyamides which provides very effective interchain cohesive forces.

Such forces do not exist in polyhydrocarbons. Increasing the chain length increases the degree of interchain cohesion and the mechanical strength increases roughly proportionately. Finally, a situation arises at very great chain lengths at which mechanical strength is no longer limited by these forces, but instead by the covalent bonds themselves. Thus at a degree of polymerization* of about 1000, both polymer types exhibit the same strengths; that is, curves I and II meet.

260   *Degree of polymerization is the average number of monomer units per polymer chain.

16-8. STEREOREGULAR POLYMERS

Man's attempts to prepare synthetic rubber had long been thwarted by his inability to make a molecule in which the geometric arrangement of atoms is *exactly* the same in each monomer unit of the polymer. For example, natural rubber is made up of units of isoprene,

$$CH_2{=}\underset{\underset{\displaystyle CH_3}{|}}{C}{-}CH{=}CH_2$$

$$n\ CH_2{=}\underset{\underset{\displaystyle CH_3}{|}}{C}{-}CH{=}CH_2 \longrightarrow \left(CH_2{-}\underset{\underset{\displaystyle CH_3}{|}}{C}{=}CH{-}CH_2\right)_n$$

which react to form the polymer

all-cis

Moreover, since the polymer has one double bond per unit, each unit is capable of existing in either a cis or trans configuration. Two forms of this polymer occur naturally, the all-cis form (natural rubber, Hevea) and the all-trans form (Gutta-percha). The all-cis form is the elastomer, and the all-trans form is hard and brittle at room temperature.

all-trans

Until the early 1950s all of man's attempts to make polyisoprene yielded polymers in which cis and trans arrangements were completely random. In 1953, a German named Karl Ziegler and an Italian named Giulio Natta, almost simultaneously, prepared the first "stereoregular" polymers. They employed an unusual catalyst made up of an aluminum compound such as aluminum triethyl, $Al[C_2H_5]_3$ mixed with titanium chloride. Since then many variations have been discovered, and it is now possible to carry out addition polymerizations to yield an all-cis, an all-trans, or a perfectly alternating cis-trans polymer. An additional advantage of the new catalysts is that they permit the use of much lower temperatures and pressures in polymerization processes.

16-9. LINEAR RIGID POLYMERS

One of the more recent trends in polymer research is the attempt to produce rigid linear chains. One such attempt involves the preparation of polyacrylonitrile (Table 16–1) followed by a closure of the ring (step 2) as shown in equation (16–2). Step 3 yields a compound containing completely delocalized electrons and therefore corresponds to a linear graphitelike structure. This latter material is black, completely infusible, insoluble, and very stable thermally. It is called "black orlon." Because of its extreme properties, it has not been useful commercially.

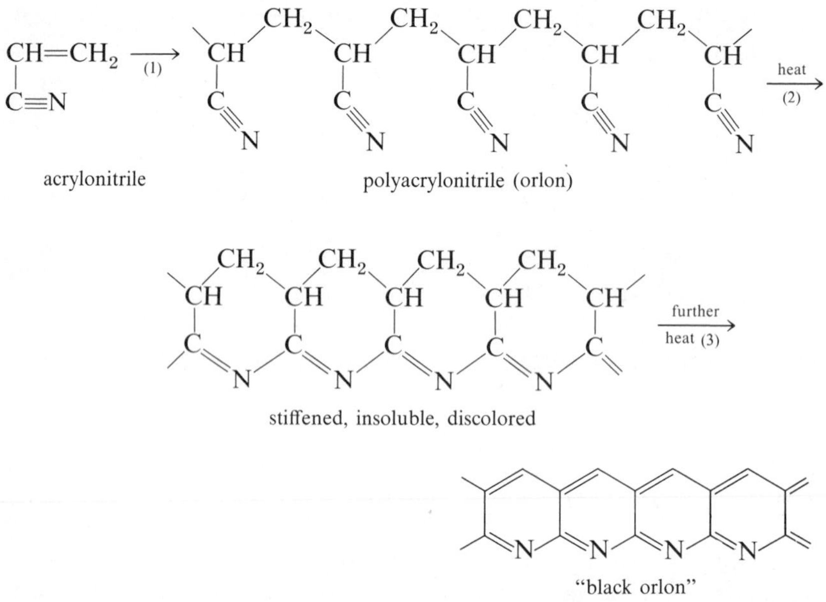

acrylonitrile   polyacrylonitrile (orlon)

stiffened, insoluble, discolored

"black orlon"

(16-2)

The stability of "black orlon" can be explained by the several contributing resonance structures that can be written

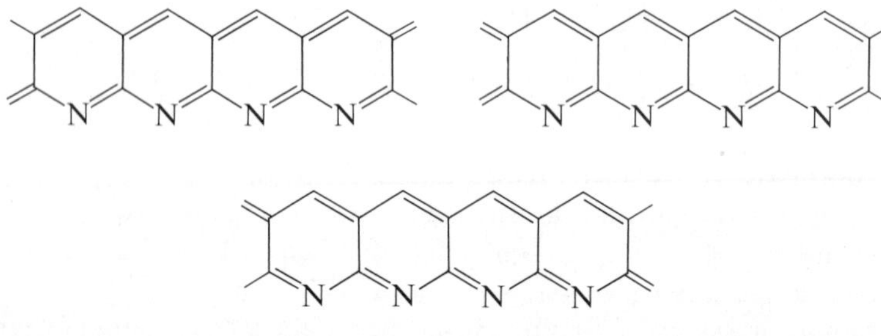

Such chains have been called "ladder" polymers. The unusual strength of such polymers is due to two factors: (1) rupture of the chain into two fragments requires the rupture of two covalent bonds in the same ring; (2) the geometry of the molecule requires that all the ring atoms lie in the same plane. This coplanarity extends the molecular chains and allows a greater degree of orientation and therefore crystallinity.

16-10. PROTEINS AND AMINO ACIDS

Proteins are one of several classes of polymers that are produced in plants and animals. Along with fats and carbohydrates, proteins are one of the three major classes of foods.

Treatment of a protein with hot acid results in a mixture of monomeric amino acids, most of which have an amino group (NH_2) on the carbon atom adjacent to the carboxy group

$$\left(-C\diagup_{OH}^{\diagup O} \quad \text{or COOH} \right)$$

$$R-\underset{NH_2}{\overset{H}{C}}-C\diagdown^{\diagup O}_{O-H}$$

R represents one of several groupings. About 20 different amino acids have been isolated by the above treatment of proteins from animal and plant sources. These are shown in Figure 16–4. Since these compounds have both acidic (—COOH) and basic (—NH_2) sites, they exist in neutral media as the internal salts

$$R-\underset{^+NH_3}{\overset{H}{C}}-C\diagdown^{\diagup O}_{O^-}$$

which accounts for the high water solubility, crystalline state, high melting points, and other saltlike properties.

All amino acids found in proteins (except glycine, the simplest) possess at least one carbon atom with four different groups attached, and hence are chiral. Probably more significantly, all amino acids from proteins exhibit the same chirality and configuration—they all possess the generalized structure

$$\underset{R}{\underset{|}{H_2N{-}C{-}H}}\overset{\overset{O\diagdown{\diagup}OH}{C}}{\underset{|}{}}$$

and most of them rotate the plane of polarized light slightly to the right.

This nearly exclusive existence of molecules of the same chirality in nature is explained by the fact that protein molecules aid in directing the construction of other protein molecules in the organism (Chapter 27). Being themselves chiral, they react preferentially with one optical isomer over the other. If, at the beginning of life on earth, a preponderance of molecules of the opposite chirality had reacted to form the first proteins, that form might have predominated. Since proteins form one of the main classes of foods, it is evident that most of the animal and plant proteins must be of the same configuration if foods are to be universally utilized by animals. Otherwise, there would exist two classes of life simultaneously on earth which could not support each other. One could speculate that both groups might have occurred early in the history of the earth during which time competition resulted in annihilation of one group.

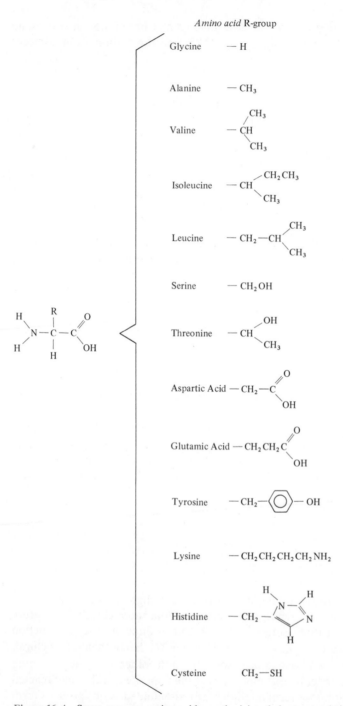

Figure 16-4. Some common amino acids emphasizing their structural similarities.

An alternative theory has been proposed recently to account for the preponderance of levo* amino acids in nature. The degradation of mixtures of dextro* and levo amino acids will always result in equal rates of decomposition

*In this case, the terms levo and dextro refer to the configuration compared to a "standard" compound—glyceraldehyde—rather than to the actual rotation of polarized light.

of the two forms provided that a chiral medium is not present. In the presence of radioactive strontium salts, however, it has been observed that the dextro amino acid was decomposed more rapidly than the levo. That the strontium exerted no chemical effect was shown by using the nonradioactive isotope. In this case, both dextro and levo isomers were decomposed at the same rate. The explanation offered is that the β particles emitted by the strontium during its radioactive decay are chiral—they all have the same magnetic spin. The result is that these β rays interact with the dextro isomer at a faster rate than with the levo isomer, rendering dextro isomers selectively more easily degradable. Large amounts of radioactive substances were probably present on the earth's surface at the time when these molecules were being formed. Although this theory is quite attractive, it will have to withstand the test of time and further experiments before it is generally accepted.

Proteins are long chains of amino acids joined to each other by **peptide linkages.** These are amides as described in Section 16–4 in which the —NH_2 group of one molecule reacts with the —COOH group of another.

amide or peptide linkage   amino acid unit

The nature of the protein will depend on the exact order and combination of amino acid molecules employed.

16–11. PROTEIN STRUCTURE

Initially, studies of protein structure focused on determinations of the exact sequences of amino acids within the polypeptide chain. This sequential arrangement is called the **primary** structure. However, the high degree of strength in some proteins and synthetic polypeptides, as well as their other properties suggested early that the polypeptide chains have very orderly arrangements or **secondary** structures. One of the outstanding scientists in the area of protein structure is Linus Pauling who, in the 1940s and 1950s made great contributions to our knowledge of this very complex subject.

Silk fibroin, stretched hair and muscle, and similar proteins (the β-keratins) have been shown by X-ray analysis to possess polypeptide chains arranged in a parallel fashion forming sheets as shown in Figure 16–5. The high degree of hydrogen bonding in this structure confers great strength on these materials and makes them very suitable for the uses to which they are put. In ordinary commercial silk, the protein chains consist of alternating glycine and alanine groups. Wild silk consists mostly of alanine.

One of Pauling's important contributions to the elucidation of protein structure was his proposed α-helix structure. He found that the properties of many native proteins as well as synthetic polypeptides could be explained by a model in which the polypeptide chain forms a coil whose shape is maintained by hydrogen

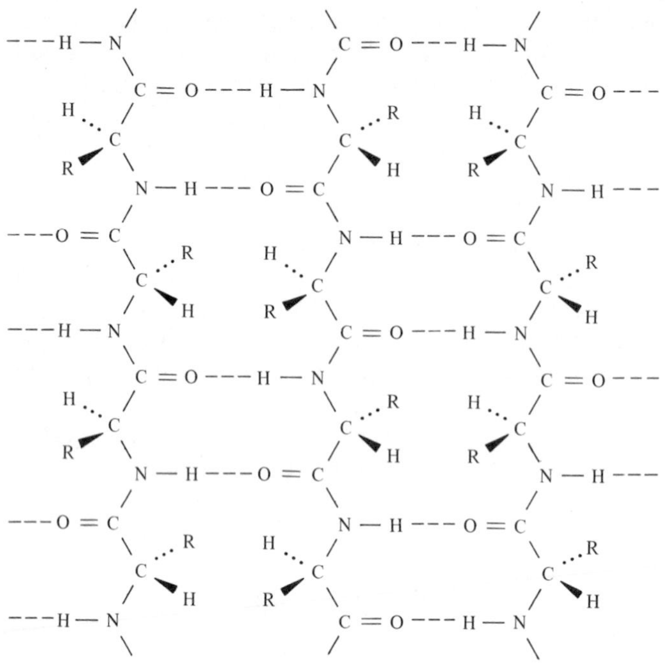

Figure 16-5. Partial structural formula of protein silk fibroin. R is CH_3 or hydrogen. Hydrogen bonds are designated by a colored dashed line.

bonding. The hydrogen bonding occurs between an ＼N—H group and a ＼C=O group three peptide linkages away. The resulting helix contains 3.6 to 3.7 amino acid units per turn (Figure 16-6). The helical structure is maintained even when the protein is dissolved in water, provided that the solution is not warmed.

The treatment of proteins by heat or acid causes a precipitation of the protein with the concomitant loss of physiological activity. The coagulation of egg white by heat is an example of this process which is called **denaturation.** In this process, it is not only the secondary structure of the protein that is affected; that is, the interchain association, but also the **tertiary** structure, which refers to the shape of the folding that is assumed as the protein is packed into a three-dimensional structure. The peptide links remain unbroken during denaturation since free amino acids are not liberated.

Some experimental results may help you to understand what happens during the denaturation process: Native protein usually has an optical rotation in the range (−) 30–60° (levorotatory). The same protein after denaturation has a rotation usually greater than (−) 100° (levorotatory). This rather large increase in levorotation is explained by assuming that the native protein contains a right-handed coil which is destroyed in denaturation. (Recall that a coil is chiral and will itself contribute to the overall optical activity (Figure 16-6). The dextrorotation of the coil opposes the levorotation of the asymmetric carbon atoms in the chain. When the coil is destroyed, the levorotation increases.

All helical protein thus far studied have been found to be similarly right-handed.

Denaturation of collagen, the protein of tendon and skin, yields gelatin with a molecular weight one third that of the original collagen. The helical structure

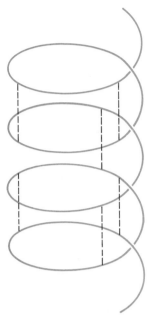

Figure 16-6. A right-handed helix. The dashed lines indicate how the α helix is held together by hydrogen bonds.

of collagen seems, from this and other evidence, to have three coils intertwined. Collagen contains the three amino acids

$$H_2NCH_2C\text{—OH}$$

glycine

proline

hydroxyproline

The nitrogen atom in the rings of the latter two amino acid units, in the peptide chain contains no hydrogen atom and so no hydrogen bonding is possible. The ability to form a regular interchain structure is therefore greatly reduced. Moreover, collagen in skin irreversibly breaks down in a first degree burn (about 145°F). This is about 60°F higher than required to cause breakdown of collagen in solution. The conclusion may be drawn that the —OH group in the hydroxyproline unit is used in strengthening tissue structure unless the substance is dissolved in water, in which case it is instead used in hydrogen bonding to water.

Globular proteins are more soluble in water and have a less regular secondary structure. Egg white, milk proteins, blood serum proteins, and most enzymes are examples of globular proteins.

The exact structures of proteins, including primary, secondary, tertiary, and quaternary (see below) structures form an area of investigation that remains under vigorous study. It is useful to summarize here the distinctions among these levels of structure complexity:

267

Primary structure is simply the exact sequence of amino acid units that appear in a given protein.

Secondary structure is the specific shape that results from hydrogen bonding as in the case of the α-helix or the sheetlike structure of silk.

Tertiary structure is the specific folding pattern that one or more protein molecules assume. Folding is not a random arrangement. Instead it is predetermined by the interactions of the amino acid R groups with one another, with the molecular framework, or with the solvent. The uniqueness of this three-dimensional structure is very closely related to the biological activity of the protein. For example, if a protein has catalytic activity, that is, an enzyme, even a small change in the tertiary structure can result in profound changes in its activity.

Interactions that produce a given tertiary structure are of several origins. Hydrogen bonding can contribute to the tertiary structure of a protein. Another form of interaction is the so-called "hydrophobic" (water-hating) interactions. If the side chain R groups are less attractive to water molecules than they are to other R groups in the protein, the chains will coalesce and exclude water much as small oil drops coalesce to form larger ones. Still another type of interaction results from electrostatic attraction between oppositely charged groups. In addition to these weak interactions, covalent bonding may occur to crosslink polypeptide chains into the tertiary structure. The only known covalent bonds of this type involve —S—S— links. Such tertiary structures have been exactly deduced in a few cases—ribonuclease, insulin—and only with great difficulty (Figure 16-7). X-ray diffraction analysis has been extremely useful in these studies.

Figure 16-7. Tertiary structure of myoglobin. [Courtesy of Magnum Photos, Inc., New York.]

Quaternary structure results when several protein units, each with its own tertiary structure, combine to form an aggregate of a higher order of complexity. Here again, the quaternary structure usually accounts for biological activity not observed with the subunit tertiary structures. The forces that hold the quaternary structure together are mainly R group surface interactions among the protein units.

16–12. DEOXYRIBONUCLEIC ACID (DNA)— MOLECULAR STRUCTURE AND GENETIC REPLICATION

The problem of explaining life has probably intrigued man longer than any other question. Although a considerable amount has been learned in the last hundred years, one of the outstanding events in this area was the proposal of the structure of deoxyribonucleic acid (DNA).

DNA is found in the nuclei of all dividing cells and is an essential component of the chromosomes. The genetic characteristics of the individual are carried by the DNA in the cells. DNA thus plays a central role in transmitting traits from one generation to the next (Chapter 27).

Until 1953, the mechanism by which genetic replication occurs was unknown. Up to that time, however, the essential chemical features of DNA were known: it consists of very high molecular weight polymers of alternating phosphate and sugar groups with one nitrogenous base attached to each sugar (Figure 16–8). The sugar units are always the same—deoxyribose—but four different nitrogenous bases are found in all DNA. It is the variation of the order in which the bases occur in the chain that allows the infinite variety possible in genetic material. The four bases are adenine, guanine, thymine, and cytosine. The first two are called purine bases and the latter two pyrimidine bases. One of the interesting early observations was the fact that the ratio of adenine to thymine and the ratio of guanine to cytosine are always very close to 1 regardless of the source of the DNA. Moreover, the sequence of bases on a single chain is not restricted. Considering the great lengths of DNA chains, the number of permutations possible for four bases is very large indeed.

In 1953, J. D. Watson and F. H. C. Crick, working in England, proposed a molecular structure that accounted for the known structural features of DNA and that suggested a mechanism for the replication of genetic material. They proposed that DNA consists of two intertwined chains with the phosphate groups on the outside of the double helix and the bases on the inside. The two chains are held together by hydrogen bonding between bases on different chains. Watson and Crick, by the use of scale molecular models, demonstrated that a smooth regular double helix with acceptable bond lengths could fit together only if certain bases were paired up by hydrogen bonding. Thus, the only pairs that can bond are adenine with thymine and guanine with cytosine. In other words, in order to form a stable double helix, if one chain has an adenine group in a given position, the other chain must have a thymine group at the corresponding position opposite the adenine. Figure 16–9 shows these two pairings, and Figure 16–10 shows diagrammatically the overall double helix structure.

269

DNA chain structure

Bases

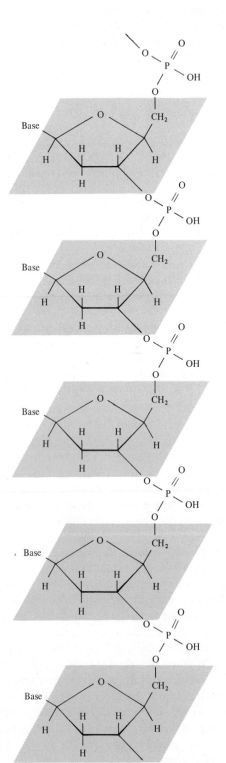

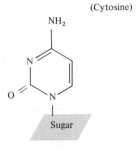

(Adenine)

(Guanine)

(Cytosine)

(Thymine)

Figure 16–8. Schematic structure of a single DNA chain. The four bases are shown at right.

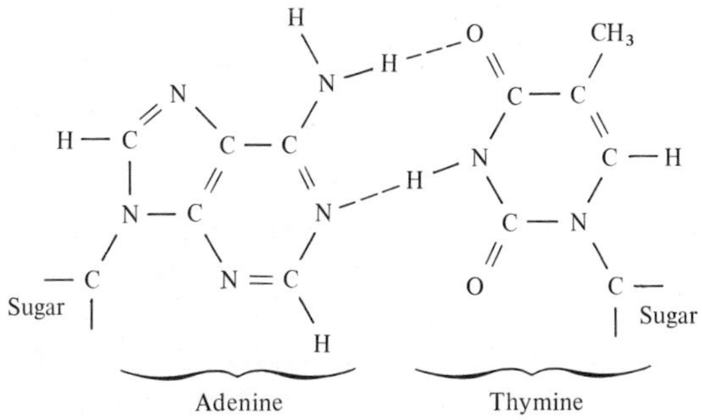

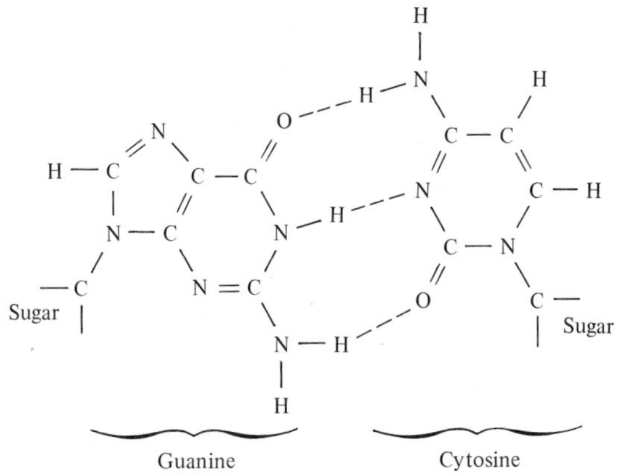

H
\
N — H --- O   CH₃
/   ‖   /
N   C — C
‖   N   =N   /   ‖
H — C   C — C   \   C — H
| ‖   \   —H --- N   /
N — C   N   C — N
/   \   /   ‖   \
— C   N = C   O   C —
Sugar | \ | Sugar
H

Adenine   Thymine

H
|
N
/
O --- H — N   H
‖   \   /
=N   C — C
H — C   C — C   ‖   ‖
| ‖   \   —H --- N   C — H
N — C   N   /
/   \   /   C — N
— C   N = C   \
Sugar | \   O   C —
N — H | Sugar
|
H

Guanine   Cytosine

Figure 16-9. Pairing of adenine-thymine and pairing of guanine-cytosine.

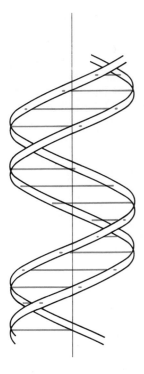

Figure 16-10. Diagrammatic representation of DNA double helix. The two ribbons represent the two phosphate-sugar chains and the horizontal rods represent the base pairs that hold the chains together. The vertical line represents the fiber axis.

The Watson-Crick model presented not only a geometrically satisfying representation that conformed to the chemical and X-ray data, but also suggested the mechanism for biological replication. One can imagine that, during replication, the single chains of the double helix become separated upon rupture of the hydrogen bonds. Each resulting chain is a complement of the other, and each can act as a template for the formation of a new double helix. Because of the rigid space requirements, only the precise complementary base can fit in a given position along the growing chain. Hence two new double helices can form from the two single chains, and they are both identical to the original double helix.

The Watson-Crick model of DNA along with Pauling's work on the structure of proteins resulted in a tremendous amount of research activity and in the growth of the new field of science called molecular biology.

PROBLEMS

1. List the ways in which the following physical properties of polymers differ from those of ordinary molecular substances:
   (a) molecular weight
   (b) viscosity
   (c) density
   (d) melting point
   (e) boiling point

2. List five examples of naturally occurring polymeric substances.

3. List five examples of synthetic polymers.

4. Draw structures of the repeating units in each of the following polymers:
   (a) polyethylene
   (b) natural rubber
   (c) nylon
   (d) polystyrene
   (e) silk
   (f) orlon
   (g) dacron
   (h) glyptal

5. How do thermoplastic and thermosetting polymers differ with respect to (a) structure, (b) melting point, (c) viscosity?

6. Why is extreme purity especially important in the synthesis of polymers?

7. Define condensation polymer and write equations for the preparation of two examples.

8. Define addition polymer and write equations for the preparation of two examples.

9. What is meant by the term crystallinity of polymers? How does a "crystalline" polymer differ from an ordinary crystalline substance?

10. What molecular feature determines the degree of crystallinity in polymers?

11. When a stretched rubber band (Section 16–6) is allowed to relax, its temperature can be observed to drop. Explain this temperature drop.

12. Why does polymer strength increase with chain length for a given polymer?

13. Why does nylon have so much greater mechanical strength than polyethylene?

14. Of what advantage is the synthesis of stereoregular polymers?

15. Of what advantage are the linear rigid polymers?

16. What are the repeating units of proteins?

17. Name and draw the structures of ten amino acids. For each, describe the R group in the generalized structure.

$$\begin{array}{c} COOH \\ | \\ H_2N-C-H \\ | \\ R \end{array}$$

18. Describe two important and commonly occurring secondary structures for proteins.

19. Give several common household examples of protein denaturation.

20. One common example of protein hydrolysis occurs during digestion. Write an equation to describe the hydrolysis of polyglycine.

21. What are the two contributors to the overall chirality of a polypeptide in the α-helical structure?

22. Describe protein tertiary structure and the various factors that account for it.

23. How does the Watson-Crick structure of DNA account for
   (a) the invariant 1 to 1 ratios of adenine to thymine and guanine to cytosine?
   (b) the seemingly infinite variety of DNA molecules in genetic material?
   (c) the transmission of genetic information from generation to generation in a given organism.

24. Correlate the chemical structures in column I with the appropriate names in column II and the correct properties in column III. (There may be more than one entry from each column.)

Column I	Column II	Column III
(a)	1. amino acid 2. polymer 3. peptide 4. thermoplastic 5. thermosetting	1. insoluble 2. dissymmetric molecule 3. soluble in water 4. insoluble in water
(b) $-(CH_2CH_2)_n-$ (n = very large)		
(c)		
(d)		

25. Explain the properties of each of the following substances on the basis of its molecular structure. **(a)** water; **(b)** polyethylene; **(c)** rubber.

26. Polymers may be classified in several ways:
   (a) addition-condensation
   (b) polyester-polyamide-, and so on
   (c) elastomer-thermosetting, thermoplastic
   (d) amorphous-crystalline
   (e) linear-crosslinked

   Library Problem: Look up and classify the following commercial products employing each of the above classifications: **(a)** nylon; **(b)** dacron; **(c)** kodel; **(d)** orlon; **(e)** lycra; **(f)** lucite; **(g)** bakelite; **(h)** acrilan.

METALLIC STRUCTURES

Metals are commonly distinguished from the nonmetals by their lustrous appearance and their relatively high conductivity for heat and electricity. In general we think of a metal as a substance capable of being beaten or forged into different shapes (malleability) and of being drawn into wires (ductility). However, the degree to which various metals possess these properties varies greatly. The control of the properties of metals by alloying with other metals has become a major study in the science of metallurgy. Chemically, metals are characterized by their relatively low ionization potentials and zero or extremely low electron affinities. Over three quarters of the elements are metals. All crystalline substances can be classified into six basic crystal systems. Yet X-ray diffraction studies have revealed that the structures of most of the metals consist primarily of three types within two of the six basic systems. This chapter will explore the structures and bonding that characterize metals.

17-1. CRYSTAL SYSTEMS

Classification of all the many crystalline substances that have been investigated into six systems according to external structure has been found possible. The classification is based on the angles between intersecting faces and the relative lengths of the sides of the crystal. The six systems are:

1. **Cubic.** All sides of equal length which intersect at 90° angles.

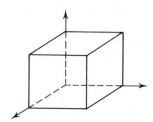

2. **Tetragonal.** One crystal dimension unequal to the other two but faces intersect at 90° angles.

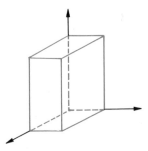

3. **Orthorhombic.** Unequal dimensions but faces intersect at 90° angles.

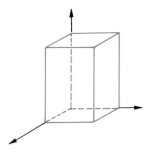

4. **Hexagonal.** Six rectangular faces intersect in pairs at 120° angles to give a regular hexagonal cross section. The hexangonal faces are perpendicular to the other six sides.

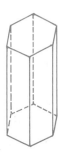

5. **Monoclinic.** Unequal dimensions. One pair of opposite faces is perpendicular to a second pair of opposite faces but not to the third pair of opposite faces.

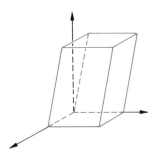

6. **Triclinic.** Unequal dimensions and no pairs of faces intersect at 90° angles.

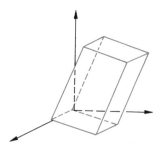

17-2.   CLOSE-PACKED STRUCTURES

Metal atoms commonly arrange themselves in what is called a close-packed structure. Such an arrangement is the most economical way to pack spheres of the same size, that is, to utilize as effectively as possible the available space. Imagine ping pong balls filling a beaker (Figure 17–1). The balls will always stack together in a manner that uses minimum space.

Now, consider a layer of spheres on top of which successive layers will stack as illustrated in Figure 17–2. The spheres of the second layer fit into the hollows where three spheres come together in the first layer. The hollows filled by the second layer are indicated by the X's in Figure 17–2. Half of the hollows in the first layer, however, will not be filled, and these positions are also indicated in Figure 17–2. A third layer of spheres can be placed in two different ways on top

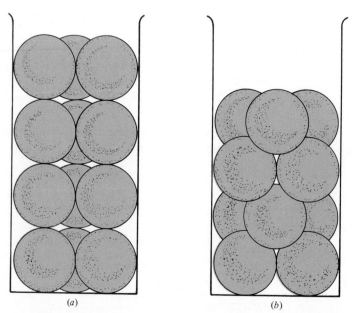

(a)	(b)

Figure 17-1. Ping pong balls will fill space as economically as possible. (a) 12 balls in a non-close-packed arrangement—not observed. (b) 12 balls in a close-packed arrangement as observed.

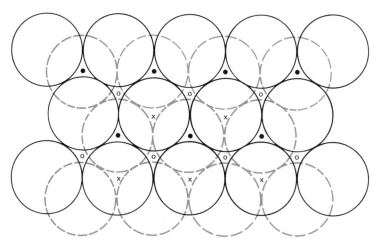

Figure 17-2. Layers of close-packed spheres building on top of each other. Dotted circles represent first layer and solid circles the second layer. x = hollows in the first layer filled by spheres in the second layer. **o** = hollows in the first layer not filled by spheres in the second layer. ● = hollows in the second layer directly over spheres in the first layer.

of the second layer. The spheres in the third layer will occupy the hollows in the second layer, but there are two different types of hollows. The solid dots (●) in Figure 17–2 indicate hollows in the second layer that lie directly above a sphere in the first layer. If the third layer of spheres occupy positions identical with the positions of spheres in the first layer, and this stacking is continued, the sequence ABABAB results. This type of close-packed arrangement is called **hexagonal close packing.** On the other hand, if the spheres of the third layer fit into the hollows in the second layer indicated by the open dots (o) in Figure 17–2, the third layer spheres have positions different from those in the first two layers. The fourth layer of spheres would become identical with the first. This stacking, when continued indefinitely, gives the sequence ABCABCABC. This type of arrangement is called **cubic close packing.**

Either of these close-packed structures will fill the maximum amount of available space namely, 76%. Thus only 24% of available space is unused when spheres of equal size are arranged in a close-packed structure. Likewise, in either of these structures, each atom has 12 nearest neighbors and the crystallographic coordination number of a close-packed arrangement is thus 12.

In treating the structures of crystalline substances, the positions of the atoms are usually defined by reference to a **unit cell.** A unit cell consists of the minimum number of atoms necessary to show the properties of symmetry characteristic of the entire crystal. In a sense, the macroscopic crystal may be imagined to be built up from countless unit cells. Figure 17–3 displays unit cells for the hexagonal and cubic close-packed structures. The relationship between the hexagonal close-packed structure and the hexagonal unit cell is fairly obvious. Not so obvious is the relationship between the cubic close-packed structure and the unit cell as it is normally described. The cubic unit cell is called a face-centered cubic arrangement because it is formed from eight atoms at the corners of a cube with six additional atoms situated in the center of each of the six faces of the cube. The cubic close-packed arrangement can indeed be described in terms of the face-centered cubic unit cell, but it is difficult to visualize this correspondence without the use of models.

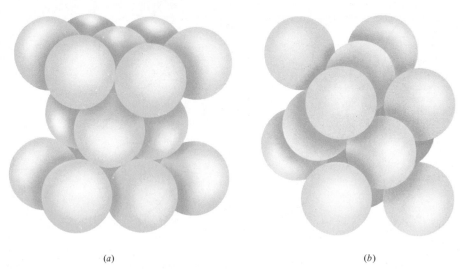

(a)   (b)

Figure 17-3.   Unit cells for close-packed structures (a) hexagonal unit cell (b) face-centered cubic unit cell.

It must be realized that the selection of the atoms comprising a unit cell is arbitrary. Thus, an atom lying at the corner of a cubic unit cell could just as well be considered to constitute the corner atom of any one of the seven adjacent unit cells (Figure 17–4). On the average, then, only one eighth of a corner atom of a cubic unit cell lies within a given unit cell. Figure 17–4 shows that the face of a cube is shared by the two adjacent unit cells. Therefore, each atom in the face of a cube in the face-centered cubic arrangement belongs, on the average, one half to one unit cell and one half to the other. The total number of atoms per unit cell for the face-centered structure is thus calculated

$$8 \text{ corner atoms} \times \tfrac{1}{8} \text{ atom/corner atom} = 1 \text{ atom}$$
$$6 \text{ face-centered atoms} \times \tfrac{1}{2} \text{ atom/face-centered atom} = 3 \text{ atoms}$$
$$\text{Total} = 4 \text{ atoms}$$

A similar analysis for the hexagonal unit cell reveals that there are six atoms per unit cell.

Avogadro's number can be determined by measuring the unit cell dimensions using X-ray diffraction. For example, copper crystallizes in a cubic close-packed

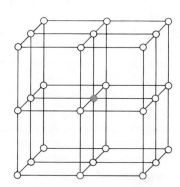

Figure 17-4.   A corner atom in a cubic arrangement is shared by eight unit cells. This can easily be seen for the shaded atom. The unit cells pictured are simple cubic unit cells.

Li H	Be H																
Na H	Mg H											Al C					
	Ca C H	Sc C H	Ti H				Fe C	Co C H	Ni C H	Cu C	Zn H						
	Sr C	Y H	Zr H		Mo H	Tc H	Ru H	Rh C	Pd C	Ag C	Cd H	In C					
		La C H	Hf H			Re H	Os H	Ir C	Pt C	Au C		Tl C H	Pb C				

Ce C H	Pr H	Nd H	Pm H	Sm H	Eu C	Gd H	Tb H	Dy H	Ho H	Er H	Tm H	Yb C	Lu H
Th C				Pu C	Am H								

Figure 17–5. Metals with close-packed crystal structures. *H* indicates hexagonal close-packing and *C* indicates cubic close-packing.

arrangement. The volume of the face-centered cubic unit cell for copper is 4.725 $\times 10^{-23}$ cm.^{3} The density of copper is 8.94 g/cm^3 and the atomic weight is 63.54. We therefore find,

$$\frac{63.54 \text{ g/g-mole} \times 4 \text{ atoms/unit cell}}{8.94 \text{ g/cm}^3 \times 4.725 \times 10^{-23} \text{ cm}^3/\text{unit cell}} = 6.02 \times 10^{23} \text{ atoms/g-mole}$$

A large number of metals crystallize in one or the other of the close-packed structures. Some of them exist in one form up to a certain temperature and change to the other structure above that temperature. Figure 17–5 shows which metals assume one or both of the close-packed arrangements.

17–3. BODY-CENTERED CUBIC STRUCTURE

Although the close-packed structures are the most efficient in terms of maximum utilization of available space, a third structure is nearly as efficient. The body-centered cubic unit cell is pictured in Figure 17–6. This structure leads to occupancy of 68% of available space by the atoms. Each atom has eight nearest neighbors and six more atoms only about 15% farther away. The coordination number is usually described as (8 + 6). There are an average of two atoms per unit cell in the body-centered arrangement because the atom in the center of the unit cell is completely contained and the corner atoms each contribute only one eighth to a given unit cell.

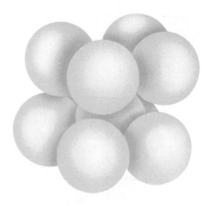

Figure 17-6. Body-centered cubic unit cell.

Figure 17–7 shows the metals which crystallize in a body-centered cubic arrangement. Notice that some of these metals also have a close-packed modification. The metals not shown in Figures 17–5 or 17–7 either have structures of a complex nature or have not yet been determined.

17-4. ALLOYS

An alloy is a solid mixture of two or more metals. Basically, three different types of alloys are recognized: (1) heterogeneous mixtures of the metals, (2) solid solutions, and (3) intermetallic compounds. The heterogeneous mixture is simply

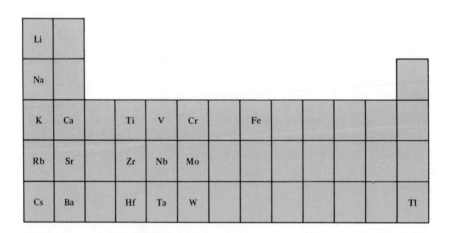

Figure 17-7. Metals with a body-centered cubic crystal structure.

a solid mixture of the metals consisting of two or more phases. A solid solution is a homogeneous mixture of solids that can vary in composition within certain solubility limits. Just as sugar has a range of solubility in water, so certain metals have a range of solubility in another metal in the solid state. The properties of a liquid solution change with the amount of solute dissolved, and, in similar fashion, the properties of solid solution alloys are altered as their composition is changed. There is a difference between a solid solution and its liquid analog, however. In a liquid solution, the solute molecules are randomly dispersed in the solvent. In a solid solution, the degree of order may vary. Changes in the degree of atomic order within a solid solution of metals can sometimes be induced by heat treatment.

Alloys that are characterized by a more or less definite stoichiometry, in contrast with the variable composition of a solid solution, are called intermetallic compounds. However, the term *compound* does not usually carry the same meaning in this context as it does when considering ordinary compounds. That is, the stoichiometries of intermetallic compounds often cannot be presently justified on a theoretical basis. Thus we have formulas such as $MgZn_2$, $LiHg$, Cu_5Sn, and Cu_3Al to give only a few examples.

Sometimes a metal near the nonmetal region of the periodic table will behave as a nonmetal toward a very reactive metal. An example of this behavior is found in the compound $NaTl$. It appears that this compound is really "sodium thallide" where the sodium atom has transferred an electron to the thallium atom. The structure of this compound is similar to that expected of a typical ionic compound.

The order-disorder relationship of the metal atoms in an alloy is usually an important consideration for both solid solutions and intermetallic compounds. An interesting example of this relationship is found in the intermetallic compound or alloy known as β-brass, which is a 50-50 mole per cent mixture of copper and zinc. The disordered phase has a body-centered cubic arrangement in which it is equally probable for a zinc or copper atom to occupy any position. Slow heating of the solid alloy between 200 and 470°C leads to a transformation into a highly ordered structure. The body-centered cubic structure now consists of unit cells in which one kind of atom occupies all of the corner positions of the unit cubes whereas the other atom occupies all of the center positions in the unit cells. In an extended body-centered cubic arrangement, the corner positions are identical with the center position as far as coordination number is concerned. Thus, each corner position plays the role of center position in some unit cell while the center position of one unit cell is a corner of another unit cell. Figure 17–8 shows a number of body-centered unit cells to illustrate the equivalence of the atoms in this type of structure. Each zinc atom thus has eight nearest neighbor copper atoms and each copper atom eight neighboring zinc atoms.

Many of the transition metals react at high temperatures with the nonmetals hydrogen, boron, carbon, silicon, or nitrogen to form what are called interstitial compounds. Although these substances are not strictly alloys, as we have defined the term, they are physically similar to alloys, and it is convenient to mention them here. Apparently, the small atoms of these nonmetals can slip into the spaces or intersticies in the structure of the metal.

Generally speaking, these compounds are much harder and have higher melting points than the pure metals. The electrical conductivity is high, like that of the metal, and the physical appearance is metallic. Often, metals that have

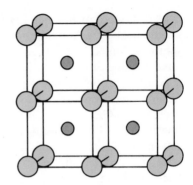

Figure 17-8. Body-centered cubic structure of the ordered phase of β-brass. ⬤ indicates the copper atoms and ⬤ the zinc atoms.

a cubic close-packed structure form interstitial compounds of the type MX, and metals with a hexagonal close-packed structure form M_2X type interstitial compounds (X stands for hydrogen, boron, carbon, silicon or nitrogen). There is a degree of variability in the stoichiometries of many of these compounds, however. Although only the smaller nonmetal atoms form these interstitial species with the transition metals, more is involved than simply a small atom filling a vacancy in the metal crystal. Some degree of bonding between the metal and nonmetal atoms must be present to account for the great thermal stability and hardness of these compounds.

The basic structures of the metals are often altered considerably by combination with the nonmetals, yet in other cases the metal retains its structure. For example, the cubic close-packed structures of scandium, lanthanum, cerium, and thorium are retained in the carbides and nitrides of these metals. It should be noted that the interstitial hydrides, borides, carbides, silicides, and nitrides described here are quite different from the ionic substances that result from the reaction of these nonmetals with very reactive metals such as sodium and calcium. The ionic compounds react readily with water and have none of the physical properties of hardness, electrical conductivity, and metallic luster associated with the interstitial compounds of the transition metals.

17-5. METALLIC BONDING

Metals are those elements with relatively few valence shell electrons. There are always unfilled orbitals in the valence shell of a metal atom, so metals have much more room for electrons in their valence shells than there are available electrons. In view of the high coordination numbers of metal atoms in their crystal structures, it is fairly obvious that ordinary electron pair bonds between all adjacent metal atoms are impossible. Likewise, because the electron affinities of metals are quite low or zero, any exchange of electrons among metal atoms to form positive and negative ions would seem to be extremely improbable. Indeed, the properties of metals indicate that there are neither strong, highly directed covalent bonds nor ions held by coulombic attractions, since metals can be deformed without breaking. Also, at least some of the electrons must be free to circulate throughout the metal in order to explain the characteristic electrical conductivity. The **metallic**

bond is obviously different from either the highly directional electron pair bond between nonmetals or the nondirectional but rigid ionic bond that causes ionic substances to be brittle and resist deformation.

The atomic orbital theory pictures resonance of an electron pair bond among all nearest neighbor metal atoms. Lithium, for example, in its body-centered cubic structure has eight closest neighbors and six more not much farther away (8 + 6 coordination); yet each lithium atom has but one valence shell electron. The bonding is pictured in terms of resonance structures involving electron pair bonds among all possible pairs of lithium atoms as well as "ionic" structures involving two bonds to one atom and none to another.

$$\begin{bmatrix} \text{Li:Li} \\ \text{Li:Li} \end{bmatrix} \quad \begin{bmatrix} \text{Li} & \text{Li} \\ \cdots & \cdots \\ \text{Li} & \text{Li} \end{bmatrix} \quad \begin{bmatrix} \overline{\text{Li}}\text{:Li} \\ \cdots & + \\ \text{Li} & \text{Li} \end{bmatrix} \quad \text{and so on}$$

The *average* bond between any two atoms is only a fraction of the conventional electron pair bond. The "extra" empty orbitals available in the valence shell make possible the contribution illustrated by the third structure and account for the movement of charge through the crystal, that is, electrical conductivity.

As one moves across a period in the periodic table, the number of valence shell electrons increases, so the degree of bonding should also increase somewhat. In general, the densities of metals do increase in moving from left to right across a given period. The electrons in the *d* orbitals of the transition metals must be included as valence shell electrons for such considerations. Of course the number of valence shell electrons alone do not determine the density of a metal; crystal structure is an important consideration along with the size of the metal atoms. Most of the densest metals have a close-packed structure as might be expected. Figures 17–9 and 17–10 show the periodic variation of density and atomic radius for the metals.

An alternate theoretical view of metallic bonding is provided by a molecular orbital treatment. The electronic energy states in a metallic crystal take on the character of bands of permitted energy called **Brillouin zones.** These Brillouin zones are composed of many closely spaced molecular orbitals so that each zone of permitted energies can contain many electrons. Between these permitted energy states are regions of forbidden energies analogous to the forbidden energies between the individual atomic orbitals for electrons in an isolated atom. The molecular orbitals within a given Brillouin zone are filled with electrons, two per orbital as usual, so that the highest energy zones contain relatively few electrons. The lower energy zones are of relatively small breadth and resemble closely in energy the inner electrons in the isolated atoms. However, the highest energy zones are associated with all the atoms in the crystal. Electrons in these energy states are no longer identified with particular atoms but belong to the crystal as a whole and are thus mobile. The molecular orbital approach has been very successful in explaining certain metallic properties, for example, electrical conductivity. The atomic orbital approach provides a somewhat more pictoral representation of the bonding in metals.

Figure 17-9. Periodic variation of the densities of metals. Densities are given in grams per cubic centimeter at 25°C.

Li 0.97	Be 1.85														
Na 0.97	Mg 1.74											Al 2.7			
K 0.87	Ca 1.54	Sc 2.99	Ti 4.54	V 6.11	Cr 7.19	Mn 7.44	Fe 7.87	Co 8.90	Ni 8.9	Cu 8.94	Zn 7.13	Ga 5.91	Ge 5.32		
Rb 1.53	Sr 2.6	Y 4.47	Zr 6.45	Nb 8.57	Mo 10.22	Tc 11.49	Ru 12.4	Rb 12.44	Pd 12.02	Ag 10.49	Cd 8.65	In 7.31	Sn 7.3	Sb 6.68	
Cs 1.87	Ba 3.5	La 6.17	Hf	Ta 16.6	W 19.3	Re 21.02	Os 22.5	Ir 22.42	Pt 21.4	Au 19.32	Hg 13.55	Tl 11.85	Pb 11.34	Bi 9.8	Po 9.40
Fr	Ra 5														

Ce 6.66	Pr 6.78	Nd 7.00	Pm	Sm 7.54	Eu 5.26	Gd 7.90	Tb 8.27	Dy 8.54	Ho 8.80	Er 9.05	Tm 9.32	Yb 6.98	Lu 9.84
Th 11.66	Pa 15.37	U 19.07	Np 20.45	Pu 19.84	Am 11.87	Cm	Bk	Cf	Es	Fm	Md	No	Lw

Figure 17-10. Periodic variation of the atomic radius for metal atoms. Radii are given in Ångstrom units (1 Å = 10^{-8} cm).

Li 1.55	Be 1.12														
Na 1.90	Mg 1.60											Al 1.43			
K 2.35	Ca 1.97	Sc 1.62	Ti 1.47	V 1.34	Cr 1.27	Mn 1.26	Fe 1.26	Co 1.26	Ni 1.25	Cu 1.24	Zn 1.28	Ga 1.38	Ge 1.41	Ge 1.37	
Rb 2.48	Sr 2.15	Y 1.80	Zr 1.60	Nb 1.46	Mo 1.39	Tc 1.36	Ru 1.34	Rh 1.34	Pd 1.37	Ag 1.44	Cd 1.54	In 1.66	Sn 1.62	Sb 1.59	
Cs 2.67	Ba 2.22	La 1.87	Hf 1.59	Ta 1.46	W 1.39	Re 1.37	Os 1.37	Ir 1.36	Pt 1.39	Au 1.44	Hg 1.57	Ti 1.71	Pb 1.75	Bi 1.70	Po 1.76
Fr	Ra														

Ce 1.65	Pr 1.65	Nd 1.64	Pm	Sm 1.66	Eu 1.85	Gd 1.61	Tb 1.59	Dy 1.59	Ho 1.59	Er 1.57	Tm 1.56	Yb 1.70	Lu 1.56

PROBLEMS

1. What is meant by close-packed structures? Describe the two types of close packing.

2. Compare the face-centered unit cell and the body-centered unit cell with respect to arrangement of the atoms and number of atoms per unit cell.

3. Name the different types of alloys. Give examples of these types.

4. Of the six basic crystal systems, which is the most symmetrical? Which is least symmetrical?

5. Verify that equal spheres of radius r packed in a cubic close-packed arrangement fill 76% of available space. (The face diagonal would be equal to $4r$ and the side of the unit cell would thus equal $4r/\sqrt{2}$).

6. Verify that equal spheres of radius r packed in a body-centered cubic arrangement fill 68% of available space. (The body diagonal of the unit cell would be equal to $4r$ and the side of the unit cell would thus equal $4r/\sqrt{3}$).

7. There is an arrangement known as the simple cubic structure (Figure 17-4). The unit cell consists simply of eight spheres at the corners of a cube. (Remember that this means there would be only one sphere on the average per unit cell.) Calculate the per cent space occupied by equal spheres of radius r in a simple cubic arrangement. Why do you suppose that no metal crystallizes with a simple cubic structure?

8. Calculate the density of aluminum using only the knowledge of its crystal structure (Figure 17-5) and atomic radius (Figure 17-10). (at. wt. Al = 27.0)

9. Calculate Avogadro's number using the information given for cesium in Figures 17-7, 17-9, and 17-10. (at. wt. Cs = 133).

10. Calculate the atomic radius for iron using the information given in Figures 17-5 and 17-9. (at. wt. Fe = 55.8)

11. Examine Figures 17-5 and 17-7. Can you give any generalization about the structures of metals according to position in the periodic table? Is there any correlation between structure and density for the metals? (Figure 17-9)

12. Make a table using Figures 17-5 and 17-7 to show which metals can exist in only one structure, which can exist in two structures, and which in all three. Is there any sort of periodic correlation in this respect?

13. What sort of entropy change is associated with the conversion of β-brass from the disordered structure to the ordered state? Is there an analogy between this type of transition and the transitions between the various physical states of matter?

14. Predict the formulas of the carbides of the following elements: Zr, Mo, Fe, Co, Ir.

15. Contrast the compounds NaH and TaH with respect to appearance, electrical conductivity in the solid state, hardness, and chemical reactivity. Rationalize the above comparisons on the basis of the structure and bonding found in these two compounds.

16. Compare and contrast covalent, ionic, and metallic bonding with respect to the basic forces involved, the directional properties of these bonds, and the physical properties of the substances utilizing these different types of bonds.

17. Examine Figure 17-9 and then try to give some explanation for the decrease in the densities of the post-transition metals.

18. What properties of metals are best described by the atomic orbital theory of metallic bonding? What properties are best described by the molecular orbital approach?

STRUCTURES OF IONIC COMPOUNDS

The nature of the ionic bond has been treated in Chapter 12. In this chapter we will describe three of the most important ionic structures and some energy relationships common to all ionic substances.

18-1. ROCK SALT—AN IMPORTANT MX STRUCTURE

The most important structure of ionic substances composed of cations and anions with the same magnitude of charge is the **rock salt** or **sodium chloride** structure. All of the group I metal halides except CsCl, CsBr, and CsI crystallize with this structure as well as all of the oxides and sulfides of group II metals except beryllium. In addition, a number of other substances approximate this structure.

To describe the sodium chloride structure, we can begin by imagining the chloride ions in an expanded face-centered cubic arrangement (Figure 17–3b) so that the ions are not in contact. The spaces between any two of the chloride ions along a side of a unit cell are then found to be surrounded by six chloride ions disposed at the corners of a regular octahedron centered on the space. Such spaces are referred to as octahedral holes. In Figure 18–1, several face-centered unit cells are shown with one of the octahedral holes marked with an X.

Careful examination of Figures 18–1 and 17–3b will reveal that there are twelve spaces between atoms along the edges of a face-centered cubic unit cell and that each of these spaces or holes is shared by four unit cells. This sharing is clearly seen in the case of the octahedral hole marked X in Figure 18–1. Thus on the average, a single unit cell will have only one fourth of each of these octahedral holes or $\frac{1}{4} \times 12 = 3$ octahedral holes per unit cell. However, in a face-centered cubic arrangement, there is one octahedral hole at the center of the unit cell, wholly contained by the unit cell. Therefore, the total number of octahedral holes per unit cell is four, the same as the average number of atoms per unit cell.

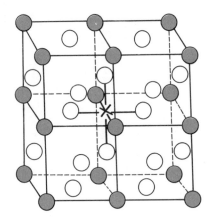

Figure 18-1. An octahedral hole in a face-centered cubic arrangement.

Return now to our face-centered cubic chloride ion structure. If all of the octahedral holes are filled with sodium ions, the rock salt structure is obtained (Figure 18–2). The cation and anion are equivalent in the rock salt structure, that is, we can describe it equally well in terms of a face-centered cubic arrangement of sodium ions with chloride ions occupying all the octahedral holes. The cation and anion are each surrounded by six ions of the opposite charge and thus the coordination number of each is six. Only cations and anions with the same magnitude of charge can crystallize with the rock salt structure since the stoichiometry is of the type MX, and electrical neutrality must be maintained.

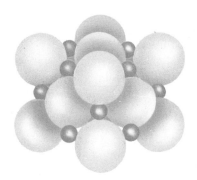

Figure 18-2. The rock salt structure. Large spheres represent the chloride ions and the small spheres the sodium ions.

18–2. FLUORITE—AN IMPORTANT MX$_2$ STRUCTURE

A structure assumed by a number of salts in which the cationic charge is twice that of the anion is the **fluorite** or **calcium fluoride** structure. Among the fluorides crystallizing in this structure are CaF_2, SrF_2, BaF_2, CdF_2, HgF_2, PbF_2, and EuF_2. Strontium chloride, $SrCl_2$, also has this structure as do the following oxides: ThO_2, CeO_2, HfO_2, ZrO_2, PoO_2, PrO_2, NpO_2, PuO_2, AmO_2, CmO_2, and UO_2.

We can describe the fluorite crystal in terms of a face-centered cubic arrangement of calcium ions with the fluoride ions occupying all of the tetrahedral holes. A tetrahedral hole is situated at a position where a corner atom of the cubic unit cell and the three atoms in the centers of the adjacent faces form a tetrahedral arrangement.

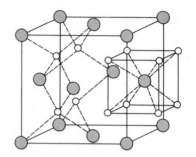

Figure 18-3. Fluorite structure: ● represent calcium ions; ○ represent fluoride ions. Each fluoride ion is tetrahedrally surrounded by calcium ions and each calcium ion is surrounded by eight fluoride ions at the corners of a cube.

Figure 18-3 illustrates the fluorite structure. Note that there are eight tetrahedral holes in each unit cell (one near each corner of the cube) and that they are all wholly contained within the unit cell. The coordination number of each fluoride ion is four because each is tetrahedrally surrounded by four calcium ions. The coordination number of the calcium ions is eight because each is surrounded by eight fluoride ions situated at the corners of a cube with respect to each calcium ion. Extra fluoride ions from adjacent unit cells are shown in Figure 18-3 to illustrate the eightfold coordination of calcium. Thus, it can be seen that, unlike the rock salt structure, the cation and anion are not structurally equivalent in the fluorite structure.

18-3. ANTIFLUORITE—AN IMPORTANT M_2X STRUCTURE

As the name implies, the antifluorite structure is one in which the roles of cation and anion are reversed from those they play in the fluorite structure.

Most of the alkali metal compounds involving dinegative anions crystallize with the antifluorite structure. For example, sodium monoxide, Na_2O, can be described in terms of a face-centered cubic arrangement of O^{2-} ions with Na^+ ions occupying the tetrahedral holes. The coordination number of O^{2-} is thus eight whereas that of Na^+ is four.

We have considered only three of the more important ionic structures. There are a number of others based on the face-centered cubic structure as well as on the hexagonal close-packed arrangement (which also has octahedral and tetrahedral holes). There are also structures based on less symmetrical arrangements. However, the general approach to a consideration of ionic structures is adequately illustrated by the examples which have been treated.

18-4. THE CALCULATION OF LATTICE ENERGIES

From a theoretical point of view, the treatment of ionic bonding is considerably easier than the treatment of bonding in covalent compounds. The attraction between anions and cations and the repulsion between ions with the same charge can be calculated using Coulomb's law. Therefore, the energy that would theoretically be released if 1 mole of an ionic solid were formed from its gaseous ions can be calculated provided the structure is known. This energy is the lattice energy of the crystal and is a direct measure of the stability of the ionic structure.

The actual calculation of lattice energies requires one to take into account several forces: (1) attractions between a given cation and its nearest neighbor anions, (2) attractions between the cation and more remote anions throughout the crystal, (3) repulsions between the cation and the other cations in the crystal, and (4) repulsions between adjacent ions regardless of their charges due to electronic interactions. Naturally, the structure of the crystal determines the geometric relationship that exists among the ions. The lattice energies given in Table 12–3 are theoretical values that were calculated by taking these effects into account.

In general, it is not possible to determine the lattice energy by direct experimental means. However, use of the conservation of energy law in a particular way known as the Born-Haber cycle has made possible the indirect experimental determination of lattice energies. Consider the reaction between solid lithium and gaseous fluorine to form one mole of lithium fluoride.

$$Li(s) + \tfrac{1}{2}F_2(g) \longrightarrow LiF(s)$$

The heat of reaction has been measured and is known to be $\Delta H = -145.7$ kcal.

Now let us break the overall reaction into a series of steps whose associated heats of reaction are known. As long as we start with solid lithium and gaseous fluorine and end with solid lithium fluoride, the sum of the heats associated with these intermediate steps must equal the overall heat of reaction. The conservation of energy assures us that this statement is true. Let us consider the following steps:

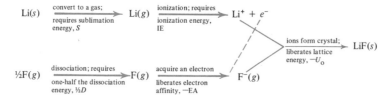

The conservation of energy requires

$$\Delta H = S + IE + \tfrac{1}{2}D - EA - U_0 \qquad (18\text{--}1)$$

Notice that nothing has been said about how the reaction actually takes place. Equation (18–1) is simply a statement of energy conservation for an overall process that has been broken into a number of steps for convenience. Substitution of experimental values for ΔH, S, IE, $\tfrac{1}{2}D$, and EA in equation (18–1) allows the calculation of U_0. It must be remembered that energy released is assigned a negative value, whereas energy absorbed is assigned a positive value. Electron affinity and lattice energy, however, are defined so that they are positive terms even though they are exothermic; therefore, the negative signs are introduced to keep them consistent with the heat of reaction which is exothermic and thus negative.

Note: The confusion that often results from such arbitrary definitions, although regrettable, is found frequently in science. One would think that heats evolved in any process could be designated by the same sign. It is a fact of scientific life, however, that with the multitudes of people and countries that participate in the development of science, differences of opinion and inconsistencies in conventions are bound to arise. We ask the student to be aware of these inconsistencies. If he is to become a practicing scientist, he may one day have a hand in their correction.

Experimental determination gives $\Delta H = -145.7$ kcal for the reaction; $S = 38.4$ kcal for Li; IE $= 124.4$ kcal for Li; $\frac{1}{2}D = 18.9$ kcal for F_2; and EA $= 79.5$ kcal for F. Substitution of these values in equation (18-1) gives

$$-145.7 = 38.4 + 124.4 + 18.9 - 79.5 - U_0$$

$$U_0 = 247.9 \text{ kcal}$$

A theoretical calculation for the lattice energy of LiF gives, $U_0 = 246.8$ kcal. The correspondence between theoretical and experimental values is thus seen to be quite good. In fact, sometimes theoretical lattice energies are used in a Born-Haber cycle, that is, equation (18-1), in conjunction with other known quantities to calculate one of the other energy terms that may be hard to determine experimentally. Electron affinities are often determined in this manner.

Table 18-1 presents data for theoretical and experimental values of the lattice energies of the alkali halides. Notice the reasonably close correspondence between the two values for the different compounds. Notice too that the experimental values are always slightly larger than the theoretical values. One reason for this fact may be that a small degree of covalent character is present even in highly ionic substances. The theoretical calculations that were used to obtain the values in Table 18-1 make no allowance for any such covalent contribution to the bonding.

Lattice energies are usually fairly large compared with the other energy terms in the Born-Haber cycle. Ionization energies are relatively large endothermic terms and, as such, represent an unfavorable energy factor in the formation of an ionic substance. Were it not for the offsetting exothermic nature of the lattice energy, most ionic substances would be energetically unstable. Stated in a slightly different way, if the ionization energy for the formation of a given cation is so large that the lattice energy of the resulting ionic crystal can not overcompensate for it, then the formation of an ionic substance is unlikely. Such cases are found in reactions between nonmetal atoms, and we have already seen that electron sharing with the formation of a covalent bond is common in these reactions.

Table 18-1. Theoretical and Experimental Lattice Energies for the Alkali Halides (kcal/mole)

Alkali Metal	Fluoride		Chloride		Bromide		Iodide	
	Theo.	Exp.	Theo.	Exp.	Theo.	Exp.	Theo.	Exp.
lithium	246.8	247.9	202.0	204.4	190.7	195.7	176.8	182.5
sodium	218.7	220.0	185.9	188.1	176.7	179.5	165.4	168.0
potassium	194.4	195.4	169.4	171.3	162.4	164.4	153.0	154.7
rubidium	185.9	187.0	164.0	164.8	157.5	158.0	148.7	149.7
cesium	178.7	182.9	155.9	160.9	151.1	155.3	143.7	147.2

18-5. SOLVATION EFFECTS

The interaction between a solvent and a molecule or ion dissolving in it is called **solvation,** and the energy change accompanying the process is called **solvation energy.** Ionic substances in general will dissolve only in polar solvents such as water. The solvation energy of the ions must be great enough to supply at least part of the lattice energy in order to break down the ionic crystal structure, otherwise the material is insoluble in the solvent. Polar solvent molecules will have relatively large attractions for the cations and anions whereas nonpolar solvent molecules will have little attractions for the ions and hence will be unable to remove the ions from the stable, crystal lattice.

The overall heat of solution of a substance is the heat effect at 1 atm and 25°C when 1 mole of the material is dissolved in a very large amount of solvent. The heat of solution of an ionic compound will thus depend upon the relative magnitudes of the lattice energy and the heats of solvation of the ions. If the heats of solvation are larger than the lattice energy, the excess energy will appear as an exothermic heat of solution. On the other hand, if the solvation energies are less than the lattice energy, the energy deficit will have to be supplied by the solvent or the surroundings, and the heat of solution will be endothermic. Many examples of both types of behavior have been observed. Potassium hydroxide has such a high exothermic heat of solution in water that considerable heating of the solution is noticeable as it is dissolved. Potassium iodide, on the other hand, produces a definite cooling effect as it is dissolved in water. In a few cases, the solvation energy almost exactly supplies the lattice energy so that the heat of solution is essentially zero. Lithium nitrate displays this type of behavior.

PROBLEMS

1. Why can it be said that the cations and anions of a rock salt structure are structurally equivalent whereas those of a fluorite or antifluorite structure are not equivalent?

2. The intermetallic compound, vanadium carbide, has the rock salt structure. **(a)** Describe the structure of this compound by analogy with a close-packed structure. **(b)** Give the simplest formula of vanadium carbide. **(c)** What is the coordination number of the vanadium atoms? the carbon atoms?

3. Titanium hydride has the fluorite structure with titanium atoms in an approximately face-centered cubic arrangement. **(a)** Give the simplest formula of titanium hydride. **(b)** Give the coordination numbers of the titanium and hydrogen atoms.

4. A carbide of iron can be described as a face-centered cubic arrangement of iron atoms with only one fourth of the octahedral holes occupied by carbon atoms. Predict the simplest formula of this compound.

5. Explain the following trends in melting points in terms of lattice energy effects
   (a) NaF, 992°C   KF, 880°C
   (b) KF, 880°C CaF_2, 1330°C
   (c) K_2SO_4, 1069°C $CaSO_4$, 1450°C

6. $CaSO_4$ and CaF_2 are both fairly insoluble in water, whereas K_2SO_4 and KF are quite soluble. Explain this behavior on the basis of lattice energies and expected solvation effects. How can you explain the fact that $MgSO_4$ is quite soluble in water?

7. Why is it possible to calculate lattice energies for many salts to a high degree of accuracy, whereas accurate covalent bond energies are extremely difficult to calculate?

8. Use the experimental value of the lattice energy of LiCl in Table 18–1 and the other necessary energy terms given in the text to calculate the electron affinity (EA) of chlorine. $\Delta H = -96.0$ kcal/mole. See Problem 9 for Cl_2 dissociation energy.

9. Calculate the expected heat of reaction for the formation of a hypothetical ionic compound, Cl^+F^-, from its elements.

$$\tfrac{1}{2}Cl_2(g) + \tfrac{1}{2}F_2(g) \longrightarrow ClF(s)$$

The dissociation energies are Cl_2, $D = 57.8$ kcal/mole: F_2, $D = 37.8$ kcal/mole. IE for chlorine is 300 kcal and EA of fluorine is 79.5 kcal. Assume that the "ionic" ClF would have $U_0 = 230$ kcal/mole.

(a) The actual ClF is a covalent compound, is gaseous at 25°C and $\Delta H = -25.6$ kcal. for the above reaction. Compare the actual ΔH with the hypothetical ΔH for the ionic formulation.

(b) In terms of energy, why is ClF covalent rather than ionic?

10. What factors determine the heat effect associated with dissolving a salt in water?

CHEMICAL
DYNAMICS

3

CHEMICAL EQUILIBRIUM

In Chapter 6, we considered an equilibrium between the liquid and vapor of a pure substance. Similarly, phase equilibria between a solid and its vapor and between a liquid and solid were discussed. The present chapter will extend this very fundamental concept of equilibrium to include chemical reactions.

19-1. THE CONCEPT OF EQUILIBRIUM

Many examples of systems in equilibrium are found in our daily lives. Common examples are a perfectly balanced see-saw, or a tug of war in which the two sides are balanced. These are examples of **static equilibria** because the equilibria exist without motion. Still another example of a static equilibrium is a person standing on a bathroom scale.

Another type of equilibrium can exist in which considerable motion takes place. This is called **dynamic equilibrium.** Obvious examples are not as plentiful; however, one can imagine two rows of baseball pitchers warming up by pitching balls at each other as fast as they can. At any given instant of time one can imagine that most of the balls will be in the air. A similar example is a snowball fight. If the pitchers do not tire, a dynamic equilibrium will be established in which, for practical purposes, there will constantly be a fixed number of balls in the air somewhere between the two rows.

As you will see in the following pages, this example of a dynamic equilibrium is a fairly good model for chemical systems in equilibrium.

19-2. CHEMICAL EQUILIBRIUM

Up to the present, we have dealt mainly with reactions that were presumed to occur completely. For example, a hydrocarbon burns in air and is converted 100%

into CO_2 and H_2O. However, if a chemical reaction is confined in a container so that products cannot escape, it is found that no reaction is absolutely 100% in its extent (although some are nearly so). Instead, the usual result is a condition in which some amount of all the reactants and products are present. Furthermore, the concentrations of all these substances do not change with time as long as external conditions of temperature and pressure are unaltered. This condition is called **chemical equilibrium.**

The reaction between carbon monoxide and water to form hydrogen and carbon dioxide is reasonably fast at 600°C. However, if 1 mole/liter each of CO and H_2O are mixed in a closed container and held at 600°C, the reaction will produce only two thirds of the expected amount of products.
Reaction:

$$CO + H_2O \longrightarrow CO_2 + H_2$$

Initial concentrations:

$$[CO] = 1.0 \qquad [CO_2] = 0.0$$
$$[H_2O] = 1.0 \qquad [H_2] = 0.0$$

Final concentrations at equilibrium:

$$[CO] = 0.33 \qquad [CO_2] = 0.67$$
$$[H_2O] = 0.33 \qquad [H_2] = 0.67$$

Use of square brackets is a conventional notation to indicate moles per liter, or molarity, of the species in the brackets.

Experiment shows that if 1 mole/liter each of the products in the above reaction are mixed (that is, CO_2 and H_2) and held at 600°C, equilibrium will be reached with the same final concentrations as those given above. In chemical notation, an equilibrium is expressed by an equation with two arrows pointing in opposite directions.

$$CO + H_2O \rightleftharpoons CO_2 + H_2$$

If one were to study the loss of reactants and the formation of products with time, graphs could be constructed. Figure 19–1 shows such a graph for the initial reaction above. After a period of time, no further changes in the concentrations of the species occur, and equilibrium is established. If the entire experiment were carried out by starting with CO_2 and H_2 (products in the reaction as written above), eventually a similar condition of unchanging concentrations would be reached. Studies of the change of concentrations of species in a chemical reaction show that equilibrium is attained in a system when the **rates** of the forward and reverse reactions become equal.

Consider the reaction exemplified by Figure 19–1. At the start of the reaction when only CO and H_2O are present, there is no reverse reaction between CO_2 and H_2. The forward reaction is rapid because there are a lot of CO and H_2O molecules present to react. However, as CO and H_2O are used up, CO_2 and H_2 molecules are formed and begin to react with each other. The reverse reaction

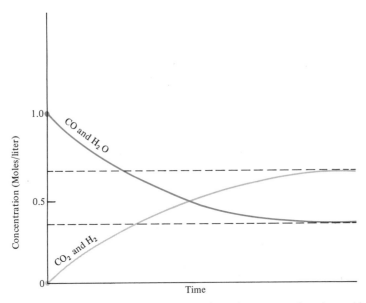

Figure 19-1. Graph of change in concentrations of reactants and products with time for the reaction: $CO + H_2O \rightarrow CO_2 + H_2$ at 600°C.

(reaction between CO_2 and H_2 to form CO and H_2O) speeds up as more molecules of CO_2 and H_2 are formed. Thus, the forward reaction slows down while the reverse reaction speeds up. Eventually the rates of these reactions will become equal, and, although both reactions continue, no overall changes in the concentrations of the species occur. At this point a dynamic equilibrium will have been established. Figure 19–2 is a graphical representation of this view of chemical equilibrium.

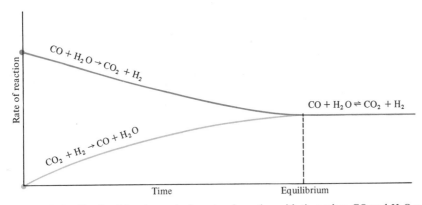

Figure 19-2. Graph of the change in the rate of reaction with time when CO and H_2O are mixed.

19-3. THE EQUILIBRIUM CONSTANT

If the CO–H_2O equilibrium that we have been discussing is attained by starting with various initial concentrations of the materials involved, different experimental

Table 19-1. Equilibrium Concentrations of Species for Various Initial Concentrations for the Reaction: $CO + H_2O \rightleftharpoons CO_2 + H_2$ at 600°C

	Initial Concentrations				Equilibrium Concentrations			
Experiment Number	[CO]	[H₂O]	[CO₂]	[H₂]	[CO]	[H₂O]	[CO₂]	[H₂]
1	1.0	1.0	0.0	0.0	0.33	0.33	0.67	0.67
2	0.0	0.0	1.0	1.0	0.33	0.33	0.67	0.67
3	1.0	1.0	1.0	1.0	0.67	0.67	1.33	1.33
4	2.0	1.0	1.0	0.0	1.25	0.25	1.75	0.75
5	10.0	1.0	0.0	0.0	9.03	0.03	0.97	0.97
6	10.0	10.0	1.0	1.0	3.67	3.67	7.33	7.33

equilibrium concentrations would be obtained. Some values are given in Table 19-1. It should be clear from an examination of Table 19-1 that the particular equilibrium concentrations that are realized depend upon the initial concentrations of the materials involved in the reaction.

A superficial consideration of these various experimental values, however, shows no relationship among the initial concentrations nor the equilibrium concentrations. A more careful consideration, though, reveals an amazing relationship among the equilibrium concentrations! The ratio of the concentrations of products multiplied together to the concentrations of reactants multipled together is very nearly a constant. No such relationship is expected nor does it exist for the initial concentrations, however. Table 19-2 illustrates this interesting result involving the concentrations of substances that are in chemical equilibrium.

Notice that the ratio of the equilibrium concentrations varies slightly, but the individual results cluster around their average value. The variation is due to experimental uncertainty. Notice too that there is no relationship among the initial concentrations, although different initial concentrations cause the actual equilibrium concentration values to be different. Still, the ratio of these equilibrium values is very nearly constant! This relationship has been verified by experiment for many different equilibrium systems. The constancy of this ratio of equilibrium concentrations has led to its being given a special name—the **equilibrium constant,** K. For the reaction we have been considering:

Table 19-2. Ratio of Concentrations of Products to Reactants for the Equilibrium Values Recorded in Table 19-1

Experiment Number	Ratio of Equilibrium Concentration $\dfrac{[CO_2][H_2]}{[CO][H_2O]}$
1	4.12
2	4.12
3	3.94
4	4.20
5	3.47
6	3.99
	average value = 3.97

$$CO + H_2O \rightleftharpoons CO_2 + H_2$$

$$K = \frac{[CO_2][H_2]}{[CO][H_2O]} = 4.0 \text{ at } 600°C$$

Several important points must be noted in using the equilibrium constant expression.

1. Convention demands that the concentrations of the products of the reaction, *as written,* be placed in the numerator and the reactant concentrations in the denominator.
2. The concentrations of the species in the numerator and in the denominator are always *multiplied* together.
3. A given equilibrium constant applies only to a given chemical reaction at a definite temperature. At a different temperature, the equilibrium constant, even for the same reaction, will have a different value.
4. Only *concentration terms* appear in an equilibrium constant expression. Various concentration scales are employed by chemists, for example sometimes mole fraction is used to express concentration. In this book, the molarity scale will usually be employed, as indicated by the square brackets.

Once the value of the equilibrium constant is known, the equilibrium concentrations that would result from given initial conditions can be calculated.

■ **Example 19-1**

Assume that 2 moles of CO and 1 mole of H_2O are mixed in a 1 liter container and held at $600°C$ until equilibrium is established. What will be the concentrations of all species at equilibrium?

□ *Solution*

$$CO + H_2O \rightleftharpoons CO_2 + H_2$$

$$K = \frac{[CO_2][H_2]}{[CO][H_2O]} = 4.0 \text{ at } 600°C$$

Initially, $[CO] = 2.0$; $[H_2O] = 1.0$; $[CO_2] = [H_2] = 0.0$. Therefore at equilibrium, $[CO] = 2.0 - x$; $[H_2O] = 1.0 - x$; and $[CO_2] = [H_2] = x$.

Therefore,

$$\frac{(x)(x)}{(2.0 - x)(1.0 - x)} = 4.0$$

$$\frac{x^2}{2.0 - 3.0x + x^2} = 4.0$$

$$3x^2 - 12x + 8 = 0$$

Use of the quadratic formula gives, $x = 0.845$. Finally, equilibrium concentrations are calculated,

$$[CO] = 2.0 - 0.845 = 1.2$$

$$[H_2O] = 1.0 - 0.845 = 0.2$$

$$[CO_2] = [H_2] = 0.845 = 0.85$$

It is customary to include in the equilibrium constant only those species whose concentrations can *change*. For example, the equilibrium that can be established at high temperatures with limestone, calcium oxide, and carbon dioxide, has species whose *concentrations* are invariant.

$$CaCO_3(s) \rightleftharpoons CaO(s) + CO_2(g)$$

The ordinary equilibrium constant (designated here as K') can be written

$$K' = \frac{[CaO][CO_2]}{[CaCO_3]} \qquad (19\text{--}1)$$

However, $CaCO_3$ and CaO are pure solids and consequently have fixed *concentrations* at a fixed temperature. Although the *amounts* of these substances can change, their concentrations cannot. The concentration of a solid is simply the number of moles of it that would occupy 1 liter. This value is fixed regardless of how much or how little of it one might have. The concentration of the CO_2, however, could easily be changed, for example, the volume of the container could be changed. Therefore, if we write equation (19–1) as

$$\frac{[CaCO_3]}{[CaO]} K' = [CO_2] \qquad (19\text{--}1a)$$

it can be seen that the left-hand side of equation (19–1a) is a constant. The equilibrium constant is then simply defined as

$$K = \frac{[CaCO_3]}{[CaO]} K' \qquad (19\text{--}2)$$

Comparing equations (19–1a) and (19–2) we have

$$K = [CO_2] \qquad (19\text{--}3)$$

Equation (19–3) is the conventional way to write the equilibrium constant for the limestone decomposition reaction. The rule is thus: *Equilibrium constant expressions do not contain pure liquids or solids because the concentrations of such substances are fixed at a given temperature.* Note that only *pure* liquids are omitted. If some solute is dissolved in a liquid, then, in general, by changing the amount of solute, the concentrations of both solute and solvent will change. If such a solvent is involved as a reactant or product in the reaction, its concentration would have to be included in the equilibrium constant expression.

19-4. THE POSITION OF EQUILIBRIUM

A chemical equilibrium can be disturbed by increasing or decreasing the concentration of one or more of the species involved in the reaction. As long as the temperature is not changed, the equilibrium that is reestablished is governed by the same equilibrium constant. Therefore, some reaction will occur to produce more products (a shift to the right) or to produce more reactants (a shift to the left). These shifts in the reaction are referred to as shifts in the position of the equilibrium.

An example will show why these shifts are necessary to maintain equilibrium.

$$CO + H_2O \rightleftharpoons CO_2 + H_2$$

$$K = \frac{[CO_2][H_2]}{[CO][H_2O]} = 4.0 \text{ at } 600°C$$

If 1 mole each of CO and H_2O is placed in a 1-liter container and equilibrium is established at 600°C, the concentrations will be fixed according to the requirements of the above expression. (Refer to experiment number 1 in Table 19–1.)

$$\frac{[0.67][0.67]}{[0.33][0.33]} = 4.0$$

If now, 1 mole each of CO_2 and H_2 is introduced into the mixture and equilibrium is reestablished at 600°C, the net effect cannot be simply a 1 mole/liter increase in the concentrations of CO_2 and H_2 because

$$\frac{[1.67][1.67]}{[0.33][0.33]} = 26$$

This result would violate the **law of chemical equilibrium,** which states that the above ratio of concentrations must be 4.0 as long as the temperature is 600°C. Reference to experiment number 3 in Table 19–1 shows what actually happens when the concentrations of CO_2 and H_2 are increased by 1 mole/liter. The new equilibrium is established so that

$$\frac{[1.33][1.33]}{[0.67][0.67]} = 4.0$$

Reaction occurs to use up some of the additional CO_2 and H_2 and produce a little more CO and H_2O to make the ratio of concentrations conform to the equilibrium constant. The equilibrium has shifted to the left in order to maintain the requirements of equilibrium, namely constancy of the equilibrium constant at a given temperature. In short, reaction will occur in an equilibrium system as necessary to maintain the concentration ratio demanded by the equilibrium constant.

Henri Le Chatelier stated this principle in 1884 in somewhat different terms. **Le Chatelier's principle** is usually formulated: *If possible, an equilibrium reaction will shift its position of equilibrium so as to oppose any change that might be imposed on the reaction.* Using this principle, a qualitative prediction for the preceding example is possible. Addition of the extra mole each of CO_2 and H_2 to the equilibrium mixture constitutes a change that the reaction can oppose by shifting the position of equilibrium to the left. Some of the additional CO_2 and H_2 are thereby used up and, in the process, some more CO and H_2O are produced.

Occasionally, a change imposed on an equilibrium requires no shift in its position to maintain the system in equilibrium. A pressure-volume alteration requires no shift if the same number of *gaseous* molecules appear on both sides of the equation. Consider once again the equilibrium given by experiment number 1 in Table 19–1.

$$\frac{[0.67][0.67]}{[0.33][0.33]} = 4.0$$

If the volume of the container is decreased by one half, the concentration of all species will double (assuming no change in temperature). But such a change requires no shift in the position of equilibrium because

$$\frac{[1.34][1.34]}{[0.66][0.66]} = 4.0$$

The increase in the concentrations of the species in this case leaves the ratio unchanged. In terms of Le Chalelier's principle, the decrease in volume at constant temperature causes an increase in pressure. However, there is no way the above reaction can oppose this pressure increase because reaction in either direction leaves the total number of gaseous molecules unchanged.

The situation is quite different for pressure changes imposed on equilibrium reactions in which different numbers of gaseous molecules appear on the two sides of the equation. Consider the gaseous reaction:

$$3\ H_2 + N_2 \rightleftharpoons 2\ NH_3$$

$$K = \frac{[NH_3][NH_3]}{[H_2][H_2][H_2][N_2]} = 6 \times 10^{-2} \text{ at } 500°C$$

or, more compactly,

$$K = \frac{[NH_3]^2}{[H_2]^3[N_2]} = 6 \times 10^{-2} \text{ at } 500°C$$

If 1 mole of NH_3 is placed in a 1 liter container and maintained at $500°C$ until equilibrium is established, the equilibrium constant is satisfied if

$$[NH_3] = 0.2$$

$$[H_2] = 1.2$$

$$[N_2] = 0.4$$

because

$$\frac{[0.2]^2}{[1.2]^3[0.4]} = 6 \times 10^{-2}$$

Now suppose the volume of the container were increased from 1 liter to 2 liters at constant temperature. A simple reduction of the concentration of each species to one half the values given above is not possible because

$$\frac{[0.1]^2}{[0.6]^3[0.2]} = 2 \times 10^{-1}$$

which violates the equilibrium law. Conformity with the equilibrium requirement means that the new concentrations are

$$[NH_3] = 0.06$$
$$[H_2] = 0.66$$
$$[N_2] = 0.22$$

because

$$\frac{[0.06]^2}{[0.66]^3[0.22]} = 6 \times 10^{-2}$$

In other words, the position of the equilibrium has shifted somewhat to the left in order to maintain constancy of the equilibrium constant. Note that the *concentrations* of all the species have decreased, as would be expected for a volume increase. However, the decrease is not proportionate because, if it had been, the equilibrium constant could not have remained constant. Le Chatelier's principle predicts qualitatively the direction of the shift as follows: An increase in volume causes a decrease in pressure. The system can oppose a pressure decrease by reacting to produce more gaseous molecules. Hence, additional NH_3 decomposes into H_2 and N_2 since two molecules of NH_3 produce a total of four molecules of H_2 and N_2 ($3 H_2 + N_2$). Conversely, for this reaction, an *increase* in the pressure would have caused a shift in the position of the equilibrium to the right, in the direction of reducing the total number of molecules in the system.

Moderate pressure changes affect only equilibria involving *gaseous* substances. Even so, we have seen that, unless the number of gaseous molecules on either side of the equation is different, pressure will have no effect on the position of equilibrium.

19-5. TEMPERATURE EFFECT ON EQUILIBRIUM

A change in the temperature causes a change in the value of the equilibrium constant. Le Chatelier's principle is useful in predicting how the equilibrium constant will change provided it is known whether the reaction is endothermic or exothermic as written. Conversely, studies of how the equilibrium constant is effected by temperature provides information about the heat of reaction, ΔH. There is, in fact, a quantitative relationship between K and temperature which

Table 19-3. The Value of K at Different Temperatures for the Reaction: $CO + H_2O \rightleftharpoons CO_2 + H_2$

Temperature (°C)	Equilibrium Constant (K)
600	4.0
700	1.6
1000	0.60

involves ΔH. Measurements of K at different temperatures are often carried out to allow calculation of ΔH for a reaction. We shall be concerned only with the qualitative relationship among these parameters, however.

Table 19-3 shows the value of K at different temperatures for the $CO-H_2O$ reaction.

Energy is liberated as CO and H_2O react to produce CO_2 and H_2, that is

$$CO + H_2O \rightleftharpoons CO_2 + H_2 + energy$$

An increase in temperature amounts to an introduction of energy into the system, and the reaction opposes this change by reacting in the direction that uses up energy, namely the endothermic reaction between CO_2 and H_2 to produce more CO and H_2O. Equilibrium will then be established at the higher temperature with a larger concentration of CO and H_2O and a smaller concentration of CO_2 and H_2. The equilibrium constant is thus smaller. A decrease in temperature causes the reaction to produce energy by reacting in the exothermic direction to give more CO_2 and H_2 at the expense of CO and H_2O, thereby increasing the value of K. We may summarize the direction in which reaction occurs with temperature changes as: increasing the temperature favors the endothermic change; decreasing the temperature favors the exothermic change. The value of the equilibrium constant will be affected accordingly, depending upon which side of the equation energy appears.

PROBLEMS

1. How does a static equilibrium differ from a dynamic equilibrium? Into which category do chemical equilibria fall?

2. What is the law of chemical equilibrium?

3. Why are pure liquids and solids not included in equilibrium constant expressions?

4. What is the effect of temperature changes on an equilibrium constant? on the equilibrium concentrations?

5. State Le Chatelier's principle. What connection is there between this principle and the law of chemical equilibrium?

6. Under what conditions do pressure changes at constant temperature affect the position of equilibrium? the equilibrium constant?

7. Fill in the following table to show how the various quantities would be affected by each of the imposed conditions after the following equilibrium is reestablished in each case.

$$3\,H_2 + N_2 \rightleftharpoons 2\,NH_3 \qquad \Delta H = -22\,\text{kcal}$$

Use the symbols: + for increase, − for decrease, 0 for no change, L for left, R for right, and I for impossible to predict.

Imposed condition	$[H_2]$	$[N_2]$	$[NH_3]$	Value of K	Position of Equilibrium
Some H_2 is added at constant volume and temperature					
Some NH_3 is added at constant volume and temperature					
The volume is doubled at constant temperature					
The temperature is increased at constant volume					
Some H_2 and NH_3 are both added at constant volume and temperature					
Some H_2 is added and temperature is decreased at constant volume					
Volume is decreased and temperature is decreased					
Volume is decreased and temperature is increased					
Some NH_3 is added, volume is increased, and temperature is increased					

8. For each of the following equilibrium reactions, predict which ones would be affected by volume changes at constant temperature, and tell how the position of equilibrium would change for an increase or decrease in volume.
   (a) $2\,SO_2(g) + O_2(g) \rightleftharpoons 2\,SO_3(g)$
   (b) $H_2(g) + I_2(g) \rightleftharpoons 2\,HI(g)$
   (c) $PCl_5(g) \rightleftharpoons PCl_3(g) + Cl_2(g)$
   (d) $3\,Fe(s) + 4\,H_2O(g) \rightleftharpoons Fe_3O_4(s) + 4\,H_2(g)$
   (e) $N_2(g) + O_2(g) \rightleftharpoons 2\,NO(g)$
   (f) $C(s) + CO_2(g) \rightleftharpoons 2\,CO(g)$

9. The equilibrium constant for the reaction

$$2\,SO_2 + O_2 \rightleftharpoons 2\,SO_3$$

decreases as the temperature increases. **(a)** Is this reaction, as written, endothermic or exothermic? Explain. **(b)** How would the equilibrium constant for the reaction $2\,SO_3 \rightleftharpoons 2\,SO_2 + O_2$ be affected by temperature changes? Explain. **(c)** What is the relationship between the equilibrium constants for the two reactions above?

10. If 1 mole of CO and H_2O are mixed in a 1-liter container and equilibrium is reached, calculate the concentrations of CO, H_2O, CO_2 and H_2 if **(a)** equilibrium is established at 700°C (see Table 19–3). **(b)** Equilibrium is established at 1000°C. **(c)** Compare your answers to parts (a) and (b) with experiment number 1 in Table 19–1. Are all of these results consistent with Le Chatelier's principle? Explain.

11. If 2.00 moles of HI are placed in a 1.00 liter container and held at 500°C until equilibrium is established, it is found that 1.56 moles of HI remain. Calculate K at 500°C for the equilibrium

$$H_2 + I_2 \rightleftharpoons 2\,HI$$

12. In carbon tetrachloride solution, the equilibrium

$$N_2O_4 \rightleftharpoons 2\,NO_2$$

has $K = 5 \times 10^{-5}$ at 25°C. If a 0.10 M solution of N_2O_4 in carbon tetrachloride is prepared at 25°C, what per cent of the N_2O_4 is dissociated into NO_2? (Note that a very small number may be neglected if it is added to or subtracted from a much larger number.)

13. One mole each of PCl_5, PCl_3, and Cl_2 are confined in a 1-liter container. Calculate the equilibrium concentrations of each species at 250°C.

$$K = \frac{[PCl_3][Cl_2]}{[PCl_5]} = 4.2 \times 10^{-2} \text{ at } 250°C$$

14. Calculate the value of K at 250°C for the reaction

$$PCl_3 + Cl_2 \rightleftharpoons PCl_5$$

from information given in problem 13.

15. At constant temperature, the molarity of a gas is proportional to its pressure (or partial pressure if in a mixture). Accordingly, equilibrium constants for gaseous reactions often utilize partial pressures of the gases involved rather than their molarities. Such an equilibrium constant using partial pressures is called a K_p. For example the reaction

$$2\,SO_2 + O_2 \rightleftharpoons 2\,SO_3$$

has

$$K_p = \frac{p_{SO_3}^2}{p_{SO_2}^2 \cdot p_{O_2}}$$

(a) Assuming all the gases are ideal in behavior, prove that the relationship between K and K_p for this reaction is, $K = K_p(RT)$, where R is the gas constant and T is the absolute temperature.

(b) Prove that for any generalized gaseous equilibrium,

$$aA + bB \rightleftharpoons cC + dD$$

as long as the gases are ideal, the relationship between K and K_p is,

$$K_p = K(RT)^{\Delta n}$$

where,

$$\Delta n = c + d - a - b$$

ELECTROLYTIC DISSOCIATION

The use of electrical conductance in the classification of matter was discussed in Chapter 12. In this chapter, we shall consider the effect upon electrical conductivity of dissolving substances in water. Historically, such studies led to the first real understanding of the nature of conducting solutions or electrolytes.

20-1. ELECTROLYTIC SOLUTIONS

During the latter part of the nineteenth century, considerable information was amassed on the electrical conductance of solutions. For instance, certain substances such as sugar or ethyl alcohol are nonconductors either in the pure state or dissolved in water. These materials are referred to as **nonelectrolytes.** On the other hand, other substances, such as sodium chloride, are good conductors of electricity both in the pure liquid state and in water solutions. More interesting, however, are substances that are nonconductors in the pure liquid state but whose water solutions are conductors. Liquid hydrogen chloride is a nonconductor, but in water this substance becomes a good conductor and is hence a **strong electrolyte.** Liquid acetic acid is also a nonconductor which becomes conducting in water, although not nearly to the same degree as hydrogen chloride solutions. Acetic acid is called a **weak electrolyte** because of the relatively poor conducting qualities of its aqueous solutions.

Based on the effect of a solute upon the conductance of water (itself a non-conductor), the following classifications can be made:

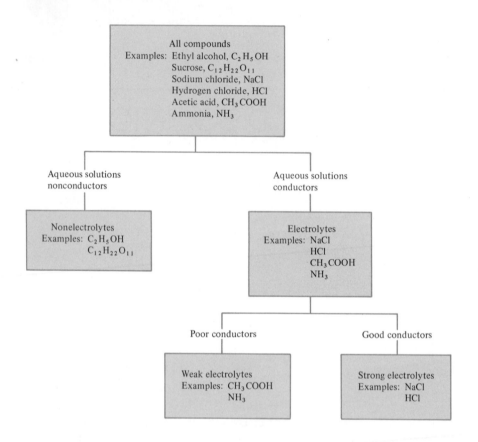

20-2. COLLIGATIVE PROPERTIES OF SOLUTIONS

A colligative property is any property that depends only on the *number* of particles present and not on their *identity*. Thus the pressure of a gas at constant volume and temperature is a colligative property, but the term is not normally applied except to liquid solutions. The properties of solutions that depend only on the number of solute particles are freezing point lowering, boiling point elevation and its related property vapor pressure lowering, and osmotic pressure.

The use of antifreeze in radiators takes advantage of the lowering of the freezing point of a liquid by the addition of some solute. For the effective elevation of the boiling point of a solvent, the solute must be nonvolatile. Otherwise, the solute may simply boil off and leave pure solvent with its original boiling point. A nonvolatile solute, however, will cause pure solvent to vaporize at a higher temperature than usual, depending upon the concentration of solute. Vapor pressure lowering of a liquid must occur if its boiling point is raised, since higher boiling points are associated with lower vapor pressures (see Chapter 6).

Osmotic pressure is a very interesting and important colligative property of a liquid. It is explained best by reference to Figure 20-1. The semipermeable

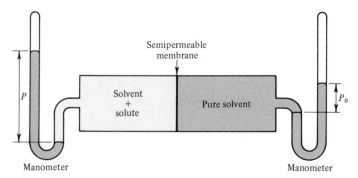

Figure 20-1. Schematic representation of the property of osmotic pressure. Osmotic pressure, $\pi = P - P_0$

membrane allows solvent molecules to pass through it but does not allow solute particles to pass. The tendency for solvent molecules to pass into the solution chamber and dilute the solution gives rise to a pressure differential across the membrane that is termed the osmotic pressure of the solution. Certain polymer films will function as a semipermeable membrane with respect to some solute particles. Osmotic pressures of several hundred atmospheres can develop in tall trees where the cellular walls in the roots function as semipermeable membranes and are largely responsible for the flow of water to the top of the tree.

The magnitude of a colligative property for dilute solutions is found to be proportional to the mole fraction of the solute. For example, the freezing point depression, ΔT_f, may be expressed for dilute solutions as,

$$\Delta T_f = K_f X_2 \tag{20-1}$$

where K_f is a proportionality constant that is called the freezing point depression constant. The value of K_f is different for each solvent. And X_2 denotes the mole fraction of the solute, that is

$$X_2 = \frac{\text{moles solute}}{\text{moles solvent} + \text{moles solute}}$$

Equation (20-1) is limited to rather small values of X_2 (that is, dilute solutions), say 0.01 or less.

Within this restriction, Equation (20-1) is found to hold for a given solvent and to have the same value of K_f for all *nonelectrolytes*. For water, $K_f = 105$, so that a freezing point depression of about 1° results when $X_2 = 0.01$. For solutions of weak electrolytes, K_f is approximately constant with a value usually only a few per cent larger than that for nonelectrolytes. However, strong electrolytes cause a large variation in K_f for a particular solvent. Jacobus van't Hoff used a different expression, as given by equation (20-2)

$$\Delta T_f = i K_f X_2 \tag{20-2}$$

The i in Equation (20-2) is called the **van't Hoff factor** and is simply the observed freezing point depression divided by the calculated depression using equation (20-1).

$$i = \frac{\Delta T_{f(\text{observed})}}{K_f X_2} \qquad (20\text{-}2\text{a})$$

Table 20–1 gives the i values for a few substances, rounded to the nearest whole number.

Table 20–1. Van't Hoff i Factor for Different Compounds

Compound	i	Classification
$C_{12}H_{22}O_{11}$	1	nonelectrolyte
C_2H_5OH	1	nonelectrolyte
CH_3COOH	1	weak electrolyte
NaCl	2	strong electrolyte
HCl	2	strong electrolyte
$MgCl_2$	3	strong electrolyte

20–3. ARRHENIUS' THEORY OF ELECTROLYTIC DISSOCIATION

In 1887, Svante Arrhenius, a young Swedish chemist, submitted a doctoral thesis in which he proposed a new and far reaching theory that accounted for the seemingly unrelated phenomena of the electrical conductance of solutions and the variation of van't Hoff i factors for various substances. In addition, his theory dealt with acid-base reactions, but we shall not be concerned with this aspect for the present.

Two essential postulates in Arrhenius' electrolytic dissociation theory were (1) substances that conduct an electric current in solution (electrolytes) are dissociated into ions when dissolved in the solvent and (2) such dissociation is not always complete, but may involve an equilibrium between dissociated and undissociated molecules. These ideas explain why sugar dissolved in water will not conduct a current whereas a hydrogen chloride solution will. The latter molecules react with water to form H^+ and Cl^- ions, which are responsible for the conductivity. Sugar molecules, on the other hand, do not dissociate into ions in solution and hence are nonconducting.

A weak electrolyte, such as acetic acid (CH_3COOH), does not react completely to form H^+ and CH_3COO^- (acetate) ions, and an equilibrium is established between the undissociated molecules and the ions. Arrhenius showed how accurate conductance data or i factors can be used to calculate the equilibrium constants. For example, a 1 M acetic acid solution at 0°C has an i factor of 1.004.

$$CH_3COOH \underset{\text{water}}{\rightleftharpoons} H^+(\text{aq}) + CH_3COO^-(\text{aq})$$

If there were no dissociation, the i factor would be 1.000. The small increase is caused by the fact that 0.4% of the neutral molecules are dissociated into ions. Therefore, at equilibrium

$$[CH_3COOH] = 1.000 - 0.004 = 0.996$$

$$[H^+] = 0.004$$

$$[CH_3COO^-] = 0.004$$

and,

$$K = \frac{[H^+][CH_3COO^-]}{[CH_3COOH]} = \frac{[0.004][0.004]}{[0.996]}$$

$$K = 1.6 \times 10^{-5}$$

Variation of the acetic acid concentration causes the i factor to vary, but the calculated equilibrium constant is a constant within experimental error. Arrhenius' theory thus not only explained a mass of experimental data but also provided a new use for such data in the calculation of the equilibrium constants governing the dissociation of weak electrolytes in solution.

20-4. DEBYE-HÜCKEL THEORY

Arrhenius attempted to extend the idea of an equilibrium between ions and undissociated molecules to strong electrolytes. He did this because accurately determined i factors for such solutions are not exactly whole numbers. A 0.001 M NaCl solution has $i = 1.97$. According to Arrhenius, complete dissociation into Na^+ and Cl^- ions should yield an i factor of 2.00. Consequently, he reasoned that some undissociated NaCl "molecules" were in equilibrium with the ions. However, attempts to calculate equilibrium constants were unsuccessful in that the "constants" vary as the concentration of the electrolyte is changed. An alternative theory is used today to explain the deviations of dilute solutions of strong electrolytes from the expected behavior.

In the early 1920s, a theory was developed by Peter Debye and Erich Hückel to explain the behavior of dilute solutions of strong electrolytes. The basic assumptions of the theory are as follows: (1) In dilute solutions, strong electrolytes are completely dissociated into ions (that is, there is no equilibrium between ions and "undissociated molecules"). (2) Except for solvation effects, the only interaction involving the ions is due to their electrical charges, that is, if the ions lost their charge, the solution would behave like one containing an equal number of nonelectrolyte particles.

Because like charges repel and opposite charges attract, a given ion, on the average, will be surrounded by more ions of the opposite charge than of the same charge. It is this **interionic attraction** that prevents the ions from functioning as free particles. The greater the ionic charge, the greater the interionic attraction; hence, the less "free" are the ions. Table 20-2 shows the effect of ionic charge on i factors.

The Debye-Hückel theory has been applied quantitatively quite successfully in explaining the properties of dilute solutions of strong electrolytes. Solutions

Table 20-2. Effect of Ionic Charge on i Factors for Salts in Aqueous Solution at 0°C (concentration 0.001 M)

Electrolyte	Cation Charge	Anion Charge	Ideal i	Observed i	Deviation (%)
NaCl	+1	−1	2.00	1.97	1.5
K_2SO_4	+1	−2	3.00	2.84	5.3
$MgSO_4$	+2	−2	2.00	1.82	9.0

of moderate to high concentrations of strong electrolytes, however, cannot be treated quantitatively by the theory, although it does provide some qualitative insight into these systems.

PROBLEMS

1. What is the difference between an electrolyte and a nonelectrolyte? Why is further classification of electrolytes necessary?

2. What sort of property is a colligative property? Name the colligative properties of liquid solutions.

3. Why was the van't Hoff i factor introduced?

4. What is a semipermeable membrane? With what colligative property is it connected?

5. Use a handbook of data or other source to help classify the following substances as nonelectrolytes, weak electrolytes, or strong electrolytes.
   (a) $BeCl_2$ (e) CCl_3COOH
   (b) CO_2 (f) C_2H_2
   (c) $Mg(OH)_2$ (g) $(NH_4)_2CO_3$
   (d) $Pb(CH_3COO)_2$ (h) HI

6. Calculate the freezing point of a solution made by dissolving 1.28 g of naphthalene ($C_{10}H_8$) in 100 g of benzene (C_6H_6). With concentration expressed in mole fraction, $K_f = 71$ for benzene. The freezing point of pure benzene is 5.5°C.

7. Calculate the boiling point elevation constant, K_b, for benzene if the solution of problem 6 boils at 80.4°C. The boiling point of pure benzene is 80.1°C. Boiling point elevation, ΔT_b, is calculated by $\Delta T_b = K_b X_2$.

8. What is the molecular weight of a nonelectrolyte if 2.0 g dissolved in 150 g of benzene causes the freezing point to be depressed by 0.40°C?

9. After a hot dog is cooked for several minutes in water it swells considerably. What colligative property is responsible for this fact? Explain.

10. What is the i factor for a certain weak electrolyte, HX, at a concentration where it is 10% ionized?

11. It is possible to formulate the compound, K_2MgCl_4, either as a coordination compound, $K_2[MgCl_4]$, or as a double salt, $MgCl_2 \cdot 2KCl$. Devise an experiment that would allow a choice between these possibilities in solution.

12. Explain the difference between the Arrhenius and Debye-Hückel explanations for the variation from whole number i factors for dilute solutions of strong electrolytes. Why is the Debye-Hückel approach favored?

13. At equal concentrations, which one of the following pairs of compounds will show the greater deviation from i factors predicted for complete dissociation?
   (a) $NaNO_3$ and Na_2SO_4
   (b) $CaCl_2$ and $MgSO_4$
   (c) KCl and CH_3COOH

21

REACTIONS THAT INVOLVE ELECTRON TRANSFER—OXIDATION AND REDUCTION

Several aspects of chemical reactions have been examined throughout this book, for example, stoichiometry, equilibrium, and chemical energy. Other topics that relate to reactions and reactivity are ionization potentials, electron affinities, atomic structure, and bonding.

The present chapter is concerned with a chemical concept that is related to all of these diverse aspects of chemistry—oxidation and reduction. The student will recall that oxidation was one of the first topics discussed in this book. In fact, the solution of the problem of combustion by Lavoisier signalled the birth of modern chemistry. At that time oxidation was shown to consist of the combination of another element with oxygen. Such a definition was useful and correlated many chemical observations. However, with time, other reactions not involving oxygen were found to be quite similar chemically to the reaction with oxygen. As more facts were added to the collective knowledge of chemistry, it became obvious that the combination of elements with oxygen can be considered a special case of a much more general type of reaction. The term **oxidation** acquired a more general meaning, which included not only combination with oxygen but also the combination with other nonmetals as well as changes in the electronic structures of atoms.

Today, *oxidation is usually defined as a loss of electrons and reduction as a gain of electrons.* Oxidation-reduction reactions therefore involve a loss of electrons by certain atoms and a gain by others. Such reactions are often referred to as **redox** reactions, for short. Because electrons cannot be left over in chemical reactions, oxidation and reduction must always occur simultaneously. This chapter will explore some of the important concepts associated with redox.

313

21-1. EXPERIMENTAL BASIS FOR ACTIVITY SERIES—
LARGE EXTENT AND SMALL EXTENT REACTIONS

A reaction in which the equilibrium favors the products can be described as occurring to a *large extent,* whereas one in which only small amounts of products are formed can be described as occurring to a *small extent.* In other words, large extent reactions are those that give relatively large amounts of products, and small extent reactions are those that give only relatively small amounts of products.

If a simple experiment is conducted as shown in Figure 21-1, examples of large extent and small extent reactions are evident. No change will ever be noted in the copper strip or the zinc sulfate solution, but if enough zinc metal is present in the copper sulfate solution, eventually almost all of the Cu^{2+} ions in solution will be replaced by Zn^{2+} ions. Furthermore, the zinc strip will disappear and copper metal will "plate out." The sulfate ions in solution are unaffected. The net large extent chemical reaction is

$$Zn + Cu^{2+} \longrightarrow Zn^{2+} + Cu$$

(The SO_4^{2-} ions are not included in the equation because they are not involved in the reaction.)

The attempt to replace Zn^{2+} ions with Cu^{2+} ions by plating out zinc metal is not successful, hence the reaction

$$Cu + Zn^{2+} \longrightarrow Cu^{2+} + Zn$$

is quite small in extent. These two reactions are the reverse of one another and illustrate a fairly obvious general rule: *A reaction that takes place to a large extent in one direction takes place only to a small extent in the opposite direction.*

An equilibrium involving the species Zn, Zn^{2+}, Cu, and Cu^{2+} will favor Zn^{2+} and Cu. Sometimes the directional tendency of a reaction is indicated by the use of unequal length arrows.

$$Zn + Cu^{2+} \rightleftharpoons Zn^{2+} + Cu$$

Such an indication also shows qualitatively the natural position of the equilibrium.

Simple observations similar to the above form the basis for the **activity series** of metals. We have seen that Zn will reduce Cu^{2+} to Cu, but Cu will not reduce

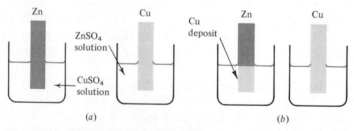

Figure 21-1. Large extent and small extent reactions. (a) A strip of zinc metal is placed in copper sulfate solution and a strip of copper metal is placed in zinc sulfate solution. (b) A short time later the zinc strip is coated with copper metal but the copper strip is unchanged.

Zn^{2+} to Zn. Therefore, we can say that Cu^{2+} ions are more easily reduced (gain electrons) in aqueous solution than Zn^{2+} ions, or that Zn atoms are more easily oxidized (lose electrons) than Cu atoms. In a general way, there is a correlation between the ionization potential of a metal and its ease of oxidation in solution. However, the correspondence is not exact because solvation of the metal ions is involved in determining the ease of oxidation in solution. Solvation effects are, of course, absent in the determination of an ionization potential of a gaseous atom.

21-2. ELECTRICAL WORK FROM CHEMICAL REACTIONS

In principle, it is always possible to obtain electrical work from a *large extent* redox reaction. Since electrons are transferred in the reaction, they will perform electrical work if they are made to travel through some sort of external circuit and drive a motor or accomplish some other task.

In practice, however, no work can be obtained if the electron loss (oxidation) and gain (reduction) occur at the same place. In the zinc-copper reaction, for example, the electrons will not be available to do work if the Cu^{2+} ions and Zn atoms are in contact as shown in Figure 21-1. The transfer of electrons from Zn atoms to Cu^{2+} ions simply occurs at the surface of the zinc strip where the Cu^{2+} ions are in contact. The oxidation and reduction processes must be physically separated to obtain work from the electrons.

Figure 21-2 illustrates several ways in which a physical separation of the oxidation and reduction processes can be achieved to give a **voltaic cell,** that is a device that converts chemical energy directly to electrical energy. The voltaic cell that utilizes the reduction of Cu^{2+} ions by Zn metal as the redox reaction which provides the electrical potential is called the Daniell cell after its inventor, John F. Daniell. This cell, developed in 1836, figured in the opening of the American West because it was used to power the telegraph in its early days. Used primarily in the gravity form (Figure 21-2*a*) for this purpose, a motion picture film based on the old west, in which numerous bottles filled with blue liquid adorn the telegraph office, shows the assistance of a competent technical advisor.

In any electrochemical process, the electrode at which *oxidation* occurs is called the **anode;** the electrode at which *reduction* occurs is called the **cathode.** In the Daniell cell, therefore, the zinc electrode is the anode and the copper electrode is the cathode. Since electrons are released at the zinc electrode, the anode is charged negatively compared to the copper cathode where the electrons go to be taken by Cu^{2+} ions. The student is cautioned to learn the definitions for anode and cathode in terms of oxidation and reduction as given in the first sentence of this paragraph and not in terms of electrical sign. In an **electrolytic cell,** an electric current is utilized to cause a reaction that is naturally small in extent to take place to a large extent. In this case the cathode is negative and the anode positive (just the opposite to voltaic cells), but reduction still occurs at the cathode and oxidation at the anode.

Let us examine the operation of the Daniell cell more closely. Figure 21-3 shows the external and internal movement of charge. Externally, electrons move through the solid conductor from anode to cathode and constitute the electrical flow. Inside the cell, there are no free electrons, and the charge is transported

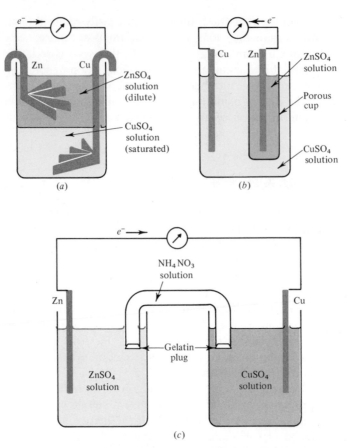

Figure 21-2. Daniell cell arrangements. (*a*) Gravity type—the solutions do not mix because the saturated $CuSO_4$ solution is much more dense than the dilute $ZnSO_4$ solution. (*b*) A porous cup prevents mixing of the solutions. (*c*) A salt bridge connects two separate containers with the two solutions.

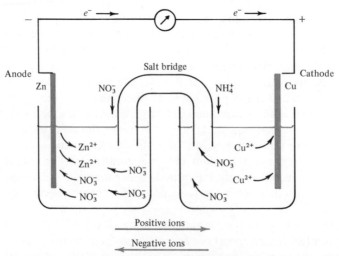

Figure 21-3. Charge flow in a voltaic cell. Externally, electrons move from anode to cathode while, internally, positive ions move from anode to cathode and negative ions move from cathode to anode.

by means of ion flow. Positive ions move from the anode toward the cathode while negative ions flow in the opposite direction. Internally as well as externally, there must be a complete circuit or there can be no difference in potential and hence no movement of charge. A salt bridge is a common way to complete the internal circuit. Negative ions in the salt bridge migrate into the anode compartment where positive ions are being put into solution and positive ions in the salt bridge migrate into the cathode compartment to replace the positive ions that are being removed from solution.

The voltage produced by a voltaic cell is affected slightly by the purity of the metals used, the concentrations of ions in solution, the pressure of any gas involved in the cell reaction, and the temperature of the cell. Because a cell will run down (its voltage will decrease) as a current is withdrawn from it, the voltage of a cell is always understood to mean its maximum voltage before any appreciable current has been withdrawn from it. In order to compare the voltages of cells under the same conditions, a set of standard conditions has been defined: (1) all metals are pure, (2) all ions are at 1 M concentration, (3) gases are at 1 atm pressure, and (4) the temperature is 25°C. A voltage measured under these conditions is called the standard cell potential, $E°$. Of the above factors, the ionic concentrations have the greatest effect. However, rather drastic changes in concentration do not produce large changes in the voltage. For example, a Daniell cell operating at standard conditions produces 1.10 volts. If the concentration of Zn^{2+} ions is made 0.1 M, the voltage becomes 1.13. A tenfold change in concentration thus causes less than a 3% change in the voltage for this particular cell.

The maximum work that can be obtained from a voltaic cell is calculated by

$$W = n\mathfrak{F}E \qquad (21–1)$$

where, n = moles of electrons transferred per mole of a given reactant; $\mathfrak{F}$ = Faraday's constant, that is, the charge carried by 1 mole of electrons, 9.65×10^4 coulombs; E = the maximum cell voltage. Under standard conditions, equation (21–1) becomes

$$W = n\mathfrak{F}E° \qquad (21–1a)$$

Using the above value for $\mathfrak{F}$, the work unit will be **joules** (J) because 1 volt is 1 J/coulomb.

Equations (21–1) and (21–1a) give the theoretical maximum value of the work that could be obtained per mole of reactant consumed from a voltaic cell at a given voltage. These expressions are valid only if the voltage is constant—a situation not always realized in practice. However, the equations do express the theoretical maximum of useful work that could be obtained from a given chemical reaction. If work is being expended, conservation of energy demands that some other form of energy decrease. The decrease in energy is in the form of chemical energy possessed by the Zn atoms and Cu^{2+} ions before reaction. As their chemical energy is lost, the electrical work becomes available.

The chemical energy of substances that can theoretically be transformed into useful work is called **free energy.** (The useful work may not always be electrical in nature.) The change in the free energy of substances involved in a chemical reaction is symbolized, ΔG, in honor of J. Willard Gibbs, who first clearly developed the ideas we are considering. At constant temperature and pressure, the free energy change attending any chemical reaction is simply the theoretical maximum useful work that could be obtained from the reaction.

$$\Delta G = -W_{max} \qquad (21\text{-}2)$$

The negative sign is introduced because positive work done by the reaction is accompanied by a decrease in the free energy of the chemical species. For a reaction that occurs only to a small extent, work would have to be done to make the reaction proceed (negative work), and the reaction would gain free energy in the process.

For the special case of a voltaic cell reaction, equations (21–1) and (21–2) give

$$\Delta G = -n\mathfrak{F}E \qquad (21\text{-}3)$$

or, considering Equation (21–1a)

$$\Delta G^\circ = -n\mathfrak{F}E^\circ \qquad (21\text{-}3a)$$

where ΔG° is called the standard free energy change.

We shall not pursue all of the consequences of these relationships. However, so that the student may gain some insight into the importance of this free energy concept, it will be noted that a relationship exists between ΔG, ΔH, and ΔS for any chemical reaction. At constant temperature,

$$\Delta G = \Delta H - T\Delta S \qquad (21\text{-}4)$$

and

$$\Delta G^\circ = \Delta H^\circ - T\Delta S^\circ \qquad (21\text{-}4a)$$

For the case of voltaic cell reactions, equations (21–3) and (21–3a) in conjunction with (21–4) and (21–4a), respectively, offer additional interrelationships. Finally, there exists an extremely important connection between ΔG° and the equilibrium constant for any chemical reaction.

$$\Delta G^\circ = -2.3RT \log K \qquad (21\text{-}5)$$

Comparing equations (21–3a) and (21–5) we obtain

$$E^\circ = \frac{2.3\,RT}{n\mathfrak{F}} \log K \qquad (21\text{-}6)$$

where R is the gas constant, and the other symbols have their usual significance. Supplying the values for these constants in the proper units and assuming 25°C ($T = 298°K$), equation (21–6) becomes

$$E^\circ = \frac{0.059}{n} \log K \qquad (21\text{-}6a)$$

Equation (21-6a) offers a powerful method for the determination of equilibrium constants of redox reactions. Provided the reaction is large in extent and can be made to function in a voltaic cell, the equilibrium constant can be determined by a measurement of $E°$ for the cell. For example, the Daniell cell has $E° = 1.10$ volts, so

$$1.10 = \frac{0.059}{n} \log K$$

$n = 2$, since each Zn atom loses and each Cu^{2+} ion gains two electrons,

$$\log K = 37.3$$
$$K = 2.0 \times 10^{37}$$

which is the equilibrium constant at 25°C for the Daniell cell reaction,

$$Zn + Cu^{2+} \rightleftharpoons Zn^{2+} + Cu$$

$$K = \frac{[Zn^{2+}]}{[Cu^{2+}]} = 2.0 \times 10^{37} \text{ at } 25°C$$

21-3. HALF-REACTIONS AND HALF-CELL POTENTIALS

Because the oxidation and reduction processes are physically separated in a voltaic cell, it is reasonable to assume that the electron loss at the anode and the electron gain at the cathode occur in separate but simultaneous steps. We can then write separate chemical equations for these steps in which electrons appear as reactants or products along with the chemical species. Such reactions are called **half-reactions** because two of them are always required to express the total cell reaction—one for electron loss (oxidation) and one for electron gain (reduction).

The Daniell cell half reactions are

Oxidation half reaction:
$$Zn \longrightarrow Zn^{2+} + 2\,e^-$$
Reduction half reaction:
$$2\,e^- + Cu^{2+} \longrightarrow Cu$$

Addition of the two half reactions gives the cell reaction. Notice that the electrons cancel. A total reaction can never show electrons because there is no creation nor destruction of electrons.

It is convenient to associate a voltage or half-cell potential with each half-reaction. In so doing, experimental reality requires that *the sum of the oxidation and reduction half-cell potentials equals the measured cell potential.* No single half-cell potential can be measured absolutely because the two half reactions always occur simultaneously. However, *relative* half-cell potentials can be measured. Consider the Daniell cell under standard conditions.

	Half Reaction	E°
Oxidation:	$Zn \longrightarrow Zn^{2+} + 2e^-$	$E^\circ_{Zn,Zn^{2+}}$
Reduction:	$2e^- + Cu^{2+} \longrightarrow Cu$	$E^\circ_{Cu^{2+},Cu}$
Cell reaction:	$Zn + Cu^{2+} \longrightarrow Zn^{2+} + Cu$	$E^\circ_{cell} = E^\circ_{Zn,Zn^{2+}} + E^\circ_{Cu^{2+},Cu} = 1.10$

Thus far, the situation is rather ambiguous because it is obvious that $E^\circ_{Zn,Zn^{2+}}$ and $E^\circ_{Cu^{2+},Cu}$ can have many different values and still sum to 1.10 volt. What is done in practice is to define the half-cell potential for one half reaction, and then measure the other half-cell potentials relative to this standard. The standard half-cell potential is that associated with the hydrogen half reaction. For the reduction of H^+ ions under standard conditions, the reduction potential is defined to be zero.

$$2\,e^- + 2\,H^+ \longrightarrow H_2 \qquad\qquad E^\circ = 0.00 \text{ volt}$$

The potential associated with an oxidation half-reaction (oxidation potential) is simply the negative of the reduction potential so that,

$$H_2 \longrightarrow 2\,H^+ + 2\,e^- \qquad\qquad E^\circ = 0.00 \text{ volt}$$

The standard hydrogen half-cell involves a gas in contact with ions in solution. To facilitate electron transfer between these substances, it is necessary to use a platinum surface coated with powdered platinum metal (Figure 21–4). The platinum electrode is not chemically involved in the reaction but functions as an inert electrode.

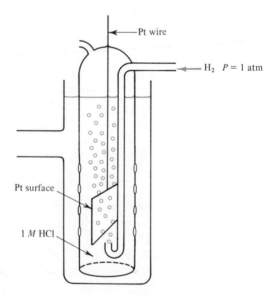

—Pt wire

—H_2 $P = 1$ atm

Pt surface

1 M HCl

Figure 21-4. The standard hydrogen half-cell. H_2 gas and H^+ ions in solution are in contact at the platinum surface. The platinum functions as an inert electrode.

If a zinc half-cell is connected to a hydrogen half-cell as shown in Figure 21–5, the cell potential under standard conditions is 0.76 volt. The zinc electrode is found experimentally to be the electron source or the anode. The half reactions are therefore,

	Half Reaction	$E°$
Oxidation (anode):	$Zn \longrightarrow Zn^{2+} + 2\,e^-$	$E°_{Zn,Zn^{2+}}$
Reduction (cathode):	$2\,e^- + 2\,H^+ \longrightarrow H_2$	$E°_{H^+,H_2}$
Cell reaction:	$Zn + 2\,H^+ \longrightarrow H_2 + Zn^{2+}$	$E°_{cell} = E°_{Zn,Zn^{2+}} + E°_{H^+,H_2} = 0.76$

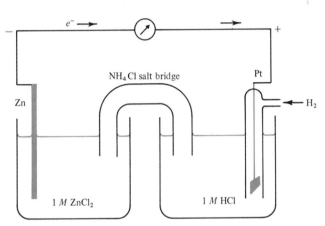

Figure 21-5. Measurement of $E°$ for a metal against the standard hydrogen half-cell.

Since $E°_{H^+,H_2} = 0.00$ $\qquad$ $E°_{Zn,Zn^{2+}} = 0.76$ volt

The Daniell cell shows, $\qquad$ $E°_{Zn,Zn^{2+}} + E°_{Cu^{2+},Cu} = 1.10$ volt

Using the value for $E°_{Zn,Zn^{2+}}$, $\qquad$ $E°_{Cu^{2+},Cu} = 1.10 - 0.76$

Finally, $\qquad$ $E°_{Cu^{2+},Cu} = 0.34$ volt

Compared with the standard hydrogen cell, the standard *oxidation* potential of zinc has been found to be 0.76 volt, and the *reduction* potential of copper has been found to be 0.34 volt. The individual values have no physical significance, but any pair of half-cell potentials will give the voltage that would be observed for the voltaic cell using those particular half reactions under standard conditions. For example, the reduction of Cu^{2+} ions is easier than the reduction of H^+ ions since $E°_{Cu^{2+},Cu} = 0.34$ volt and $E°_{H^+,H_2} = 0.00$ volt. Therefore a voltaic cell using these half-cells would have a potential of 0.34 volt. The hydrogen half-cell would be the anode and the copper half-cell the cathode.

Half-cell potentials may be listed either as oxidation potentials or as reduction potentials. Present usage favors listing reduction potentials, but tables of oxidation potentials are found in some reference books. One must be quite clear about the particular process to which a half-cell potential applies before using the value. Most tables give the half reaction along with the potential, so that there will be no confusion. Table 21-1 gives reduction potentials for a number of substances. Remember that an oxidation potential is easily obtained simply by putting a minus sign in front of the reduction potential.

Table 21-1. Standard Reduction Potentials

Half Reaction	$E°$ (volts)
$e^- + Li^+ \longrightarrow Li$	-3.05
$e^- + Na^+ \longrightarrow Na$	-2.71
$2\,e^- + Mg^{2+} \longrightarrow Mg$	-2.37
$3\,e^- + Al^{3+} \longrightarrow Al$	-1.66
$2\,e^- + Zn^{2+} \longrightarrow Zn$	-0.76
$2\,e^- + Fe^{2+} \longrightarrow Fe$	-0.44
$2\,e^- + Ni^{2+} \longrightarrow Ni$	-0.25
$2\,e^- + Sn^{2+} \longrightarrow Sn$	-0.14
$2\,e^- + Pb^{2+} \longrightarrow Pb$	-0.13
$2\,e^- + 2\,H^+ \longrightarrow H_2$	0.00
$2\,e^- + Cu^{2+} \longrightarrow Cu$	0.34
$2\,e^- + I_2 \longrightarrow 2\,I^-$	0.54
$e^- + Ag^+ \longrightarrow Ag$	0.80
$2\,e^- + Br_2 \longrightarrow 2\,Br^-$	1.07
$4\,e^- + O_2 + 4\,H^+ \longrightarrow 2\,H_2O$	1.23
$2\,e^- + Cl_2 \longrightarrow 2\,Cl^-$	1.36
$3\,e^- + Au^{3+} \longrightarrow Au$	1.50
$5\,e^- + MnO_4^- + 8\,H^+ \longrightarrow Mn^{2+} + 4\,H_2O$	1.51
$2\,e^- + F_2 \longrightarrow 2\,F^-$	2.87

Use of the standard potentials is illustrated by the following example:

■ **Example 21-1**

Calculate the voltage of a cell using magnesium and lead electrodes operating under standard conditions and write the cell reaction.

□ *Solution:*
Reference to Table 21–1 shows that Pb^{2+} is more easily reduced than Mg^{2+} ($E°_{Pb^{2+},Pb}$ less negative than $E°_{Mg^{2+},Mg}$). The Pb^{2+} ion will therefore be reduced at the cathode and Mg will be oxidized at the anode

	Half Reaction	$E°$
Anode:	$Mg \longrightarrow Mg^{2+} + 2\,e^-$	2.37
Cathode:	$2\,e^- + Pb^{2+} \longrightarrow Pb$	-0.13
Cell reaction:	$Mg + Pb^{2+} \longrightarrow Mg^{2+} + Pb$	$E°_{cell} = 2.37 + (-0.13) = 2.24$ volt

Whenever the $E°$ values of two half-cells are added to give an overall $E°$, the numbers are never changed even though the half reactions may have to be multiplied by some factor to balance the total reaction. As an example, the cell using magnesium and silver electrodes would have the standard voltage calculated as follows

$$
\begin{array}{lr}
 & E° \\
Mg \longrightarrow Mg^{2+} + 2\,e^- & 2.37 \\
(2\times)\; e^- + Ag^+ \longrightarrow Ag & 0.80 \\
\hline
Mg + 2\,Ag^+ \longrightarrow 2\,Ag + Mg^{2+} & 3.17
\end{array}
$$

The silver half reaction must be multiplied by 2 so that the electrons will cancel and the total reaction will balance. The $E°$ value for the silver half reaction, however, is unchanged, because it is an **intensive** quantity. Intensive quantities, like density, color, and temperature do not depend on the amount of material present. **Extensive** quantities, like weight, volume, and kinetic energy, do depend on the amount of substance being considered.

A standard notation has been devised to represent voltaic cells. The cell described in Example 21–1 is represented

$$Mg\,|\,Mg^{2+}\,\|\,Pb^{2+}\,|\,Pb$$

Convention demands that the anode be placed on the left. The vertical line between Mg and Mg^{2+} indicates the phase boundary between the solid Mg anode and the solution. The double lines represent a salt bridge, and the Pb^{2+} ions are separated from the Pb cathode by a vertical line. Standard conditions are understood unless the contrary is indicated. If the ionic concentrations are not $1\,M$, for example, they are given as subscripts on the ion symbol, that is, $Pb^{2+}_{(0.1)}$ means $0.1\,M\;Pb^{2+}$ ion. An inert electrode must be included. The Zn-H_2 cell previously discussed is represented by

$$Zn\,|\,Zn^{2+}\,\|\,H^+\,|\,H_2\,|\,Pt$$

21-4. USE OF $E°$ TO PREDICT THE RELATIVE EXTENT OF A REACTION

A redox reaction that occurs to a large extent will always have a positive cell potential because, in prinicple, the reaction could be used to provide electrical work. A redox reaction that is small in extent will always have a negative cell potential because energy would have to be supplied to cause the reaction to take place to an appreciable extent. The calculated $E°$ for a redox reaction can therefore be used as a measure of the relative extent of the reaction. A positive $E°$ indicates that the reaction will take place to a large extent as written, whereas a negative $E°$ indicates that the reaction will occur only slightly or to a small extent as written.

Actually, these criteria apply strictly only if the reaction is run under standard conditions, but we have seen that rather large changes in the concentration of the species involved affects the voltage only slightly. Therefore a rough rule may be stated: *Redox reactions with $E°$ greater than 0.05 volt will occur to a large extent under most conditions, whereas those with $E°$ less than -0.05 volt will occur only to a small extent under most conditions.* Between 0.05 volt and -0.05 volt, the reaction conditions (concentration, pressure, temperature), if they are far removed from standard, can invalidate a prediction of the direction of reaction. Several examples will be given to illustrate the use of this criterion.

■ **Example 21-2:**

Can nickel be used to produce hydrogen from 1 M HCl?

☐ *Solution:*

The question is really one of determining if the reaction

$$2 H^+ + Ni \longrightarrow Ni^{2+} + H_2$$

is large or small in extent. From Table 21-1

	Half Reaction	$E°$
	$2 e^- + 2 H^+ \longrightarrow H_2$	0.00
	$Ni \longrightarrow Ni^{2+} + 2 e^-$	0.25
Total reaction	$2 H^+ + Ni \longrightarrow Ni^{2+} + H_2$	0.25

Since the total $E°$ is positive, the reaction is large in extent as written and nickel can be used to produce hydrogen from the acid.

■ **Example 21-3:**

Is iodine stable in water or will the reaction

$$2 I_2 + 2 H_2O \longrightarrow 4 I^- + 4 H^+ + O_2$$

take place?

☐ *Solution:*

From Table 21-1

	Half Reaction	$E°$
	$(2\times)2 e^- + I_2 \longrightarrow 2 I^-$	0.54
	$2 H_2O \longrightarrow 4 H^+ + O_2 + 4 e^-$	-1.23
Total reaction	$2 H_2O + 2 I_2 \longrightarrow 4 I^- + 4 H^+ + O_2$	-0.69

The reaction is small in its extent, so iodine is stable in water. Notice that although the iodine reduction half reaction must be multiplied by 2 before adding the two half reactions, the $E°$ for the iodine half-cell is not changed in any way.

■ **Example 21-4**

Is the reaction

$$Pb + 2\,H^+ \longrightarrow Pb^{2+} + H_2$$

large or small in extent?

☐ *Solution:*

From Table 21-1

	Half Reaction	*E°*
	$2\,e^- + 2\,H^+ \longrightarrow H_2$	0.00
	$Pb \longrightarrow Pb^{2+} + 2\,e^-$	0.13
Total reaction	$Pb + 2\,H^+ \longrightarrow Pb^{2+} + H_2$	0.13

The reaction is thus seen to be large in extent under most conditions, yet lead will not react appreciably with most acids. The reason is that, although the reduction of H^+ ions by Pb atoms is favored, the reduction is very slow in most acids. Although most reactions which take place to a large extent are also fairly rapid, it should not be assumed that this will always be the case. In HCl solutions, the Pb^{2+} ions form insoluble $PbCl_2$, which coats the surface of the remaining lead metal and prevents the reaction from taking place readily. Insoluble $PbSO_4$ causes a similar situation in H_2SO_4. An $E°$ calculation really indicates only where the position of equilibrium is reached but says nothing about how long it may take to reach equilibrium.

21-5. OXIDATION NUMBER AS AN ELECTRONIC BOOKKEEPING DEVICE

The **oxidation number** of an atom is a number that can be assigned to the atom through the application of one or more arbitrary rules. In most instances, the oxidation number has no physical significance. However, an increase in the oxidation number of an atom is associated with the loss of electrons by the atom and a decrease in its oxidation number indicates a gain of electrons. Changes in the oxidation number of an atom therefore give an indication of oxidation or reduction of the atom and thereby serve as a bookkeeping device for electrons. Additional uses of oxidation numbers include the systematization of the chemistry of the elements and a basis for the nomenclature of compounds. The term **oxidation state** is synonymous with oxidation number.

The rules for the assignment of oxidation numbers are

1. An uncombined atom, or an atom in a molecule of an element is given an oxidation number of zero, for example, in Na, H_2, O_3, S_8, all atoms have zero oxidation numbers.

325

2. A monatomic ion has an oxidation number identical to its charge, for example, Na^+ is $+1$; Cl^- is -1; Al^{3+} is $+3$; and S^{2-} is -2.

3. In a compound with another element, fluorine always has an oxidation number of -1, for example, in KF, F is -1; in OF_2, each F is -1.

4. In a compound with an element other than fluorine or another oxygen atom, oxygen is assigned an oxidation number of -2, for example, in H_2O, H_3PO_4, Na_2O, and OsO_4, each O atom is -2.

5. When oxygen is bonded to itself and another element, as in peroxides, each O atom has an oxidation number of -1, for example, in H_2O_2 and Na_2O_2, each O atom is -1. The superoxide ion, O_2^-, is an exception, and each O atom is assigned an oxidation number of $-\frac{1}{2}$.

6. Hydrogen atoms are given an oxidation number of $+1$ in compounds with other elements except the reactive metals of groups I and II and aluminum. In combination with the latter elements, hydrogen has an oxidation number of -1, for example, in H_2O, HCl, and CH_4, all H atoms are $+1$; in NaH, CaH_2, and AlH_3, all H atoms are -1.

7. The algebraic sum of the oxidation numbers of all atoms in a species equals the net charge on the species. This rule will usually provide the oxidation number of atoms other than those specifically covered by the other rules, for example, in H_2SO_3, S is $+4$ because each O is -2 and each H is $+1$, $(-6 + 2 + S) = 0$, so $S = 4$; in NO_3^-, $(N - 6) = -1$ so $N = +5$; in NH_3, N is -3; in Fe_3O_4, Fe is $+2\frac{2}{3}$, $(3\ Fe - 8) = 0$ so $Fe = +2\frac{2}{3}$.

Occasionally the assignment of an oxidation number is not covered by one of the above rules. An example is the cyanide ion, CN^-. In such cases, the element with the higher electron affinity is usually given the more negative oxidation number. Nitrogen can be given the oxidation number -3 (as in ammonia) and carbon therefore would have to be $+2$.

Use of oxidation numbers allows an alternate definition of a redox reaction as one in which the oxidation numbers of some atoms change. The reaction

$$\overset{+4-2}{2\ SO_2} + \overset{0}{O_2} \longrightarrow \overset{+6-2}{2\ SO_3}$$

is thus clearly seen to be a redox reaction because the oxidation numbers of sulfur and oxygen change. It is also immediately obvious that the S atoms are oxidized and the O atoms (from the O_2 molecules) are reduced since the oxidation number of sulfur increases and the oxidation number of oxygen (in O_2) decreases. On the other hand, the reaction

$$\overset{+4-2}{CO_2} + \overset{+1-2}{H_2O} \longrightarrow \overset{+1+4-2}{H_2CO_3}$$

is not a redox reaction because there is no change in the oxidation number of any atom. A redox reaction such as

$$\overset{0}{Cl_2} + \overset{-2+1}{2\ OH^-} \longrightarrow \overset{-1}{Cl^-} + \overset{+1-2}{ClO^-} + \overset{+1-2}{H_2O}$$

where the same kinds of atoms are both oxidized and reduced is called a **disproportionation** reaction. In this reaction, some of the Cl atoms are reduced to Cl^- ions and some of the Cl atoms are oxidized to the $+1$ state in the hypochlorite ion, ClO^-. The H and O atoms are, of course, neither oxidized nor reduced since their oxidation numbers do not change.

21-6. OXIDATION NUMBER AS A PERIODIC PROPERTY

Certain useful generalizations can be made about the expected oxidation states of an atom and its position in the periodic table.

1. The highest *possible* oxidation state of an atom is equal to its group number in the periodic table. An atom may not be able to attain this state, and those that do are usually combined with fluorine or oxygen. The exceptions to this generalization are found in group I transition metals—the copper-silver-gold family. Copper has a very common $+2$ state and gold a $+3$ state in addition to the expected $+1$ states. Silver has a stable $+1$ state and an unstable $+2$ state.

2. The representative elements of group I have only a $+1$ oxidation number in compounds and those of group II have only a $+2$ oxidation number in compounds. Except for mercury, which has $+1$ as well as $+2$ oxidation numbers in compounds, transition elements of group II also have only the $+2$ state in compounds. Group III transition metals all have only the $+3$ state in compounds, and aluminum behaves as if it were a transition metal of group III in this respect.

3. The representative elements of groups III through VII and the noble gases (group VIII) are characterized by decreasing stability of the higher oxidation states, relative to their lower oxidation states, in moving from the lighter to the heavier atoms. For example, nitrogen has a very stable $+5$ oxidation number, but bismuth, the heaviest member of the nitrogen family has no stable $+5$ state and exists primarily in the $+3$ state in compounds. This generalization does not mean that the highest states never exist in the heavier elements (IO_4^- is quite stable), but simply is an indication of trends in stability. The oxidation numbers of atoms in these groups vary by two units, for example, HOCl, HOClO, $HOClO_2$, and $HOClO_3$ are the oxyacids of chlorine, and it can be seen that the oxidation numbers of chlorine in these compounds are $+1$, $+3$, $+5$, and $+7$, respectively.

4. The transition metals of a given family are characterized by increasing stability of their higher oxidation states, relative to their lower oxidation states, in moving from the lighter to the heavier atoms. Thus, OsO_4 is a stable compound, RuO_4 is explosive, and FeO_4 apparently does not exist. In contrast with the representative elements, the transition elements display variations in oxidation number which differ by one unit. The following compounds of vanadium illustrate this behavior, namely, VCl_2, VCl_3, VCl_4, and V_2O_5, where the oxidation numbers of vanadium are $+2$, $+3$, $+4$, and $+5$, respectively.

5. The rare earth metals of the inner transition series are characterized by

327

a stable $+3$ oxidation state. A few of these elements have other oxidation numbers in compounds (Ce^{4+} is the most important), but none are as stable as the $+3$ state.

21-7. OXIDATION NUMBER AS A BASIS FOR NOMENCLATURE

Compounds containing elements that have more than one common oxidation state are numerous. Some system must be employed to indicate the particular oxidation state of the element in naming a particular compound. The traditional method uses the suffix **-ous** to indicate the lower oxidation state and **-ic** to indicate the higher state. Thus we have ferrous chloride ($FeCl_2$) and ferric chloride ($FeCl_3$); nitrous acid (HNO_2) and nitric acid (HNO_3); sulfurous acid (H_2SO_3) and sulfuric acid (H_2SO_4). This system is adequate if only two common oxidation numbers are encountered. However, for an element that commonly shows more than two oxidation states, the procedure is highly ambiguous.

Several alternative systems are currently used to treat the problem of naming compounds of elements with multiple oxidation numbers. Binary compounds may be named by using the prefixes: **di-, tri-, tetra-, penta-, hexa-,** to indicate two, three, four, five, or six atoms of each element in the molecule, for example, iron dichloride ($FeCl_2$) and iron trichloride ($FeCl_3$). Such prefixes are not usually applied to a metal atom but only to nonmetal atoms. A method more dependent on the use of oxidation numbers leads to the naming of these two compounds as iron(II) chloride and iron(III) chloride, respectively. Occasionally, a compound cannot be unambiguously named by indicating the oxidation number in this manner. As an example, nitrogen(IV) oxide does not differentiate between NO_2

Table 21-2. Comparison of Systems of Nomenclature

Compound	Suffix -ic and -ous	Prefix di-, tri-	Indication of Oxidation Number
Cu_2O	cuprous oxide	—	copper(I) oxide
CuO	cupric oxide	—	copper(II) oxide
VO	—	vanadium oxide	vanadium(II) oxide
V_2O_3	—	vanadium trioxide	vanadium(III) oxide
VO_2	—	vanadium dioxide	vanadium(IV) oxide
V_2O_5	—	vanadium pentoxide	vanadium(V) oxide
IF	—	iodine fluoride	iodine(I) fluoride
IF_3	—	iodine trifluoride	iodine(III) fluoride
IF_7	—	iodine heptafluoride	iodine(VII) fluoride
$HOIO_2$	iodic acid	—	—
$HOIO_3$	periodic acid	—	—
N_2O	nitrous oxide	dinitrogen oxide	nitrogen(I) oxide
NO	nitric oxide	nitrogen oxide	nitrogen(II) oxide
N_2O_2	—	dinitrogen dioxide	—
NO_2	—	nitrogen dioxide	nitrogen(IV) oxide
N_2O_4	—	dinitrogen tetroxide	—
N_2O_5	—	dinitrogen pentoxide	nitrogen(V) oxide

and N_2O_4, and these substances are always referred to as nitrogen dioxide and dinitrogen tetroxide, respectively.

Ternary compounds such as oxyacids in which the central atom can have more than two oxidation states are dealt with in a special way. An oxidation number higher than that of the -ic compound is indicated by the prefix **per,** and an oxidation number lower than that of the -ous compound is given by the prefix **hypo.** Examples of this usage are found in the halogen oxyacids: hypochlorous acid (HOCl), chlorous acid (HOClO), chloric acid ($HOClO_2$), and perchloric acid ($HOClO_3$). In naming the anions derived from oxyacids, the -ic ending is replaced by **-ate** and the -ous ending by **-ite,** for example, OCl^- is the hypochlorite ion, ClO_2^- the chlorite ion, ClO_3^- is called the chlorate ion, and ClO_4^- the perchlorate ion. Table 21-2 on previous page shows a comparison of the various naming schemes for several compounds.

PROBLEMS

1. What is oxidation and reduction?
2. What is the difference between a voltaic cell and an electrolytic cell? Into which category does the Daniell cell fall?
3. Define the terms anode and cathode.
4. What is the function of a salt bridge?
5. Physically, what is the free energy change associated with a process?
6. What is the difference between a reduction potential and an oxidation potential?
7. Give some reasons for the use of oxidation numbers.
8. What is a disproportionation reaction? Give an example.
9. Sketch a diagram of a voltaic cell operating under standard conditions using aluminum and copper electrodes. Indicate on your diagram **(a)** all necessary solutions and their concentrations, **(b)** the anode and cathode, **(c)** the direction of electron flow, **(d)** the direction of ion flow.
10. Write the anode and cathode half reactions and the total reaction for the cell of problem 9 and calculate the maximum voltage.
11. Use standard notation to indicate voltaic cells utilizing the following reactions:
    (a) $Fe + Cu^{2+} \longrightarrow Fe^{2+} + Cu$
    (b) $Fe + Cl_2 \longrightarrow Fe^{2+} + 2\,Cl^-$
    (c) $3\,Sn + 2\,Au^{3+} \longrightarrow 3\,Sn^{2+} + 2\,Au$
12. Calculate $E°$, $\Delta G°$, and K for each of the reactions of problem 11.
13. Write the anode and cathode half reactions and the cell reaction for the following:
    (a) $Zn\,|\,Zn^{2+}\,\|\,Br^-\,|\,Br_2\,|\,Pt$
    (b) $Ni\,|\,Ni^{2+}\,\|\,H^+\,|\,H_2\,|\,Pt$
    (c) $Sn\,|\,Sn^{2+}\,\|\,Pb^{2+}\,|\,Pb$
14. Calculate $E°$ for each cell in problem 13.
15. Predict whether each of the following redox reactions proceeds to a large or small extent on the basis of an $E°$ calculation.
    (a) $2\,Cl_2 + 2\,H_2O \longrightarrow 4\,Cl^- + 4\,H^+ + O_2$
    (b) $Cl_2 + 2\,F^- \longrightarrow 2\,Cl^- + F_2$
    (c) $2\,Ag + Cu^{2+} \longrightarrow 2\,Ag^+ + Cu$
    (d) $H_2 + Cl_2 \longrightarrow 2\,H^+ + 2\,Cl^-$

329

    (e) $Al + 3 Na^+ \longrightarrow 3 Na + Al^{3+}$
    (f) $Li + Na^+ \longrightarrow Na + Li^+$
    (g) $2 Li + F_2 \longrightarrow 2 Li^+ + 2 F^-$

16. List each of the reations of problem 15 from largest to least in extent.

17. Calculate the equilibrium constant at 25°C for each of the reactions in problem 15.

18. Suppose it were decided to make the silver half-cell the new standard, that is, to define $e^- + Ag^+ \longrightarrow Ag$, $E° = 0.00$ volt. How would the standard reduction potentials in Table 21–1 be changed? Would the voltage of any voltaic cell be changed? Explain.

19. A voltaic cell has the following half reactions:
Anode half-reaction:

$$Ag \longrightarrow Ag^+ + e^-$$

Cathode half-reaction:

$$3e^- + M^{3+} \longrightarrow M$$

The cell has $E° = 1.00$ volt. Using this information, calculate the following:
    (a) The standard reduction potential of M.
    (b) The maximum electrical work per mole of M produced.
    (c) The maximum electrical work per mole of Ag used.
    (d) $\Delta G°$ for the cell reaction.
    (e) K for the cell reaction.

20. Of the following reactions, which ones are redox and which are not?
    (a) $PCl_3 + 3 H_2O \longrightarrow 3 HCl + H_3PO_3$
    (b) $Cu(H_2O)_4^{2+} + 4 NH_3 \longrightarrow Cu(NH_3)_4^{2+} + 4 H_2O$
    (c) $C + O_2 \longrightarrow CO_2$
    (d) $HSO_4^- + HS^- \longrightarrow SO_4^{2-} + H_2S$
    (e) $6 H^+ + 2 MnO_4^- + 5 H_2C_2O_4 \longrightarrow 2 Mn^{2+} + 10 CO_2 + 8 H_2O$
    (f) $2 NO + O_2 \longrightarrow 2 NO_2$
    (g) $2 H_2O_2 \longrightarrow 2 H_2O + O_2$

21. Assign oxidation numbers to the underlined atom in each species.

(a) $H_3\underline{As}O_3$	(h) $\underline{N}H_2OH$	(o) $Na\underline{H}$
(b) $\underline{Br}F_5$	(i) $\underline{Fe}_3O_4$	(p) $\underline{O}_2F_2$
(c) $\underline{Cl}_2O_7$	(j) $I\underline{Cl}_2^-$	(q) $\underline{S}_8$
(d) $\underline{Cl}O_4^-$	(k) $\underline{S}_4O_6^{2-}$	(r) $\underline{Al}_2Cl_6$
(e) $Na\underline{Cl}O_4$	(l) $\underline{S}_2O_6^{2-}$	(s) $Na_2\underline{O}_2$
(f) $H\underline{N}_3$	(m) $\underline{Mn}O_4^{2-}$	(t) $\underline{H}_2^+$
(g) $\underline{N}_2H_4$	(n) $\underline{Mn}_2O_3$	(u) $\underline{C}NO^-$

22. Predict the more stable species in each of the following pairs.
    (a) CCl_4 and $PbCl_4$ (f) Eu_2O_3 and EuO
    (b) H_3PO_4 and H_3BiO_4 (g) XeF_6 and ArF_6
    (c) $CuCl_3$ and $AuCl_3$ (h) Sc_2O_3 and LaO_2
    (d) $MgCl_2$ and $BaCl_3$ (i) $AlCl$ and $AlCl_3$
    (e) BeF and $BeSO_4$ (j) Cs_2O and CsO

23. Name the following compounds using the -ic and -ous suffixes.
    (a) Ce_2O_3 and CeO_2 (c) SnF_2 and $SnCl_4$
    (b) $TlCl_3$ and $TlCl$ (d) $Hg_2(NO_3)_2$ and HgS

24. Name the following compounds using prefixes to indicate the number of each kind of atom in the molecule.
    (a) ClO_2 (d) S_2Cl_2 (g) N_4S_4
    (b) XeF_4 (e) N_2O_3 (h) $AsCl_5$
    (c) SCl_2 (f) P_4S_3

25. Name the following compounds by indicating the oxidation number of the proper atom in parenthesis.

   (a) SF_6 (d) CeO_2 (g) ClO_2

   (b) SCl_2 (e) $AsCl_3$ (h) XeO_3

   (c) Hg_2Cl_2 (f) Cl_2O

26. How are differences in structure between Li_2O, Na_2O_2, and KO_2 indicated in naming? between BaO_2 and PbO_2?

27. Write formulas for the following compounds.

   (a) silver bromide   (j) phosphorus(III) bromide

   (b) krypton(IV) fluoride   (k) cuprous nitrate

   (c) perbromic acid   (l) zinc carbonate

   (d) ferric nitrate   (m) calcium hypochlorite

   (e) sodium chlorite   (n) diphosphorus tetraiodide

   (f) uranium trioxide   (o) aluminum oxide

   (g) platnium(II) oxide   (p) sulfur(VI) fluoride

   (h) mercury(II) sulfide   (q) chromium(VI) oxide

   (i) thallous sulfate   (r) plumbic chloride

ACIDS AND BASES

During the seventeenth century, the major problems confronting chemistry were

1. The nature of elementary substances.
2. Combustion.
3. Acids and bases.

By the middle of that century, Robert Boyle had recognized and studied all three subjects. The first two have been discussed earlier in this book. The latter problem involved the observation that many substances could be placed into one of two groups—acids or bases. The members of the acid group all had the common properties of a sour taste and the ability to turn litmus, a vegetable dye, red. Bases, on the other hand, were bitter to the taste and turned litmus blue. Moreover, acids and bases bore a complementary relationship to each other in that they neutralized each other's properties. The products of their reaction were a salt and water. The present state of knowledge about acids and bases is much broader. Acids and bases are no longer defined as simply as stated above.

Many theories of acids and bases have been suggested over the past several hundred years. This chapter will take up the acid-base system proposed by J. N. Brønsted and T. M. Lowry in some detail, because it is probably the best approach at an elementary level to a systematic treatment of acids and bases. Several important acid-base equilibria will also be considered.

22-1. BASIC BRONSTED-LOWRY DEFINITIONS

Bronsted in Denmark and Lowry in England proposed, almost simultaneously, in 1923 the following definitions:

1. An acid is any substance that is capable of giving up or donating a proton (H^+).
2. A base is any substance that is capable of accepting a proton.
3. Every acid has its conjugate base, that is, the fragment that remains after the acid has lost a proton.

4. Every base has its conjugate acid, that is, the species formed when the base gains a proton.

Table 22–1 lists some common acids and their conjugate bases. Of course, each acid is also the conjugate acid of the base.

Table 22-1. Some Conjugate Acid-Base Pairs

Acid	Base
HCl	Cl^-
H_2SO_4	HSO_4^-
H_3O^+	H_2O
HSO_4^-	SO_4^{2-}
NH_4^+	NH_3
H_2O	OH^-

Many substances can function both as an acid and a base (amphoterism), for example, H_2O and HSO_4^-.

In the Brønsted-Lowry scheme, every reaction between an acid and a base produces another acid and base—the conjugates of the reactants.

$$HCl + H_2O \longrightarrow H_3O^+ + Cl^-$$
$$\text{acid}_1 \quad \text{base}_2 \qquad \text{acid}_2 \quad \text{base}_1$$

Notice that the pairs, acid$_1$-base$_1$ and base$_2$-acid$_2$ are conjugates. This relationship between the reactants and products of an acid-base reaction always holds true. We can consider Brønsted-Lowry acid-base reactions as **proton transfer** reactions.

22-2. RELATIVE ACID-BASE STRENGTH

The strength of an acid is measured by its ability to donate a proton to some standard base. Usually, water is chosen as the standard base because it is so frequently used as a solvent. The strength of an acid, HX, is then determined by the extent of the reaction,

$$HX + H_2O \rightleftharpoons H_3O^+ + X^-$$

If the reaction tends to occur essentially completely to the right, the acid is said to be **strong.** If the reaction is small in extent, the acid is said to be **weak.** Differences in the strengths of weak acids can be measured by the equilibrium constants of the acids in the above reaction.

Strong acids react *completely* with water to produce the hydronium ion (H_3O^+) and the conjugate base of the acid. In such cases, it is obviously impossible for the acid to exist in an aqueous solution (except possibly in very small concentration). In fact, the strongest acid that can exist in large concentration in any solvent is the conjugate acid of the solvent as a base. Similarly, the strongest base that can exist in large concentration in a solvent is the conjugate base of the solvent as an acid. For water, this means that H_3O^+ is the strongest acid and OH^- is the strongest base that can exist in large concentrations. If any stronger acid or base is placed in water, it will simply react with water, virtually completely. This phenomenon is known as the **leveling effect** of a solvent, that is, any strong acid or base will be *leveled* to the strength of the conjugate acid or base of the solvent.

A list of acids in descending order of their strength as proton donors toward water along with their conjugate bases listed directly across from them is called an acid-base table. The acids and bases are arranged in this fashion in Table 22-1. An interesting relationship exists in an acid-base table—the acid strength *decreases* going down the table but the base strength *increases* going down the table. Thus, we find that HCl is a strong acid and H_2O is a very weak acid, whereas Cl^- is a very weak base but OH^- a very strong one.

A little thought will show that this inverse relationship between the strengths of conjugate acids and bases is quite logical. A strong acid readily gives up a proton; therefore, its conjugate base has little ability to hold on to (or accept) a proton. Conversely, a weak acid is a poor proton donor, indicative that the conjugate base has a high affinity for the proton. Table 22-2 is an acid-base table for a number of substances. The K_a for each acid (a type of equilibrium constant) is included in the table for use later on.

Table 22-2. Acid-Base Table: Acid-Base Strengths Are Compared in Water Solutions

Acid Name	Acid Formula	Base Formula	K_a (25°C)
perchloric	$HClO_4$	ClO_4^-	∞
hydroiodic	HI	I^-	∞
hydrochloric	HCl	Cl^-	∞
nitric	HNO_3	NO_3^-	∞
sulfuric	H_2SO_4	HSO_4^-	∞
hydronium ion	H_3O^+	H_2O	55.5
oxalic	$H_2C_2O_4$	$HC_2O_4^-$	5.6×10^{-2}
sulfurous	H_2SO_3	HSO_3^-	1.7×10^{-2}
hydrogen sulfate ion	HSO_4^-	SO_4^{2-}	1.0×10^{-2}
phosphoric	H_3PO_4	$H_2PO_4^-$	5.9×10^{-3}
hydrofluoric	HF	F^-	6.7×10^{-4}
hydrogen oxalate ion	$HC_2O_4^-$	$C_2O_4^{2-}$	5.2×10^{-5}
acetic	$CH_3COOH(HOAc)$	$CH_3COO^-(OAc^-)$	1.8×10^{-5}
carbonic	H_2CO_3	HCO_3^-	4.5×10^{-7}
hydrogen sulfide	H_2S	HS^-	1.1×10^{-7}
hydrogen sulfite ion	HSO_3^-	SO_3^{2-}	6.2×10^{-8}
dihydrogen phosphate ion	$H_2PO_4^-$	HPO_4^{2-}	6.2×10^{-8}
hydrocyanic	HCN	CN^-	7.2×10^{-10}
ammonium ion	NH_4^+	NH_3	5.6×10^{-10}
hydrogen carbonate ion	HCO_3^-	CO_3^{2-}	4.7×10^{-11}
hydrogen phosphate ion	HPO_4^{2-}	PO_4^{3-}	4.8×10^{-13}
hydrosulfide ion	HS^-	S^{2-}	1.0×10^{-14}
water	H_2O	OH^-	1.8×10^{-16}
hydroxide ion	OH^-	O^{2-}	less than 10^{-33}

Increasing Acid Strength (arrow pointing up between H_3PO_4 and $H_2PO_4^-$ region)

Increasing Base Strength (arrow pointing down)

22-3. RELATIVE EXTENT OF REACTION

The relative extent of an acid-base reaction is easy to predict using an acid-base table. A stronger acid and base will react to produce a weaker acid and base to a relatively large extent. A weaker acid and base, however, will react only to a small extent to produce a stronger acid and base. Stated in a slightly different way, the position of equilibrium will favor the weaker acid and base. An acid-base table affords a ready comparison of the relative strengths of acids and bases.

Consider the reaction between the acid, ammonium ion, NH_4^+, and the base, carbonate ion, CO_3^{2-}. Will this reaction occur to a small or large extent? The products of the reaction will, of course, be the conjugate acid of CO_3^{2-} and the conjugate base of NH_4^+,

$$NH_4^+ + CO_3^{2-} \longrightarrow HCO_3^- + NH_3$$

Reference to Table 22-2 shows that NH_4^+ is a stronger acid than HCO_3^- and that CO_3^{2-} is a stronger base than NH_3. The stronger acid and base will react to a large extent and the reaction will take place as written. Unequal length arrows are sometimes employed to show the direction in which the equilibrium is favored.

$$NH_4^+ + CO_3^{2-} \rightleftharpoons HCO_3^- + NH_3$$

The reaction between hydrocyanic acid, HCN, and the acetate ion, which will be abbreviated OAc^-, will produce cyanide ion, CN^-, and acetic acid, HOAc. However, for this reaction, we find that the reactants are weaker than the products. At equilibrium, therefore, the reaction lies to the left.

$$HCN + OAc^- \rightleftharpoons HOAc + CN^-$$

The stronger acid and base will *always* appear on the same side of an acid-base equation because of the conjugate relationship between reactants and products. A simple device to give a quick prediction of the relative extent a given acid-base reaction using an acid-base table is the following:

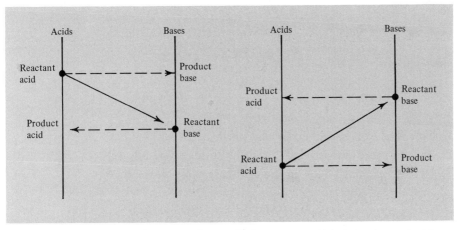

If the reactant acid is *above* the reactant base, the reaction will favor the *product* conjugate base and acid.

If the reactant acid is *below* the reactant base, the reaction will occur only slightly and the equilibrium will favor the reactant base and acid.

Moreover, the greater the separation between the acid and base the greater the displacement of the equilibrium.

22-4.   THE PRINCIPAL ACID-BASE REACTION

In any mixture of acids and bases there are always several equilibrium acid-base reactions. The reaction between the strongest acid and base that are present in relatively large concentrations in the mixture is called the principal acid-base reaction. This reaction will largely determine the acid-base properties of the solution. Usually, the other acid-base reactions can be neglected.

Several steps are helpful in writing the principal acid-base reaction, particularly in a solution with a number of acids and bases.

1. List all acids and bases initially present in relatively large concentration.
2. Pick the strongest acid and base using an acid-base table if necessary.
3. Write the reaction between this acid and base and use unequal length arrows to indicate the relative extent of the reaction. (Even if the principal acid-base reaction is small in extent, it will be the most important reaction in the system.)

■ Example 22-1

Write the principal acid-base reaction in a solution prepared by mixing approximately equal volumes of 0.1 M $NaHC_2O_4$ and 0.1 M Na_2SO_3.

□ *Solution*

In this solution there will initially be Na^+, $HC_2O_4^-$, and SO_3^{2-} ions and H_2O molecules.

1.

Acids Present Initially	Bases Present Initially
$HC_2O_4^-$	$HC_2O_4^-$
H_2O	H_2O
	SO_3^{2-}

Note that Na^+ is neither an acid nor a base. Why?

2. The strongest acid is $HC_2O_4^-$ and the strongest base is SO_3^{2-}.
3. $HC_2O_4^-$ lies above SO_3^{2-} in the acid-base table so we have

$$HC_2O_4^- + SO_3^{2-} \rightleftharpoons HSO_3^- + C_2O_4^{2-}$$

as the principal acid-base reaction in this solution. Notice that the principal acid-base reaction is not the only acid-base reaction. Many other acid-base equilibria exist in the solution, namely

$$HC_2O_4^- + HC_2O_4^- \longrightarrow H_2C_2O_4 + C_2O_4^{2-}$$

$$HC_2O_4^- + H_2O \rightleftharpoons H_3O^+ + C_2O_4^{2-}$$

$$H_2O + HC_2O_4^- \rightleftharpoons H_2C_2O_4 + OH^-$$

$$H_2O + H_2O \rightleftharpoons H_3O^+ + OH^-$$

$$H_2O + SO_3^{2-} \rightleftharpoons HSO_3^- + OH^-$$

■ **Example 22-2**

Some 1 M NaOAc is mixed with some 1 M $(NH_4)_2HPO_4$. Write an equation for the principal acid-base equilibrium.

☐ *Solution*

Initially, there will be Na^+, OAc^-, NH_4^+, HPO_4^{2-}, and H_2O in the mixture.

1. *Acids Initially Present*	*Bases Initially Present*
NH_4^+	OAc^-
HPO_4^{2-}	HPO_4^{2-}
H_2O	H_2O

2. The strongest acid is NH_4^+ and the strongest base is HPO_4^{2-}.
3. NH_4^+ lies below HPO_4^{2-} in the acid-base table, so the principal acid-base reaction is,

$$NH_4^+ + HPO_4^{2-} \rightleftharpoons H_2PO_4^- + NH_3$$

In this case, the strongest acid and base come from the same original solution. This result simply means that mixing NaOAc solution with $(NH_4)_2HPO_4$ solution has a negligible effect on the acid-base properties of the latter solution. Notice that there are eight other acid-base equilibria existing in the solution—all smaller in extent than the principal reaction.

There is one instance in which the rules for writing the principal acid-base reaction require a slight modification. In solutions where an acid and its conjugate base are the strongest acid and base present respectively, a trivial equation will result if the normal procedure is followed. For example, a solution containing only HOAc and NaOAc (in addition to H_2O) has as the strongest acid HOAc and as the strongest base OAc^-. But the equation

$$HOAc + OAc^- \rightleftharpoons HOAc + OAc^-$$

is obviously useless even though correct. In such cases, the principal acid-base reaction is taken to be either the reaction between the strongest acid and the next strongest base,

$$HOAc + H_2O \rightleftharpoons H_3O^+ + OAc^-$$

or the reaction between the strongest base and next strongest acid,

$$OAc^- + H_2O \rightleftharpoons HOAc + OH^-$$

22-5. K_w, pH, AND pOH

Water is both an acid and a base (although very weak as either), and hence will undergo acid-base reaction.

$$H_2O + H_2O \rightleftharpoons H_3O^+ + OH^-$$

The equilibrium constant for this reaction is,

$$K = \frac{[H_3O^+][OH^-]}{[H_2O]^2} = 3.2 \times 10^{-18} \text{ at } 25°C$$

Because water is usually the solvent, its concentration is large and will never vary much from the value of pure water, namely 55.5 moles/liter. Consequently, a simpler expression is used in place of the ordinary equilibrium constant

$$[H_2O]^2K = [H_3O^+][OH^-]$$

and

$$[H_2O]^2K = (55.5)^2(3.2 \times 10^{-18}) = 1.0 \times 10^{-14}$$

The combination, $[H_2O]^2K$ is called the water constant, K_w.

$$K_w = [H_3O^+][OH^-] = 1.0 \times 10^{-14} \text{ at } 25°C. \tag{22-1}$$

This equilibrium always exists in any water solution although it is rarely the principal acid-base equilibrium. Consequently, if $[H_3O^+]$ is known, $[OH^-]$ can be calculated and vice-versa.

In aqueous solutions, the acidity is determined by the concentration of H_3O^+ ions. In order to express this concentration using simple numbers, the pH scale was defined by S. P. L. Sorensen in 1909:

$$pH = -\log [H_3O^+] \quad \text{or} \quad [H_3O^+] = 10^{-pH} \tag{22-2}$$

An analogous definition is also used to express the OH^- ion concentration

$$pOH = -\log [OH^-] \quad \text{or} \quad [OH^-] = 10^{-pOH} \tag{22-3}$$

Taking logarithms of both sides of equation (22–1),

$$\log K_w = \log [H_3O^+] + \log [OH^-] \tag{22-1a}$$

and introducing the value of log K_w at 25°C,

$$-14 = \log [H_3O^+] + \log [OH^-] \qquad (22\text{--}1b)$$

Finally, substitution of equations (22–2) and (22–3) into (22–1b) yields

$$\text{pH} + \text{pOH} = 14 \quad \text{at } 25°C. \qquad (22\text{--}4)$$

In pure water

$$[H_3O^+] = [OH^-]$$

so that substitution into equation (22–1) gives,

$$[H_3O^+]^2 = 1.0 \times 10^{-14}$$

$$[H_3O^+] = 1.0 \times 10^{-7}$$

Pure water at 25°C thus has pH = pOH = 7.00. It is often said that a neutral solution has a pH = 7, and acidic solutions have pHs less than 7 whereas basic solutions have pHs greater than 7. Aqueous solutions are neutral when $[H_3O^+] = [OH^-]$. At 25°C this condition gives a pH of 7, but at a different temperature, the pH of a solution in which $[H_3O^+] = [OH^-]$ is not 7. For example, at 100°C

$$K_w = [H_3O^+][OH^-] = 1.0 \times 10^{-12}$$

and $\qquad\qquad [H_3O^+] = [OH^-] = 1.0 \times 10^{-6}$

A neutral aqueous solution at 100°C would therefore have a pH = 6.

22-6.   WEAK ACIDS AND BASES—K_a AND K_b

The principal acid-base equilibrium in a solution containing only a weak acid, say acetic acid, is

$$\text{HOAc} + H_2O \rightleftharpoons H_3O^+ + \text{OAc}^-$$

and the equilibrium constant is

$$K = \frac{[H_3O^+][\text{OAc}^-]}{[\text{HOAc}][H_2O]} = 3.2 \times 10^{-7} \text{ at } 25°C$$

Again, we find the solvent, water, appearing as a reactant in a reaction and in the equilibrium constant expression. The large, nearly constant, concentration of water can be combined with the equilibrium constant to give a simpler expression.

$$[H_2O]K = \frac{[H_3O^+][\text{Ac}^-]}{[\text{HAc}]}$$

$$[H_2O]K = (55.5)(3.2 \times 10^{-7}) = 1.8 \times 10^{-5}$$

339

The term $[H_2O]K$ is called the acid constant, K_a, for the weak acid.

$$K_a = \frac{[H_3O^+][Ac^-]}{[HAc]} = 1.8 \times 10^{-5} \text{ at } 25°C$$

The K_a for any weak acid refers to the expression,

$$K_a = \frac{[H_3O^+][\text{conjugate base}]}{[\text{acid}]}$$

Table 22–2 lists the K_a values at 25°C for all the acids given.

In similar fashion, K_b refers to the reaction between a weak base and water as an acid

$$\text{base} + H_2O \rightleftharpoons \text{conjugate acid} + OH^-$$

$$K_b = \frac{[\text{conjugate acid}][OH^-]}{[\text{base}]}$$

Consider the weak base, F^-, as a specific example.

$$F^- + H_2O \rightleftharpoons HF + OH^-$$

$$K_b = \frac{[HF][OH^-]}{[F^-]} = 1.5 \times 10^{-11} \text{ at } 25°C$$

Table 22–2 does not list any K_b values; however, it is very easy to calculate K_b if the K_a for the conjugate acid is known. The following derivation will show how such a calculation is done.

Consider the two simultaneous equilibria,

$$HX + H_2O \rightleftharpoons H_3O^+ + X^-$$
$$X^- + H_2O \rightleftharpoons HX + OH^-$$

and their respective constants,

$$K_a = \frac{[H_3O^+][X^-]}{[HX]}$$

$$K_b = \frac{[HX][OH^-]}{[X^-]}$$

Multiplication of these constants yields

$$K_a \times K_b = \frac{[H_3O^+][\cancel{X^-}]}{\cancel{[HX]}} \times \frac{[\cancel{HX}][OH^-]}{[\cancel{X^-}]} = [H_3O^+][OH^-]$$

or,

$$K_a \times K_b = K_w \qquad (22\text{–}5)$$

The K_a and K_b are the acid and base constants for a **conjugate** acid-base pair. Using equation (22-5), the K_b for any base listed in Table 22-2 is readily calculated.

Several problems will illustrate the principles which have been developed.

■ Example 22-3

Calculate $[H_3O^+]$ in 0.10 M NH_4Cl at 25°C.

□ *Solution*

The principal acid-base reaction in this solution is

$$NH_4^+ + H_2O \rightleftharpoons H_3O^+ + NH_3$$

and

$$K_a = \frac{[H_3O^+][NH_3]}{[NH_4^+]} = 5.6 \times 10^{-10}$$

At equilibrium

$$[NH_4^+] = 0.10 - [H_3O^+]$$
$$[H_3O^+] = [NH_3]$$

so that

$$\frac{[H_3O^+]^2}{0.10 - [H_3O^+]} = 5.6 \times 10^{-10}$$

Assuming that $[H_3O^+]$ is small and is negligible compared with 0.10, the expression can be simplified to,

$$\frac{[H_3O^+]^2}{0.10} = 5.6 \times 10^{-10}$$
$$[H_3O^+]^2 = 5.6 \times 10^{-11}$$
$$[H_3O^+]^2 = 56 \times 10^{-12}$$
$$[H_3O^+] = 7.5 \times 10^{-6}$$

Indeed, 7.5×10^{-6} is small compared with 0.10, so the simplification achieved by dropping the $[H_3O^+]$ term in the denominator is justified. The effect of dropping the $[H_3O^+]$ term is to make the final $[H_3O^+]$ value *larger* than it would be if the term were retained in the denominator so that, if the final result is still much smaller than the constant term, the dropping procedure is justified.

■ Example 22-4

Calculate the pH at 25°C of a solution that is 0.10 M $NaHC_2O_4$ and 0.010 M $Na_2C_2O_4$.

□ *Solution*

Because the strongest acid and base are conjugate, the principal acid-base reaction will be written using the strongest acid and next strongest base.

$$HC_2O_4^- + H_2O \rightleftharpoons H_3O^+ + C_2O_4^{2-}$$

$$K_a = \frac{[H_3O^+][C_2O_4^{2-}]}{[HC_2O_4^-]} = 5.2 \times 10^{-5}$$

At equilibrium,

$$[HC_2O_4^-] = 0.10 - [H_3O^+]$$

$$[C_2O_4^{2-}] = 0.010 + [H_3O^+]$$

and,

$$\frac{[H_3O^+](0.010 + [H_3O^+])}{0.10 - [H_3O^+]} = 5.2 \times 10^{-5}$$

Assuming $[H_3O^+]$ is small, we can simplify the expression to

$$\frac{[H_3O^+](0.010)}{0.10} = 5.2 \times 10^{-5}$$

and

$$[H_3O^+] = 5.2 \times 10^{-4}$$

$$pH = -\log [H_3O^+]$$

$$= -\log (5.2 \times 10^{-4})$$

$$= 3.28$$

■ **Example 22-5**

Calculate the pH at 25°C of 0.10 M Na_2SO_3.

□ *Solution*

The principal acid-base reaction is

$$SO_3^{2-} + H_2O \rightleftharpoons HSO_3^- + OH^-$$

$$K_b = \frac{[HSO_3^-][OH^-]}{[SO_3^{2-}]} = \frac{K_w}{K_a}$$

$$\frac{K_w}{K_a} = \frac{1.0 \times 10^{-14}}{6.2 \times 10^{-8}} = 1.6 \times 10^{-7}$$

so

$$\frac{[HSO_3^-][OH^-]}{[SO_3^{2-}]} = 1.6 \times 10^{-7}$$

At equilibrium

$$[HSO_3^-] = [OH^-]$$

$$[SO_3^{2-}] = 0.10 - [OH^-]$$

$$\frac{[OH^-]^2}{0.10 - [OH^-]} = 1.6 \times 10^{-7}$$

Simplifying

$$\frac{[OH^-]^2}{0.10} = 1.6 \times 10^{-7}$$

$$[OH^-]^2 = 1.6 \times 10^{-8}$$

$$[OH^-] = 1.3 \times 10^{-4}$$

$$pOH = -\log [OH^-]$$

$$= -\log (1.3 \times 10^{-4})$$
$$= 3.89$$

$$pH + pOH = 14.00$$
$$pH = 10.11$$

Often it is necessary that the pH of a solution be fixed at a certain value and maintained near that value even though strong acids and bases may be added to the solution. A solution capable of resisting changes in its pH is called a **buffer.** Biological systems usually require a definite pH for proper functioning. Human blood, for example, has a pH close to 7.4, and variation from this value by more than about 0.1 pH unit will result in rapid death.

To resist the effect of some added acid or base, a solution must have relatively high concentrations of both a base and an acid to react with the added substances. A conjugate acid-base pair can exist in the same solution because the products of their reaction are identical with reactants. The principal acid-base reaction of such a solution must be written between the acid (or base) and the next strongest base (or acid). Consider the generalized weak acid, HX, in solution.

$$HX + H_2O \rightleftharpoons H_3O^+ + X^-$$

$$K_a = \frac{[H_3O^+][X^-]}{[HX]}$$

Taking logarithms of both sides,

$$\log K_a = \log [H_3O^+] + \log \frac{[X^-]}{[HX]}$$

and defining,

$$-\log K_a = pK_a$$

there results,

$$-pK_a = -pH + \log \frac{[X^-]}{[HX]}$$

$$\log \frac{[X^-]}{[HX]} = pH - pK_a \qquad (22\text{-}6)$$

Equation (22-6) is useful in determining the proper ratio of concentrations of conjugate base to acid to fix the pH of a solution at some desired value.

■ **Example 22-6**

How could a solution with pH = 10.00 be prepared using the conjugate acid-base pair, $NH_4^+ - NH_3$?

□ *Solution*

For NH_4^+

$$K_a = 5.6 \times 10^{-10}$$

$$pK_a = -\log (5.6 \times 10^{-10})$$

$$pK_a = 9.25$$

Using equation (22-6)

$$\log \frac{[NH_3]}{[NH_4^+]} = 10.00 - 9.25$$

$$\log \frac{[NH_3]}{[NH_4^+]} = 0.75$$

so that

$$\frac{[NH_3]}{[NH_4^+]} = 5.6$$

A solution in which the concentration of NH_3 is 5.6 times the concentration of NH_4^+ will therefore be buffered at a pH = 10.00. For example, 1 mole of NH_4Cl dissolved in enough 5.6 M NH_3 to give 1 liter of solution would have the proper concentration ratio.

Having now seen how to prepare a particular buffer in the Example 22-6, let us investigate how this solution would resist changes in its pH. Suppose 0.10 mole HCl is added to 1 liter of the buffer solution. What would happen to its pH? (Note that if 0.10 mole of HCl were added to a liter of pure water, the pH would drop to 1.00). The 0.10 mole of HCl will become 0.10 mole of H_3O^+ in the solution and will react with the strongest base present, namely, NH_3.

$$H_3O^+ + NH_3 \longrightarrow NH_4^+ + H_2O$$

Assuming that this reaction occurs nearly to completion, the concentrations of species in the solution after the reaction are

$$[NH_3] = 5.6 - 0.1 = 5.5$$

$$[NH_4^+] = 1.0 + 0.1 = 1.1$$

Equation (22–6) can now be used to calculate the new pH for the slightly different concentration ratio of conjugate base to acid,

$$\log \frac{5.5}{1.1} = pH - 9.25$$

or

$$pH = 9.25 + \log 5.0$$
$$= 9.95$$

We see that the addition of quite a large amount of a strong acid only lowers the pH of the buffered solution by 0.05 pH units.

Let us now see how the addition of 0.10 mole of OH^- ions (for example, NaOH) would change the pH of the buffer. The principal acid-base reaction is

$$OH^- + NH_4^+ \longrightarrow H_2O + NH_3$$

Assuming a virtually complete reaction, the concentrations after reaction would be

$$[NH_3] = 5.6 + 0.1 = 5.7$$
$$[NH_4^+] = 1.0 - 0.1 = 0.9$$

and the pH would be calculated as

$$\log \frac{5.7}{0.9} = pH - 9.25$$

$$pH = 9.25 + 0.80$$
$$pH = 10.05$$

Again, we see that our buffer system fairly successfully withstands the effects of the addition of a relatively large amount of a strong base.

For an efficient buffer, the ratio of concentrations of the conjugate base and acid should not be too far from unity—usually 10/1 or 1/10 is about the maximum acceptable variation. If a greater variation is necessary to fix the pH at the desired value, a different conjugate acid-base pair should be chosen so that pK_a of the acid will be closer to the desired pH. As an example, if a $pH = 5.00$ were desired, the $NH_4^+ - NH_3$ acid-base pair would require

$$\log \frac{[NH_3]}{[NH_4^+]} = 5.00 - 9.25$$

$$\log \frac{[NH_3]}{[NH_4^+]} = -4.25$$

$$\frac{[NH_3]}{[NH_4^+]} = 5.6 \times 10^{-5}$$

Such a concentration would be unrealistic because there would simply be too little NH_3 present to react with an added acid. For a $pH = 5.00$, a much better buffer would result from the use of the HOAc-OAc⁻ acid-base pair, where $pK_a = 4.74$. In this case

$$\log \frac{[OAc^-]}{[HOAc]} = 5.00 - 4.74$$

$$\log \frac{[OAc^-]}{[HOAc]} = 0.26$$

$$\frac{[OAc^-]}{[HOAc]} = 1.8$$

which is a very reasonable concentration ratio.

22-8. MORE COMPLEX ACID-BASE EQUILIBRIA

Sometimes information is needed about the concentration of a species that does not appear in the principal acid base reaction. A common problem involves the calculation of the pH of a solution in which water is not a reactant in the principal acid base reaction. A specific example will illustrate this type of problem.

■ **Example 22-7**

Calculate the pH of 1.0 M NaHCO₃.

□ *Solution*

A consideration of the acids and bases present in this solution reveals that HCO_3^- ion is both the strongest acid and strongest base. Consequently, the principal acid-base reaction must be

$$HCO_3^- + HCO_3^- \rightleftharpoons H_2CO_3 + CO_3^{2-}$$

$$K = \frac{[H_2CO_3][CO_3^{2-}]}{[HCO_3^-]^2}$$

Note that K is *not* the K_a for the acid HCO_3^- since it is reacting with the base HCO_3^- and not with water. However, the above reaction primarily controls the concentrations of these species because it is the principal acid-base reaction. The concentration of H_3O^+ or OH^- must be determined before the pH can be calculated. Ultimately, therefore, some equilibrium involving H_3O^+ or OH^- will have to be considered, but the principal acid-base reaction determines the concentrations of the species that control the H_3O^+ and OH^- ion concentrations. The value of K is needed and must be calculated from known equilibrium constants.

Two equilibria, which between them contain the species appearing in the principal acid-base equilibrium, are

$$H_2CO_3 + H_2O \rightleftharpoons H_3O^+ + HCO_3^-$$

$$K_{a_1} = \frac{[H_3O^+][HCO_3^-]}{[H_2CO_3]} = 4.5 \times 10^{-7}$$

and

$$HCO_3^- + H_2O \rightleftharpoons H_3O^+ + CO_3^{2-}$$

$$K_{a_2} = \frac{[H_3O^+][CO_3^{2-}]}{[HCO_3^-]} = 4.7 \times 10^{-11}$$

Examination of these equilibrium constant expressions reveals that

$$\frac{K_{a_2}}{K_{a_1}} = \frac{\dfrac{[H_3O^+][CO_3^{2-}]}{[HCO_3^-]}}{\dfrac{[H_3O^+][HCO_3^-]}{[H_2CO_3]}} = \frac{[H_2CO_3][CO_3^{2-}]}{[HCO_3^-]^2}$$

which is the necessary constant, K.

$$K = \frac{K_{a_2}}{K_{a_1}} = \frac{4.7 \times 10^{-11}}{4.5 \times 10^{-7}} = 1.0 \times 10^{-4}$$

Referring back to the principal acid-base reaction, we find that at equilibrium

$$[H_2CO_3] = [CO_3^{2-}]$$

$$[HCO_3^-] = 1.0 - [H_2CO_3]$$

so that

$$\frac{[H_2CO_3]^2}{(1.0 - [H_2CO_3])^2} = 1.0 \times 10^{-4}$$

Neglecting $[H_2CO_3]$ in the denominator

$$[H_2CO_3]^2 = 1.0 \times 10^{-4}$$

$$[H_2CO_3] = 1.0 \times 10^{-2}$$

Either the expression for K_{a_1} or K_{a_2} may now be used to calculate $[H_3O^+]$ because the correct concentrations of H_2CO_3 and CO_3^{2-} are known.

Using K_{a_2}

$$\frac{[H_3O^+][CO_3^{2-}]}{[HCO_3^-]} = 4.7 \times 10^{-11}$$

$$\frac{[H_3O^+](1.0 \times 10^{-2})}{1.0 - 1.0 \times 10^{-2}} = 4.7 \times 10^{-11}$$

$$[H_3O^+] = 4.7 \times 10^{-9}$$

347

and

$$pH = \log 4.7 \times 10^{-9}$$
$$pH = 8.3$$

Example 22-7 appears at first to be fairly complicated. However, the alternative to this method based on the principal acid-base reaction is to consider all possible acid-base equilibria involving the HCO_3^- and H_3O^+ ions, namely,

$$HCO_3^- + H_2O \rightleftharpoons H_3O^+ + CO_3^{2-}$$

$$HCO_3^- + H_2O \rightleftharpoons H_2CO_3 + OH^-$$

$$H_2O + H_2O \rightleftharpoons H_3O^+ + OH^-$$

with five unknown concentrations (HCO_3^-, H_3O^+, CO_3^{2-}, H_2CO_3, and OH^-). Three equilibrium constants relate these unknown species, K_a for HCO_3^-, K_b for HCO_3^-, and K_w. However, three independent equations are insufficient to solve five unknowns. Two more independent equations can be derived by taking into account the conservation of mass and the conservation of charge. Mass conservation requires that,

$$[HCO_3^-] + [CO_3^{2-}] + [H_2CO_3] = 1.0$$

because all the carbon-containing species come from the original 1.0 molar HCO_3^- ion. Charge conservation requires that the sum of the concentrations of all positive charges equals the sum of the concentrations of all negative charges,

$$[Na^+] + [H_3O^+] = [HCO_3^-] + [OH^-] + 2[CO_3^{2-}]$$

or

$$[Na^+] = 1.0 = [HCO_3^-] + [OH^-] + 2[CO_3^{2-}] - [H_3O^+]$$

Solution of the five equations simultaneously will yield essentially the same answer as that obtained by working from the principal acid-base reaction. The solution of several simultaneous equations will convince one that any method which avoids such a calculation is greatly to be preferred, particularly if the simpler method yields essentially the same answer.

Occasionally, a given solution will have two acids or bases so nearly the same in strength that it is not possible to write a single principal acid-base reaction. When such cases arise, one has little recourse except to solve the necessary simultaneous equations. Fortunately, the circumstances where a single principal acid-base reaction cannot be written are rare.

22-9.   PERIODIC VARIATIONS IN ACID-BASE STRENGTH

Acids that are binary compounds are called **hydroacids,** for example, HCl and H_2S. The **oxyacids** are ternary compounds in which the hydrogen atoms that can be lost as protons are always bonded to an O atom which, in turn, is bonded to some central atom, for example, HOClO and H_2SO_4.

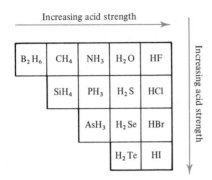

Figure 22-1. Variation of acid strength of nonmetal hydrides.

Figure 22–1 illustrates the change in strength of the hydroacids, and other nonmetal hydrides not usually considered as acids, with their position in the periodic table. The general rule is as follows: *Within a given group, a proton is more easily removed from a large atom than from a small one.* In a qualitative way this rule can be justified on the basis of Coulomb's law because the attraction between the proton and an incipient anion is weaker the larger the anion. Moving across a period, *acid strength increases as the electron affinity of the non-metal increases.* This latter effect is more pronounced than the size factor as is evidenced by the fact that of two hydroacids along a diagonal, the one to the right is stronger, for example, HF is much stronger than H_2S as an acid. Many of the nonmetal hydrides have basic properties, for example, NH_3 and H_2O. Not unexpectedly, base strength increases opposite to increasing acid strength. It should be mentioned that CH_4, SiH_4, and B_2H_6 do not function as bases because there is no place in these molecules for a proton to bond.

The increase in the strength of oxyacids with increasing nonmetal character (increasing electron affinity) of the central atom has already been discussed in Section 15–2. Oxyacids of a given nonmetal can vary greatly in acid strength, however. Consider the chlorine oxyacids and their K_a values at 25°C.

Acid	K_a
HOCl	3.2×10^{-8}
HOClO	1.1×10^{-2}
HOClO_2	∞
HOClO_3	∞

The drastic change in acid strength for these compounds can be rationalized in terms of the formal charges on the central atom, that is

$H:\ddot{O}:\ddot{C}l:$ ⌒0 $H:\ddot{O}:\ddot{C}l:\ddot{O}:$ ⌒+1 $H:\ddot{O}:\ddot{C}l:\ddot{O}:$ ⌒+2 $H:\ddot{O}:\ddot{C}l:\ddot{O}:$ ⌒+3

349

The increase in formal charge on the Cl atom causes increased attraction for electrons. Consequently, the H-O bond will be weakened as electrons are withdrawn from the oxygen atom.

It was pointed out in Section 15–5 that, if an H atom in an oxyacid is bonded directly to the central atom, it will not be available as a proton. Phosphoric, phosphorous, and hypophosphorous acids have three, two, and one replaceable H atoms, respectively.

phosphoric acid
$K_a = 6 \times 10^{-3}$

phosphorous acid
$K_a = 5 \times 10^{-2}$

hypophosphorous acid
$K_a = 1 \times 10^{-2}$

The formal charge on the P atom is $+1$ in each of these compounds, and their acid strengths in terms of loss of the first proton are roughly the same.

22–10. ACIDIC PROPERTIES OF HYDRATED METAL IONS

Small cations of $+2$ or $+3$ charge will withdraw enough electron density from the O atoms of hydrated water molecules to confer definite acidic character on the hydrated ion. This property of hydrated Be^{2+} and Al^{3+} ions was mentioned in Sections 15–2 and 15–3. The hydrated Al^{3+} ion in solution undergoes the reaction,

$$Al(H_2O)_6^{3+} + H_2O \rightleftharpoons H_3O^+ + Al(H_2O)_5(OH)^{2+}$$

$$K_a = \frac{[H_3O^+][Al(H_2O)_5(OH)^{2+}]}{[Al(H_2O)_6^{3+}]} = 1.3 \times 10^{-5} \text{ at } 25°C$$

This K_a value shows $Al(H_2O)_6^{3+}$ to be about as strong an acid as acetic acid. Other ions with $+3$ charges have roughly this same strength unless they are quite large. Very small cations with a $+2$ charge like Be^{2+} are also noticeably acidic in solution. The Mg^{2+} is much less acidic in solution because of its larger size. The positive electric field around the larger ion is less intense, which results in less polarization of the electrons in the hydrated water molecules toward the metal ion. Except for the tiny Li^+ ion, metal ions with only a $+1$ charge do not confer measurable acidic properties on their water solutions. Table 22–3 lists a few K_a values for some hydrated metal ions.

The amphoteric nature of the insoluble, hydrated hydroxides of some $+2$ and $+3$ metal ions was also discussed in Chapter 15. In general, the same factors that lead to acidity of the hydrated ions (high charge and small size of the metal ion) also cause amphoterism in the hydroxide. For the representative metal ions, this generalization is strictly followed. However, there are exceptions among the transition metal ions. For instance, $Fe(H_2O)_6^{3+}$ is quite acidic, yet $Fe(H_2O)_3(OH)_3$

Table 22-3. K_a Values for
Some Hydrated Metal Ions

Hydrated Ion	K_a (25°C)
$Fe(H_2O)_6^{3+}$	6.3×10^{-3}
$Cr(H_2O)_6^{3+}$	1.0×10^{-4}
$Al(H_2O)_6^{3+}$	1.3×10^{-5}
$Fe(H_2O)_6^{2+}$	4.0×10^{-9}
$Ca(H_2O)_6^{2+}$	1.0×10^{-13}

reacts only as a base and shows none of the amphoteric behavior of $Al(H_2O)_3(OH)_3$.

PROBLEMS

1. What is the relationship between a conjugate acid-base pair?
2. How is acid strength measured in the Brønsted-Lowry system?
3. What is the leveling effect of a solvent?
4. What is meant by the principal acid-base reaction in an acid-base system?
5. What is pH? Why was this term defined?
6. Define the following special equilibrium constants: K_a, K_b, K_w. How do they differ from an ordinary equilibrium constant?
7. The K_a for water is given in Table 22-2 as 1.8×10^{-16}. This is not K_w, which has the value 1.0×10^{-14}. Explain.
8. By means of a calculation, show why the K_a for H_3O^+ is listed in Table 22-2 as 55.5.
9. Write equations showing the principal acid-base equilibrium that would exist in each of the following solutions. Use unequal length arrows to indicate the relative position of equilibrium.
   (a) $0.1\ M$ $NaHSO_4$ (c) $0.1\ M$ Na_2HPO_4
   (b) $0.1\ M$ Na_3PO_4 (d) pure water
10. Write a chemical equation for the principal acid-base reaction that would occur in the following solutions. Where equilibria are established, use unequal length arrows to indicate the direction in which reaction is large in extent.
   (a) gaseous HCl is bubbled into water
   (b) gaseous HF is bubbled into $1\ M$ NaOH
   (c) $0.1\ M$ HCl and $0.1\ M$ KOH are mixed
   (d) $0.1\ M$ HOAc and $0.1\ M$ NH_3 are mixed
   (e) a carbonic acid solution is added to $0.1\ M$ $NaHSO_3$
   (f) $0.1\ M$ $(NH_4)_2HPO_4$ and $0.1\ M$ $Na_2C_2O_4$ are mixed
   (g) $0.1\ M$ $NaHC_2O_4$ is added to $0.1\ M$ $NaHSO_3$
   (h) some NH_4Cl is dissolved in a NaHS solution
   (i) $0.1\ M$ $HClO_4$ and $0.1\ M$ NaOAc are mixed
   (j) $0.1\ M$ $NaHC_2O_4$ and $0.1\ M$ NaH_2PO_4 are mixed

351

(k) 0.1 M $NaHCO_3$ is added to 0.1 M $NaHSO_4$

(l) some NaH_2PO_4 and some Na_3PO_4 are dissolved in water

11. Write the conjugate acid and conjugate base for each of the following species.
   (a) H_2Y^- **(c)** CO_3^{2-} **(e)** HS^-
   (b) PH_3 **(d)** H_2O **(f)** H_2S

12. A 0.1 M solution of a weak acid, HX, is 10% ionized at 25°C. Calculate K_a for HX and K_b for X^- at 25°C.

13. Determine $[H_3O^+]$ in 0.20 M HCN at 25°C.

14. Calculate $[OH^-]$ in 0.10 M NH_3 at 25°C.

15. What is the pH of the solution in problem 13? problem 14?

16. Determine the pH and pOH at 25°C for the following:
   (a) 0.010 M HCl   **(b)** 0.010 M NaOH   **(c)** 0.010 M HOAc

17. Calculate the pH of 1.0 M Na_3PO_4 at 25°C.

18. Calculate $[H_3O^+]$ and $[OH^-]$ for the following solutions at 25°C.
   (a) pure water   **(c)** pOH = 2.50
   (b) pH = 4.00   **(d)** pH = 11.50

19. A solution is prepared by adding 0.10 mole of NaF to sufficient 1.0 M HF to make 1 liter at 25°C. What is $[H_3O^+]$ in this solution? the pH? Is the solution buffered?

20. Calculate the pH of a buffer prepared by adding 0.010 mole of NH_4Cl and 0.050 mole of NH_3 to enough water to make 1.0 liter of solution.

21. What concentration of HPO_4^{2-} would be necessary in a solution with $[H_2PO_4^-] = 0.50$ to fix the pH at 7.0?

22. Calculate the pH of the solutions of problem 16 after 0.05 mole of HCl is added to 1 liter of each.

23. Calculate the pH of 1 M NaH_2PO_4.

24. Choose the stronger acid of the following pairs.
   (a) NH_3 and H_2O **(e)** HOI and HOCl
   (b) H_2S and HF   **(f)** H_3BO_3 and H_2CO_3
   (c) H_2S and H_2Se **(g)** H_2SeO_3 and H_2SeO_4
   (d) CH_4 and PH_3 **(h)** H_3PO_2 and HNO_2

25. Choose the stronger base of the following pairs.
   (a) NH_3 and H_2S **(c)** HF and H_2O
   (b) H_2S and H_2O **(d)** AsH_3 and PH_3

26. Calculate the pH of 1.0 M $Fe(NO_3)_3$.

27. Arrange the following ions in order of decreasing acidity of their hydrated ions. Mg^{2+}, Ga^{3+}, Al^{3+}, Li^+, Sr^{2+}

28. Of the ions listed in problem 27, only two have amphoteric hydroxides. Which ones are they? Explain.

EQUILIBRIUM IN THE SERVICE
OF ANALYTICAL CHEMISTRY

The chemical analysis of a mixture usually involves two steps: (1) the separation of the components of the mixture and (2) the identification of each component and/or the determination of the amount of each component.

Occasionally, one or more components in a mixture can be identified without prior separation by means of a specific chemical test called a spot test or by some instrumental method such as nuclear magnetic resonance, infrared, or X-ray diffraction spectroscopy. More often, however, some sort of prior separation is necessary before the final qualitative or quantitative analysis* can be carried out. Selective precipitation of certain species is a general method of separation as well as a method for the quantitative determination of a substance.

Titration is the determination of the concentration of a given solution by adding measured amounts of a standard solution (a solution of exactly known concentration) until the desired chemical reaction has taken place. Acid-base and redox titrations are widely used as analytical tools.

The equilibrium principles underlying quantitative precipitation reactions and acid-base and redox titrations will be treated in this chapter.

23-1.   QUANTITATIVE PRECIPITATIONS—K_{sp}

A quantitative precipitation is one in which the species that is precipitated from solution has a residual concentration in the solution not exceeding 10^{-5} molar at equilibrium. The equilibrium under consideration is established between a solid salt and a saturated solution of its ions. For example, silver chloride (AgCl) in equilibrium with its ions in solution can be expressed

---

*Qualitative analysis is the identification of the chemical species present in an unknown, whereas quantitative analysis involves the determination of the amount of a species present.

353

$$AgCl \rightleftharpoons Ag^+ + Cl^-$$

$$K = \frac{[Ag^+][Cl^-]}{[AgCl]}$$

Since the AgCl is a pure solid, its concentration is fixed at a constant temperature and is therefore a constant which can be combined with K.

$$[AgCl]K = [Ag^+][Cl^-]$$

The term $[AgCl]K$ is called the solubility product constant, K_{sp}.

$$K_{sp} = [Ag^+][Cl^-] = 1.7 \times 10^{-10} \text{ at } 25°C$$

Table 23–1 lists K_{sp} values for some salts.

Table 23–1. K_{sp} Values for Some Salts at 25°C

Salt	K_{sp}	Salt	K_{sp}	Salt	K_{sp}
bromides		fluorides		sulfates	
PbBr$_2$	4.6×10^{-6}	BaF$_2$	2.4×10^{-5}	BaSO$_4$	1.5×10^{-9}
AgBr	5.0×10^{-13}	CaF$_2$	3.9×10^{-11}	CaSO$_4$	2.4×10^{-5}
carbonates		PbF$_2$	4×10^{-8}	Ag$_2$SO$_4$	1.2×10^{-5}
BaCO$_3$	1.6×10^{-9}	hydroxides		sulfides	
PbCO$_3$	1.5×10^{-15}	Fe(OH)$_2$	1.8×10^{-15}	CdS	1.0×10^{-28}
Ag$_2$CO$_3$	8.2×10^{-12}	Fe(OH)$_3$	6×10^{-38}	CuS	8×10^{-37}
chlorides		Mg(OH)$_2$	8.9×10^{-12}	FeS	4×10^{-19}
PbCl$_2$	1.6×10^{-5}	Zn(OH)$_2$	4.5×10^{-17}	PbS	7×10^{-29}
Hg$_2$Cl$_2$	1.1×10^{-18}	iodides		MnS	7×10^{-16}
AgCl	1.7×10^{-10}	PbI$_2$	8.3×10^{-9}	HgS	1.6×10^{-54}
chromates		AgI	8.5×10^{-17}	NiS	3×10^{-21}
PbCrO$_4$	2×10^{-16}	oxalates		Ag$_2$S	5.5×10^{-51}
Ag$_2$CrO$_4$	1.9×10^{-12}	BaC$_2$O$_4$	1.5×10^{-8}	ZnS	2.5×10^{-22}
		CaC$_2$O$_4$	1.3×10^{-9}		
		MgC$_2$O$_4$	8.6×10^{-5}		

If it were necessary to remove the Cl$^-$ ion from a solution, the Ag$^+$ ion would be a good precipitating agent, for example, as AgNO$_3$, since relatively little excess Ag$^+$ ion would be needed to precipitate AgCl quantitatively. The K_{sp} expression for AgCl shows that to reduce the final Cl$^-$ ion concentration to 10^{-5} molar, the excess Ag$^+$ ion concentration need only be

$$[Ag^+] = \frac{1.7 \times 10^{-10}}{10^{-5}}$$

$$[Ag^+] = 1.7 \times 10^{-5}$$

If the Ag$^+$ ion concentration were raised to some moderate level, say $[Ag^+] = 1 \times 10^{-2}$, then the Cl$^-$ ion remaining in solution at equilibrium would be even less.

$$[Cl^-] = \frac{1.7 \times 10^{-10}}{1 \times 10^{-2}}$$

$$[Cl^-] = 1.7 \times 10^{-8}$$

After collecting, washing, drying and weighing the AgCl, the original concentration of Cl^- ion in the solution could be calculated.

On the other hand, a more soluble chloride, TlCl, would not be suitable for the quantitative precipitation of Cl^-. The following calculation shows the problem in attempting to have $[Cl^-] = 1 \times 10^{-5}$.

$$K_{sp} = [Tl^+][Cl^-] = 1.9 \times 10^{-4} \text{ at } 25°C$$

$$[Tl^+] = \frac{1.9 \times 10^{-4}}{1 \times 10^{-5}}$$

$$[Tl^+] = 19$$

It would be impossible to raise the excess Tl^+ ion concentration to 19 moles/liter since no thallium salt is soluble enough.

Often times, when solutions containing the ions of a sparingly soluble salt are mixed, some precipitation will occur even if the precipitation is not quantitative.

■ **Example 23–1**

10 ml each of 0.1 M TlNO$_3$ and 0.1 M NaCl are mixed. Will TlCl precipitate?

□ *Solution*

If no reaction should occur, the concentrations of the ions would be

$$[Tl^+] = \frac{10 \text{ ml} \times 0.1 \text{ } M}{20 \text{ ml}} = 5 \times 10^{-2} \text{ } M$$

$$[Cl^-] = \frac{10 \text{ ml} \times 0.1 \text{ } M}{20 \text{ ml}} = 5 \times 10^{-2} \text{ } M$$

At equilibrium

$$[Tl^+][Cl^-] = 1.9 \times 10^{-4}$$

and if

$$[Tl^+] = 5 \times 10^{-2}$$

then the maximum Cl^- ion concentration can be only

$$[Cl^-] = \frac{1.9 \times 10^{-4}}{5 \times 10^{-2}}$$

$$[Cl^-] = 4 \times 10^{-3}$$

which is much smaller than the available Cl^- ion concentration of 5×10^{-2} molar. Consequently, some TlCl will precipitate.

Another way to predict whether or not precipitation will occur is to calculate what is referred to as the ion product, i.p. The i.p. has the same form as the K_{sp} but applies to any condition—equilibrium or nonequilibrium. The i.p. for the above example is calculated to be

$$\text{i.p.} = [Tl^+][Cl^-]$$
$$= (5 \times 10^{-2})(5 \times 10^{-5})$$
$$= 2.5 \times 10^{-3}$$

Comparison of this value with the K_{sp} for TlCl shows that the i.p. is larger than the K_{sp}. The solution would thus have more ions than allowed by the equilibrium; hence, precipitation would occur to reduce the ion concentrations to their equilibrium values. An i.p. is calculated assuming no reaction occurs. It is then compared with the K_{sp} for the particular salt. Three conditions can arise: (1) the i.p. is larger than the K_{sp}, indicating that precipitation will occur; (2) the i.p. is smaller than the K_{sp}, indicating that precipitation will not occur; (3) the i.p. is equal to the K_{sp}, indicating that a slight increase in the concentration of either ion will cause precipitation to commence.

For salts whose formulas have more than one cation and anion, the ion concentrations must be raised to their appropriate power in the K_{sp} expression.

$$CaF_2 \rightleftharpoons Ca^{2+} + 2F^- \qquad K_{sp} = [Ca^{2+}][F^-]^2$$
$$Ag_2CrO_4 \rightleftharpoons 2Ag^+ + CrO_4^{2-} \qquad K_{sp} = [Ag^+]^2[CrO_4^{2-}]$$

Several examples will illustrate calculations for such systems.

■ **Example 23-2**

Calculate the K_{sp} for CaF_2 if 2.2×10^{-4} mole of CaF_2 dissolves in 1 liter of water at 25°C.

□ *Solution*

$$CaF_2 \rightleftharpoons Ca^{2+} + 2F^-$$

For every mole of CaF_2 dissolved, 1 mole of Ca^{2+} and 2 moles of F^- go into solution. At equilibrium,

$$[Ca^{2+}] = 2.2 \times 10^{-4}$$
$$[F^-] = 4.4 \times 10^{-4}$$

$$K_{sp} = [Ca^{2+}][F^-]^2$$
$$= (2.2 \times 10^{-4})(4.4 \times 10^{-4})^2$$
$$= 4.3 \times 10^{-11}$$

■ **Example 23-3**

Calculate the molar solubility of Ag_2CrO_4 at 25°C if $K_{sp} = 1.9 \times 10^{-12}$

□ *Solution*

$$Ag_2CrO_4 \rightleftharpoons 2 Ag^+ + CrO_4^{2-}$$
$$K_{sp} = [Ag^+]^2[CrO_4^{2-}] = 1.9 \times 10^{-12}$$

At equilibrium, molar solubility of $Ag_2CrO_4 = [CrO_4^{2-}]$ and $[CrO_4^{2-}] = \frac{1}{2}[Ag^+]$

$$[Ag^+]^2(\tfrac{1}{2}[Ag^+]) = 1.9 \times 10^{-12}$$
$$[Ag^+]^3 = 3.8 \times 10^{-12}$$
$$[Ag^+] = 1.6 \times 10^{-4}$$

and the molar solubility of $Ag_2CrO_4 = [CrO_4^{2-}] = 8.0 \times 10^{-5}$

If an ion is added to a solution containing two or more different ions that form sparingly soluble salts with the first ion, the least soluble salt will precipitate first. As an example, consider the addition of Ag^+ ion to a solution containing Cl^- and CrO_4^{2-} ions each at a concentration of 0.01 M. Both AgCl and Ag_2CrO_4 are sparingly soluble salts, but AgCl is less soluble. The Ag^+ ion concentration when Ag_2CrO_4 *just starts* to precipitate is calculated as follows

$$Ag_2CrO_4 \rightleftharpoons 2 Ag^+ + CrO_4^{2-}$$
$$[Ag^+]^2[CrO_4^{2-}] = 1.9 \times 10^{-12}$$

if

$$[CrO_4^{2-}] = 1 \times 10^{-2}$$

then Ag_2CrO_4 just starts to precipitate when

$$[Ag^+]^2 = \frac{1.9 \times 10^{-12}}{1 \times 10^{-2}}$$
$$= 1.9 \times 10^{-10}$$
$$[Ag^+] = 1.4 \times 10^{-5}$$

But the maximum Cl^- ion concentration in the presence of Ag^+ at this concentration is calculated as

$$AgCl \rightleftharpoons Ag^+ + Cl^-$$
$$[Ag^+][Cl^-] = 1.7 \times 10^{-10}$$

if

$$[Ag^+] = 1.4 \times 10^{-5}$$

then

$$[Cl^-] = \frac{1.7 \times 10^{-10}}{1.4 \times 10^{-5}}$$

$$[Cl^-] = 1.2 \times 10^{-5}$$

In other words, the Cl^- ion has been quantitatively precipitated as AgCl before any Ag_2CrO_4 begins to precipitate. Furthermore, Ag_2CrO_4 is deep red whereas AgCl is white. The concentration of Cl^- ion in a solution can be determined by titrating it with a standard solution of $AgNO_3$. A small amount of CrO_4^{2-} ion is added to the unknown solution at the start of the titration, and, when the red Ag_2CrO_4 appears, the volume of standard $AgNO_3$ solution used to this point allows calculation of the original Cl^- ion concentration. The CrO_4^{2-} ion serves as a **precipitation indicator.**

23-2. METAL SULFIDE PRECIPITATIONS

The sulfides of many metals are highly insoluble; yet there is a wide variation in the K_{sp} values of these compounds. Because the S^{2-} ion is a fairly strong base, it is easy to control its concentration in solution by appropriate adjustment of the pH. The *selective* precipitation of metal sulfides is thus possible.

An easy way to introduce S^{2-} ions into a solution is by means of the weak acid, H_2S. This acid is gaseous but dissolves in water to give an approximately $0.1\ M\ H_2S$ solution at $25°C$. H_2S is easily generated by the action of HCl on FeS or by heating a mixture of paraffin and sulfur. However, H_2S is very toxic, and a safer source for it in the laboratory is the reaction between thioacetamide and water.

$$H_3C\overset{\overset{\displaystyle S}{\|}}{C}-NH_2 + H_2O \longrightarrow H_2S + H_3C\overset{\overset{\displaystyle O}{\|}}{C}-NH_2$$

This reaction is extremely slow at $25°C$, but proceeds rapidly when the solution is heated. The H_2S is formed right in the solution and very little escapes into the air. The horrible rotten egg odor is fortunately a ready indicator of the presence of H_2S long before lethal concentrations are reached. The other product in the above reaction, acetamide, remains in the solution and does not interfere with subsequent reactions.

The principal acid-base reaction in a solution of H_2S is

$$H_2S + H_2O \rightleftharpoons H_3O^+ + HS^- \tag{23-1}$$

$$K_a = \frac{[H_3O^+][HS^-]}{[H_2S]} = 1.1 \times 10^{-7} \tag{23-2}$$

This equilibrium will control the acid-base properties of the solution, but the S^{2-} ion does not appear in it. Consequently, we must consider the extremely slight reaction

$$HS^- + H_2O \rightleftharpoons H_3O^+ + S^{2-} \tag{23-3}$$

$$K_a = \frac{[H_3O^+][S^{2-}]}{[HS^-]} = 1.0 \times 10^{-14} \tag{23-4}$$

which controls the S^{2-} ion concentration. Multiplication of equations (23-2) and (23-4) yields

$$\frac{[H_3O^+]^2[S^{2-}]}{[H_2S]} = 1.1 \times 10^{-21} \tag{23-5}$$

Equation (23-5) is an overall equilibrium expression for the composite equation obtained by adding equations (23-1) and (23-3)

$$H_2S + 2 H_2O \rightleftharpoons 2 H_3O^+ + S^{2-} \tag{23-6}$$

The student is cautioned that equation (23-6) does not represent a *real* equilibrium because two H_3O^+ ions are not produced for every one S^{2-} ion. The concentration of H_3O^+ is much greater than twice the S^{2-} ion in a saturated solution of H_2S. Some simple calculations will show why this is true. Using the principal acid-base reaction (23-1), the $[H_3O^+]$ can be calculated because

$$[H_3O^+] = [HS^-]$$

Substitution into equation (23-2) gives

$$\frac{[H_3O^+]^2}{[H_2S]} = 1.1 \times 10^{-7}$$

and since the solution is saturated with H_2S

$$\frac{[H_3O^+]^2}{0.1} = 1.1 \times 10^{-7}$$

$$[H_3O^+]^2 = 1.1 \times 10^{-8}$$

$$[HS^-] = [H_3O^+] = 1.1 \times 10^{-4}$$

The concentration of S^{2-} must be calculated from equation (23-4)

$$\frac{(1.1 \times 10^{-4})[S^{2-}]}{(1.1 \times 10^{-4})} = 1.0 \times 10^{-14}$$

$$[S^{2-}] = 1.0 \times 10^{-14}$$

It is therefore seen that $[H_3O^+]$ is *ten billion* times that of $[S^{2-}]$ in a saturated solution of H_2S at 25°C.

As long as no assumption is made about the H_3O^+ or S^{2-} ion concentrations, equation (23-5) can be used to calculate either the H_3O^+ or the S^{2-} ion concentration if one of these concentrations is known. Because the solution is usually kept saturated with H_2S (that is, $[H_2S] = 0.1$), equation (23–5) can be written as

$$[H_3O^+]^2[S^{2-}] = 1.1 \times 10^{-22} \tag{23–5a}$$

for any aqueous solution saturated with H_2S at 25°C. The use of expression (23–5a) is best illustrated by some examples.

■Example 23-4

A solution contains Fe^{2+}, Pb^{2+}, and Cu^{2+} all at concentrations of 0.1 M. If this solution is buffered to a pH = 1.0 and saturated with H_2S at 25°C., what substances will precipitate?

☐Solution

If pH = 1.0, then $[H_3O^+] = 1 \times 10^{-1}$ and because

$$[H_3O^+]^2[S^{2-}] = 1.1 \times 10^{-22}$$

$$[S^{2-}] = \frac{1.1 \times 10^{-22}}{(1 \times 10^{-1})^2}$$

$$[S^{2-}] = 1.1 \times 10^{-20}$$

The i.p.'s will all be the same in this case,

$$\text{i.p.} = [M^{2+}][S^{2-}]$$
$$= (0.1)(1.1 \times 10^{-20})$$
$$= 1.1 \times 10^{-21}$$

Comparison of the i.p. with the appropriate K_{sp}'s in Table 23–1 shows that PbS and CuS would precipitate whereas FeS would not, that is, the Fe^{2+} would remain in solution at 0.1 M.

■ Example 23-5

What is the lowest pH at which quantitative precipitation of FeS will occur in a solution saturated with H_2S?

☐ Solution

For quantitative precipitation

$$[Fe^{2+}] = 1 \times 10^{-5}$$

and since at equilibrium

$$[Fe^{2+}][S^{2-}] = 4 \times 10^{-19}$$

therefore

$$[S^{2-}] = \frac{4 \times 10^{-19}}{1 \times 10^{-5}}$$

$$[S^{2-}] = 4 \times 10^{-14}$$

Because the solution is saturated with H_2S

$$[H_3O^+]^2[S^{2-}] = 1.1 \times 10^{-22}$$

$$[H_3O^+]^2 = \frac{1.1 \times 10^{-22}}{4 \times 10^{-14}}$$

$$[H_3O^+]^2 = 28 \times 10^{-10}$$

$$[H_3O^+] = 5.3 \times 10^{-5}$$

so that

$$pH = 4.3$$

The preceding examples illustrate how control of the pH can be used to separate metal ions by selective precipitation of their sulfides. The same principle can, of course, be used with other precipitating agents, for example, CO_3^{2-} and $C_2O_4^{2-}$. Such separations are important analytical techniques.

23-3. ACID-BASE TITRATIONS

The determination of the concentration of an unknown solution of an acid or base by titration with a suitable standard is a common analytical procedure (Figure 23–1). Acid-base titrations usually fall into one of two catagories: (1) the titration of a strong acid by a strong base (or a strong base by a strong acid), and (2) the titration of a weak acid by a strong base (or a weak base by a strong acid). A plot of the pH of the solution as a function of the volume of standard solution added is known as a titration curve. In practice, the pH is usually determined by a pH meter, which is a device that records the potential of a cell that is sensitive to changes in the H_3O^+ ion concentration. The instrument is calibrated so that the cell voltage is read directly as pH.

Figure 23–2 shows a titration curve for the titration of an unknown HCl solution with 0.10 M NaOH. Because all strong acid-strong base reactions in water undergo the same principal acid-base reaction, namely

$$H_3O^+ + OH^- \longrightarrow 2\,H_2O$$

the pH at the point where all the acid and base have exactly reacted—the end-point—should be 7.00 at 25°C. Ideally, this point should occur exactly half way up the steeply rising portion of the curve (Figure 23–2). Near the end-point, the curve rises so steeply with a very small volume of standard solution added,

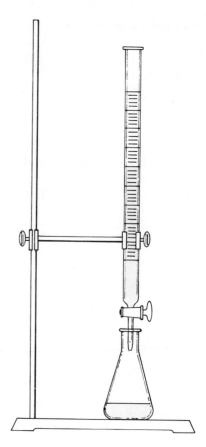

Figure 23-1. Titration of an unknown with a reagent of known concentration. Volume of the standard solution added is accurately measured with a buret.

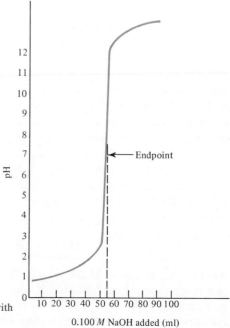

Figure 23-2. Titration of 20.0 ml of an unknown HCl solution with standard NaOH solution at 25°C.

that little error is encountered in determining the volume of standard solution added at the end-point. This volume of standard solution (in liters) multiplied by the known concentration gives the moles of standard acid or base added and thus allows determination of the original concentration of the unknown.

■ **Example 23-6**

Determine the concentration of 20.0 ml of HCl solution if titration with 0.100 M NaOH gives the titration curve of Figure 23-2.

□ *Solution*

The principal acid-base reaction is

$$H_3O^+ + OH^- \longrightarrow 2\,H_2O$$

so that the moles of OH^- used equals the moles of H_3O^+ originally present. From Figure 23-2, it is seen that the end-point occurs when 55.0 ml of 0.100 M NaOH has been added

$$\text{moles } OH^- \text{ used} = 0.0550 \text{ liter} \times 0.100 \text{ mole/liter}$$
$$= 0.00550 \text{ mole}$$

so that

$$\text{moles } H_3O^+ \text{ originally present} = 5.50 \times 10^{-3} \text{ mole}$$

and

$$\text{original HCl concentration} = \frac{5.50 \times 10^{-3} \text{ mole}}{0.0200 \text{ liter}} = 0.275 \ M$$

The titration of a strong base with a strong acid gives a curve similar to that in Figure 23-2 except that it is inverted (Figure 23-3). Calculation using such a curve is similar to that in the above example.

When a weak acid is titrated with a strong base, the end-point will occur at a pH greater than 7. Consider, for example, the titration of a solution of acetic acid by a standard NaOH solution.

The principal acid-base reaction during the addition of NaOH solution will be

$$HOAc + OH^- \longrightarrow H_2O + OAc^-$$

Notice that a base stronger than water, namely, OAc^- ion, is a product (see Table 22-2, the acid-base table in Chapter 22). At the end-point, therefore, the principal acid-base equilibrium will be

$$OAc^- + H_2O \rightleftharpoons HOAc + OH^-$$

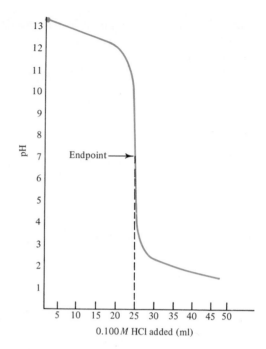

Figure 23-3. Titration of 25 ml of 0.100 M NaOH with standard HCl solution at 25°C.

and the OH^- ions produced by this equilibrium will cause the pH of the solution to be greater than 7. Figure 23–4 is a titration curve for the titration of an unknown HOAc solution by standard NaOH solution. Calculation of the concentration of the unknown solution from the volume of standard solution added at the end-point is carried out exactly as for the strong acid–strong base titration.

The acid constant, K_a, can be calculated from a titration curve such as that in Figure 23–4. At any point during the titration before the end-point is reached,

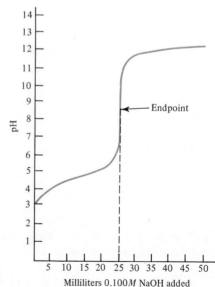

Figure 23-4. Titration of 25.0 ml of an unknown HOAc solution with a standard NaOH solution at 25°C.

relatively large amounts of both the weak acid and its conjugate base will be in the solution. The principal acid base equilibrium in this buffer region of the titration may be written

$$HOAc + H_2O \rightleftharpoons H_3O^+ + OAc^-$$

$$K_a = \frac{[H_3O^+][OAc^-]}{[HOAc]}$$

At the point where half of the HOAc has been converted into OAc^- (half-way to the end-point), the condition

$$[HOAc] = [OAc^-]$$

arises. Consequently, at this half-neutralization point the terms [HOAc] and $[OAc^-]$ cancel in the K_a expression, and

$$K_a = [H_3O^+]$$

This result means that regardless of the initial concentration of the weak acid, the half-neutralization point of a titration curve always occurs at the same pH because

$$K_a = [H_3O^+]$$

so that

$$-\log K_a = -\log [H_3O^+]$$

or

$$pK_a = pH$$

The pK_a for a given weak acid is, of course, a constant that is independent of the initial concentration of the acid.

■ **Example 23-7**

The end-point in the titration of 0.270 g of a weak acid, HX, with 0.100 M NaOH occurs after 30.0 ml of base is added. The pH after 15.0 ml of base is added is 5.00. Calculate (a) the molecular weight and (b) the K_a of the weak acid.

□ *Solution*

(a) At end-point, moles OH^- added = moles HX originally present. Thus,

$$\text{moles } OH^- = 0.0300 \text{ liter} \times 0.100 \text{ mole/liter} = 0.00300 \text{ mole}$$

$$\text{molecular weight of HX} = 0.270 \text{ g}/0.00300 \text{ mole}$$

$$\text{molecular weight of HX} = 90.0$$

365

(b) At the half-neutralization point, $pH = pK_a = 5.00$

$$-\log K_a = 5.00$$

$$K_a = 1.0 \times 10^{-5}$$

The titration of a weak base with a strong acid will yield a titration curve similar to that for the weak acid–strong base case except that the curve is inverted. The pH at the end-point will be less than 7 at 25°C because the conjugate acid of the weak base will be formed during the titration. The pOH at the half neutralization point will be equal to the pK_b for the weak base. Figure 23–5 is a typical titration curve for a weak base-strong acid titration.

It is not necessary to determine the pH throughout a titration if only the amount or concentration of the substance being titrated is desired. In such cases, only the end-point is necessary. The use of an **acid-base indicator** is a common procedure for simply determining an end-point. The function of such indicators is the same as that of a precipitation indicator discussed in the previous section, namely, to show when equivalent amounts of reactants have been added.

Acid-base indicators are weak acids whose conjugate bases have a different color. If we use the formula, HIn, to stand for any weak acid indicator, then in a water solution the following equilibrium will exist.

$$HIn + H_2O \rightleftharpoons H_3O^+ + In^-$$
$$\text{color 1} \qquad\qquad\qquad \text{color 2}$$

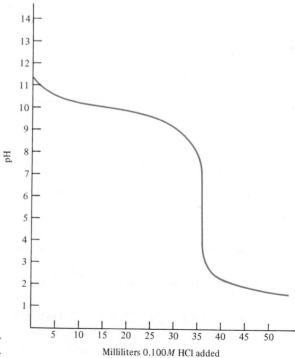

Figure 23–5. Titration of 25.0 ml of an unknown solution of a weak base with standard HCl solution at 25°C.

Milliliters 0.100M HCl added

The position of this equilibrium will determine the color of the solution, and the position of this equilibrium will be determined by two things—the H_3O^+ ion concentration in the solution and the K_a of the indicator. Normally, the concentration of weak acid must be about ten times that of the conjugate base to see only color 1, and the concentration of In^- must be ten times that of HIn to see only color 2. When the concentrations of the two species lie between these values, an intermediate color produced from the mixture of the two colors will be observed.

Usually, indicators will change from one color through the intermediate mixture of colors to the other color over a pH range of about 2 units. The following calculation will show why this is so.

$$K_a = \frac{[H_3O^+][In^-]}{[HIn]}$$

To see only color 1

$$[HIn] = 10[In^-]$$

so that

$$K_a = \frac{[H_3O^+][In^-]}{10[In^-]}$$

$$K_a = \frac{[H_3O^+]}{10}$$

and

$$\log K_a = \log [H_3O^+] - \log 10$$

or

$$-\log K_a = -\log [H_3O^+] + \log 10$$

thus

$$pK_a = pH + 1$$

Therefore the pH at which color 1 is seen is

$$pH = pK_a - 1$$

On the other hand, to see color 2,

$$[In^-] = 10[HIn]$$

so that

$$K_a = \frac{[H_3O^+](10[HIn])}{[HIn]}$$

$$K_a = [H_3O^+]10$$

and

$$\log K_a = \log [H_3O^+] + \log 10$$

or

$$-\log K_a = -\log [H_3O^+] - \log 10$$

thus

$$pK_a = pH - 1$$

Therefore the pH at which color 2 is seen is

$$pH = pK_a + 1$$

The pK_a for a given indicator is, of course, a fixed number, and so every acid-base indicator has its own pH range over which is changes color. It is important to choose an indicator that will change colors over a pH range at or very near the pH corresponding to the end-point of the titration. Table 23–2 lists some acid-base indicators with the approximate pH ranges over which they are effective.

Table 23–2. Acid-Base Indicators

Indicator	Color	pH
methyl violet	green	1 or lower
	blue	2
	violet	3 or higher
methyl orange	pink	3 or lower
	orange	4
	yellow	5 or higher
bromthymol blue	yellow	6 or lower
	green	7
	blue	8 or higher
phenolphthalein	colorless	8 or lower
	pale pink	9
	red	10 or higher
malachite green	green	11 or lower
	pale green	12
	colorless	13 or higher

In addition to changing color near the pH corresponding to the end-point, the indicator must change sharply to be useful in titrations, Although bromthymol blue changes colors at the end-point pH of strong acid–strong base reactions, it is not used in practice because there is no sharp color change. Phenolphthalein is the indicator of choice for such titrations because the difference between a pink and colorless solution is very easy to see.

23–4. REDOX TITRATIONS

The use of redox titrations in quantitative analysis is similar to the use of precipitation and acid-base titrations except that a redox reaction is involved. The permanganate ion, MnO_4^-, is a good oxidizing agent which is commonly used to determine the amount of a substance that will undergo oxidation. The MnO_4^- ion has another advantage in that it also serves as the **redox indicator.** The deep purple color of the MnO_4^- ion serves to indicate when one drop in excess has been added. As long as any reducing agent is present to react with MnO_4^-, its color is destroyed and the analyst knows that the end-point has not been reached. The first discernible persistent purple color in the solution indicates that the end-point has been reached.

A common procedure for the analysis of iron ore involves dissolving a weighed sample of ore in HCl solution and then titrating the Fe^{2+} ions with standard $KMnO_4$ solution. The equation for the reaction is

$$8 H_3O^+ + MnO_4^- + 5 Fe^{2+} \longrightarrow 5 Fe^{3+} + Mn^{2+} + 12 H_2O \qquad (23-7)$$

■ Example 23–8

Calculate the per cent Fe in a sample if 2.00 g of it requires 25.0 ml of 0.100 M $KMnO_4$ to titrate to the end-point (at. wt. Fe = 55.8).

☐ *Solution*

From equation (23–7) we see that

$$\text{moles } Fe^{2+} = 5 \times \text{moles } MnO_4^- \text{ used}$$

$$\text{moles } MnO_4^- \text{ used } = 0.0250 \text{ liter} \times 0.100 \text{ mole/liter} = 0.00250 \text{ mole}$$

so

$$\text{moles } Fe^{2+} = 5 \times 0.00250 = 0.0125 \text{ mole}$$

and

$$\text{weight Fe in sample} = 0.0125 \text{ mole} \times 55.8 \text{ g/mole} = 0.698 \text{ g}$$

finally

$$\frac{0.698 \text{ g Fe}}{2.00 \text{ g total}} \times 100 = 34.9\% \text{ Fe in sample}$$

Standard $KMnO_4$ solutions are often prepared by titrating the solution against a known amount of $Na_2C_2O_4$, a material which can be obtained in a state of high purity. The titration is carried out in an acidic solution and the reaction is

$$6 H_3O^+ + 2 MnO_4^- + 5 H_2C_2O_4 \longrightarrow 10 CO_2 + 2 Mn^{2+} + 14 H_2O \quad (23-8)$$

■ Example 23–9

Calculate the concentration of a $KMnO_4$ solution if 20.0 ml of it is required to titrate 0.268 g of $Na_2C_2O_4$ to the end-point.

□ *Solution*

According to equation (23–8)

$$\text{moles of KMnO}_4 = \tfrac{2}{5} \times \text{moles Na}_2\text{C}_2\text{O}_4 \text{ used}$$

$$\text{moles Na}_2\text{C}_2\text{O}_4 = \frac{0.268 \text{ g}}{134 \text{ g/mole}} = 0.00200 \text{ mole}$$

so that

$$\text{moles of KMnO}_4 = \tfrac{2}{5} \times 0.00200 = 0.00080 \text{ mole}$$

finally

$$\text{concentration of KMnO}_4 = \frac{0.000800 \text{ mole}}{0.0200 \text{ liter}} = 0.0400 \text{ } M$$

Examples 23–8 and 23–9 demonstrate the need for a balanced equation for the reaction under consideration. Up to this point in our study, chemical equations have been balanced simply by inspection. Many redox reactions, however, have such complex stoichiometries that balancing by inspection is very difficult if not impossible. A systematic method for balancing difficult redox equations will be presented in Section 23–5.

23–5.   BALANCING DIFFICULT REDOX EQUATIONS

Since oxidation and reduction always occur simultaneously, a redox equation can be broken into two parts—an oxidation half and a reduction half. The use of half reactions has already been discussed in Chapter 21 in conjunction with electrochemical cells. At this point, we shall use the idea of half reactions to help in balancing redox equations.

The assignment of oxidation numbers to all atoms in an unbalanced equation is frequently the best thing to do first. In this way, the oxidation and reduction half reactions can be separated. Let us use the Fe^{2+}–MnO_4^- reaction considered above as an example.

$$Fe^{2+} + \overset{(+7)}{MnO_4^-} \longrightarrow Fe^{3+} + Mn^{2+} \qquad \text{(unbalanced)}$$

Oxidation half reaction: $\quad\quad Fe^{2+} \longrightarrow Fe^{3+}$ \hfill (unbalanced)

Reduction half reaction: $\quad\quad MnO_4^- \longrightarrow Mn^{2+}$ \hfill (unbalanced)

The next step involves balancing the atoms of all elements that appear in the half reactions. The oxidation half reaction is already balanced in terms of atoms, but not in terms of the charges.

Oxidation half reaction: $\quad Fe^{2+} \longrightarrow Fe^{3+}$ (atoms balanced, charges unbalanced)

The reduction half reaction is unbalanced in terms of atoms because four oxygen atoms are on the left and none are on the right. To balance hydrogen and oxygen atoms, the species H_3O^+, OH^-, and H_2O are used as necessary because water is always understood to be the solvent and therefore to be in excess. However, in an acidic solution, only H_3O^+ and H_2O may be used, whereas OH^- and H_2O are used in basic solutions. It is therefore necessary to know the conditions of acidity under which the reaction is carried out. The reaction we are considering occurs in acidic solution, so the reduction half reaction must be balanced using H_3O^+ and H_2O.

Reduction half reaction: $8\ H_3O^+ + MnO_4^- \longrightarrow Mn^{2+} + 12\ H_2O$
(atoms balanced, charges unbalanced)

The charges in each half reaction are next balanced using electrons.

Oxidation half reaction: $Fe^{2+} \longrightarrow Fe^{3+} + e^-$ (atoms and charges balanced)

Reduction half reaction: $5\ e^- + 8\ H_3O^+ + MnO_4^- \longrightarrow Mn^{2+} + 12\ H_2O$
(atoms and charges balanced)

The final step consists in multiplying through each half reaction by whatever factor is necessary to make the number of electrons the same in each half reaction so that they will cancel and not appear in the final equation. Addition gives

Oxidation half reaction:
$$5\ Fe^{2+} \longrightarrow 5\ Fe^{3+} + 5\ e^-\ (\text{multiply through by 5})$$
Reduction half reaction:
$$5\ e^- + 8\ H_3O^+ + MnO_4^- \longrightarrow Mn^{2+} + 12\ H_2O$$
$$\overline{}$$

Total balanced reaction:
$$5\ Fe^{2+} + 8\ H_3O^+ + MnO_4^- \longrightarrow 5\ Fe^{3+} + Mn^{2+} + 12\ H_2O$$

At this point, a recheck should be made to see that the equation is balanced with respect to both atoms and charges. Any species that appears on both sides of the equation should be removed.

The steps involved in balancing redox equations by the method given above can be summarized.

1. Assign oxidation numbers to the atoms and write skeleton oxidation and reduction half reactions.
2. Balance the atoms in each half reaction. Use H_3O^+ and H_2O to balance oxygen and hydrogen in acidic solutions; use OH^- and H_2O in basic solutions.
3. Use electrons to balance the charges in each half reaction. The oxidation half reaction must show electrons as products and the reduction half reaction must show electrons as reactants.
4. Equate the number of electrons gained and lost by multiplying each half reaction by the necessary factor.
5. Add the two half reactions. Cancel any species that appear on both sides of the total equation. Make a final check to see that the equation is balanced with respect to both atoms and charges.

Another example of a redox reaction in acidic solution will be given using the $MnO_4^- - H_2C_2O_4$ reaction.

$$\overset{(+7)}{MnO_4^-} + \overset{(+3)}{H_2C_2O_4} \longrightarrow \overset{(+4)}{Mn^{2+} + CO_2}$$

1. $H_2C_2O_4 \longrightarrow CO_2$
   $MnO_4^- \longrightarrow Mn^{2+}$

2. $2 H_2O + H_2C_2O_4 \longrightarrow 2 CO_2 + 2 H_3O^+$
   $8 H_3O^+ + MnO_4^- \longrightarrow Mn^{2+} + 12 H_2O$

3. $2 H_2O + H_2C_2O_4 \longrightarrow 2 CO_2 + 2 H_3O^+ + 2 e^-$
   $5 e^- + 8 H_3O^+ + MnO_4^- \longrightarrow Mn^{2+} + 12 H_2O$

4. $10 H_2O + 5 H_2C_2O_4 \longrightarrow 10 CO_2 + 10 H_3O^+ + 10 e^-$
   $10 e^- + 16 H_3O^+ + 2 MnO_4^- \longrightarrow 2 Mn^{2+} + 24 H_2O$

5. $10 H_2O + 5 H_2C_2O_4 + 16 H_3O^+ + 2 MnO_4^- \longrightarrow$
   $10 CO_2 + 10 H_3O^+ + 2 Mn^{2+} + 24 H_2O$
   The final, balanced equation after removing the H_2O and H_3O^+ appearing on both sides is

$$5 H_2C_2O_4 + 6 H_3O^+ + 2 MnO_4^- \longrightarrow 10 CO_2 + 2 Mn^{2+} + 14 H_2O$$

An example of a reaction in basic solution will be given. This reaction is a disproportionation so that both half reactions will contain the atom undergoing disproportionation:

$$P_4 \longrightarrow H_2PO_2^- + PH_3$$

1. $P_4 \longrightarrow H_2PO_2^-$
   $P_4 \longrightarrow PH_3$

2. $8 OH^- + P_4 \longrightarrow 4 H_2PO_2^-$
   $12 H_2O + P_4 \longrightarrow 4 PH_3 + 12 OH^-$

3. $8 OH^- + P_4 \longrightarrow 4 H_2PO_2^- + 4 e^-$
   $12 e^- + 12 H_2O + P_4 \longrightarrow 4 PH_3 + 12 OH^-$

4. $24 OH^- + 3 P_4 \longrightarrow 12 H_2PO_2^- + 12 e^-$
   $12 e^- + 12 H_2O + P_4 \longrightarrow 4 PH_3 + 12 OH^-$

5. $24 OH^- + 4 P_4 + 12 H_2O \longrightarrow 12 H_2PO_2^- + 4 PH_3 + 12 OH^-$
   The final, balanced equation after reduction of coefficients is thus

$$3 OH^- + P_4 + 3 H_2O \longrightarrow 3 H_2PO_2^- + PH_3$$

PROBLEMS

1. What is the function of an indicator in titrations? What are the characteristics of acid-base indicators? Redox indicators?

2. Calculate the molar solubilities of the following substances in the solutions as indicated at 25°C. Use necessary K_{sp} values from Table 23-1. **(a)** $BaSO_4$ in pure water **(b)** $BaSO_4$ in 0.010 M Na_2SO_4 **(c)** PbF_2 in 0.10 M $Pb(NO_3)_2$ **(d)** Ag_2CO_3 in pure water.

3. Calculate the K_{sp} of a sparingly soluble salt, MX_2, if 1.0×10^{-3} mole dissolves in 1 liter of water.

4. How many grams of CaF_2 will dissolve in 500 ml of water at 25°C?

5. Na_2SO_4 is slowly added with stirring to a solution in which $[Ag^+] = [Ba^{2+}] = [Ca^{2+}] = 0.010$. In what order will Ag_2SO_4, $BaSO_4$, and $CaSO_4$ precipitate? What concentration of SO_4^{2-} ion is necessary for the quantitative precipitation of each of these salts? Do you think quantitative precipitation of all three would be possible? Explain.

6. 25.0 ml of 0.10 M $AgNO_3$ is required to titrate 35.0 ml of an unknown NaCl solution to the end-point. What is the concentration of the NaCl solution?

7. A solution contains $[Fe^{2+}] = [Mn^{2+}] = [Cu^{2+}] = 0.10$. What will precipitate if this solution is buffered at a pH $= 1.0$ and saturated with H_2S at 25°C?

8. What is the lowest pH at which the solution of problem 7 will precipitate all three sulfides quantitatively?

9. Is a quantitative separation of Ni^{2+} and Zn^{2+} ions possible by means of the selective precipitation of their sulfides from a solution in which they are present in approximately equal conditions? Explain.

10. Calculate the concentrations of H_3O^+, HS^-, S^{2-}, and H_2S in a solution saturated with H_2S at 25°C if the pH $= 18.00$.

11. 30.0 ml of 0.100 M HCl is required to titrate 20.0 ml of NaOH solution to the end-point at 25°C. **(a)** Calculate the concentration of the NaOH solution **(b)** What is the pH of the solution at the end-point?

12. **(a)** If 0.112 g of $NaHC_2O_4$ is titrated to the end-point with 0.0500 M NaOH, what volume of the standard solution is required? **(b)** What would be the pH at the end-point if the $NaHC_2O_4$ were originally dissolved in 10.0 ml of water?

13. What is the concentration of the unknown HOAc solution that gave the titration curve in Figure 23-4?

14. Refer to the titration curve in Figure 23-5.
    (a) What is the concentration of the unknown weak base?
    (b) What is K_b for this base?
    (c) What is the molecular weight of the base if 0.600 g of it was used for the titration?

15. 30.0 ml of 0.100 M HCl is titrated with 0.100 M NaOH at 25°C. Calculate the pH values for every 5 ml of NaOH added up to 55 ml, make a table of these values, and plot the resulting titration curve.

16. Which of the indicators listed in Table 23-2 would be suitable for use in the titrations of Figures 23-2, 23-3, 23-4, and 23-5?

17. Estimate the pH of each of the following solutions, which have been tested as indicated.
    (a) methyl orange and bromthymol blue are both yellow.
    (b) methyl violet is green and methyl orange is pink.
    (c) phenolphthalein is colorless.
    (d) bromthymol blue is green.

18. Estimate the K_a's for each of the indicators listed in Table 23-2.

19. A certain indicator has $K_a = 1 \times 10^{-5}$. Over what approximate pH range will this indicator change colors?

20. An iron ore sample, after proper preliminary treatment, required 20.0 ml of 0.0500 M

$KMnO_4$ to titrate to the end-point. If the ore is 20.0% Fe, what was the weight of the sample analyzed?

21. An impure sample of $Na_2C_2O_4$ weighing 0.200 g required 22.0 ml of 0.0200 M $KMnO_4$ to titrate to the end-point in acidic solution. Calculate the per cent $Na_2C_2O_4$ in the sample.

22. What volume of 0.075 M $K_2Cr_2O_7$ would be required to titrate 5.00×10^{-3} mole of Fe^{2+} to the end-point? The unbalanced equation for the reaction in acidic solution is

$$Cr_2O_7^{2-} + Fe^{2+} \longrightarrow Cr^{3+} + Fe^{3+}$$

23. Balance the following redox equations for reactions in acidic solution.
   (a) $Cu + NO_3^- \longrightarrow Cu^{2+} + NO$
   (b) $H_2O_2 + MnO_4^- \longrightarrow Mn^{2+} + O_2$
   (c) $I^- + H_2SO_4 \longrightarrow H_2S + I_2$
   (d) $Cr_2O_7^{2-} + Cl^- \longrightarrow Cl_2 + Cr^{3+}$
   (e) $Cu + H_2SO_4 \longrightarrow Cu^{2+} + SO_4^{2-} + SO_2$
   (f) $P_4 + HClO \longrightarrow H_2PO_4^- + Cl^-$

24. Balance the following redox equations for reactions in basic solution.
   (a) $Al + OH^- \longrightarrow Al(OH)_4^- + H_2$
   (b) $MnO_4^- + ClO_2^- \longrightarrow MnO_2 + ClO_3^-$
   (c) $Zn + NO_3^- \longrightarrow Zn(OH)_4^{2-} + NH_3$
   (d) $Cl_2 \longrightarrow ClO_3^- + Cl^-$
   (e) $MnO_4^- + Br_2 \longrightarrow MnO_2 + BrO_3^-$
   (f) $Cl_2O^{2-} + AsO_2^- \longrightarrow Cl^- + AsO_4^{3-}$

CHEMICAL KINETICS—THE APPROACH TO EQUILIBRIUM

The preceeding chapters have dealt with various aspects of chemical equilibria. Equilibrium is one of the cornerstones of the science of chemistry. As has become evident in the preceeding chapters, very few chemical processes occur to completion. Instead, in an isolated system, most chemical reactions proceed to a point of equilibrium which is determined by the free energies of reactants and products.

An aspect of equilibrium that must now be considered is the rate at which equilibrium is approached. This study of the rates of chemical reactions is called **chemical kinetics.** Rates of chemical reactions cover an extremely broad range. The detonation of explosives and the reaction of photographic film under the influence of light are two examples of extremely fast reactions. The rusting of iron and the radioactive decay of carbon-14 are examples of very slow reactions. All of these examples, it might be noted, are examples of reactions that proceed essentially to completion.

Chemical reaction rates can be changed dramatically by changing experimental conditions such as concentrations, solvent, and temperature. The study of the influence of these factors on reaction rates and the development of theories to explain these phenomena will be the objectives of this chapter. The use of reaction rate data in understanding detailed mechanisms of reactions will form the subject of the following chapter. But before examining these facets of reaction kinetics, it will be useful to survey some of the actual experimental problems that the chemist encounters in these studies.

24-2. EXPERIMENTAL DETERMINATION OF REACTION RATES

Accurate determination of the rate of a chemical reaction often requires considerable care. The rate of the reaction:

$$A \longrightarrow B$$

is defined as the rate of decrease of the concentration of A or as the rate of increase of the concentration of B with time. The rate of the above simple reaction is usually given in terms of moles per liter per unit time. In actual practice the concentration of either A or B may be measured at varying time intervals (Figure 24-1). The concentration measurement may, of course, be made by any suitable analytical procedure capable of giving the needed accuracy. Usually, when a rate study is carried out in solution, samples are removed periodically and the reaction is stopped by a suitable quenching method, such as sudden cooling, before analysis. If the reaction is sufficiently slow, it is often not necessary to quench the reaction before analysis.

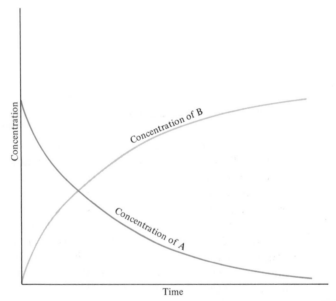

Figure 24-1. A plot of concentration versus time for the concentrations of reactant A and product B in the reaction A ⟶ B.

In cases of gas reactions, especially when the number of molecules changes as a result of reaction, it is often possible to use pressure change as a measure of reaction rate. For example, the decomposition of N_2O_5 according to the equation

$$2\,N_2O_5 \longrightarrow 4\,NO_2 + O_2$$

involves the conversion of two reactant molecules of gas into five product molecules. The relation between pressure and concentration is shown in Figure 24-2. Thus the variation of pressure with time affords a direct measure of the concentrations and thus of the reaction rate.

An autocatalytic reaction is a reaction in which one of the products catalyzes the reaction (Section 24-5). Such a reaction has an induction period and gives

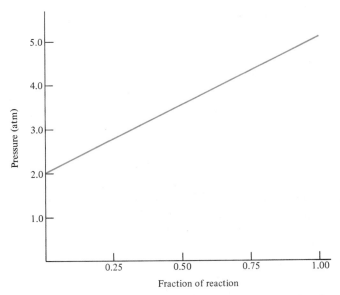

Figure 24-2. Variations of pressure with fraction of reaction: $2\,N_2O_5 \longrightarrow 4\,NO_2 + O_2$.

an S-shaped rate curve, as shown in Figure 24-3. Examples of such autocatalytic reactions abound in nature. The acid catalyzed hydrolysis of esters is one.

$$\underset{\text{an ester}}{R-\overset{\overset{\textstyle O}{\|}}{C}-O-R'} + H_2O \xrightarrow{\text{acid}} \underset{\text{an acid}}{R-\overset{\overset{\textstyle O}{\|}}{C}-OH} + \underset{\text{an alcohol}}{HOR'}$$

Biochemical reactions are commonly autocatalytic, such as the conversion of trypsinogen into trypsin with the latter catalyzing the reaction. A nonchemical example is the growth of populations.

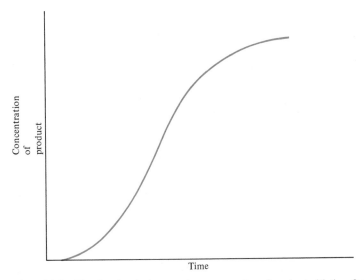

Figure 24-3. Plot showing the increase in concentration of product with time for a typical autocatalytic reaction.

377

24-3.   NATURE OF RATE BEHAVIOR—ORDER OF REACTION

The mathematical expression that relates the rate of a chemical reaction to the concentrations of the reactants is called the **rate law.** The rate law for the hypothetical reaction,

$$2\,A + B \longrightarrow C + D$$

has the form

$$\text{rate} = k[A]^m[B]^n$$

where m and n are exponents that *must be determined experimentally,* and k is a proportionality constant called the **specific rate constant,** or sometimes called simply the **rate constant.** Because there is a tendency for the student to assume that the values of m and n are 2 and 1, as would be the case with the equilibrium constant expression, it is imperative to note that such is *not* usually the case in the rate expression. The sum $m + n$ is called the **order** of the overall reaction. The order of the reaction with respect to A is m, and the order with respect to B is n.

The most direct method of determining the order of a reaction is to determine the initial reaction rates. The initial rate is significant because all concentrations are known with certainty at the beginning of a reaction (Figure 24–4). Table 24–1 shows some typical rate data for different initial concentrations of A and B. These data show that the rate doubles when either [A] or [B] is doubled. Thus the rate

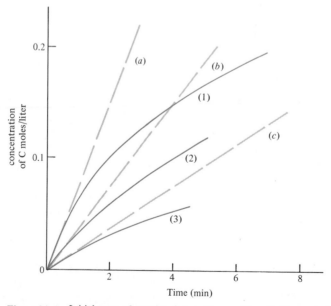

Figure 24-4. Initial rates of reaction $2\,A + B \longrightarrow C + D$. Curves 1, 2, and 3 represent the actual plots of concentration of product C with time. Curves a, b, and c represent the tangents to curves 1, 2, and 3, respectively, at time = 0 sec; that is, curves a, b, and c represent the respective initial rates. The conditions under which the three determinations were made are given in Table 24–1.

Table 24-1. Initial Rates of the Reaction $2 A + B \longrightarrow C + D$

Experiment No.	Initial Concentrations (moles/liter) [A]	[B]	Curve No.	Initial Rate* (moles/liter min)
1	0.2	0.1	1	0.08
2	0.1	0.1	2	0.04
3	0.1	0.05	3	0.02

*The initial rates are determined graphically as the slopes of the curves a, b, or c: slope = $\Delta[C] \div \Delta$time

is directly proportional to [A] and [B], and the rate law for this reaction is

$$\text{rate} = k[A][B]$$

Thus the reaction is second order overall, first order in A and first order in B.

24-4. EFFECT OF TEMPERATURE ON RATE

We have seen that the rate of a chemical reaction is strongly influenced by the concentration of reactants. The exact dependence of rate on concentration is given by the rate law. Another important variable that affects the rate of a chemical reaction is temperature. A rule of thumb that has long been employed by chemists is that every 10°C rise in temperature increases the rate by a factor of two to three.

The actual behavior of reaction rate with temperature depends on the type of reaction. The most common variation of rate with temperature is the one that obeys the above rule of thumb and is shown in Figure 24-5. This type of variation will be discussed more fully below. Two other types of behavior of rate with temperature are given. In an explosion, an abrupt rise in reaction rate occurs

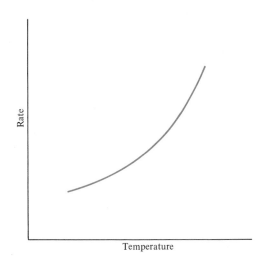

Temperature

Figure 24-5. Variation of reaction rate with temperature for an ordinary chemical reaction. This relationship obeys the rule of thumb that states that the rate doubles or triples for every 10°C rise in temperature.

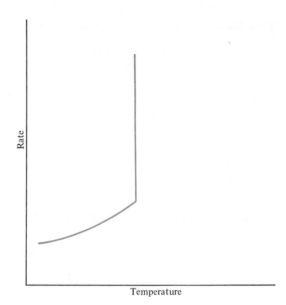

Figure 24-6. Variation of reaction rate with temperature for an explosion. The discontinuity in the curve represents the ignition temperature where the rate increases abruptly.

at the ignition temperature as shown in Figure 24-6. Figure 24-7 is typical of enzymatic reactions. Enzymes are catalysts (Section 24-5) that accelerate reactions in biological systems. As the temperature of these systems is raised, the rate increases until enzyme breakdown begins to occur, when the reaction slows down.

By far the most commonly encountered variation of rate with temperature is the one shown in Figure 24-5. This rate behavior was observed in the latter half of the nineteenth century. In 1887, van't Hoff proposed, on the basis of

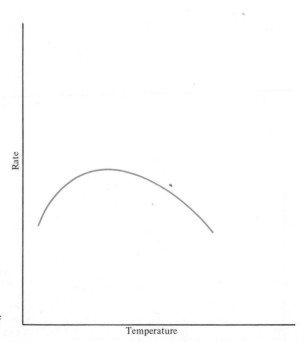

Figure 24-7. Variation of reaction rate with temperature for an enzymatic reaction.

theoretical arguments, that the logarithm of the rate constant should be inversely proportional to the negative of the absolute temperature. S. Arrhenius, in 1889, extended this idea and offered a model for chemical reactions that could explain this relationship, which has become known as the Arrhenius rate law:

$$\log k \propto -\frac{1}{T}$$

Arrhenius' work on the rates of several reactions led him to the final form of the equation

$$\ln k = -\frac{E_a}{RT} + \text{a constant}$$

in which ln is the natural logarithm ($= 2.303 \times \log_{10}$), E_a is called the **activation energy,** and R is the gas constant. It is reasonable to assume that the rate of reaction should depend on the frequency of collisions of reacting molecules. Because molecules move faster the higher the temperature, an increase of reaction rate with temperature is to be expected.

Arrhenius, however, argued that this dependence of rate on temperature was too great to be attributed solely to increased molecular velocities, that is, more frequent molecular collisions. Instead, collisions that lead to reaction must involve "activated" molecules, and the proportion of activated molecules is increased rapidly when the temperature is raised. Since translational molecular energies are distributed as shown in Figure 24-8, any sample of molecules will have at least a few activated molecules. A rise in temperature produces a disproportionate increase in the number of activated molecules, as shown in Figure 24-8.

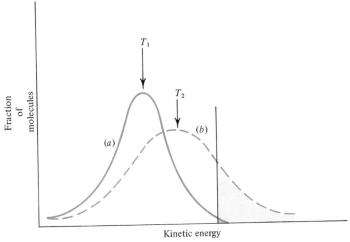

Figure 24-8. Distribution of molecular kinetic energies for a gas at temperature T_1 (curve *a*) and at a higher temperature T_2 (curve *b*). Recall that $T \propto$ KE. The shaded area represents those molecules that have sufficient energy to be activated.

Molecules can be activated by means other than by increasing their translational kinetic energies. We have seen (Section 10–5) that electronic excitations are produced in atoms and molecules by ultraviolet and visible light. Similarly, molecules may be excited to various vibrational energy levels. This excitation can be brought about by irradiation with energies as low as those in the infrared region of the electromagnetic spectrum.

Examples of vibrations in the water molecule are shown by the colored arrows.

Each vibration of a given molecule has its own fixed energy, and thus vibrational energy levels are quantized as are electronic energy levels.

The activation energy E_a, according to Arrhenius' theory, is the energy that molecules must possess in order to be activated. Once a molecule possesses the activation energy, it can form product molecules spontaneously upon collision. This process is described by the diagram in Figure 24–9. As shown, reaction of normal molecules requires first that they be activated by some process such as collision with a hot surface or by irradiation. Once activated, the molecule can either cross over the energy barrier to become products, or return to the original state by transferring its energy to another molecule.

The concept of activation is quite reasonable if one considers the alternative. If molecules could react without activation, then all chemical reactions would be

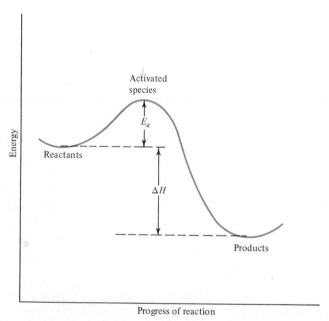

Figure 24-9. Energy diagram for the conversion of reactant molecules through activated molecules to products. E_a is the activation energy and ΔH is the energy released in the overall reaction.

instantaneous and the universe would have long ago reached the condition of equilibrium.

The rate of a chemical reaction is related to its activation energy. The greater the activation energy, the slower the reaction. It is important to contrast the difference between the quantities E_a and ΔH for a chemical reaction. These two quantities are demonstrated in Figure 24-9. From previous chapters, we have seen that the equilibrium position is influenced greatly by the energy of reaction, which is the difference between the energies of reactants and products. In the example of Figure 24-9, this corresponds to ΔH. This factor affords a measure of the extent of reaction, but it gives no indication of the velocity at which this equilibrium position is attained. Thus in Figure 24-10 are shown two reaction diagrams that have the same ΔH, but different E_a's. Under the same conditions, these reactions will attain the same equilibrium positions but they will do so at different rates.

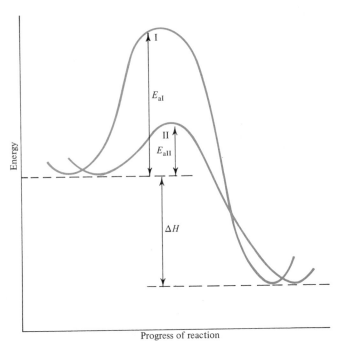

Figure 24-10. Comparison of two reactions that have the same ΔH but different E_as.

24-5. CATALYSIS

The two reactions shown in Figure 24-10 are good examples of the effect of a catalyst on a reaction. A **catalyst** is defined as a substance that affects the rate of a reaction but is itself unspent at the end of the reaction. In terms of the activation concept, catalysts act to reduce the activation energy by providing a new pathway for the reaction. This may occur in many ways.

The reaction between gaseous oxygen and hydrogen is energetically extremely favorable to the formation of water. However, the two gases can be mixed and kept in contact for very long periods without any reaction taking place because the activation energy of the reaction is extremely high. The addition of a small amount of finely divided platinum, palladium, or nickel to the mixture results in a tremendous increase in the rate that can lead to an explosion. The finely divided metal acts as a catalyst by adsorbing hydrogen on its surface. The adsorption process apparently weakens the H-H bond enough to allow its easy reaction with oxygen molecules.

Many other types of catalysts are known. In some, the catalyst actually undergoes reaction but is, in a later step, converted back into its original form. An example is the use of nitric oxide (NO) as a catalyst in the oxidation of sulfur dioxide (SO_2) to sulfur trioxide (SO_3)

$$2 SO_2 + O_2 \xrightarrow{\text{NO}} 2 SO_3$$

The simple uncatalyzed reaction occurs very slowly. By undergoing reaction with O_2, the nitric oxide forms nitrogen dioxide (NO_2) at a reasonable speed, and the nitrogen dioxide can react with SO_2 fairly rapidly.

$$NO + \tfrac{1}{2}O_2 \longrightarrow NO_2$$
$$NO_2 + SO_2 \longrightarrow SO_3 + NO$$

Thus the overall reaction rate is increased, and the nitric oxide is regenerated. For this reason, catalysts are frequently employed in much less than stoichiometric amounts.

24-6. COLLISION THEORY

We can now consider a generalized theory that coordinates the observed behavior of reaction rates. The simplest such theory is the so-called collision theory, which states that reaction occurs as the result of collisions between reactant molecules, and that, to be effective, the collisions must be sufficiently energetic to supply the activation energy. As we have seen in Chapter 5, molecules under ordinary conditions are in a state of chaotic motion and collisions occur with extremely high frequency. For a gas at standard conditions, there are approximately 10^{28} collisions/cm^3 each second. Normally very few collisions are sufficiently energetic to be effective.

The factors that determine reaction rates, according to this theory, are three:

1. **Collision frequency.** Since molecular collisions are necessary for reactions to occur, any factor that increases the frequency of collisions will increase the rate. Concentration is therefore a major factor, because doubling the concentration of one reactant, in a reaction that requires a collision between two molecules, doubles the rate (Table 24-1). In gaseous reactions, the same effect is achieved by increasing the pressure.

2. **Orientation factor.** The proportion of effective collisions is also influenced in most cases by the molecular orientation at the instant of collision. This factor is of no importance in the reaction of simple spherical atoms or ions,

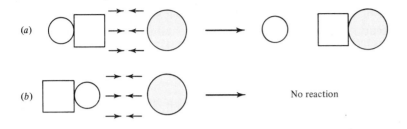

Figure 24–11. Symbolic model of a chemical reaction: (a) and (b). In many chemical reactions, orientation of reactant molecules at the instant of collision is important. A chemical example of the model is shown in (c) and (d).

but it is quite important for most molecules. The effect is shown in Figure 24–11 in which the collision of two molecules must occur in a fixed orientation for reaction to be possible.

3. **Energy factor.** The energy requirement for reaction has already been discussed. The point to be emphasized here is that temperature is the major factor that determines the proportion of activated molecules. Of course, the actual energy required for activation depends on the nature of the reaction itself; but for a given reaction, temperature is the most important controlling factor.

We may examine these three factors by comparing two reaction types—ionic and covalent. When an ionic crystalline substance is dissolved in water, the ion-water interactions are sufficiently strong to account for the disruption of the crystal lattice. What happens when solutions of different ionic substances are mixed? If, for example, a solution of sodium chloride is mixed with a solution of potassium bromide, the resulting solution simply contains all four of the ions Na^+, K^+, Cl^-, Br^-. Unless something else is done, it cannot really be said that a chemical reaction has occurred. Evaporation of the water would leave behind a mixture of all four possible crystal combinations: NaCl, KCl, NaBr, and KBr. This mixture can be considered a reaction product because new substances have been formed.

In many cases, however, the crystal lattice is so strong that the substance is insoluble in water. An example is barium sulfate ($BaSO_4$). Thus if a solution of sodium sulfate (Na_2SO_4) is mixed with a solution of barium nitrate [$Ba(NO_3)_2$], the combination of Ba^{2+} and SO_4^{2-} ions forms such a strong crystal lattice that precipitation occurs. The resulting solution then contains mostly Na^+ ions and NO_3^- ions. This reaction is expressed by equation (24–1), in which dissolved ionic substances are written as free ions and undissolved ionic substances are written in the "molecular" form followed by an arrow ($\downarrow$).

$$[2\,Na^+ + SO_4^{2-}] + [Ba^{2+} + 2\,NO_3^-] \longrightarrow BaSO_4\downarrow + 2[Na^+ + NO_3^-] \quad (24\text{–}1)$$

According to the collision theory, this reaction occurs by the combination of oppositely charged ions. Electrostatic attractions increase the collision frequency over that expected for neutral molecules and reduce the required activation energy. No orientation requirement exists because monatomic ions are spherical. Rates of such ionic reactions are consequently very rapid.

Reactions of covalent compounds are normally much slower than those of ionic compounds. In the former, the simple collision of molecules is not sufficient for reaction to occur; covalent bonds must be broken in the original molecules and others must be formed in the product molecules. An example given previously was the oxidation of sulfur dioxide to the trioxide.

24-7. KINETICS AND EQUILIBRIUM

A final point to be made is a further relationship between reaction kinetics and equilibrium. When a reaction is in a state of equilibrium, no change in the concentrations of reactants or products can be observed. This does not mean, however, that reactions are not occurring, since molecules continue to collide, some with the required orientation and energy. Under these conditions, the rates of forward and reverse reactions are equal, otherwise the equilibrium would shift. For the reaction at equilibrium,

$$A + B \underset{k_r}{\overset{k_f}{\rightleftharpoons}} C + D,$$

the equilibrium constant can be expressed as the ratio of k_f/k_r. At equilibrium, therefore

$$\frac{[C][D]}{[A][B]} = k_{eq} = \frac{k_f}{k_r}$$

PROBLEMS

1. Define or explain briefly each of the following terms:
   (a) order of reaction   (d) heat of reaction
   (b) rate law   (e) catalyst
   (c) activation energy

2. In a given sample of a gas, would you expect a higher percentage of the molecules to be in excited electronic or in excited vibrational energy levels? Explain.

3. Suggest a method for following the rate of each of the reactions given below:

(a) $C_2H_4\ (g) + H_2\ (g) \longrightarrow C_2H_6\ (g)$

(b) —$CH_2F + H_2O \longrightarrow$ —$CH_2OH + HF$

(c) $+ Br_2 \xrightarrow[\text{solution}]{CCl_4}$

4. The rate constant of a first order reaction is 2×10^{-6} sec^{-1}*, and the initial concentration of reactant is 0.1 M. What is the initial rate in moles per liter per second, in moles per cubic centimeter per second, and in moles per cubic centimeter per minute?

5. The initial rate of a second order reaction is 5.7×10^{-7} moles/liter sec, when the initial concentrations of the two reactants are 0.2 M. What is the rate constant in liters per mole per second, in cubic centimeters per mole per minute, and in cubic centimeters per molecule per second?

6. The decomposition of NO_2 is a second order reaction with the following k's:

$k(\text{cm}^3/\text{mole sec})$	$T(^\circ K)$
522	592
755	603
1700	627
4020	652

Calculate the Arrhenius activation energy, E_a.

7. Which condition normally has the greater effect on reaction rate, concentration or temperature? Explain or justify your answer. (*Hint:* Use the rate law).

8. Calculate the order of the reaction

$$A + B \longrightarrow C + 2\ D,$$

from the data given.

Experiment No.	Initial Concentrations		Relative Initial Rate
	[A]	[B]	
1	0.2	0.2	4
2	0.2	0.1	1
3	0.4	0.1	1

*sec^{-1} = 1/sec, and is read "per second," or "reciprocal second."

387

9. Write the rate law for the reaction described in problem 8.

10. Make a plot of pressure versus fraction of reaction as in Figure 24–2 for the reaction of hydrogen with chlorine to produce hydrogen chloride. Can pressure be used to follow this reaction? Explain.

11. For a hypothetical reaction that has a rate of 10^{-4} moles/liter sec. at 25°C, and for which the rate doubles for every 10°C rise in temperature, make an accurate plot of rate versus temperature between 25 and 100°C.

12. From what source do reacting molecules acquire activation energy? Does this help to explain why reactions proceed at faster rates when the temperature is raised? Explain.

13. What kinds of information about chemical reactions are given by (a) activation energies, (b) heats of reaction.

14. What are the factors that govern the rates of chemical reactions according to the collision theory?

15. According to the collision theory, which of the reactions below should occur at the faster rate. Cite reasons for your choice.

(a) $\overset{+}{Na}\overset{-}{Cl} + \overset{+}{Ag}NO_3^- \xrightarrow[\text{solution}]{\text{water}} \overset{+}{Ag}Cl^- \downarrow + \overset{+}{Na}NO_3^-$

(b) $H\!-\!\overset{\displaystyle H}{\underset{\displaystyle H}{C}}\!\text{-----}H + Cl_2 \longrightarrow H\!-\!\overset{\displaystyle H}{\underset{\displaystyle H}{C}}\!\text{-----}Cl + HCl$

16. Draw an energy diagram (in the format shown in Figure 24–9) for a typical equilibrium process. Label the energies of activation of the forward and reverse reactions and the heat of reaction. Discuss the relative rates, rate constants, and position of equilibrium.

17. Using the energy diagram in problem 16 show how a catalyst would affect (a) the shape of the curve, (b) the rates of the forward and reverse reactions, and (c) the position of equilibrium.

CHEMICAL REACTIVITY

25-1. STATISTICAL NATURE OF CHEMICAL REACTIONS

Let us examine the statistics of chemical reactions. Although molecular weights are equal for all molecules in a pure sample, excluding isotopic differences, not all molecules in the sample possess the same amounts of energy. The result is that molecular energy is always distributed about a mean value. Chemists, however, are able to operate as though energies were fixed. This is because of the extremely large numbers of individual molecules involved in even the smallest operation. A teaspoon of water, for example, contains approximately 10^{23} molecules.

When dealing with such unbelievably large numbers, the laws of probability work perfectly. Although no one could predict exactly what would happen if one molecule of hydrogen were to collide with one of oxygen, under any given set of conditions, any chemist can predict exactly what will happen if 1 liter of each gas is made to combine. Thus, the study of chemistry is greatly simplified.

Why does a particular reaction occur? This question is perhaps best answered in terms of bond stabilities and bond energies discussed in Chapter 12. Chemical processes tend to proceed in the direction of greatest stability; that is, chemical reactions tend to form products of lowest free energy (ΔG). This does not necessarily mean only energetic stability, ΔH, but also involves configurational stability, ΔS. A substance that reacts to form a molecule containing a more random atomic arrangement will tend to proceed even if it forms equally energetic molecules (Figure 25–1). An example of this randomness is the preferential formation of hexane ($CH_3CH_2CH_2CH_2CH_2CH_3$) over cyclohexane, which has similar bond energies.

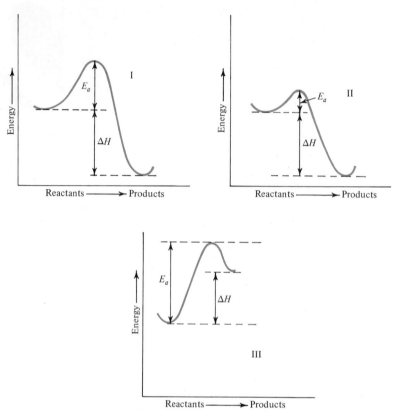

Figure 25–1. Energy diagrams (E_a = energy of activation; ΔH = heat released in the reaction). I, an exothermic reaction with a relatively large energy of activation (slow reaction); II, an exothermic reaction with low energy of activation (fast reaction); and III, an endothermic reaction with a high energy of activation.

This preference for open chain compounds is due to the greater randomness, that is, configurational stability, of linear molecules over that of cyclic molecules. The measure of this randomness of a system has been defined as the *entropy* of the system. *In general, systems tend toward greater entropy* (Chapter 6). Energy factors also play an important role in determining the products of a chemical reaction. The summation of all these terms, called the **free energy** (ΔG), is the quantity that allows the prediction of the direction of reaction: A reaction always proceeds of its own accord with a decrease in free energy. Recall that free energy is related to heat of reaction, ΔH, and entropy, ΔS, by the equation

$$\Delta G = \Delta H - T\Delta S$$

25-2. MECHANISM—A MODEL FOR CHEMICAL REACTIONS

Many important and interesting questions arise from the equation

$$CH_4 + 2\,O_2 \longrightarrow CO_2 + 2\,H_2O + energy$$

The first one, of course, is the driving force (ΔG) which has been examined in Section 25-1. Another question might be the sequence of steps by which each reactant molecule is converted into products. For example, the oxygen molecule might break apart into two atoms first, then one of these atoms could extract a hydrogen atom from the methane molecule; and this could be followed by several similar steps. The exact sequence of events that occur in the conversion of reactant molecules into product molecules is called the **mechanism** of the reaction. Unfortunately, the reaction of methane with oxygen is an exceedingly difficult reaction to study because it is so fast, as are most burning reactions.

25-3. REACTION OF ALKYL HALIDES WITH BASE

An entirely different kind of reaction occurs when ethyl chloride (CH_3CH_2Cl) is heated with sodium hydroxide in solution. The products of this reaction are ethyl alcohol (CH_3CH_2OH) and sodium chloride.

$$NaOH + CH_3CH_2Cl \longrightarrow CH_3CH_2OH + NaCl$$

The overall process involves the replacement of a chlorine atom by a hydroxy group (OH) on a carbon atom. If potassium hydroxide is employed in place of sodium hydroxide, the same ethyl alcohol forms along with potassium chloride. The nature of the metal part of the hydroxide is, in fact, immaterial.

If a different hydrocarbon chloride, such as tertiary butyl chloride

$$\begin{array}{c} CH_3 \\ | \\ CH_3-C-Cl \\ | \\ CH_3 \end{array}$$

is used in place of ethyl chloride, the same type of product, namely tertiary butyl alcohol,

$$\begin{array}{c} CH_3 \\ | \\ CH_3-C-OH \\ | \\ CH_3 \end{array}$$

is formed in small amounts along with isobutylene,

$$\begin{array}{c} CH_3 \\ | \\ CH_3-C=CH_2 \end{array}$$

391

as the major product. Considering only the alcohol for a moment, although the reaction appears to be the same as before, the process by which it occurs is quite different. For example, in the reaction of ethyl chloride, an increase in the concentrations of either ethyl chloride or sodium hydroxide results in an increase in the rate of the reaction. The rate of the corresponding reaction of tertiary butyl chloride does not depend on the concentration of sodium hydroxide, but only on the concentration of butyl chloride. That is, the use of different concentrations of sodium hydroxide will not affect the rate of the latter reaction, provided that the same concentration of butyl chloride is employed. The rate laws for these two reactions are:

$$rate_{(C_2H_5Cl)} = k_1[C_2H_5Cl][OH^-]$$

$$rate_{(C_4H_9Cl)} = k_2[C_4H_9Cl]$$

Thus the ethyl chloride reaction is second order and the butyl chloride reaction is first order. Since the effect of an increase in concentration is to increase the probability of collision, it must be concluded that the reaction step that controls the reaction rate—**rate determining step**—is different in these two reactions. In the reaction of ethyl chloride, the rate determining step involves the collision of ethyl chloride with hydroxide ions. The rate determining step in the reaction of tertiary butyl chloride does not involve such a collision.

These reactions also exhibit other differences that suggest that they proceed by different paths. For example, the rate of the ethyl chloride reaction is not greatly different when the reaction is carried out in water solution than when it is carried out in acetone solution. The reaction of tertiary butyl chloride proceeds at a much slower rate in acetone solution than it does in water solution. A summary of the differences between the two reactions is given in Table 25–1. An explanation of these differences can be found in the mechanisms of the reactions. Mechanisms that conform with all of the above data are given in the following paragraphs.

The reaction of ethyl chloride with sodium hydroxide occurs in a single step that involves attack by hydroxide ion at the carbon atom containing the chlorine atom, and simultaneous release of chloride ion.

Table 25–1.   A Comparison of the Factors that Influence the Reaction of Ethyl Chloride and tert-Butyl Chloride with Alkali

Experimental Condition	Effect on the Rate of Reaction	
	$CH_3CH_2Cl \rightarrow CH_3CH_2OH$	$(CH_3)_3C-Cl \rightarrow (CH_3)_3C-OH$
1. NaOH vs. KOH	1. no effect	1. no effect
2. rate law	2. rate $= k[C_2H_5Cl][OH^-]$	2. rate $= k[C_4H_9Cl]$
3. changing solvent from water to acetone	3. slight effect on rate	3. rate is reduced drastically
4. byproducts	4. very small amounts	4. $CH_3-\overset{\overset{\textstyle CH_3}{\vert}}{C}=CH_2$ is the major product in the reaction

$$OH^- + \quad \overset{H}{\underset{CH_3}{\overset{|}{C}}}-Cl \longrightarrow \left\{ HO{\cdots}\overset{H}{\underset{CH_3}{\overset{|}{C}}}{\cdots}Cl \right\}^- \longrightarrow HO-\overset{H}{\underset{CH_3}{\overset{|}{C}}}H + Cl^-$$

"activated state"

This mechanism obeys second order kinetics. Note that at one point in the process carbon is bound at least partially to both the hydroxide ion and the chloride ion. During the reaction, the molecule is "inverted" like an umbrella in a high wind. Since the reaction depends on the collision of OH^- with CH_3CH_2Cl, it follows that a reduction in the concentration of either would result in a reduction in the probability of collision, hence in the rate of the reaction.

The reaction of tertiary butyl chloride, on the other hand, occurs in two steps:

1. $CH_3 \overset{CH_3}{\underset{CH_3}{\overset{|}{C}}}-Cl \xrightarrow{\text{(slow)}} CH_3 \overset{CH_3}{\underset{CH_3}{\overset{|}{C^+}}} + Cl^-$ planar

2. $CH_3 \overset{CH_3}{\underset{CH_3}{\overset{}{C^+}}} + OH^- \xrightarrow{\text{(fast)}} CH_3 \overset{CH_3}{\underset{CH_3}{\overset{}{C}}}-OH$

Step 1 is very slow compared to step 2, and any reduction in OH^- concentration, which would slow down step 2 only, would not slow it down enough to make it slower than step 1. The overall formation of product cannot be faster than the slowest step in the sequence—the rate determining step—hence the rate is unaffected by the concentration of OH^-, and the reaction is first order. The great influence of the solvent in this reaction is due to its ability to stabilize the intermediate ion,

$$CH_3 \overset{CH_3}{\underset{CH_3 \quad CH_3}{\overset{|}{C^+}}}$$

Because this ion is extremely short-lived, a solvent of high polarity like water will enhance its formation and consequently increase the rate of reaction. A solvent like acetone, which is relatively nonpolar, inhibits the formation of such ions and, hence, slows down the reaction.

The formation of isobutylene as the major product also conforms to this mechanism. The ion formed in step 1 has an alternative to collision with an OH^- ion; namely, it can lose a H^+ from one of the methyl groups.

$$CH_3 \overset{CH_3}{\underset{\underset{H}{\overset{|}{\underset{H}{\overset{|}{C}}}}{\overset{|}{C^+}}}} \quad OH^- \longrightarrow CH_3 \overset{CH_3}{\underset{\underset{H}{\overset{}{\underset{H}{\overset{||}{C}}}}}{\overset{}{C}}} + H_2O$$

This H^+ is attracted by an OH^- ion to form the very stable water molecule, which provides the impetus for the tertiary butyl chloride to follow this path to iso-butylene rather than the one leading to the alcohol.

The reason why tertiary butyl chloride reacts by a different mechanism can be simply explained by the bulky CH_3 groups that prevent backside attack by the hydroxide ion. Such hindrance to backside attack is much less important in ethyl chloride.

These two mechanisms can be further tested by resorting to another feature of chemical reactions—their stereochemistry. The second-order reaction involves an inversion of the configuration of the carbon atom at which the substitution reaction occurs. Thus if an optically active alkyl halide were reacted, the second-order reaction should result in an inversion of configuration, and consequently in an inversion of the optical rotation. If the first-order mechanism operates, on the other hand, the intermediate ion has a plane of symmetry. Attack by the hydroxide ion can occur equally well from either side and the equal formation of both enantiomers would result. The optical activity of the sample would thus disappear.

Examples of each mechanism have been verified as shown in the sequences below.

$$OH^- + \quad \overset{\overset{C_6H_{13}}{\underset{H}{\vphantom{|}}}}{\underset{CH_3}{C}}{-}Br \longrightarrow HO{-}\overset{\overset{C_6H_{13}}{\underset{H}{\vphantom{|}}}}{\underset{CH_3}{C}} \quad + \; Br^-$$

optically pure   inverted optical rotation

optically pure   optically inactive mixture

25-4. REACTION OF HYDROGEN WITH CHLORINE—A CHAIN REACTION

The reaction of hydrogen and chlorine gases is another interesting reaction and one that poses entirely different problems and follows an entirely different mechanism.

Some of the facts about the reaction

$$H_2 + Cl_2 \longrightarrow 2\,HCl + energy$$

are as follows:

1. Reaction is very slow in the dark near room temperature.
2. Reaction is rapid in the dark when the gases are heated to relatively high temperatures.

3. Reaction is very rapid (explosive) when the mixture of gases is exposed to sunlight or ultraviolet light at room temperature.

4. When the reaction is carried out in the light, thousands of molecules of HCl are produced for each photon of light energy absorbed by the mixture.

5. The presence of even small amounts of oxygen gas inhibits the reaction for a period of time after which the reaction occurs normally; the length of this *induction* period depends on the amount of oxygen present.

These facts cannot be accounted for by a reaction mechanism that involves ions because ionic reactions are known to be indifferent to light. In fact, reactions that are light sensitive normally involve species with unpaired electrons. A mechanism that accounts for the above facts and which is generally accepted as explaining this reaction involves the following sequence of steps:

$$:\ddot{C}l:\ddot{C}l: \xrightarrow{\text{heat or light}} 2 :\ddot{C}l \cdot \qquad \text{(initiation step)} \qquad (25\text{--}1)$$

$$:\ddot{C}l \cdot + H:H \longrightarrow H:\ddot{C}l: + H \cdot \qquad \text{(propagation step)} \qquad (25\text{--}2)$$

$$H \cdot + :\ddot{C}l:\ddot{C}l: \longrightarrow H:\ddot{C}l: + :\ddot{C}l \cdot \qquad \text{(propagation step)} \qquad (25\text{--}3)$$

The first step (25–1) is the cleavage of the chlorine-chlorine bond into two equal atoms; that is, the pair of bonding electrons is split so that each chlorine atom contains an unpaired electron. Naturally, the free atoms are extremely reactive and will, in fact, react with nearly anything that they strike. Step (25–2) is an example of the reaction that occurs when a chlorine atom collides with a hydrogen molecule to produce a molecule of hydrogen chloride and a hydrogen atom that is itself quite reactive because of its unpaired electron. This hydrogen atom can then collide with a chlorine molecule (step 25–3) to produce another molecule of hydrogen chloride and another chlorine atom. The initial situation is now restored and steps (25–2) and (25–3) can repeat over and over again. It is this cycling of the propagation steps that accounts for the name **chain reaction** used to describe this type of process.

Closer examination reveals that there are other collisions that can occur besides the ones shown above. For instance, the chlorine atom may collide with a chlorine molecule:

$$:\ddot{C}l \cdot + :\ddot{C}l:\ddot{C}l: \longrightarrow :\ddot{C}l:\ddot{C}l: + :\ddot{C}l \cdot$$

Such an occurrence, however, represents no net change and is therefore non-productive. By a similar process, a hydrogen atom may collide with a hydrogen molecule to give a nonproductive reaction. Of course, a chlorine may collide with another chlorine atom or a hydrogen atom with another hydrogen atom, which ends the chain process. Such an occurrence is made unlikely by the very small number of such atoms present at any given time. This and other **chain termination** steps are as follows:

$$:\overset{..}{\underset{..}{Cl}}\cdot \: + \: \cdot\overset{..}{\underset{..}{Cl}}: \longrightarrow \: :\overset{..}{\underset{..}{Cl}}:\overset{..}{\underset{..}{Cl}}: \qquad \text{(termination step)} \qquad\qquad (25\text{--}4)$$

$$:\overset{..}{\underset{..}{Cl}}\cdot \: + \: \cdot H \longrightarrow \: H:\overset{..}{\underset{..}{Cl}}: \qquad \text{(termination step)} \qquad\qquad (25\text{--}5)$$

$$H\cdot \: + \: \cdot H \longrightarrow \: H:H \qquad \text{(termination step)} \qquad\qquad (25\text{--}6)$$

Because there is a finite possibility of such chain termination collisions, chain reactions do not proceed indefinitely. But the chain process accounts for the large number of product molecules produced per photon of light absorbed. A few chlorine atoms produced by a few photons (equation 25–1) leads to the production of a very large number of hydrogen chloride molecules (equations 25–2 and 25–3).

Table 25–2. Bond Dissociation Energies (kcal/mole)

$X:Y \longrightarrow X\cdot + \cdot Y$	$\Delta H = $ *Bond Dissociation Energy*
H—H	103
H—Cl	103
Cl—Cl	58
CH_3—H	101
CH_3—Cl	80

Still another aspect of this reaction mechanism must be examined. Why is it proposed that the initiating step is the rupture of the chlorine molecule instead of the hydrogen molecule? Table 25–2 contains the answer to this question in the dissociation energies of the two molecules. Because the dissociation energy of Cl_2 is so much smaller than that of H_2—58 kcal/mole compared with 103 kcal/mole for H_2—it is assumed that only chlorine molecules will be broken apart. Also, although a hydrogen atom is produced in step (25–2), this step is not as unfavorable as first appears because the dissociation energy required is supplied by the bond energy of formation of the hydrogen chloride molecule.

$$:\overset{..}{Cl}\cdot \: + \: H:H \: \longrightarrow \: H:\overset{..}{\underset{..}{Cl}}: \: + \: H\cdot$$

$$\underset{103 \text{ kcal}}{} \qquad\qquad \underset{103 \text{ kcal}}{}$$

By the same reasoning, step (25–3) is energetically quite favorable.

An energy diagram of the initiation process is represented in Figure 25–2. Reactions of species with unpaired electrons have nearly zero activation energy, as seen from the reverse reaction which has nearly zero activation energy.

One last fact about this reaction requires explanation: the retarding effect of oxygen gas. You will recall (Section 15–6) that molecular oxygen has two unpaired electrons. It can react with a chlorine atom to form a new molecule that is quite stable and therefore not sufficiently reactive to bring about the propagation steps.

$$:\overset{..}{\underset{..}{O}}—\overset{..}{\underset{.}{O}}: \: + \: \cdot\overset{..}{\underset{..}{Cl}}: \: \longrightarrow \: :\overset{..}{\underset{.}{O}}—\overset{..}{\underset{.}{O}}:$$
$$\underset{\underset{..}{\overset{|}{\underset{..}{Cl}}}}{}$$

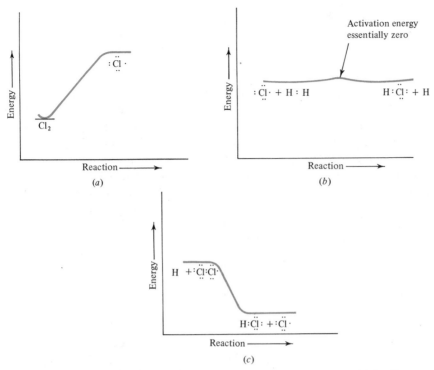

Figure 25-2. Energy diagrams for chain reaction steps: (a) initiation step (25-1), (b) propagation step (25-2), (c) propagation step (25-3).

Only after all the oxygen has been thus consumed can the propagation steps (25-2) and (25-3) proceed. Since the slowest step in the chain reaction is the initiation step (25-1), the reaction experiences an induction period (delay in initiation) when oxygen is present.

25-5. TAILOR MADE MOLECULES—THE USE OF SYNTHESIS

At this point let us consider some chemical reactions of great interest to the synthesis chemist. These will provide examples of the use of chemical reactions in the design of molecules of given specifications. Benzene can be made to react with chlorine gas if the reaction is carried out in the presence of iron powder. Whereas iron powder is necessary if this reaction is to proceed at a reasonable rate, light has no effect on the reaction. The reaction is represented by:

The reaction involves the replacement of one hydrogen atom in the benzene ring by a chlorine atom. Methane also reacts with chlorine in an apparently similar way

397

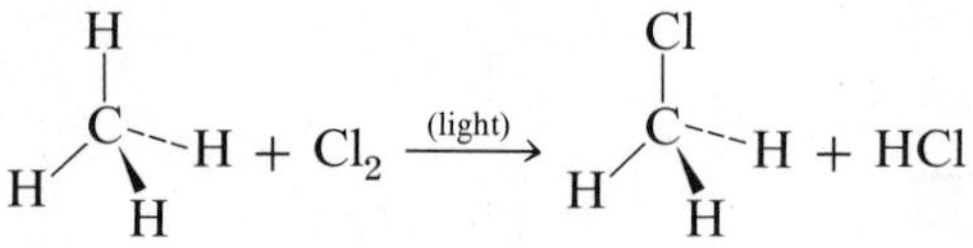

except that, in this case, light energy is required to cause the reaction to occur at a reasonable rate and iron powder has no effect. The very fact that the requirements are different for the two reactions is sufficient evidence that they occur through different mechanisms. The mechanisms will not be discussed, however, because the details are not necessary to the argument that follows.

The substance toluene has the structural formula shown, which bears a close similarity both to benzene and to methane. With this substance we find that

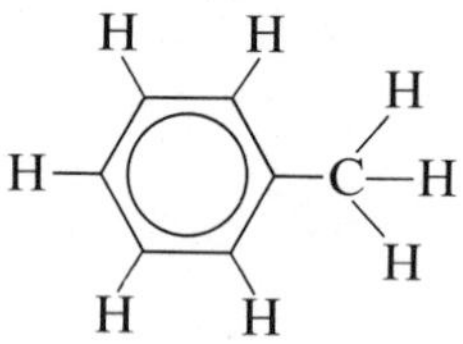

toluene

chlorine can react in either of two ways. Chlorine can substitute for one of the ring hydrogen atoms or for one of the hydrogen atoms on the CH$_3$ part of the molecule. As might be expected, if the reaction is carried out in the dark in the presence of iron powder, reaction occurs only in the ring, and if no iron is present and the reaction mixture is exposed to sunlight, then the reaction occurs only in the CH$_3$ group.

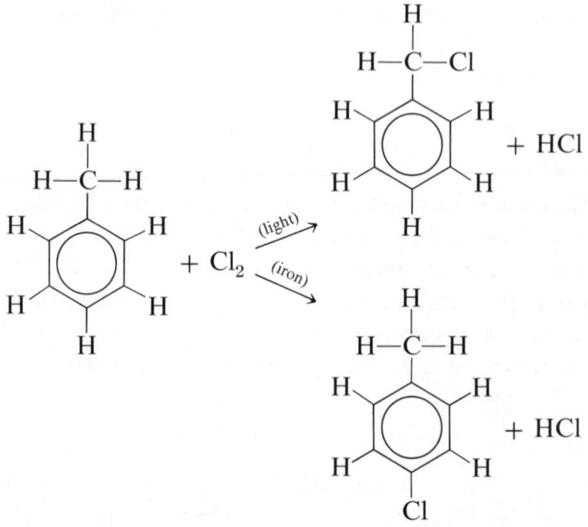

This is a simple example of the versatility that the chemist has in tailormaking molecules. It is an example of how he can direct a given reaction to occur in one of two locations in a particular molecule. Not only are there hundreds of such specific reactions available to the chemist, but also many chemists devote nearly all their efforts to working out new synthetic methods. When you realize that these reactions may be applied to thousands of molecules in any of a variety of ways, it becomes apparent that the chemist has an enormous "bag of tricks" with which to create new molecules for many purposes.

25-6. CLASSIFICATION OF REACTIONS OF COVALENT MOLECULES

The reaction of chlorine with either kind of hydrocarbon as described in Section 25–5 is often referred to as a substitution reaction. The process involves the substitution of a chlorine atom for a hydrogen atom attached to carbon. For the convenience of the practicing chemist (not to mention of the student) the hundreds of known chemical reactions may be classified into a reasonably small number of general types of reactions, of which substitution is one. This section will be devoted to a description of one or two illustrative examples of each of the following general types of reaction:

1. Substitution
2. Addition
3. Elimination
4. Oxidation
5. Rearrangement

1. Substitution Reactions □ The examples studied in Sections 25–3 and 25–5 are typical of the hundreds of substitution reactions that are known. No other examples will be given here.

2. Addition Reactions □ Compounds that contain double or triple covalent bonds frequently undergo reactions in which these extra electron pairs can be used to add other atoms. An example of such an addition reaction is the reaction of bromine with ethylene.

This reaction occurs quite readily and requires neither light nor heat. One has simply to mix the two substances and reaction occurs almost immediately. Since bromine has a reddish brown color and the other substances are colorless, the progress of the reaction can easily be followed by observing the disappearance of color.

The reaction between bromine and ethylene could occur in any of several ways. One is by direct "broadside" collision of the two molecules.

399

Another is the dissociation of the bromine molecule into ions ($Br^+ + Br^-$) followed by attack of Br^+ on the carbon-carbon double bond, then by attachment of Br^-.

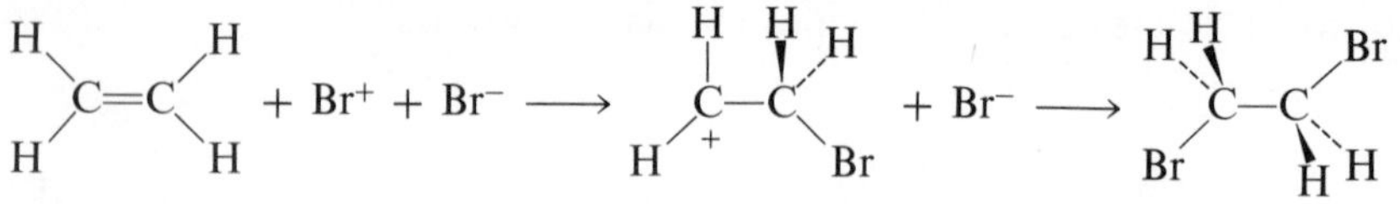

The accepted mechanism is a compromise between these two. Bromine is not known to dissociate spontaneously into ions as required by the second mechanism above. Moreover, if the first mechanism operated, a molecule such as cyclopropene would react to form the *cis*-dibromide (both bromine atoms on the same side of the ring).

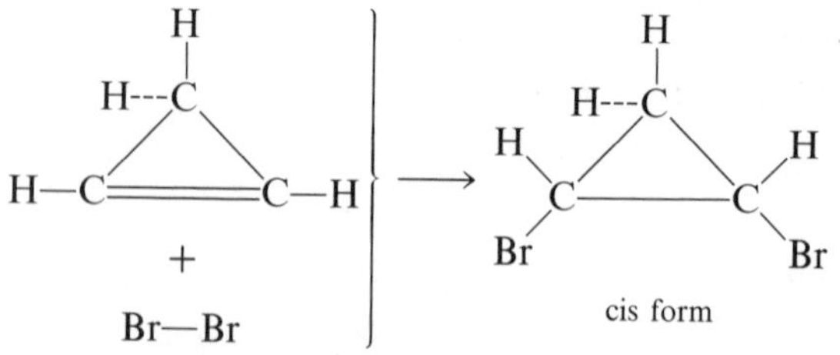

cis form

Actually only the trans form is produced.

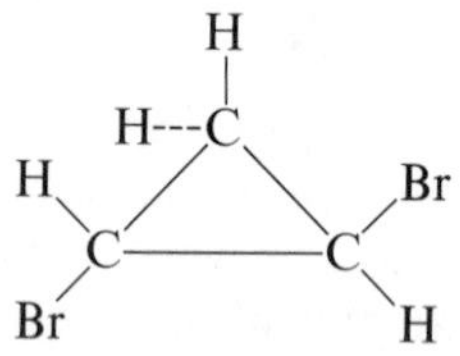

trans form

That ions are indeed involved is suggested by the fact that, if the reaction is carried out with some sodium chloride added to the mixture, some of the mixed chloride-bromide is formed, but no dichloride.

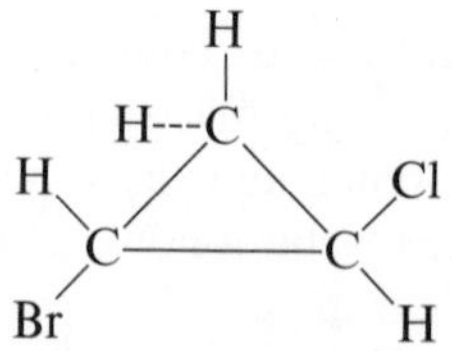

These results can be rationalized by the scheme shown in equation (25–7) in which dissociation of the bromine molecule into ions is facilitated by the approach of the highly electron rich double bond, which causes a polarization of the Br—Br molecule into a nearly ionic species (see equation 25–7).

The transient intermediate is then attacked by Br^- from the side opposite the attached Br^+, otherwise only a useless collision to yield Br_2 would occur. If other negative ions are present, such as Cl^-, they can also attack.

$$\underset{H}{\overset{H}{\text{C}}}=\underset{H}{\overset{H}{\text{C}}} + \overset{+}{\text{Br}}-\overset{-}{\text{Br}} \longrightarrow \left[\begin{array}{c} \underset{H}{\overset{H}{\text{C}}} \\ \text{Br}^{+} \\ \underset{H}{\overset{}{\text{C}}} \end{array} \right] + \text{Br}^{-} \qquad (25\text{-}7)$$

transient intermediate

Another addition reaction is the oxidation of carbon-carbon double bonds. If potassium permanganate is used as an oxidizing agent in water solution at room temperature, the product is entirely the cis form.

A plausible explanation for this specificity is the formation of an intermediate addition compound with the MnO_4^- ion which undergoes subsequent reaction with water to form the dihydroxy compound.

The *trans*-dihydroxy compound may be prepared if the oxidizing agent is peroxyformic acid. In this reaction, an intermediate is formed similar to the addition of bromine, in which the second OH group must attack from the opposite side of the molecule.

401

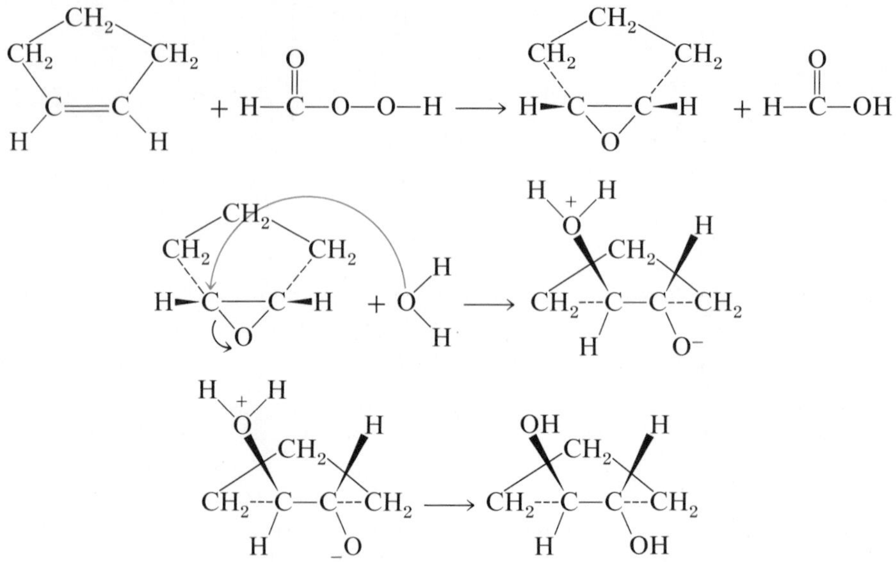

Without going into mechanistic details, the following reactions are presented as typical of the remaining types.

3. Elimination Reactions □ Alcohols are compounds in which one hydrogen atom of a hydrocarbon is replaced by a hydroxy group. One of the reactions characteristic of alcohols is that, when they are heated in the presence of an acid such as sulfuric acid, they lose a molecule of water to form a carbon-carbon double bond.

$$CH_3-CH_2-\underset{\underset{OH}{|}}{CH}-CH_3 \xrightarrow{\text{acid}} \underset{\text{major product}}{CH_3-CH=CH-CH_3} + H_2O +$$

$$\underset{\text{minor product}}{CH_3-CH_2-CH=CH_2}$$

Elimination of a molecule of water occurs predominantly so that the carbon-carbon double bond formed has the smallest number of hydrogens attached (most stable product). Analogous compounds with a chlorine, bromine, or iodine atom in place of a hydroxy group can also undergo the elimination reaction to form the carbon-carbon double bond and a molecule of HCl, HBr, or HI. In this reaction, the compound is usually heated in a mixture with sodium hydroxide.

$$CH_3-\underset{\underset{Cl}{|}}{\overset{\overset{CH_3}{|}}{C}}-CH_3 + OH^- \xrightarrow[\text{solvent}]{\text{alcohol}} CH_3-\overset{\overset{CH_3}{|}}{C}=CH_2 + H_2O + Cl^-$$

4. Oxidation Reactions □ Hydrocarbon groupings attached to a benzene ring may be selectively oxidized* without affecting the ring simply by heating with an acidic solution of $KMnO_4$. Regardless of the length or complexity of the hydrocarbon chain, oxidation degrades it down to the COOH group.

---

*Oxidation here is used in its limited sense to mean combination with oxygen. Note that since oxygen has more electron affinity than carbon, combination with oxygen involves a relative loss of electrons by carbon. Hence the definition used here is a special case of the more general definition.

$$+ \; KMnO_4 \; \xrightarrow[H_3O^+]{heat} \; + \; CO_2 + H_2O +$$

$$Mn^{2+} + \ldots$$

If there is more than one hydrocarbon chain attached to the ring, each is oxidized to a COOH group. This reaction has proved useful in determining the exact positions of various groupings on benzene rings because the oxidation products are all well known and easily identified.

5. *Rearrangement Reactions* □ Some chemical reactions occur in which all the atoms present in the original molecule are present in the product, but in rearranged positions. In the overall reaction, atoms are neither added nor taken from the original molecule.

One example of the many known rearrangement reactions was discovered by a chemist named Claisen and the reaction bears his name.

Compound A is converted into B simply by heating. A and B are both liquids. The progress of the reaction can be followed easily because B has a boiling point about 20°C higher than that of A. By boiling the mixture until the desired boiling point is reached, it is easy to tell when the reaction is complete.

The overall process involves an exchange between one hydrogen atom on the ring and the allyl ($-CH_2CH=CH_2$) group on the oxygen atom. Several mechanisms can be suggested; for example, the allyl group becomes detached† from the oxygen atom and later becomes attached to the carbon atom of the ring. Some brilliant experiments finally led to the accepted mechanism in which the allyl group never becomes detached from the molecule. If in compound A, the allyl group bears the radioactive isotope ¹⁴C in position 3, as shown in compound C,

†The electrons in the detached ion $CH_2-CH=CH$ are actually delocalized over three atoms so that the positive charge is spread to two atoms. This is shown by the resonance structures $(+)CH_2-CH=CH_2$, $CH_2=CH-CH_2(+)$ and accounts for the two possible products.
*Denotes the radioactive isotope ¹⁴C.

then detachment of the allyl group from the oxygen atom prior to attachment to the ring carbon atom would lead to the mixture of products D and E.

Actually only compound E is formed. The correct mechanism must therefore account for the selectivity of attachment as well as for the absence of even a small amount of compound D in the product.

The solution is shown in the intermediate product F in which carbon 3 of the allyl group becomes attached to the ring carbon atom before carbon 1 becomes detached from the oxygen atom:

(F)

The activated state F can be represented as a resonance hybrid of the two forms The dashed lines in F are used to denote partial bonds.

In a subsequent step, the hydrogen atom is lost from the carbon atom and becomes attached to oxygen.

PROBLEMS

1. What is meant by the term reaction mechanism?

2. A certain reaction A + B ⟶ C + D, has the following properties at 25°C: $\Delta H = 27.3$ kcal/mole, $\Delta S = 75$ cal/mole deg.
   (a) Will the reaction proceed in the desired direction?
   (b) At what temperature will the reaction proceed in the direction opposite to that at 25°?

3. Assuming that you are writing for an intelligent person who knows nothing about

chemistry, explain how chemists are able to determine the mechanisms of chemical reactions. You need not define the word mechanism in your answer.

4. In the reactions described in Section 25–3 what conclusions can be drawn from each of the following facts?
   (a) Substituting KOH for NaOH has no effect on the reaction provided that the same number of moles are used in each case.
   (b) Elimination to form the carbon-carbon double bond does not occur with ethyl chloride but does occur with tertiary butyl chloride.
   (c) The rate of reaction of tertiary butyl chloride to form the alcohol is independent of the [OH⁻] concentration.

5. The rate of reaction of tertiary butyl chloride to form isobutylene depends on the [OH⁻] concentration. What does this suggest about the mechanism of the elimination reaction mechanism?

6. In the reaction of hydrogen with chlorine, why is the initiation step believed to be the rupture of the chlorine molecule rather than the rupture of the hydrogen molecule?

7. Would you expect the reverse reaction: namely, $H:\overset{..}{\underset{..}{Cl}}: \xrightarrow{light} H\cdot + \cdot\overset{..}{\underset{..}{Cl}}:$ to occur to any appreciable extent under the conditions described? Explain.

8. Why are the termination reactions—steps (25–4), (25–5), and (25–6)—considered to be rare?

9. How does oxygen act to delay the onset of the H_2–Cl_2 reaction?

10. Calculate the heat of reaction, ΔH for each step in the mechanism given in steps 25–1 through 25–6.

11. Write a mechanism for the chlorination of methane: $CH_4 + Cl_2 \longrightarrow CH_3Cl + HCl$. Assume a mechanism similar to that for the H_2–Cl_2 reaction.

12. Calculate the heat of reaction of each step of the mechanism proposed in problem 11.

13. Under what conditions could benzene be used as a solvent for the chlorination of methane?

14. Classify each of the following reactions as substitution, addition, elimination, oxidation, or rearrangment.
   (a) $CH_3CH_2CH_2CH_2{-}Br + NaOH \longrightarrow CH_3CH_2CH_2CH_2{-}OH + NaBr$

   (b)

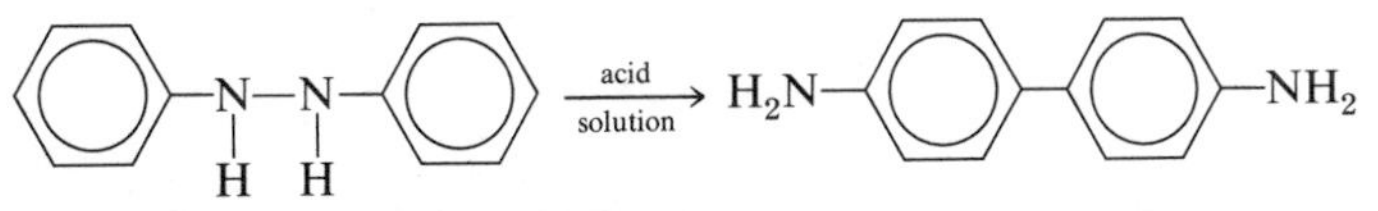

   (c) $CH_3{=}CH_2 + H_2O \xrightarrow[solution]{acid} CH_3CH_2OH$

   (d) $HCl + NaOH \longrightarrow H_2O + NaCl$
   (e) $2\,H_2 + O_2 \longrightarrow 2\,H_2O$

   (f)

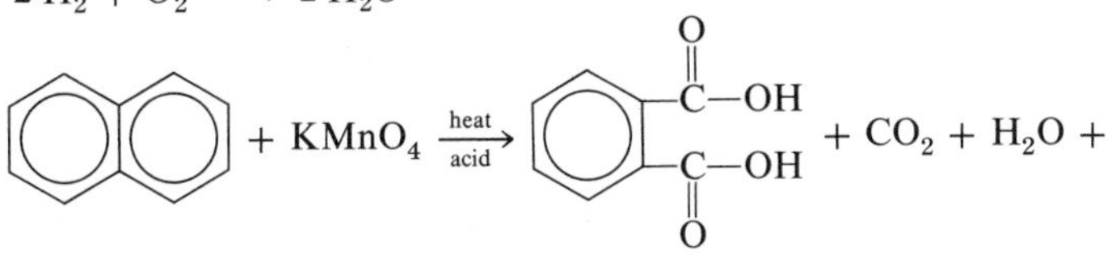

15. A certain compound (A), on treatment with hot $KMnO_4$ in an acidic solution yielded only benzoic acid

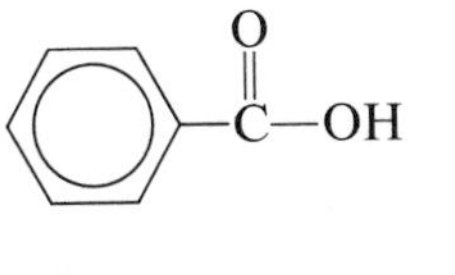

Which of the following compounds might A have been?

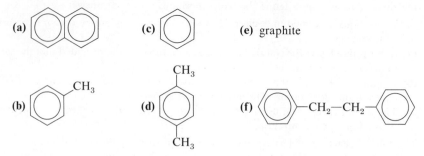

(a)   (c)   (e) graphite

(b)   (d)   (f)

16. A certain compound X reacts with bromine in the dark in the absence of catalyst and with no evolution of gaseous byproducts. Which of the following are possible compounds X?

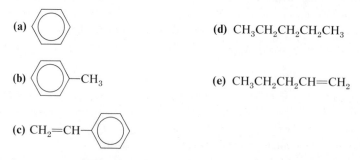

(a)

(d) $CH_3CH_2CH_2CH_2CH_3$

(b) —CH_3

(e) $CH_3CH_2CH_2CH=CH_2$

(c) $CH_2=CH$—

17. A certain compound, B, of formula C_8H_9Cl reacts with aqueous NaOH to form compound C ($C_8H_{10}O$). Treatment of C with hot H_2SO_4 yields compound D (C_8H_8), which adds bromine readily at room temperature to form E ($C_8H_8Br_2$). Compounds B, C, D, and E all react with hot acidic $KMnO_4$ solution to produce the same product—benzoic acid,

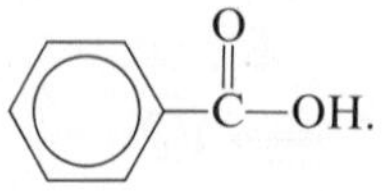

$$\begin{array}{c} O \\ \parallel \\ \text{—C—OH.} \end{array}$$

Write structures for B, C, D, and E and write equations for all the reactions described.

ORGANIC CHEMICALS

Several examples of the importance of classification in science have been offered in the previous chapters. One example is periodicity of the elements. Were it not for periodic variations of the properties of the elements, the study of chemistry would be indeed difficult if not impossible.

The branch of chemistry called organic chemistry is a very broad field that includes about three million known compounds. Organic compounds are those that contain carbon. The compounds of all the other chemical elements number about 100,000. The study of such a vast area as organic chemistry would be truly difficult were it not for its classification into easily grasped categories. In fact, the classification of organic compounds has been so thorough that it is generally considered easier to comprehend than inorganic chemistry.

Many examples of organic chemicals have already been encountered in previous chapters.

It will be the purpose of this chapter to present a survey of the ways in which organic chemicals are classified.

26-1. WHY ORGANIC CHEMISTRY?

This question is actually twofold: (1) why is the term **organic** used and (2) why is it a separate branch of chemistry? The first question is answered by the observation that, early in the development of chemistry, many compounds were isolated from living organisms, both plant and animal. These compounds were typically very complex compared to those of inorganic origin, which, incidentally, resulted in a lag in the development of organic chemistry. Not only were these organic substances more complex in nature, but also they almost always occurred as complex mixtures.

Knowledge of organic substances is evidenced in the oldest literature. For example, the Bible makes several references to the intoxicating effect of wine produced by fermentation. In one of the Proverbs of Solomon, the action of acetic acid on chalk to produce carbon dioxide is alluded to. The Egyptians also knew many organic substances such as concoctions for embalming and dyeing. In the fifth century A.D., distillation was first described; and in the eighth century, the Arabian alchemists had learned to concentrate acetic acid and alcohol by distillation. Still another milestone in organic chemistry was in the thirteenth century when alcohol was first converted into ether by the action of sulfuric acid.

$$CH_3CH_2OH + HOCH_2CH_3 \xrightarrow{H_2SO_4} CH_3CH_2-O-CH_2CH_3 + H_2O$$

alcohol   ether

Organic substances came to be considered capable of originating only from living systems and, hence, to be endowed with a vital force. Until the early nineteenth century, it was believed that man could not synthesize organic compounds in the laboratory from purely inorganic substances! In 1828, Friedrich Wöhler, while conducting some experiments with ammonium cyanate, an inorganic substance, found that, on heating, it was converted into urea

$$NH_4OCN \xrightarrow{heat} H_2N-\overset{\overset{\textstyle O}{\|}}{C}-NH_2$$

(Note that this is a rearrangement reaction.) Urea is an organic substance found in the urine. This discovery paved the way for dispelling the notion of a "vital force."

In the following few years, many known organic substances were synthesized in the laboratory. The term *organic* was retained, however, probably because of convenience and custom, but its use eventually came to be applied to compounds of carbon.

This brings up the second question: Why is organic chemistry still a separate branch of chemistry? The answer is that carbon has the unique ability to form virtually limitless chains, and this ability results in a tremendous proliferation of compounds. For example, consider the number of isomers possible for each of the molecular formulas in the table:

Number of Carbon Atoms in the Molecule	Number of Isomers Possible
C_7H_{16}	9
$C_{10}H_{22}$	75
$C_{20}H_{42}$	366,319
$C_{30}H_{62}$	4,111,846,763
$C_{40}H_{82}$	62,491,178,805,831

The problem of comprehending such staggering numbers of isomers is mediated by the fact that the saturated hydrocarbons, that is, compounds of carbon and hydrogen that have only single bonds, all behave nearly identically in most chemical reactions.

This similarity may be further emphasized by considering the idea of functional groups in organic chemistry.

26-2. FUNCTIONAL GROUPS

A functional group is a grouping of atoms that behaves chemically more or less independently of the rest of the molecule. Thus, the compounds in Table 26-1 are all organic carboxylic acids because they all contain the carboxy functional group

$$-\overset{\overset{\displaystyle O}{\|}}{C}-OH$$

The first three acids in Table 26-1 are nearly identical in physical properties. They have very similar odors and tastes and are very similar in acid strength.

The last three, while containing additional functional groups, namely, ⬡—

and —OH, still possess the characteristic properties of acidity—taste, reaction toward litmus—in addition to the properties of the other functional groups. In short, any compound containing a carboxy group will exhibit the properties of this group in addition to those of the other functional groups present.

Other functional groups are the carbon-carbon double bond (C=C), the triple bond (—C≡C—), the hydroxy group (—OH) characteristic of alcohols, the amine group (NH_2) characteristic of organic bases, the carbonyl group (C=O) characteristic of aldehydes and ketones, and several others (Table 26-2). About a dozen

Table 26-1. Some Organic Carboxylic Acids

formic acid	$H-\overset{\overset{\displaystyle O}{\|}}{C}-OH$ or HCOOH
acetic acid	$CH_3-\overset{\overset{\displaystyle O}{\|}}{C}-OH$ or CH_3COOH
propionic acid	$CH_3CH_2-\overset{\overset{\displaystyle O}{\|}}{C}-OH$ or CH_3CH_2COOH
benzoic acid	⬡—$\overset{\overset{\displaystyle O}{\|}}{C}$—OH or ⬡—COOH
citric acid	$HO-\overset{\overset{\displaystyle O}{\|}}{C}-CH_2-\overset{\overset{\displaystyle OH}{\|}}{\underset{\underset{\displaystyle O}{\|}}{\underset{C-OH}{C}}}-CH_2-\overset{\overset{\displaystyle O}{\|}}{C}-OH$ or $HOOCCH_2C(OH)(COOH)CH_2COOH$
lactic acid	$CH_3-\overset{\overset{\displaystyle OH}{\|}}{C}H-\overset{\overset{\displaystyle O}{\|}}{C}-OH$ or $CH_3CH(OH)COOH$

Table 26–2. Simple Organic Functional Groups

Functional Group Name	Structural Formula*
alcohol (hydroxy group)	$-\ddot{\underset{\cdot\cdot}{O}}H$
aldehyde (carbonyl group)	$-\overset{\overset{\cdot\cdot O \cdot\cdot}{\parallel}}{C}-H$
ketone (carbonyl group)	$-\overset{\overset{:O}{\parallel}}{C}-$
acid (carboxy group)	$-\overset{\overset{:O}{\parallel}}{C}-\ddot{O}H$
ester (carbalkoxy group)	$-\overset{\overset{:O}{\parallel}}{C}-\ddot{O}R$
amide	$-\overset{\overset{:O}{\parallel}}{C}-\ddot{N}H_2$
acid chloride	$-\overset{\overset{:O}{\parallel}}{C}-\ddot{C}l:$
acid anhydride	$-\overset{\overset{:O}{\parallel}}{C}-\ddot{O}-\overset{\overset{O:}{\parallel}}{C}-$
nitrile (cyano group)	$-C\equiv N:$
nitro	$-N\overset{\nearrow \overset{\cdot\cdot}{O}\cdot}{\underset{\searrow \ddot{\underset{\cdot\cdot}{O}}:}{}}$
nitroso	$-\ddot{N}=\ddot{O}:$
amino	$-\ddot{N}H_2$
ethylene	$\overset{\diagdown}{C}=\overset{\diagup}{C}$
acetylene	$-C\equiv C-$

*R denotes any hydrocarbon grouping.

functional groups exist in various combinations in all organic compounds. It is not the intention of this book to discuss in detail the physical and chemical behavior of the organic functional groups, but only to emphasize that they are limited in number and that knowledge of their behavior allows an understanding of the behavior of literally thousands of substances.

Instrumental analyses, especially those involving spectroscopic methods, have been extremely useful in the characterization and proof of structures of many complex compounds. The use of these methods is made possible by the fact that the absorption of electromagnetic energy by a functional group in a molecule is virtually independent of the presence of other groups. Thus, each group exhibits its own contributions to the overall spectrum of the molecule.

Nearly all of the functional groups in Table 26–2 can be thought of as derivatives of either water or ammonia (Figure 26–1). Thus the replacement of one

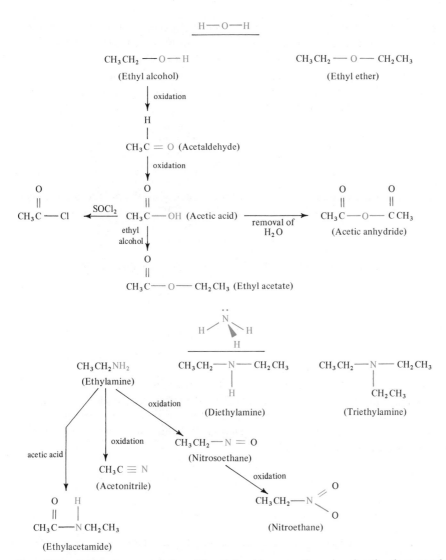

Figure 26-1. Schematic representation of the relationship among the various functional groups related to water and ammonia.

hydrogen atom in water by a hydrocarbon radical like ethane results in ethyl alcohol. Similarly ethyl alcohol can be oxidized to acetaldehyde, which may in turn be oxidized to acetic acid. Figure 26-1 shows schematically these relationships.

26-3. CLASSIFICATION BY REACTION TYPE

Chapter 25 dealt with some chemical reactions. In that chapter, reactions were classified according to whether they involved substitution, addition, elimination, oxidation, condensation, or rearrangement. This is only one of many ways to classify chemical reactions. Reactions may also be classified according to whether

411

ions are involved or not, or according to mechanism type. Following are one or two characteristic reactions of each of the more important functional groups— carboxy, hydroxy, aldehyde, ketone, ester, amide, and amine.

As implied above, each functional group undergoes certain characteristic reactions. For example, compounds that possess a carboxy group are acidic and therefore neutralize bases, for example

$$CH_3\overset{\overset{\textstyle O}{\|}}{C}{-}OH + Na^+OH^- \longrightarrow CH_3\overset{\overset{\textstyle O}{\|}}{C}{-}O^-Na^+ + H_2O$$

They also turn blue litmus red, taste sour, and react with alcohols to form esters and with amines to form amides (Chapters 16 and 25).

$$R{-}\overset{\overset{\textstyle O}{\|}}{C}{-}OH + CH_3CH_2OH \xrightarrow[\text{heat}]{\text{acid}} R{-}\overset{\overset{\textstyle O}{\|}}{C}{-}O{-}CH_2CH_3 + H_2O$$

$$\text{an ester}$$

$$R{-}\overset{\overset{\textstyle O}{\|}}{C}{-}OH + CH_3CH_2NH_2 \xrightarrow{\text{heat}} R{-}\overset{\overset{\textstyle O}{\|}}{C}{-}\overset{\overset{\textstyle H}{|}}{N}{-}CH_2CH_3 + H_2O$$

$$\text{an amide}$$

Alcohols possess the hydroxy group, OH, attached directly to a saturated carbon atom. Thus they differ structurally from the carboxylic acids. The hydroxy groups in alcohols are covalently bound to carbon and therefore are neutral; that is, free OH^- does not exist. Typical reactions besides ester formation are the reaction with alkali metals

$$2\, CH_3CH_2OH + 2\, Na \longrightarrow 2\, CH_3CH_2O^-Na^+ + H_2$$

and the elimination of water to form alkenes (Chapter 25).

$$CH_3CH_2OH \xrightarrow[\text{heat}]{H_2SO_4} CH_2{=}CH_2 + H_2O$$

Aldehydes and ketones are similar in having the carbonyl group, $C{=}O$. This group exists as a resonance hybrid of the structures

$$\overset{\overset{\textstyle :\ddot{O}}{\|}}{-C-} \qquad \overset{\overset{\textstyle :\ddot{O}:^-}{|}}{\underset{+}{-C-}}$$

which suggests that the oxygen has a resultant negative charge and the carbon has a resultant positive charge. For this reason, the carbonyl group undergoes reactions with reagents that have unshared electron pairs, for example, amines.

$$CH_3{-}\overset{\overset{\textstyle :\ddot{O}}{\|}}{C}{-}CH_3 + :NH_3 \longrightarrow [CH_3{-}\overset{\overset{\textstyle ^-:\ddot{O}:}{|}}{\underset{\underset{+}{NH_3}}{C}}{-}CH_3] \longrightarrow CH_3{-}\overset{\overset{\textstyle O}{\|}}{\underset{:NH}{C}}{-}CH_3 + H_2O$$

In addition to this type of addition reaction, aldehydes, but not ketones, are easily oxidized to carboxylic acids.

$$2 \; CH_3\overset{\overset{\displaystyle O}{\|}}{C}\!\!-\!\!H + O_2 \longrightarrow 2 \; CH_3\overset{\overset{\displaystyle O}{\|}}{C}\!\!-\!\!OH$$

Esters and amides may both be hydrolyzed to the carboxylic acid and either the alcohol or the amine. These reactions are the reverse of the reactions shown above for the formation of these compounds.

$$CH_3\overset{\overset{\displaystyle O}{\|}}{C}\!\!-\!\!OCH_2CH_3 + H_2O \xrightarrow[\text{heat}]{\text{acid}} CH_3\overset{\overset{\displaystyle O}{\|}}{C}\!\!-\!\!OH + CH_3CH_2OH$$

$$CH_3\overset{\overset{\displaystyle O}{\|}}{C}\!\!-\!\!\overset{\overset{\displaystyle H}{|}}{N}CH_2CH_3 + H_3O^+ \xrightarrow[\text{heat}]{\text{acid}} CH_3\overset{\overset{\displaystyle O}{\|}}{C}\!\!-\!\!OH + CH_3CH_2\overset{+}{N}H_3$$

Here, the amine is obtained as the ammonium salt, which may be neutralized with sodium hydroxide to the free amine.

$$CH_3CH_2\overset{+}{N}H_3 + Na^+OH^- \longrightarrow CH_3CH_2NH_2 + Na^+ + H_2O$$

Amides are *not* basic because the unshared electron pair on the nitrogen atom is involved in resonance structures such as

$$CH_3\!\!-\!\!\overset{\overset{\displaystyle \cdot\cdot\overset{\cdot}{O}\cdot}{\|}}{C}\!\!-\!\!\overset{\overset{\displaystyle \cdot\cdot}{}}{\underset{\underset{\displaystyle H}{|}}{N}}\!\!-\!\!CH_2CH_3 \quad \text{and} \quad CH_3\!\!-\!\!\overset{\overset{\displaystyle :\overset{\cdot\cdot}{O}:^-}{|}}{C}\!\!=\!\!\overset{+}{\underset{\underset{\displaystyle H}{|}}{N}}\!\!-\!\!CH_2CH_3$$

Amines on the other hand, possess an unshared electron pair and are, therefore, proton acceptors and undergo the typical reactions of bases. In addition, they undergo many other reactions. An example is alkylation, in which a hydrocarbon group may replace one of the hydrogen atoms.

$$CH_3CH_2NH_2 + CH_3CH_2Cl \longrightarrow CH_3CH_2\!\!-\!\!\overset{\overset{\displaystyle H}{|}\, +}{\underset{\underset{\displaystyle H}{|}}{N}}\!\!-\!\!CH_2\!\!-\!\!CH_3 + Cl^-$$

$$H_2O + CH_3CH_2\!\!-\!\!\overset{\overset{\displaystyle H}{|}}{\underset{}{N}}\!\!-\!\!CH_2CH_3 + Na^+Cl^- \longleftarrow \!\!\!\!\! \rule{0pt}{2.5em}\!\!\! Na^+OH^-$$

26-4. SYNTHESIS

Wohler's original discovery ushered in the era of synthetic organic chemistry. Today over 90% of all the known organic compounds have been made synthetically in laboratories.

Synthetic chemistry has provided chemists with essentially two goals: One has been to tailormake molecules with particular uses. Examples are the drugs, plastics, and explosives of our modern technological age. A second goal has been the sheer challenge of synthesizing complex molecules from simple ones. The synthesis of a molecule using unambiguous reactions and the production of a substance identical to a complex, naturally occurring molecule affords the ultimate proof of the latter's structure.

PROBLEMS

1. Write structures for all the possible isomers of C_6H_{12}.
2. Identify and name the functional groups in the following compounds.

(a) 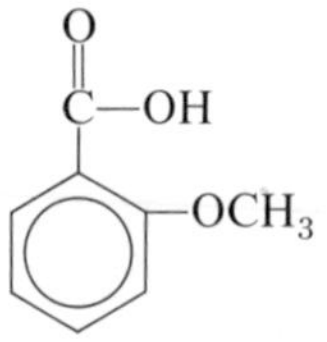

(b) Citric acid (Table 26–1)
(c) Lactic acid (Table 26–1)

(d) 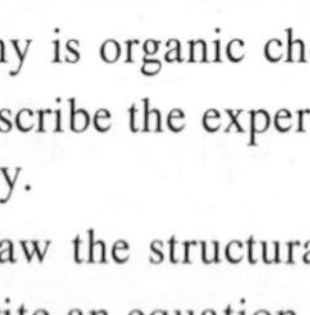

(e) 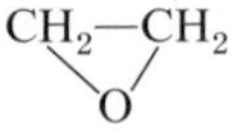

(f) 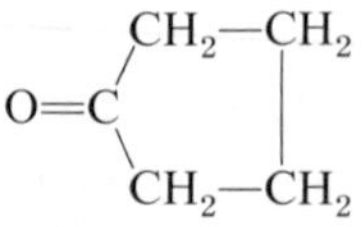

(g) 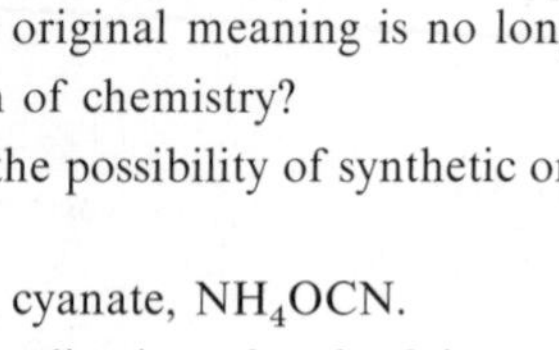

3. Explain why the term "organic" was first used to describe carbon compounds. Why is this term still used today even though its original meaning is no longer valid.
4. Why is organic chemistry a separate branch of chemistry?
5. Describe the experiment that demonstrated the possibility of synthetic organic chemistry.
6. Draw the structural formula of ammonium cyanate, NH_4OCN.
7. Write an equation to show the complete neutralization of each of the carboxylic acids in Table 26–1 with sodium hydroxide.
8. Cite typical reactions that would distinguish between the compounds in each of the following pairs.

(a) $CH_3\overset{\overset{O}{\|}}{C}{-}OH$ and CH_3CH_2OH

(b) $CH_3\overset{\overset{O}{\|}}{C}{-}OH$ and $CH_3\overset{\overset{O}{\|}}{C}{-}H$

(c) $CH_3\overset{\overset{O}{\|}}{C}CH_3$ and $CH_3\overset{\overset{O}{\|}}{C}{-}O{-}CH_2CH_3$

(d) $CH_3CH_2\overset{\overset{O}{\|}}{C}{-}H$ and $CH_3\overset{\overset{O}{\|}}{C}CH_3$

(e) CH_3CH_2OH and $CH_3\overset{\displaystyle O}{\overset{\|}{C}}-H$

(g) $CH_3CH_2NH_2$ and $CH_3\overset{\displaystyle O}{\overset{\|}{C}}-\overset{\displaystyle H}{\underset{|}{N}}-CH_2CH_3$

(f) CH_3CH_2OH and $CH_3CH_2NH_2$

9. Classify each of the reactions in Section 26–3 according to the classification given in Section 25–6.

10. From the structural formula of the compounds given below, predict and explain how each would react with (a) H^+ and (b) $:NH_3$; that is, show which atom in the compound will be attacked by each of these reagents. (*Hint:* draw out all the contributing resonance structures.)

(a) $CH_3-\overset{\displaystyle :\ddot{O}}{\overset{\|}{C}}\underset{\underset{\displaystyle \ddot{O}}{}}{\diagdown}\overset{\displaystyle :\ddot{O}}{\overset{\|}{C}}-CH_3$

(c) $CH_3-\overset{\displaystyle :\ddot{O}}{\overset{\|}{C}}-H$

(b) $CH_3-\overset{\displaystyle :\ddot{O}}{\overset{\|}{C}}\underset{\underset{\displaystyle \ddot{O}}{}}{\diagdown}CH_2CH_3$

(d) $CH_3-C\equiv N:$

THE CHEMISTRY OF LIFE

27-1. ORGANIC COMPOUNDS OF LIVING MATTER

Organic compounds found in nature consist primarily of the elements carbon, hydrogen, oxygen, and nitrogen. In fact, about 99% of all living substances are composed of these four elements (recall that hydrogen and oxygen also form water, which is one of the principal components of living matter).

The organic compounds that occur in living organisms may be grouped into four major classes—fats, carbohydrates, nucleic acids, and proteins (Figure 27–1). Other kinds of substances are also necessary to life, namely, hormones, vitamins, and inorganic salts. These will be described as the need arises in the following discussions. Proteins and nucleic acids have been discussed in Chapter 16. The next two sections will take up fats and carbohydrates.

27-2. FATS

The naturally occurring fats and oils are called **lipids** and consist of esters of glycerine, a trihydroxy alcohol, and any of several long

$$
\begin{array}{ll}
CH_2OH & CH_2-O-\overset{\overset{\displaystyle O}{\|}}{C}-(CH_2)_7-CH=CH-(CH_2)_7-CH_3 \\[2em]
CHOH & CH-O-\overset{\overset{\displaystyle O}{\|}}{C}-(CH_2)_7-CH=CH-(CH_2)_7-CH_3 \\[2em]
CH_2OH & CH_2-O-\overset{\overset{\displaystyle O}{\|}}{C}-(CH_2)_7-CH=CH-(CH_2)_7-CH_3
\end{array}
$$

glycerine   glyceryl trioleate

416

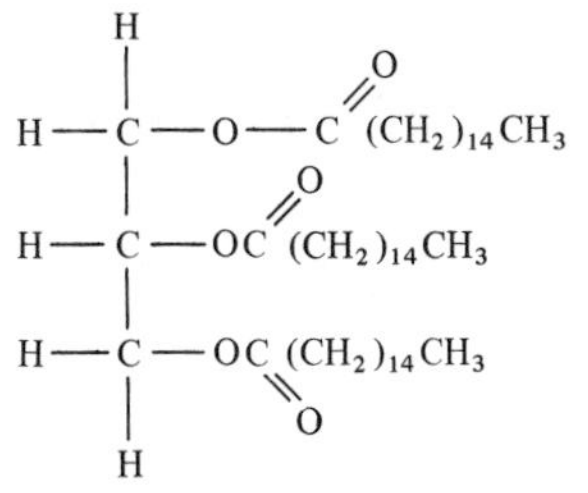

Palmitin, an example of a fat, is an ester of one molecule of glycerol (a trialcohol) and three molecules of palmitic acid.

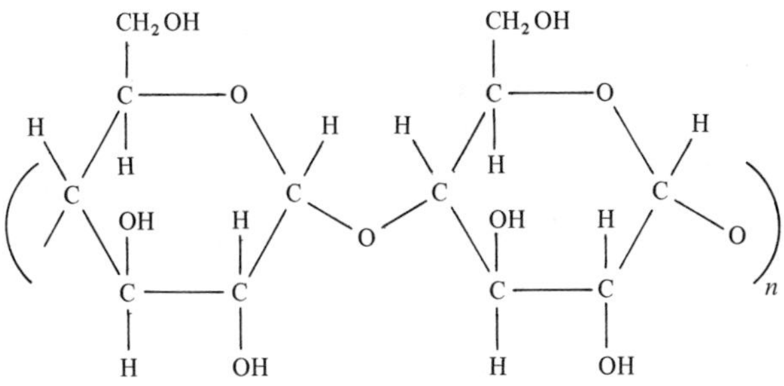

A carbohydrate is typically a polymer of glucose or similar $C_6H_{10}O_5$ unit, two of which are shown at left.

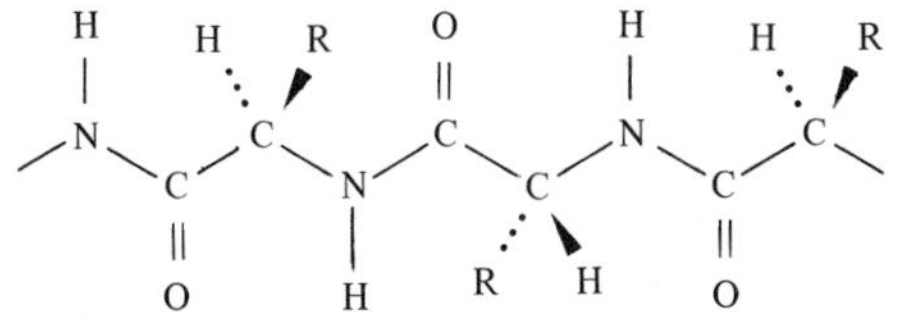

Proteins are polymers of amino acid units (Chapter 16).

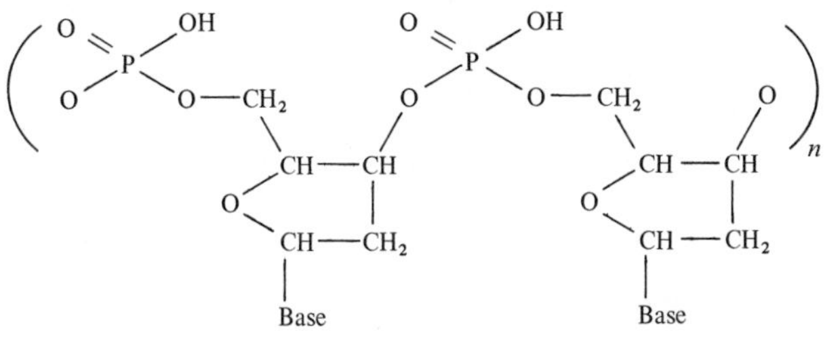

Nucleic acids contain a backbone which is a polymeric phosphate ester of a five-carbon sugar. The bases shown at left may be any of four nitrogenous compounds (Chapter 16).

Figure 27-1. The four great classes of organic compounds found in living organisms.

chain carboxylic acids. An example is the lipid formed from glycerine and oleic acid which is called glyceryl trioleate. Fats containing oleic acid comprise nearly 50% of the fats found in the human body. The more important natural lipids are milk fat, lard, tallow, corn oil, olive oil and peanut oil. The more important carboxylic acids (also called fatty acids) that occur in nature are listed as follows:

myristic acid	$CH_3(CH_2)_{12}COOH$
palmitic acid	$CH_3(CH_2)_{14}COOH$

417

stearic acid $CH_3(CH_2)_{16}COOH$

palmitoleic acid $CH_3(CH_2)_5CH=CH(CH_2)_7COOH$

oleic acid $CH_3(CH_2)_7CH=CH(CH_2)_7COOH$

linoleic acid $CH_3(CH_2)_3CH_2CH=CHCH_2CH=CH(CH_2)_7COOH$

Treatment of fats with strong bases causes them to undergo hydrolysis (called saponification) to form glycerine and the salt of the acid.

$$
\begin{array}{l}
CH_2-O-\overset{\overset{\displaystyle O}{\|}}{C}-R \\[2mm]
CH-O-\overset{\overset{\displaystyle O}{\|}}{C}-R \quad +\ 3\ NaOH \ \longrightarrow \\[2mm]
CH_2-O-\overset{\overset{\displaystyle O}{\|}}{C}-R
\end{array}
\qquad
\begin{array}{l}
CH_2OH \\[2mm]
CHOH \quad +\ 3\ Na\overset{+}{O}\overset{-}{}-\overset{\overset{\displaystyle O}{\|}}{C}-R \\[2mm]
CH_2OH
\end{array}
$$

a soap

$\Big(R-\overset{\overset{\displaystyle O}{\|}}{C}-O-$ may be

any of the *listed*

fatty acids $\Big)$

These salts have been used for ages as soaps because of their ability to form emulsions* of grease and water. Because of their very polar ionic ($Na^{+-}OOC$) end, they are able to dissolve in water; because of their long hydrocarbon chain, they are able to dissolve in droplets of oil or grease. The soap molecule is therefore able to bring oil and water together into dispersions that can be rinsed away.

In the body, a reaction analogous to saponification occurs during digestion. The body cannot tolerate such a highly basic substance as sodium hydroxide, however, and the hydrolysis is catalyzed by an **enzyme.** Enzymes are proteins that catalyze specific reactions in the body. The enzymes that catalyze lipid hydrolysis are called lipases.

$$
\begin{array}{l}
CH_2-O-\overset{\overset{\displaystyle O}{\|}}{C}-R \\[2mm]
CH-O-\overset{\overset{\displaystyle O}{\|}}{C}-R \quad +\ 3\ H_2O \ \xrightarrow{\ lipase\ } \\[2mm]
CH_2-O=\overset{\overset{\displaystyle O}{\|}}{C}-R
\end{array}
\qquad
\begin{array}{l}
CH_2OH \\[2mm]
CHOH \quad +\ 3\ R-\overset{\overset{\displaystyle O}{\|}}{C}-OH \\[2mm]
CH_2OH
\end{array}
$$

In the body, fats serve several functions including (1) insulation against heat loss, (2) protection against shock, (3) support for the various organs, and (4) a stored energy source. In this latter capacity, fats are the most effective of foods. Oxidation of fats yields over twice as much energy per unit mass as either carbohydrates or proteins. Fats are synthesized in the body from fats in the diet or from other food sources.

*An emulsion is a dispersion of fine particles or globules in a liquid.

Carbohydrates derive their name from the fact that many of them have the formula $C_nH_{2n}O_n$, or $C_n(H_2O)_n$. Although this name is not an accurate description of these substances, it has been retained to the present time.

Carbohydrates include the sugars, starches, and cellulose. These compounds have diverse functions in living organisms, including structural tissue (cellulose), synthesis of proteins and lipids by plants (sugars), and energy sources for animals (starch and sugars).

The carbohydrates occur as units containing usually five or six carbon atoms. Carbohydrates that contain one such unit are called **monosaccharides.** The most important monosaccharides are glucose and fructose.

glucose   fructose

These formulas include the stereochemistry of the compounds. Written in this format, the groups or atoms attached to the vertical column of carbon atoms are restricted to the directions shown. The more descriptive formula implied here would be

The student should verify these formulas with molecular models. It should be noted that glucose possesses four asymmetric carbon atoms and is therefore one of $2^4 = 16$ stereoisomers. Fructose has three asymmetric carbon atoms and is therefore one of $2^3 = 8$ stereoisomers. Note also that glucose and fructose are constitutional isomers.

These sugars and others like them exist mainly in a cyclic form in which one of the hydroxy oxygen atoms is linked to the carbon atom that originally bore the double-bonded oxygen (see formulas on page 420).

Glucose is also known as dextrose (dextrorotatory—rotates polarized light to the right) and occurs in many plants and fruits. Ripe grapes may contain as much as 30% of glucose. Fructose is also called levulose because it rotates polarized light toward the left (levorotatory). It occurs in fruits and honey. Fructose exists in combination with glucose as the common table sugar, sucrose, which occurs

glucose   cyclic form of glucose

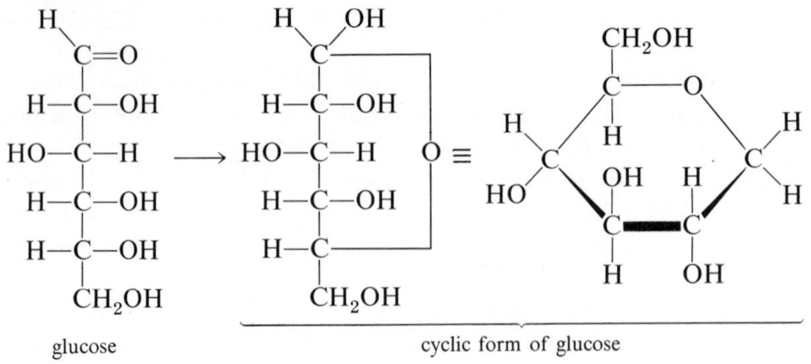

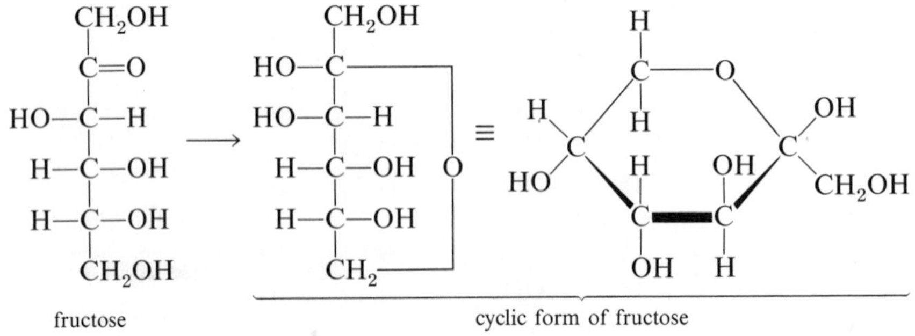

fructose   cyclic form of fructose

in sugar cane and sugar beets. Fructose is the sweetest of all the sugars and is twice as sweet as sucrose.

Sucrose is an example of a **disaccharide;** that is, it consists of two simple units—one glucose and one fructose—linked as shown below.

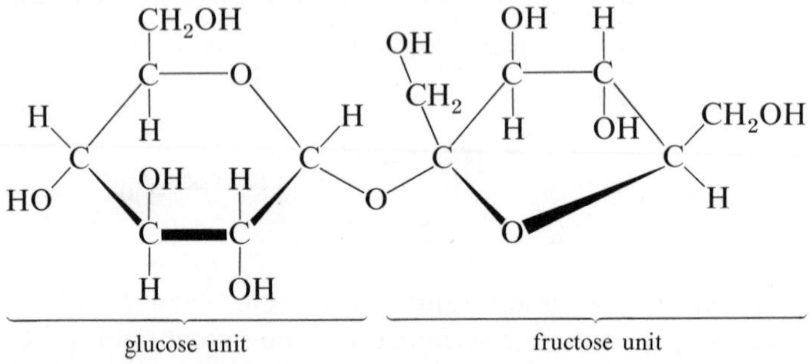

glucose unit   fructose unit

Such disaccharides are formed from two monosaccharide units by the elimination of one water molecule between them. Disaccharides may be made up of the same units or of different units, as in the case with sucrose.

Polysaccharides are polymeric carbohydrates that consist of monosaccharide units linked through the loss of one water molecule between each pair of monosaccharide units.

Starch, $(C_6H_{10}O_5)_n$, on hydrolysis yields only glucose, indicating that it is made up entirely of repeating glucose units. It occurs widely in plants, and is concentrated mainly in seeds and roots.

Cellulose has the same empirical formula as starch—$(C_6H_{10}O_5)_n$—and also yields only glucose on hydrolysis. Cellulose and starch have quite different properties, however. Cellulose serves as a structural material in most plants. The source of the purest cellulose is cotton.

An interesting structural difference distinguishes these two substances. Starch is digestible and is at best partially water soluble. Cellulose is completely insoluble in water and is not digestible. The difference lies in the linkage that joins the monosaccharide units. In cellulose, the C-O-C linkages project on alternating sides of the ring to give a more or less linear structure to the polymer. In starch, the linkages are all on the same side, which causes the polymeric chain to curl. Thus, cellulose is structurally useful because of the strength of its fibers, and starch forms granules.

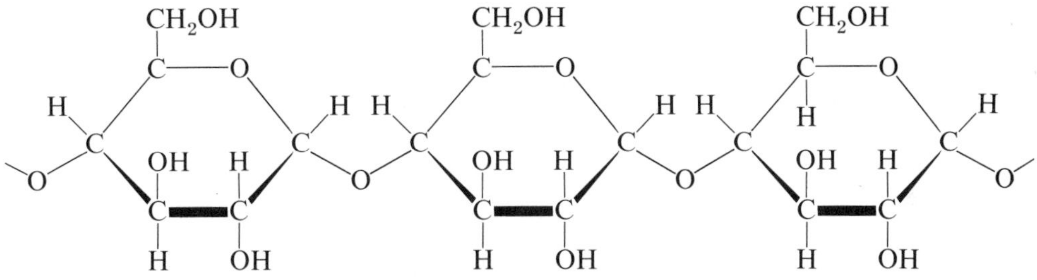

starch

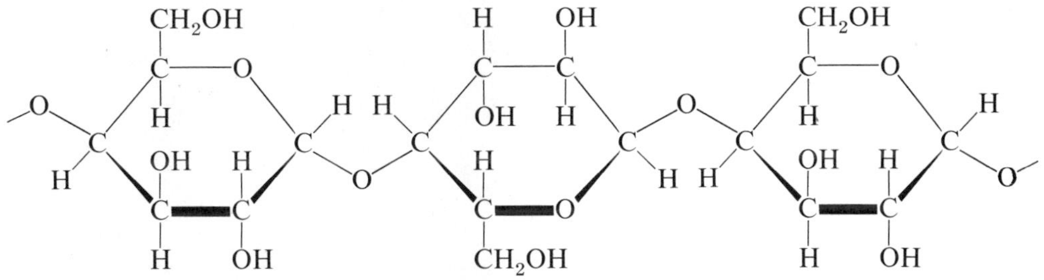

cellulose

27-4. CELLS—THE BASIC UNITS OF LIFE

The ideas embodied in Darwinian evolution have led scientists to speculate that life as we understand it originated one to two billion years ago in a very simple unicellular form, probably resembling the PPLO (Pleuropneumonia-like organism). The PPLO resembles higher cells but is very small—about 1000 Å in diameter (Figure 27-2). Still simpler entities are the viruses, which contain nucleic acids surrounded by a protein coat. Viruses, however, must use a cell's synthetic machinery for their replication. Hence they cannot be considered a living form.

The proposal that the cell is the fundamental unit of all living things was put in its present form in 1838 by Schleiden and Schwann in Germany. Their "cell theory" simply states that all plants and animals are constructed from small fundamental units called cells (Figure 27-3, 27-4, and 27-5). Each cell is sur-

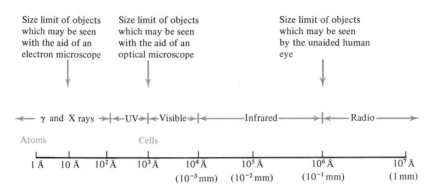

Size limit of objects which may be seen with the aid of an electron microscope

Size limit of objects which may be seen with the aid of an optical microscope

Size limit of objects which may be seen by the unaided human eye

←— γ and X rays →|←UV→|←Visible→|←————Infrared————→|←— Radio —→

Atoms   Cells

1 Å   10 Å 10^2 Å 10^3 Å 10^4 Å 10^5 Å 10^6 Å 10^7 Å
                            (10^{-3} mm)   (10^{-2} mm)   (10^{-1} mm)   (1 mm)

Figure 27-2. Comparison of the sizes of entities ranging from atoms to cells with wavelength regions of the electromagnetic spectrum. In order to compress the scale to proportion, a logarithmic scale is used.

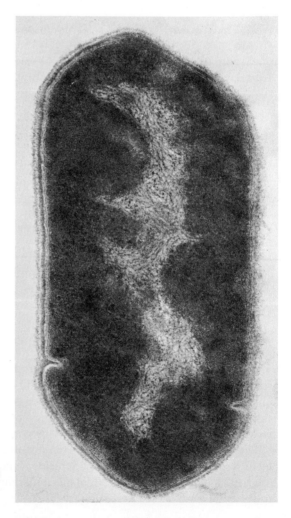

Figure 27-3. Bacterial cell. Light region in the center is DNA. (*Courtesy of Professor John V. Betz, Botany Dept., University of South Florida.*)

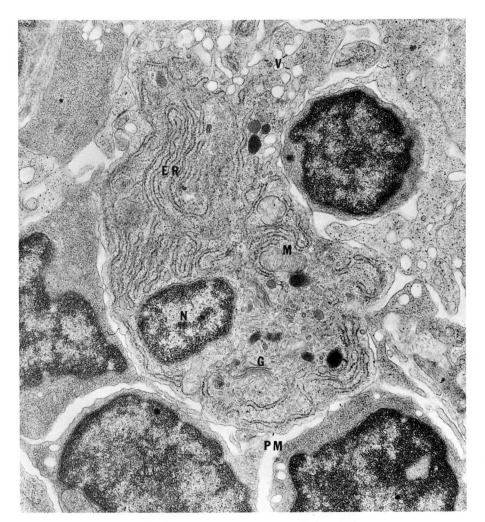

Figure 27-4. *Lepomis macrochirus*—blue gill plasma cell (central cell) surrounded by lymphocytes (14,000×). **ER** = granular endoplasmic reticulum, **G** = golgi apparatus, **M** = mitochondrion, **N** = nucleus, **PM** = plasma membrane, **V** = vesicles. (*Courtesy of Professor Kenneth Muse, Zoology Dept., North Carolina State University.*)

rounded by a membrane and has a nucleus surrounded by a double membrane with pores.

Cells reproduce themselves by splitting into two "daughter" cells, each complete with nucleus and all the other structural features of the original parent cell.

In complex organisms, there may be literally thousands of different kinds of cells each performing a different function. A human being has up to five million million (5×10^{12}) cells, which all originated from a single cell. The fertilized egg cell, therefore, contains all the information necessary for the production of these many cell types and the consequent growth and development of the adult.

Research on cell structure and function over the past 150 years has produced a wealth of descriptive information. But only within the last few decades has any real understanding been acquired about the detailed chemical reactions that occur within the living cell and how they relate to cell structure and function. The

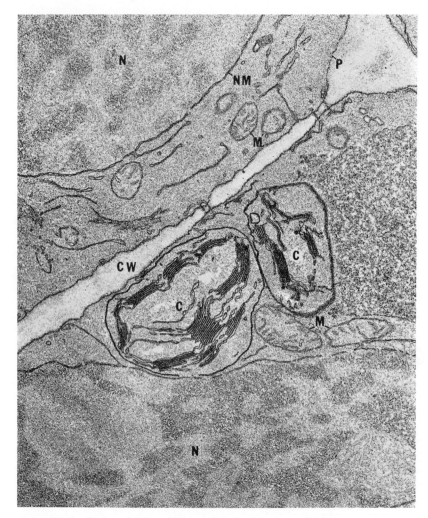

Figure 27-5. Chrysanthemum leaf cell (16,700×). **N** = nucleus, **NM** = nuclear membrane, **M** = mitochondrion, **P** = plasmalemma, **CW** = cell wall, **C** = chloroplast. (*Courtesy of Professor Robert K. Reid, Botany Dept., North Carolina State University.*)

problem is made quite difficult because chemical investigations necessarily disrupt the cell and change the very delicate balances within it.

27-5. INTERMEDIARY METABOLISM

The cell obtains its food by absorption through its external membranes. Once inside the cell, the food molecules are transformed extensively. Foods do not contain all of the molecules required by the cell. Instead these molecules must be synthesized in the cell by decomposing the food molecules and resynthesizing the needed ones. Such transformations usually occur through complex series of many intermediate steps. The intermediate molecules are termed **metabolites** and many have no function other than as intermediates in the overall process. The term **intermediary metabolism** refers to the sum total of all the chemical reactions that occur in the transformation of food molecules into cellular building blocks.

Whereas some organisms like *Escherichia coli,* can synthesize all the compounds necessary for its metabolism, most organisms are not capable of this; some of the molecules required in the various metabolic pathways must be furnished to the organism in its food supply. Certain amino acids are among those molecules that many organisms cannot synthesize. These are called **growth factors.** In addition, some small organic molecules are required in trace amounts. These are the **vitamins** (vital molecules).

27-6. ENERGY TRANSFORMATIONS IN THE CELL—METABOLISM

Metabolism describes essentially two broad operations: (1) **catabolism**—the decomposition of food molecules for the production of other metabolites with the generation of chemically usable energy—and (2) **anabolism**—the biosynthesis of proteins, carbohydrates, fats, and nucleic acids necessary in carrying out the functions of the cell. Anabolism will be discussed in Section 27-7.

The conbustion of a fuel like coal involves a release of energy in the form of heat and light. The released energy

$$C + O_2 \longrightarrow CO_2 + energy$$

results from the rupture of many carbon-carbon covalent bonds of relatively high potential energy and the formation of carbon-oxygen bonds, which have a relatively low potential energy.

Cells, likewise, carry out an energy-liberating operation in which food molecules, for example, carbohydrates and fats, combine with oxygen through a series of intermediate steps, with the ultimate production of carbon dioxide, water, and energy.

$$C_6H_{12}O_6 + 6\,O_2 \longrightarrow 6\,CO_2 + 6\,H_2O + energy$$

In this overall reaction, the higher energy covalent bonds in glucose and oxygen are converted into lower energy bonds in carbon dioxide and water. Approximately 690 kcal of energy is liberated from the conversion of 1 mole of glucose into carbon dioxide and water.

Such oxidation in the cell is controlled. This differs very significantly from the uncontrolled combustion of coal. Part of the energy liberated during metabolism is trapped and stored in even higher energy bonds for subsequent use by the cell as an energy source for the synthesis of more complex molecules needed by the cell. Thus the cell is able to do work that involves an increase in the order of the system, that is, a decrease in entropy, which the burning of coal cannot do. This is not, however, a violation of the second law of thermodynamics because it is the total entropy of the system plus its surroundings that must spontaneously increase. There is a greater entropy increase external to the cell which more than compensates for the entropy decrease within the cell.

Fuel oxidation in the cell differs in another way from a simple combustion process; it is rigorously controlled. The cell is unable to utilize heat as such in processes like muscle contraction. Moreover, the cell operates at a constant temperature; therefore, no temperature differential can exist like that required to perform simple mechanical work such as expansion. The energy is converted into useful forms through the use of enzymes. The breakdown of fats to glycerine and fatty acids, proteins to amino acids, polysaccharides to simple carbohydrates, and nucleic acids to nucleotides can also occur extracellularly, for example, as in the stomach and intestines.

Oxidative processes within the cell can be divided into three major stages. The first is a breakdown of simple sugars, fatty acids, and amino acids into simpler fragments that are more or less common denominators for the processes that are to follow. The second stage involves the ultimate conversion of carbon in these intermediate compounds into carbon dioxide. The third is the respiratory pathway: the transfer of electrons from highly hydrogen-rich intermediates during their conversion to carbon dioxide. The electrons are eventually used to convert hydrogen ions and oxygen to water. This process occurs within the cell in tiny granules called **mitochondria,** and the energy released is trapped in a form that is usable by the cell instead of being lost as heat. Each of the above three stages involves a labyrinth of chemical transformations. The energy liberated throughout these steps is stored in the form of relatively unstable (high energy) covalent bonds in special molecules to be used in the subsequent energy consuming processes of the cell.

The formation of these high energy molecules is a very interesting and important process in all cells. In the late 1930s, two biochemists, H. M. Kalekar in Denmark and V. A. Belitser in Russia, made an important deduction based on a puzzling experimental result. When a suspension of ground kidney or muscle was incubated at 37°C with glucose in the presence of oxygen gas, the concentration of phosphate ions present in the suspension decreased as the glucose was consumed. They correctly interpreted this result as an incorporation of phosphate into organic compounds in the system. The particular compound formed was shown to be adenosine triphosphate (ATP). The symbol $\sim$ designates "high energy" bonds.

ATP consists of a purine base component—adenine—attached to the five-carbon sugar—D-ribose—together forming a nucleotide, which is universally found in cells. The formation of ATP molecules is referred to as **oxidative phosphorylation.** Upon demand, as in muscle contraction, ATP molecules release their energy to the muscle by severing their terminal phosphate groups producing adenosine diphosphate (ADP).

The conversion of one molecule of glucose into carbon dioxide and water involves the conversion of approximately 38 molecules of ADP to 38 molecules of ATP. On a mole basis

$$\text{ADP} + \text{PO}_4^{3-} + 12 \text{ kcal} \longrightarrow \text{ATP}$$

and

$$38 \times 12 \text{ kcal} = 456 \text{ kcal}$$

which equals the energy stored in the form of high energy bonds in ATP per mole of glucose oxidized. Since the overall oxidation of glucose liberates 690 kcal/mole, this process is about 66% efficient.

Of the energy released in the oxidation of glucose, about 90% arises from the third stage: the combination of hydrogen from glucose with oxygen supplied by the lungs and carried to each cell by the blood.

27-7. BIOSYNTHETIC METABOLISM—PHOTOSYNTHESIS

Although a variety of foods is employed by the many kinds of cells, ultimately the fundamental foods derive from one fundamental chemical process—**photosynthesis.** This is the series of chemical reactions by which green plants convert atmospheric carbon dioxide and water vapor into carbohydrate molecules and, in the process, transform the sun's energy into chemical energy.

$$6 \text{ CO}_2 + 6 \text{ H}_2\text{O} + \text{energy} \longrightarrow \text{C}_6\text{H}_{12}\text{O}_6 + 6 \text{ O}_2$$

Thus, the very stable molecules—CO_2 and H_2O—are converted into less stable (higher energy content) molecules like glucose ($C_6H_{12}O_6$).

Photosynthesis is chemically the reverse of respiration or combustion. In the latter, carbohydrates are combined with oxygen to produce carbon dioxide and water.

$$\text{C}_6\text{H}_{12}\text{O}_6 + 6 \text{ O}_2 \longrightarrow 6 \text{ CO}_2 + 6 \text{ H}_2\text{O} + \text{energy}$$

This relationship between photosynthesis and respiration is shown diagramatically in Figure 27-6. Thus, these two processes support each other and provide an energy and material balance in the world of living things.

It was demonstrated in the 1930s by C. B. van Niel that the oxygen gas evolved in the photosynthetic process comes from the water molecules. His experiment involved a certain bacteria that absorbs H_2S instead of H_2O in its photosynthetic process. The expelled product is elemental sulfur instead of oxygen.

$$6 \text{ CO}_2 + 6 \text{ H}_2\text{S} \longrightarrow \text{C}_6\text{H}_{12}\text{O}_6 + 6 \text{ S}$$

Most of the energy absorbed during photosynthesis goes to split the water molecules into oxygen and hydrogen atoms. The hydrogen atoms react with

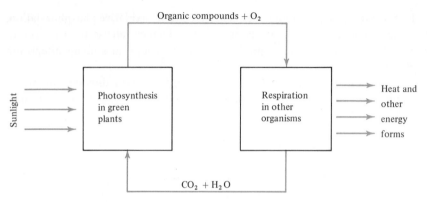

Figure 27-6. Schematic representation of the complementary relationship between photosynthesis and respiration.

carbon dioxide molecules through a long series of intermediate steps called carbon dioxide fixation. The reaction sequence involves many intermediate products and finally ends up in carbohydrates, fats, carboxylic acids, amino acids, and other organic substances.

The chemical pathways involved in photosynthesis have been largely worked out by Dr. Melvin Calvin at the University of California. In this work, which spanned the period 1946 to 1960, many new techniques were employed, including the use of radioactive isotopes to trace the metabolic pathways. In one technique, carbon dioxide containing $_{6}^{14}C$ was fed to the plant, and the labeled carbon was traced through the various intermediates in the biosynthetic pathways leading to carbohydrates.

27-8. BIOSYNTHETIC METABOLISM—PROTEINS AND NUCLEIC ACIDS

In contrast to the fats and carbohydrates, which are relatively simple molecules, proteins and nucleic acids contain an additional complication—they contain *information* in their structure. Thus, the exact order in which the building blocks are put together in the formation of these polymers determines their function. As we have seen (Chapter 16), these molecules are able to use this information in the synthesis of exact duplicates of themselves.

The DNA molecule is the master informational unit in the cell. It contains all the genetic information required for the differentiation of each type of cell in the organism. Nearly all the DNA molecules are to be found in the cell nucleus. The structure of DNA was described in Chapter 16. Its double helix structure provides the mechanism for self-duplication by untwining and allowing each strand of the double helix to reproduce the complementary strand (Figure 27-7). The information contained in the DNA structure is that required to specify the amino acid sequence in the biosynthesis of the various proteins used in the cell.

In protein synthesis, information in the DNA is transferred to the cell fluid outside the nucleus—the **cytoplasm**—where protein synthesis principally occurs,

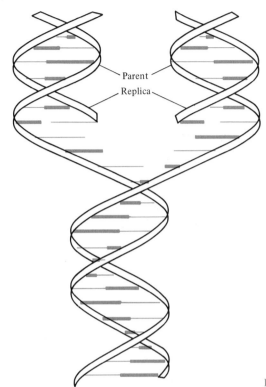

Parent

Replica

Figure 27-7. Diagram of DNA replication.

by way of so-called "messenger" molecules. These messenger molecules are similar to DNA except that they are single-stranded, they contain uracil in place of thymine, and their backbone is composed of D-ribose instead of 2-deoxyribose. These ribonucleic acid molecules are appropriately called **"messenger RNA"** or m-RNA. m-RNA is a copy of DNA and is synthesized enzymatically in the nucleus with DNA as template. The m-RNA then diffuses into the cytoplasm where it becomes attached to the **ribosome***.

The other component of the synthetic scheme is **transfer RNA** (t-RNA). There are at least 20 different t-RNAs, each capable of forming a linkage with one specific amino acid. This product is called the "charged" t-RNA. Protein synthesis then occurs on the surface of the ribosome by bringing together an m-RNA and two charged t-RNAs. The information in the m-RNA is now translated by forming bonds with the proper pair of t-RNAs on adjacent sites from those present by virtue of the complementary nucleotide units in each. The peptide bond is then formed enzymatically between the two adjacent amino acid units. The dipeptide now remains attached to one of the t-RNAs, the uncharged one leaves, and the next properly charged t-RNA becomes attached to the new m-RNA unit. After the required repetitions, each selecting and attaching the specific amino acid unit needed at that point in the chain, the finished protein results (Figure 27–8).

---

*Ribosomes are small granules composed of RNA and protein. Protein synthesis occurs in the ribosomes of the cell.

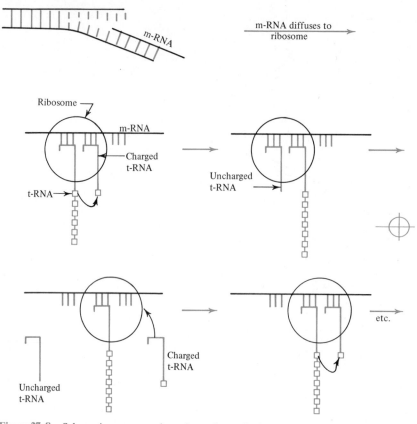

Figure 27-8. Schematic representation of protein synthesis. (□ = amino acid residues)

PROBLEMS

1. Refer to the structures of the fatty acids given in Section 26–2. Suggest why unsaturated fats are considered more digestible than saturated fats.

2. Using the linear formulas, draw all the steroisomers of glucose.

3. Redraw the linear structures below in the cyclic form.

(a)   (b)

4. What are the four major groups of organic compounds that occur in living organisms? Describe them.

5. Why is the cell considered to be the fundamental unit of all living things?

6. What is meant by the term intermediary metabolism?

7. What essential difference is there between the oxidation that occurs during the combustion of coal and the oxidation that takes place within cells?

8. How is the energy that is produced during food oxidation in the cell stored for later use?

9. Explain how photosynthesis in plants is related chemically to respiration in animals.

10. Explain the role of DNA, m-RNA, and t-RNA in protein synthesis in the cell.

11. To the chemist, a high bond energy represents a very stable bond. In biochemical processes, however, "high energy" bonds are found in molecules that contain a high degree of potential energy. Is this consistent? Explain.

12. Discuss the meaning of the term oxidation as it is used in this chapter.

13. In an overall way, discuss the ultimate fate of that part of the sun's energy that is absorbed in photosynthesis.

NUCLEAR REACTIONS

The early history of radioactivity was presented in Chapter 9. The student should remember the work of Bequerel and the Curies; the detection and identification of α, β, and γ radiations; and the discovery of isotopes through the work of Thomson, Rutherford, and Soddy. A review of Chapter 9 is suggested if these facts have been forgotten because the present chapter will build upon these ideas and extend the consideration of chemical dynamics to the heart of the atom—the nucleus.

28-1. RADIOACTIVITY IN NATURE

Recall that Rutherford and Soddy first unraveled the fate of radioactive elements found in nature. These unstable elements decay into other elements (daughter elements), which in turn decay until finally a stable nucleus is achieved. Virtually every naturally occurring unstable element decays by elimination either of an α particle (α decay) or a β particle (β decay) from its nucleus. Release of γ radiation usually accompanies decay by either mode.

The stability of a nucleus is often expressed in terms of its **half-life**—the time it takes for one half of any given number of the particular nuclei to decay. All radioactive decay reactions follow first-order kinetics. For example, the β decay of an unstable isotope of thallium into lead can be represented

$$^{210}_{81}\text{Tl} \longrightarrow ^{210}_{82}\text{Pb} + ^{0}_{-1}\text{e}$$

where the subscripts give the charge on the species (atomic number for atoms, Z) and the superscripts give the mass number, A, or sum of neutrons and protons in the nucleus. Both the charge and mass numbers must balance in the reaction. For an atom, then, Z represents the number of protons and $A - Z$ the number of neutrons in the nucleus. The nuclear particles—neutrons and protons—are often referred to as **nucleons.** A particular nucleus of given atomic number and mass number is called a **nuclide.**

The rate of the above reaction can be expressed as loss of thallium nuclei per second = $k \times$ number of thallium nuclei present at any instant or,

$$-\frac{\Delta N_t}{\Delta t} = k N_t \qquad (28\text{-}1)$$

where, N_t = the number of thallium nuclei at any time t and k = the first-order rate constant.

By the use of calculus, it is possible to express equation (28-1) in a different and often more useful form, namely

$$2.3 \log \frac{N_0}{N_t} = kt \qquad (28\text{-}2)$$

where, N_0 = number of nuclei originally present (that is, at time $t = 0$) and t = time elapsed. When a substance has passed through a half-life, then, $t = t_{1/2}$, and $N_t = N_0/2$. Substituting these values into equation (28-2),

$$2.3 \log \frac{N_0}{N_0/2} = kt_{1/2}$$

$$2.3 \log 2 = kt_{1/2}$$

finally

$$t_{1/2} = \frac{0.69}{k} \qquad (28\text{-}3)$$

Equation (28-3) is a fundamental relationship between the half-life and specific rate constant for any reaction following a first-order rate law. Since all radioactive decay reactions obey a first-order rate law, equation (28-3) is particularly useful in this connection. The half-life is seen by this equation to be a constant for a particular substance, completely independent of the actual number of nuclei one starts with. Half-lives of radioactive nuclei vary enormously. For $^{210}_{81}\text{Tl}$ decaying by the above equation, $t_{1/2} = 1.3$ min, whereas the decay of $^{238}_{92}\text{U}$ to $^{234}_{90}\text{Th}$ by α emission has $t_{1/2} = 4.5 \times 10^9$ years!

Three radioactive decay series exist in nature. The **4n series,** which consists of elements with mass numbers evenly divisible by 4, originates with $^{232}_{90}\text{Th}$ and ends with the stable $^{208}_{82}\text{Pb}$. A **4n + 2 series** starts with $^{238}_{92}\text{U}$ and ends with $^{206}_{82}\text{Pb}$ and has only nuclei with mass numbers divisible by 4 with a remainder of 2. There is also a **4n + 3 series,** which starts with $^{235}_{92}\text{U}$ and ends with $^{207}_{82}\text{Pb}$ and contains only nuclides with mass numbers divisible by 4 with a remainder of 3. Interestingly enough, although no 4n + 1 series now exists in nature, all the necessary members of such a series have been produced by artificial means. It is believed that such a series may have once existed but has long since disappeared because no member of the scheme had a sufficiently long half-life to survive over the years except the last—the stable $^{209}_{83}\text{Bi}$ isotope. Incidently, no radioactively stable isotope exists for any element of atomic number greater than 83. The decay

$$^{232}_{90}\text{Th} \xrightarrow[1.4 \times 10^{10}\text{ years}]{\alpha} {}^{228}_{88}\text{Ra} \xrightarrow[6.7\text{ years}]{\beta} {}^{228}_{89}\text{Ac} \xrightarrow[6.1\text{ hr}]{\beta} {}^{228}_{90}\text{Th} \xrightarrow[1.9\text{ years}]{\alpha}$$

$$^{224}_{88}\text{Ra} \xrightarrow[3.6\text{ days}]{\alpha} {}^{220}_{86}\text{Rn} \xrightarrow[55\text{ sec}]{\alpha} {}^{216}_{84}\text{Po}
\begin{cases} \xrightarrow[0.16\text{ sec}]{99.99\%\ \alpha} {}^{212}_{82}\text{Pb} \xrightarrow[11\text{ hr}]{\beta} \\ \xrightarrow[0.16\text{ sec}]{\beta} {}^{216}_{85}\text{At} \xrightarrow[54\text{ sec}]{\alpha} \end{cases}$$

$$^{212}_{83}\text{Bi}
\begin{cases} \xrightarrow[61\text{ sec}]{66.3\%\ \beta} {}^{212}_{84}\text{Po} \xrightarrow[3 \times 10^{-7}\text{ sec}]{\alpha} \\ \xrightarrow[61\text{ sec}]{\alpha} {}^{208}_{81}\text{Tl} \xrightarrow[3.1\text{ min}]{\beta} \end{cases} {}^{208}_{82}\text{Pb}$$

Figure 28–1. The naturally occuring 4n series of radioactive decay. The symbol above the arrow shows the mode of decay and the half-life for each nuclide is given below the arrow.

schemes for the three naturally occurring series and the hypothetical $4n + 1$ series are presented in Figures 28–1 through 28–4.

The members of the naturally occurring series have reached a steady state so that their amounts are constant. The ratio of the amounts found in minerals has been used to estimate the age of the earth. For example, rocks containing uranium are found to have roughly equal molar ratios of $^{238}_{92}\text{U}$ and $^{206}_{82}\text{Pb}$. The longest-lived member of the $4n + 2$ series by far is $^{238}_{92}\text{U}$ itself with a half-life of 4.5×10^9 years, so it appears that about one half of the original $^{238}_{92}\text{U}$ has been converted into $^{207}_{82}\text{Pb}$. Assuming there has been no other source of this isotope of lead, the uranium has passed through approximately one half-life since it was formed. The age of the rock and, presumably, of the earth itself is thereby indicated to be 4 to 5 billion years.

When a radioactive element has passed through about ten half-lives, it is effectively gone. Because the longest-lived member of the hypothetical $4n + 1$

$$^{238}_{92}\text{U} \xrightarrow[4.5 \times 10^9\text{ years}]{\alpha} {}^{234}_{90}\text{Th} \xrightarrow[25\text{ days}]{\beta} {}^{234}_{91}\text{Pa} \xrightarrow[1.1\text{ min}]{\beta} {}^{234}_{92}\text{U} \xrightarrow[2.7 \times 10^5\text{ years}]{\alpha}$$

$$^{230}_{90}\text{Th} \xrightarrow[8.3 \times 10^4\text{ years}]{\alpha} {}^{226}_{88}\text{Ra} \xrightarrow[1.6 \times 10^3\text{ years}]{\alpha} {}^{222}_{86}\text{Rn} \xrightarrow[3.8\text{ days}]{\alpha} {}^{218}_{84}\text{Po}
\begin{cases} \xrightarrow[3.1\text{ min}]{99.96\%\ \alpha} \\ \xrightarrow[3.1\text{ min}]{\beta} \end{cases}$$

$$\begin{array}{l} {}^{214}_{82}\text{Pb} \xrightarrow[27\text{ min}]{\beta} \\ {}^{218}_{85}\text{At} \xrightarrow[2\text{ sec}]{\alpha} \end{array} {}^{214}_{83}\text{Bi}
\begin{cases} \xrightarrow[20\text{ min}]{99.96\%\ \beta} {}^{214}_{84}\text{Po} \xrightarrow[1.5 \times 10^{-4}\text{ sec}]{\alpha} \\ \xrightarrow[20\text{ min}]{\alpha} {}^{210}_{81}\text{Tl} \xrightarrow[1.3\text{ min}]{\beta} \end{cases} {}^{210}_{82}\text{Pb} \xrightarrow[22\text{ years}]{\beta}$$

$$^{210}_{83}\text{Bi}
\begin{cases} \xrightarrow[4.8\text{ days}]{\sim 100\%\ \beta} {}^{210}_{84}\text{Po} \xrightarrow[1.4 \times 10^2\text{ days}]{\alpha} \\ \xrightarrow[4.8\text{ days}]{\alpha} {}^{206}_{81}\text{Tl} \xrightarrow[4.2\text{ min}]{\beta} \end{cases} {}^{206}_{82}\text{Pb}$$

Figure 28–2. The naturally occuring $4n + 2$ series of radioactive decay. The symbol above the arrow shows the mode of decay and the half-life for each nuclide is given below the arrow.

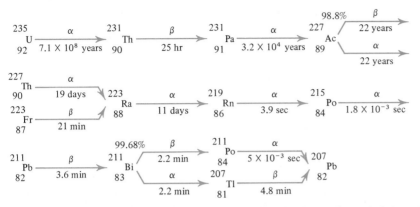

Figure 28-3. The naturally occuring $4n + 3$ series of radioactive decay. The symbol above the arrow shows the mode of decay and the half-life for each nuclide is given below the arrow.

series is $^{237}_{93}\mathrm{Np}$ with $t_{1/2} = 2.2 \times 10^6$ years, it is easy to see why this series no longer exists in nature (assuming it once did). The age of the earth is sufficient for $^{237}_{93}\mathrm{Np}$ to have passed through several thousand half-lives. Of course, it is quite possible for a very unstable element with a very short half-life to exist in nature if it is a member of one of the natural radioactive series. The only exceptions are those elements that appear as minor branched disintegration products. Astatine (At) is found in both the $4n$ and $4n + 2$ series as a branched product of the decay of polonium (Po) isotopes. However, the production of astatine accounts for far less than 1% of the decay products of polonium in either case. Such small traces have not yet been unequivocally detected in nature.

Only a very few radioactive nuclides that are not members of one of the three series are found in nature. One of the most interesting of these is the carbon isotope, $^{14}_6\mathrm{C}$. This species undergoes β decay with a half-life of 5770 years.

$$^{14}_6\mathrm{C} \longrightarrow {}_{-1}^{0}\mathrm{e} + {}^{14}_7\mathrm{N}$$

241
Pu $\xrightarrow[\text{13 years}]{\beta}$ 241
Am $\xrightarrow[\text{4.7} \times 10^2 \text{ years}]{\alpha}$ 237
Np $\xrightarrow[\text{2.2} \times 10^6 \text{ years}]{\alpha}$ 233
Pa $\xrightarrow[\text{27 days}]{\beta}$
94 · 95 · 93 · 91

233
U $\xrightarrow[\text{1.6} \times 10^5 \text{ years}]{\alpha}$ 229
Th $\xrightarrow[\text{7.3} \times 10^3 \text{ years}]{\alpha}$ 225
Ra $\xrightarrow[\text{15 days}]{\beta}$ 225
Ac $\xrightarrow[\text{10 days}]{\alpha}$
92 · 90 · 88 · 89

221
Fr $\xrightarrow[\text{4.8 min}]{\alpha}$ 217
At $\xrightarrow[\text{2} \times 10^{-2} \text{ sec}]{\alpha}$ 213
Bi · 98% β 47 min → 213
Po $\xrightarrow[\text{4.2} \times 10^{-6} \text{ sec}]{\alpha}$ / α 47 min → 209
Tl $\xrightarrow[\text{2.2 min}]{\beta}$
87 · 85 · 83 · 84 · 209 · 81

209
Pb $\xrightarrow[\text{3.3 hr}]{\beta}$ 209
Bi
82 · 83

Figure 28-4. The hypothetical $4n + 1$ series of radioactive decay. The symbol above the arrow shows the mode of decay and the half-life for each nuclide is given below the arrow.

The reason that this nuclide exists in nature is that it is continuously formed by the bombardment of nitrogen in the upper atmosphere by neutrons which are a component of cosmic radiation.

$$\frac{1}{0}n + \ _{7}^{14}N \longrightarrow \ _{6}^{14}C + \ _{1}^{1}H$$

In a nuclear reaction, the charges and mass numbers must balance just as in the decay of an unstable nucleus. The amount of $_{6}^{14}C$ produced by the above reaction is very small, but its reasonably long half-life has led to a steady-state concentration. All carbon found in nature, therefore, contains a small, but detectable, constant fraction of this isotope. About one carbon atom out of every 10^{12} is the $_{6}^{14}C$ isotope. Because carbon is found in all living cells, a feeble radioactivity is associated with cells due to the β decay of the $_{6}^{14}C$ isotope. As long as the cell is alive, the amount of $_{6}^{14}C$ in the total carbon content remains constant because it is continuously replenished by the natural production of $_{6}^{14}C$. When the cell dies, however, its carbon cycle is broken, and the $_{6}^{14}C$ content of the cell slowly begins to diminish. By comparing the radioactivity of some old wooden object, for instance, with that of living trees, the number of half-lives through which the object has passed since the tree was cut down can be calculated. In this way, old objects can be very accurately dated. This method of **radiocarbon dating** is limited to about 55,000 to 60,000 years maximum (about ten half-lives for $_{6}^{14}C$), but objects of archeological importance are often only several thousand years old.

28-2.   ARTIFICIAL NUCLEAR REACTIONS

Given the new α and β particles, scientists in the first few years of the century were armed with a great new weapon with which to examine further the properties of atomic nuclei. α particles emitted from radium have a velocity of about 10,000 miles/sec, that is, 20,000 times greater than the velocity of a rifle bullet. On a comparable mass basis the energies of α particles are over 400 million times as great as the bullet. Ernest Rutherford, who determined the nature of these particles, was the first to use them as projectiles. During the course of this work, Rutherford and his associates examined the scattering of α and β particles by a wide variety of substances. The scattering of α particles by metal foils and the consequences of the results were discussed in Section 8-6.

The scattering of α particles by gases also yielded very intriguing results. The apparatus used by Rutherford is shown schematically in Figure 28-5. When hydrogen gas was passed through the apparatus, occasional collision between the high velocity α particles and hydrogen nuclei produced hydrogen nuclei of velocities about four times as great as that of the α particles. By inserting sufficient mica sheets between the thin silver sheet and the zinc sulfide scintillation screen, these hydrogen nuclei (protons) could be stopped so that no scintillations were observed. Knowledge of the thickness of mica required to stop the protons afforded a measure of their energy. When oxygen was passed through the apparatus, only occasional protons were observed, and they were probably due to impurities in the radiation source. The use of air, however, produced three or four times as many scintillations. The increased number of scintillations, shown to be due to protons, was ascribed to the presence of nitrogen in the air, and,

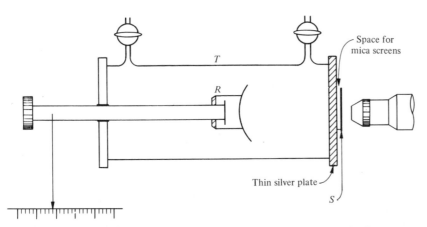

Figure 28-5. Rutherford's apparatus for studying the scattering of α particles by gases: α particles emitted by the radioactive source R are allowed to bombard the nuclei of gas which is passed through the cell T. The resulting particles pass through the thin silver plate striking the zinc sulfide scintillating screen S. The space between the silver screen and S may be filled with sheets of mica which absorb the radiation before it strikes S.

in fact, an even greater activity was observed with pure nitrogen gas. Moreover, it took a considerably greater thickness of mica to stop these protons than it did to stop the protons formed when hydrogen was used. Rutherford concluded that the protons produced in the bombardment of nitrogen originated in the nitrogen nucleus when it was struck by a fast-moving α particle. This nuclear disintegration occurred with such violence that the proton fragments were of much higher velocities than would result from the simple elastic collision of an α particle with a hydrogen nucleus.

Rutherford and his associates examined all the elements up to atomic weight 40 (calcium) with the exception of helium, neon, and argon. The heavier elements, that is, those above phosphorus, failed to give this effect, presumably because of the improbability of collision of the positively charged α particle with highly positively charged nuclei due to the high electrostatic repulsions. The most active elements were boron, nitrogen, fluorine, sodium, aluminum, and phosphorus. Thus **artificially induced atomic disintegrations** were observed for the first time.

Examples of the reactions discussed can be written using the notation given previously:

$$^4_2\text{He} + {}^{14}_7\text{N} \longrightarrow {}^1_1\text{H} + {}^{17}_8\text{O}$$

$$^4_2\text{He} + {}^{10}_5\text{B} \longrightarrow {}^1_1\text{H} + {}^{13}_6\text{C}$$

$$^4_2\text{He} + {}^{19}_9\text{F} \longrightarrow {}^1_1\text{H} + {}^{22}_{10}\text{Ne}$$

An alternate notation for nuclear reactions is often used in which the heavy nucleus under bombardment is represented followed by the bombarding particle and the light product particle enclosed in parentheses. Using a p to represent the proton, the above equations are represented, respectively by

$$^{14}_7\text{N}(\alpha, p) \qquad {}^{10}_5\text{B}(\alpha, p) \qquad {}^{19}_9\text{F}(\alpha, p)$$

The heavy product nucleus can be deduced by balancing the charges and mass numbers.

437

28-3. DISCOVERY OF THE NEUTRON

During the ten years that followed Rutherford's announcement in 1922 of artificial disintegration of atomic nuclei, many similar experiments were carried out by other workers. In particular, several scientists observed that some of the light elements, notably boron and beryllium, emit very high energy radiations that appeared to be similar to γ rays. Later, it was shown that the radiation excited in beryllium on bombardment by α particles had a penetrating power much greater than any γ radiation that had ever been observed in the radioactive elements. These radiations were soon shown to have other striking properties. For example, they were able to eject protons with considerable velocities from substances containing hydrogen, such as paraffin (a hydrocarbon with the approximate formula $C_{40}H_{82}$).

In 1932, James Chadwick, a former student of Rutherford, attempted a synthesis of these apparently irreconcilable observations, If the radiations were, in fact, electromagnetic, theoretical calculations of their interactions with matter required that the ejected protons have velocities quite different from those observed. If the radiation consisted of particles, however, the ejection of protons from paraffin could be explained nicely. But the particles, if they existed, could not be detected in the usual ways. They were unaffected by electric or magnetic fields and did not leave a track when introduced into a cloud chamber.* However, when a sheet of paraffin was placed in the path of the radiation, a large number of tracks were observed which were shown to be due to protons.

Chadwick showed that the interesting behavior of the radiation could be explained by assuming that it consisted of neutral particles of mass nearly equal to that of the proton. Rutherford had suggested the existence of such a particle, which he called a **neutron,** in 1920. Chadwick calculated the mass by comparing the velocity imparted to a proton with that imparted to a nitrogen nucleus. If an elastic collision is assumed, the velocity imparted by a moving object to a stationary object is inversely proportional to the sum of the two masses; that is

$$V_H \propto \frac{1}{(M_n + M_H)}$$

where V_H = velocity of the proton, and M_H and M_n are the masses of the proton and neutron, respectively, and

$$V_N \propto \frac{1}{(M_n + M_N)}$$

where V_N and M_N are the velocity and mass, respectively, of a nitrogen atom. Thus

$$\frac{V_H}{V_N} = \frac{(M_n + M_N)}{(M_n + M_H)}$$

---

*Wilson cloud chamber.** This apparatus consists of a drum filled with saturated water vapor through which radiation is allowed to pass. The path of a charged particle can be traced because it leaves a fog track in its wake. The close approach of a moving α-particle, for example, can occasionally cause ionization of nearby water molecules, which in turn attract other water molecules to them forming tiny droplets. The resulting fog track provides a visualization of the path of the α particle.

If we know the atomic masses of hydrogen and nitrogen and measure their velocities when bombarded by neutrons, the equation can readily be solved for the neutron mass, M_n.

$$M_n = \frac{V_N M_N - V_H M_H}{V_H - V_N}$$

The radiation was thus calculated to consist of particles of mass very nearly equal to that of the proton. Their high penetrating power suggested that they had no net charge. Chadwick supposed that the neutron was a proton and an electron in close combination, which is one of the currently held views on the nature of the neutron.

The sequence of events in the field of nuclear physics in the 1920s could only be surpassed in sheer excitement by those in the 1930s. And toward the end of the 1930s the intrigues of war entered the scene.

Enrico Fermi, in Italy, was the first to realize the significance of the penetrating power of neutrons in the study of nuclear structure. Rutherford had not been able to bombard elements heavier than potassium with α particles because of the high positive nuclear charges. Fermi now saw the neutron as a particle that would not be repelled by even the heaviest nuclei and might therefore be used to cause

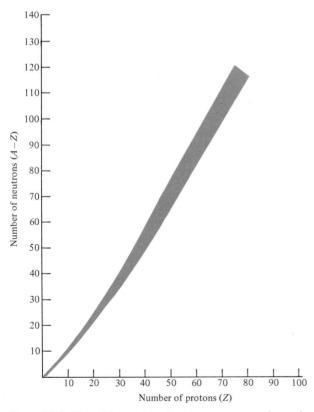

Figure 28-6. Plot of the number of neutrons versus atomic number for stable nuclides. Color region denotes the zone of nuclear stability.

disintegrations of these heavy elements. During the early thirties, he examined all of the elements from nitrogen to uranium (atomic numbers 7 to 92) and found all of them capable of yielding radioactive products when bombarded with neutrons. The results of Fermi's fascinating idea will be discussed in Section 28-7 after some necessary fundamental concepts are developed.

The relative numbers of neutrons and protons in a nuclide are found to be an index of its stability. This measure is referred to as the neutron-proton ratio. Through element number 20 (calcium), this ratio is roughly unity, that is, about equal numbers of neutrons and protons. Beyond element number 20, the ratio slowly increases until it reaches a value of about 1.5. Figure 28-6 shows a plot of the number of neutrons versus atomic number for stable nuclides. Upset of the neutron-proton ratio generally leads to nuclear instability. For example, neutron capture by a nucleus increases the ratio and the resulting nuclide will often decay in such a way as to decrease the ratio.

28-4. DECAY OF ARTIFICIAL RADIOACTIVE NUCLIDES

Decay of α and β emission has been discussed in connection with natural radioactivity. Artificially produced radioactive nuclides also decay by these modes, although α decay is normally observed with nuclei of mass numbers 210 or higher and atomic numbers 83 or higher. Loss of a β particle increases the atomic number by one unit and leaves the mass number unchanged; hence, β decay always decreases the neutron-proton ratio.

Two other decay processes are also observed with artificial radioactive nuclides. Positron emission consists of the ejection of a positive particle with a mass identical with that of the electron. Positrons were first observed by Carl Anderson in 1932 in a cloud chamber in which he was studying cosmic rays. K capture involves the capture of one of the 1s (K shell) orbital electrons of an atom by its nucleus. Loss of a nuclear positive charge in positron emission or gain of a negative charge by a nucleus in K capture, without a change in mass number, both lead to an increase in the neutron-proton ratio. Some examples of these two types of decay are the following:

$$\ce{^{19}_{10}Ne} \longrightarrow \ce{^{19}_{9}F} + \ce{^{0}_{+1}e}$$

$$\ce{^{15}_{8}O} \longrightarrow \ce{^{15}_{7}N} + \ce{^{0}_{+1}e}$$

$$\ce{^{55}_{26}Fe} + \ce{^{0}_{-1}e} \xrightarrow{K} \ce{^{55}_{25}Mn}$$

$$\ce{^{106}_{47}Ag} + \ce{^{0}_{-1}e} \xrightarrow{K} \ce{^{106}_{46}Pd}$$

28-5. ENERGY CHANGES IN NUCLEAR REACTIONS

One of Lavoisier's greatest contributions to chemistry was the law of conservation of mass. During chemical reactions matter is neither created nor destroyed. The energies absorbed or released during chemical changes are primarily due to differences in bonding energies on going from reactants to products.

The study of nuclear reactions now brought complications into the picture.

For example, neutrons outside the nucleus are known to decay into protons and electrons according to the equation

$$\ce{^1_0 n} \longrightarrow \ce{^1_1 H} + \ce{^0_{-1} e} + \text{energy}$$

But the sum of the masses of a proton and an electron does not quite equal the mass of the neutron:

$\ce{^1_0 n}$: mass = 1.00866 atomic mass units (amu)

$\ce{^1_1 H}$: mass = 1.00728 amu

$\ce{^0_{-1} e}$: mass = 0.0005486 amu

The difference (ΔM) is approximately 0.0008 amu or about 1.3×10^{-27} g. Einstein's equivalence of mass and energy states that the amount of energy, E, produced when a given amount of mass, m, is lost is given by the equation $E = mc^2$, where c is the velocity of light. Taking this into account, the destruction of 0.0008 amu of matter in the above reaction should liberate 2.8×10^{-14} cal/neutron, or 1.7×10^7 kcal/mole of neutrons, which is very close to the observed quantity of energy released. It must be emphasized that the release of energy in such a process is quantitatively different from that observed in ordinary chemical reactions. For example, the complete combustion of 1 g of carbon releases only 7.8 kcal; less than one millionth the amount obtained in the nuclear reaction.

$$C + O_2 \longrightarrow CO_2 + \text{energy}$$

An energy unit useful for nuclear processes is **million electron volts** (Mev). One electron volt (ev) is the kinetic energy acquired by an electron when it moves through an electrical potential difference of 1 volt. In terms of the equivalence of mass expressed in atomic mass units and energy, it can be shown by the Einstein equation that

$$931 \text{ Mev} = 1.000 \text{ amu}$$

If 1 mole of particles possessed 1 Mev of energy, this would be equivalent to 2.3×10^7 kcal. The million electron volt is thus seen to be a fairly large energy unit and therefore quite suitable for the energies encountered in nuclear reactions.

28-6. BINDING ENERGY AND THE MASS DEFECT

As a result of the work of nuclear physicists during the first 30 years of the twentieth century, it became established that atoms are composed of protons and neutrons in the nucleus and "planetary" electrons beyond the nucleus. Thus all the observed nuclear transformations by which radioactive atoms disintegrated into daughter atoms and elementary particles could be understood. But in nearly all such nuclear transmutations the two sides of the equation do not balance exactly in their mass because photons of energy (γ rays) are nearly always released as the result of the destruction of a small amount of mass.

Moreover if one sums up the masses of the protons, electrons, and neutrons that combine to form a given atom, this sum is never equal to the actual atomic mass of the nuclide formed. An example is the most stable isotope of sodium, $^{23}_{11}\text{Na}$. The sum of 11 protons, 12 neutrons and 11 electrons is

$$11 \times {}^{1}_{1}\text{H} = 11(1.00728) = 11.08008$$

$$12 \times {}^{1}_{0}n = 12(1.00866) = 12.10392$$

$$11 \times {}_{-1}^{0}e = 11(0.0005486) = \underline{0.00603}$$

$$\text{Sum} = 23.19003$$

$$\text{Actual atomic mass of } {}^{23}_{11}\text{Na} = \underline{22.9898}$$

$$\text{Mass defect} = 0.2002 \text{ amu}$$

$$\text{Binding energy} = 0.2002 \times 931 = 186 \text{ Mev}$$

$$\text{Binding energy per nucleon} = 186/23 = 8.1 \text{ Mev}$$

The difference between the sum of the masses of the constituent particles and that of the actual atom is called the **mass defect.** The mass defect may be thought of as the mass that would be converted into energy if the correct numbers of particles could combine in exactly the right way. As such, the mass defect represents the stability of the atom compared with the separated protons, neutrons, and electrons. The energy equivalent of the mass defect is called the **binding energy.** The total binding energy divided by the number of nucleons is useful for the comparison of nuclear stabilities. The binding energy per nucleon shows an interesting variation with atomic mass. As shown in Figure 28–7 the binding

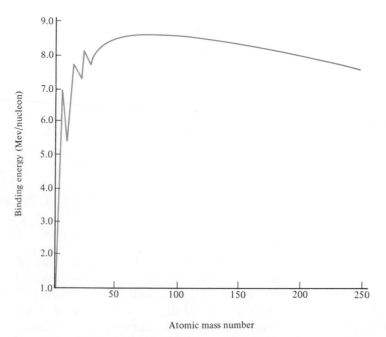

Figure 28-7.   Variation of nuclear binding energy with atomic mass.

energy reaches a maximum at atomic mass number 75 and then drops with the heavier atoms.

An examination of the 274 naturally occurring stable isotopes leads to the following observations:

Isotopes with even numbers of protons and even numbers of neutrons = 162
Isotopes with odd numbers of protons and even numbers of neutrons = 52
Isotopes with even numbers of protons and odd numbers of neutrons = 56
Isotopes with odd numbers of protons and odd numbers of neutrons = 4

Or, in other terms:

78% of all isotopes contain even numbers of neutrons.
80% contain even numbers of protons.
88% of the earth's crust is made up of elements with even numbers of protons and neutrons.
0.0007% of the earth's crust is made up of stable elements with odd numbers of protons and neutrons. Those are ^2_1H, ^6_3Li, $^{10}_5B$, and $^{14}_7N$.
98% of the earth's crust is made up of elements with even numbers of neutrons.

The apparent stability of nuclides with even numbers of protons and neutrons has led to a theory of nuclear structure in which energy levels are presumed to exist in the nucleus somewhat analogous with the energy levels for the orbital electrons of atoms. It is assumed that the nucleons fill their energy levels in regular fashion and that the completion of a given energy level results in a nuclide of enhanced stability.

28-7. NUCLEAR FISSION

The experiments of Enrico Fermi led him, toward the end of the 1930s, to the realization that, if slow neutrons were used to bombard a heavy atom like $^{238}_{92}U$, the possibility existed that a neutron might be captured. The resulting nucleus might be less stable due to the increased neutron-proton ratio and β emission should occur. The resulting nuclide would be the hitherto undiscovered element number 93.

$$^{238}_{92}U + {}^1_0n \longrightarrow {}^{239}_{92}U \longrightarrow {}^{\ 0}_{-1}e + {}^{293}_{93}X$$

By repetition or by the capture of several neutrons, perhaps a whole series of **transuranium** elements could be synthesized.

Fermi attempted this experiment and the result was startling! At first he believed that element 93 had been produced, but the element produced was shown not to be an isotope of uranium nor was it an isotope of any element near uranium. Up to that time, only α and β particles and neutrons had been observed to be emitted during radioactive decay, and the inability to find among the decay products an element near uranium in the periodic table was quite confusing. Even if several α particles had been ejected, the product should be between 84 and 92. Curie and Savitch, in 1938, repeated the experiment and concluded that the element behaved like an alkaline earth; that is, it belonged in group II of the periodic table. Hahn and Strassman in Germany very soon afterwards again repeated the experiment with painstaking care and came to the unambiguous conclusion that the product was barium, atomic number 56.

It should be recalled that World War II was in its embryonic stages in 1938. Professor Hahn, a chemist in Germany, had the collaboration of a very able physicist, Lise Meitner, an Austrian-born Jew. When Hitler's army invaded Austria and declared it part of the German Reich, Professor Meitner immediately became subject to the German racial laws. She lost no time in severing her many years' association with Dr. Hahn, and, with the help of Niels Bohr in Denmark, managed to escape to Sweden. There, she heard of the news that Hahn and Strassmann had demonstrated that at least one of the products of the neutron bombardment of uranium was barium.

Dr. Meitner conceived the notion that the nucleus had split into two nuclei of roughly equal mass. Together with her nephew, O. R. Frisch, she worked out the problems inherent in such a theory and performed experiments that confirmed it. The major problem was in the large energies that theory demanded must be evolved. This was observed as calculated—about 200 Mev per nucleus split. They gave the process the name **fission** because of its apparent similarity to biological cell division (Figure 28–8). Later, it was found that only the $^{235}_{92}U$ isotope in the natural uranium was fissioned by the fairly low energy neutrons employed in these early experiments.

The discovery of nuclear fission had enormous consequences for the world. Within a matter of months, nearly all the nuclear scientists in the world were engaged in a race to achieve the practical utilization of this great energy release. By this time, Dr. Fermi had fled Italy and had resumed his research at Columbia University from which he was soon to move to the University of Chicago. There he and his colleagues built the first self-sustaining nuclear reactor.

A fission reaction becomes self-sustaining when there are enough fissionable nuclei present for one or more of the neutrons released in each fission to strike another nucleus and continue the reaction. A chain reaction is thereby established. The minimum amount of fissionable material necessary for such a reaction is called the **critical mass.** Although $^{238}_{92}U$ is fissioned by high energy (fast) neutrons,

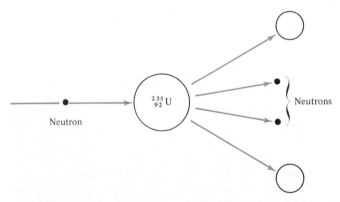

Figure 28–8. Fission of $^{235}_{92}U$ nucleus by bombardment of a neutron yields two new nuclei of masses approximately one half that of the uranium, and several additional neutrons.

it is not split by low energy (slow) neutrons. However, the $^{235}_{92}$U isotope is capable of undergoing fission by a slow neutron. A chain reaction does not occur in nature because the $^{238}_{92}$U isotope accounts for more than 99% of the isotopic composition of uranium. Although an occasional fission of a $^{235}_{92}$U nucleus can occur due to stray neutrons, there is (fortunately!) simply not a high enough concentration of the fissionable nuclide (subcritical mass) to give rise to a self-sustaining reaction.

One effort in the early work on the atomic bomb was directed toward the separation of pure $^{235}_{92}$U from natural uranium. A diffusion method was developed which accomplished this goal, but it is a fairly slow and laborious process. Atomic fuel for reactors can be produced from the abundant $^{238}_{92}$U isotope by the capture of slow neutrons and the subsequent decay into a plutonium isotope.

$$^{238}_{92}\text{U} + {}^{1}_{0}n \xrightarrow[\text{slow}]{} {}^{239}_{92}\text{U} \longrightarrow {}^{239}_{93}\text{Np} + {}^{-0}_{1}e$$
$$\longrightarrow {}^{239}_{94}\text{Pu} + {}^{0}_{-1}e$$

The $^{239}_{93}$Np isotope has a half-life of only 2.3 days, but the $^{239}_{94}$Pu isotope has $t_{1/2} = 2.4 \times 10^4$ years. This latter nuclide is fissioned by slow neutrons and thus can be used conveniently as an atomic fuel.

28-8. FUSION PROCESSES

The energy of the sun and other stars is thought to arise from nuclear **fusion** reactions in which lighter nuclei are combined to form a heavier nucleus. The solar fusion reaction involves the fusion of protons to α particles with the liberation of positrons and vast amounts of energy.

$$4\,{}^{1}_{1}\text{H} \longrightarrow {}^{4}_{2}\text{He} + 2\,{}^{0}_{+1}e$$

The source of the energy in this process is the binding energy of the α particle— about 28 Mev per α particle. For each *gram* of hydrogen consumed in the reaction, *162 million* kcal is released. Less than 20 million kcal is released in the fission of 1 g of $^{238}_{92}$U. The fusion process is thus seen to be capable of liberating much more energy per unit weight of fuel than the fission process.

The mechanism of a fusion reaction such as the one above is quite complex, however. One mechanism that has been suggested for the solar fusion reaction is the following:

Step 1.	$^{12}_{6}\text{C} + {}^{1}_{1}\text{H} \longrightarrow {}^{13}_{7}\text{N}$
Step 2.	$^{13}_{7}\text{N} \longrightarrow {}^{13}_{6}\text{C} + {}^{0}_{+1}e$
Step 3.	$^{13}_{6}\text{C} + {}^{1}_{1}\text{H} \longrightarrow {}^{14}_{7}\text{N}$
Step 4.	$^{14}_{7}\text{N} + {}^{1}_{1}\text{H} \longrightarrow {}^{15}_{8}\text{O}$
Step 5.	$^{15}_{8}\text{O} \longrightarrow {}^{15}_{7}\text{N} + {}^{0}_{+1}e$
Step 6.	$^{15}_{7}\text{N} + {}^{1}_{1}\text{H} \longrightarrow {}^{12}_{6}\text{C} + {}^{4}_{2}\text{He}$

$$4\,{}^{1}_{1}\text{H} \longrightarrow {}^{4}_{2}\text{He} + 2\,{}^{0}_{+1}e$$

445

This mechanism is reasonable because simple, two-body collisions are assumed in steps 1, 3, 4, and 6 involving hydrogen—the principal element in the sun. The spontaneous decay of unstable nuclides in steps 2 and 5 is reasonable because the half-lives are known; $^{13}_{7}N$, $t_{1/2} = 10$ min and $^{15}_{8}O$, $t_{1/2} = 2.1$ min. The $^{12}_{6}C$ isotope has been identified spectroscopically as present on the sun and its function as a catalyst in the proposed scheme is reasonable.

Man-made fusion reactions have not utilized the fusion of pure hydrogen to helium. The activation energy of this reaction is too high for it to be practical. However, hydrogen bombs have been exploded, probably using hydrogen and a lithium isotope. The reaction

$$^{1}_{1}H + ^{7}_{3}Li \longrightarrow 2\,^{4}_{2}He$$

produces 8.7 Mev per α particle rather than the 28 Mev per α particle in the fusion of pure hydrogen. Nevertheless, it can be seen that the reaction is quite exothermic. A very high activation energy is required for this reaction, however, and the detonation of a small fission device has been the most successful technique yet devised for supplying the necessary high temperatures. High energy laser and electron beams are also under study as possible sources of the activation energy. Needless to say, the development of controlled fusion processes presents a staggering problem to the nuclear engineer.

28-9. MODERN EXPERIMENTAL METHODS OF PRODUCTION AND DETECTION OF NUCLEAR PARTICLES

Early workers in nuclear phenomena utilized the α and β particles only at the energies obtained from naturally occurring radioactive decay. The acceleration of charged particles provides a much wider range of experimental conditions. Acceleration is readily accomplished by means of an electric field, but acceleration to speeds required to carry out various nuclear reactions requires special techniques. One such instrument, which was developed in the early 1930s, is the **cyclotron.**

Figure 28-9 is a schematic representation of the **dees** of a cyclotron. The dees are D-shaped hollow boxes that are placed between the poles of a large magnet. The entire apparatus is inside an evacuated chamber. A radiofrequency oscillator is connected to the dees so that the electrical polarity of one dee is opposite the other and is alternately switched from positive to negative. A charged particle, for example, an α particle, is generated at the center and is attracted to the negatively charged dee and repelled by the positive dee. Under the influence of the magnetic field, the particle will travel a curved path. The polarity of the dees is now reversed so that, as the particle emerges from the first dee, it is attracted by the other dee and repelled by the first dee. While inside a dee, the particle experiences no electrical forces. The speed of the particle is thus increased each time it crosses the gap between the dees. As its speed increases, it will travel a path of increasing radius of curvature until it finally reaches the outer edge of a dee and leaves the instrument to strike the target.

A later variation of the cyclotron is the **synchrotron.** This instrument utilizes

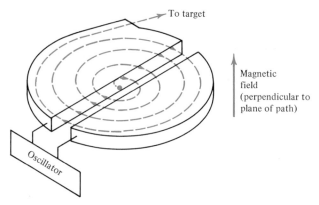

To target

Magnetic field (perpendicular to plane of path)

Oscillator

Figure 28-9. Schematic representation of the dees of a cyclotron.

the same basic principle of the cyclotron to accelerate the particles. However, instead of an ever increasing radius of curvature, the particles are made to follow a path of constant radius by progressively increasing the magnetic field strength. The physical size of the magnets may therefore be greatly reduced since the magnetic field is necessary only around the ring in which the particles move.

No magnetic fields are employed in the **linear accelerator.** Figure 28-10 illustrates the principle of the linear accelerator, in which the particles travel in a straight path. A series of tubes called drift tubes are suspended in an evacuated tank. A high frequency oscillator is connected to the drift tubes so that, as a positive particle enters the first drift tube, the tube is negatively charged. As the particle emerges from the drift tube, the polarity is switched to positive while that of the second drift tube is made negative. In this way, the particle is accelerated as it moves along its path. Because the particle moves faster and faster, the length of succeeding drift tubes is increased so that the times when the particle crosses the gaps between the drift tubes remain synchronized with the oscillator frequency. Electrons possessing thousands of million electron volts can be produced using linear accelerators several miles in length.

A number of instruments are available for the quantitative detection of radiation. Many of these devices operate on the principle that radiation leads to ionization of materials and the ions thus produced are detected. The Wilson

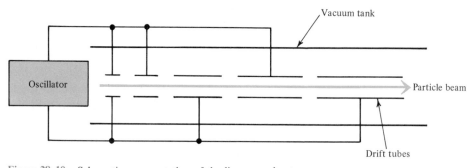

Vacuum tank

Oscillator

Particle beam

Drift tubes

Figure 28-10. Schematic representation of the linear accelerator.

447

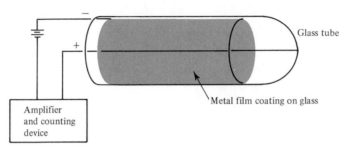

Figure 28-11. Schematic representation of a Geiger counter.

cloud chamber has already been mentioned in this regard. Water vapor and the molecules in air are ionized by radiation, and the condensation of water vapor on these ions causes a fog track of the path of the original radiation. These fog tracks are usually photographed so that they can be studied with regard to the length of the paths and the effects of external magnetic and electric fields on the radiation.

The detection of the ionization produced in a substance when radiation passes through it is usually done by electronic means, however. In the Geiger counter (Figure 28-11), a gas such as argon is contained at low pressure in a tube within which a high electrical potential difference is maintained. A single α or β particle entering the tube will cause ionization of some of the argon atoms. These few electrons will be rapidly accelerated toward the negative electrode and will strike many more argon atoms and produce an avalanche of ionization. The electrical impulse thus produced is then amplified electronically.

Scintillation counters are widely used for the detection of γ rays. One such instrument employs sodium iodide containing a small amount of thallium iodide. When a γ ray is stopped by this material, a weak flash of light is emitted whose intensity is proportional to the energy of the γ ray. A photomultiplier converts this light into an electrical impulse whose size is proportional to the intensity of the light. By this means, the energy of the impinging γ ray can be determined.

28-10. EXTENSION OF THE PERIODIC TABLE

The transuranium elements have been prepared by methods similar to Fermi's original notion. Glenn Seaborg prepared plutonium (at. no. 94) by the bombardment of $^{238}_{92}U$ with deuterons ($^{2}_{1}H$), an isotope of hydrogen. The subsequent experiments during the early 1940s brought a host of new discoveries that became legend within just a few years.

To date at least one isotope of each element up to atomic number 105 has been prepared synthetically by variations of Fermi's idea. The preparation of these heavy elements is currently being investigated by Albert Ghiorso and his collaborators at the University of California and by a Russian group under the leadership of G. N. Flerov.

At least one isotope of each of the elements from 92 to 105 is shown in Table 28-1 along with some of its properties.

Table 28–1. Some Isotopes of the Transuranium Elements

Name	Half-Life	Source
uranium	$^{238}_{92}U$ $\quad$ 4.5×10^9 years	pitchblende and other ores
neptunium	$^{237}_{93}Np$ $\quad$ 2.2×10^6 years	$^{238}_{92}U + {}^1_0n \longrightarrow {}^{237}_{92}U + 2\,{}^1_0n$ $\quad\quad\quad\quad\quad \rightarrow {}^{237}_{93}Np + {}^{\;\;0}_{-1}e$
plutonium	$^{239}_{94}Pu$ $\quad$ 2.4×10^4 years	traces in pitchblende and from reaction: $^{238}_{92}U + {}^1_0n \longrightarrow {}^{239}_{93}Np + {}^{\;\;0}_{-1}e$ $\quad\quad\quad\quad\quad \rightarrow {}^{239}_{94}Pu + {}^{\;\;0}_{-1}e$
americium	$^{241}_{95}Am$ $\quad$ 4.6×10^2 years	$^{239}_{94}Pu + {}^1_0n \longrightarrow {}^{240}_{94}Pu \xrightarrow{+{}^1_0n} {}^{241}_{94}Pu \longrightarrow {}^{241}_{95}Am + {}^{\;\;0}_{-1}e$
curium	$^{242}_{96}Cm$ $\quad$ 1.6×10^2 days	$^{241}_{95}Am + {}^1_0n \longrightarrow {}^{242}_{95}Am \longrightarrow {}^{242}_{96}Cm + {}^{\;\;0}_{-1}e$
berkelium	$^{249}_{97}Bk$ $\quad$ 3.1×10^2 days	$^{249}_{96}Cm$ (formed by neutron capture of $^{242}_{96}Cm$) $\longrightarrow {}^{249}_{97}Bk + {}^{\;\;0}_{-1}e$
californium	$^{252}_{98}Cf$ $\quad$ 2.6 years	$^{243}_{95}Am$ + prolonged ${}^1_0n$ bombardment $\longrightarrow {}^{252}_{98}Cf$
einsteinium	$^{254}_{99}Es$ $\quad$ 2.7×10^2 days	first identified in the debris from the "Mike" thermonuclear explosion in Eniwetok in 1952 along with fermium
fermium	$^{254}_{100}Fm$ $\quad$ 3.3 h	(See einsteinium above) and $^{254}_{99}Es \longrightarrow {}^{254}_{100}Fm + {}^{\;\;0}_{-1}e$
mendelevium	$^{256}_{101}Md$ $\quad$ 1.5 h	$^{253}_{99}Es + {}^4_2He \longrightarrow {}^{256}_{101}Md + {}^1_0n$
nobelium	$^{254}_{102}No$ $\quad$ 55 sec	$^{242}_{96}Cm + {}^{12}_6C \longrightarrow {}^{254}_{102}No$
lawrencium	$^{256}_{103}Lw$ $\quad$ 45 sec	$^{98}Cf + ({}^{10}_5B + {}^{11}_5B) \longrightarrow {}^{256}_{103}Lr$
104	4 min	$^{249}_{98}Cf + {}^{12}_6C \longrightarrow {}^{257}_{104}X + 4\,{}^1_0n$
105		$^{249}_{98}Cf + {}^{15}_7N \longrightarrow {}^{260}_{105}Z + 4\,{}^1_0n$

PROBLEMS

1. Explain the chemical relationship between a radioactive element and its daughter element when α decay and β decay occur.

2. What modes of decay occur with artificial radioactive nuclides that are not normally observed in naturally occurring radioactive nuclides?

3. What is the difference between a fission reaction and a fusion reaction? What do they have in common?

4. In what region of the periodic table are found the elements with the most stable nuclei, that is, the highest binding energy per nucleon?

5. Fill in the gaps in the following equations for nuclear reactions.

(a) $^{28}_{14}Si + ^{1}_{0}n \longrightarrow ^{28}_{13}Al +$ (e) $^{40}_{19}K \longrightarrow ^{40}_{20}Ca +$

(b) $^{25}_{12}Mg + ^{4}_{2}He \longrightarrow ^{1}_{1}H +$ (f) $^{9}_{4}Be + ^{4}_{2}He \longrightarrow ^{12}_{6}C +$

(c) $^{27}_{13}Al + \longrightarrow ^{28}_{13}Al + ^{1}_{1}H$ (g) $^{224}_{88}Ra \longrightarrow ^{4}_{2}He +$

(d) $^{209}_{83}Bi + ^{2}_{1}H \longrightarrow ^{210}_{84}Po +$

6. Write balanced equations for the following nuclear transformations.

(a) $^{14}_{7}N(\alpha, p)$ (e) $^{107}_{47}Ag(n, 2n)$

(b) $^{27}_{13}Al(\alpha, n)$ (f) $^{40}_{18}Ar(\alpha, p)$

(c) $^{23}_{11}Na(n, \alpha)$ (g) $^{24}_{12}Mg(d, \alpha)$

(d) $^{35}_{17}Cl(n, \alpha)$ (h) $^{7}_{3}Li(p, n)$

7. Explain why $^{227}_{89}Ac$ with a half-life of 22 years is found in nature whereas $^{99}_{43}Tc$ with a half-life of 2.1×10^5 years is not.

8. Cite evidence for the existence of a $4n + 1$ decay series in the past.

9. The half-life of a certain nuclide is 19 sec. Identify this nuclide from among the three possibilities whose decay constants are $^{10}_{6}C$, $k = 3.64 \times 10^{-2}$ sec^{-1}; $^{9}_{3}Li$, $k = 4.06$ sec^{-1}; $^{14}_{8}O$, $k = 9.07 \times 10^{-3}$ sec^{-1}.

10. Three stable isotopes of oxygen are $^{16}_{8}O$, $^{17}_{8}O$, and $^{18}_{8}O$. Use the neutron-proton ratio to explain which of the unstable isotopes $^{14}_{8}O$ and $^{19}_{8}O$ is a positron emitter and which is a β emitter.

11. Fragments from a fossilized creature gave 3.8 disintegrations per minute per gram of carbon. If the steady-state level of $^{14}_{6}C$ in living organisms gives 15.3 disintegrations per minute per gram of carbon, how long ago did the creature live?

12. Explain why there are very few unstable nuclides of low atomic number whereas all nuclides with atomic numbers larger than 83 are unstable.

13. (a) Compute the mass defect, binding energy, and binding energy per nucleon for the deuteron, $^{2}_{1}H$. ($^{2}_{1}H = 2.01410$ amu)

(b) How do you account for the stability of the deuteron in view of your answer in part (a)?

14. A third isotope of hydrogen is tritium, $^{3}_{1}H$. This isotope is unstable; predict its decay behavior.

15. Outline Fermi's method for producing transuranium elements. Why did he fail?

16. Show by means of nuclear equations how $^{238}_{92}U$ can be converted into readily fissionable atomic fuel.

17. Explain why a self-sustaining fission reaction of $^{235}_{92}U$ does not occur in nature.

18. Calculate the energy that would be released in the fusion process

$$^{3}_{1}H + ^{2}_{1}H \longrightarrow ^{4}_{2}He + ^{1}_{0}n$$

($^{3}_{1}H = 3.01605$ amu, $^{2}_{1}H = 2.01410$ amu, $^{4}_{2}He = 4.00260$ amu)

19. Explain the difficulties encountered in the discovery of the neutron. Why is the neutron a better bombarding particle for heavy atoms than the α particle?

20. Explain the differences and similarities between the cyclotron and the linear accelerator.

21. Outline several possible methods for the production of isotopes of element number 110.

CHEMISTRY IN THE SERVICE OF MAN

4

TECHNOLOGICAL PROCESSES

Would there be a science of chemistry without a chemical technology? The question has two meanings: It asks if chemical science could exist without the feedback and stimulus from the industry. It also asks the more philosophical question of whether the science would be pursued were it not for the possibility of any kind of profitable return from its application to real problems.

The first question is relatively easy to answer. Certainly chemical technology has played a major role in stimulating basic research, especially in modern times. But it should not be forgotten that behind each scientist of any historical age was the underlying impetus of the utility of his science. The ancients were concerned with practical matters of dyeing, warfare, glasses and soap making, medicinal substances, and other chemical problems; the alchemists were concerned with eternal life and gold. Knowledge gained in such technical pursuits fed the growing infant chemistry until it had developed in the eighteenth century into a mature science capable of sustaining its own growth.

As to the second question, it is difficult to speculate on the possibility of a strictly pure science with no utility whatever because such has probably never existed. Certainly the motivation of scientists takes many forms.

There can be no question that pure science and technology are very closely coupled and that their mutual effects are profound. The present chapter discusses only a few industrial processes. The particular processes have been included not only for their importance but also to show the diversity of pursuits in which chemistry plays a part.

29-1. NYLON

Ever since the recognition of polymers, subject to exploitation by synthetic chemists, one of the primary goals has been the imitation of natural products such

453

as rubber, wool, and silk. One of the outstanding polymer chemists in America was Wallace H. Carothers who worked for the DuPont Company during the 1930s. One of his principal tasks was the synthesis of a substitute for silk. By the time of his death, Carothers and his coworkers had laid the foundation for the production of the now common fibers and plastics.

His most noteworthy product, of course, was nylon, his silk substitute. Recall that silk is a polypeptide (Chapter 16). The production of a synthetic material with similar properties would thus be approached by attempts to prepare polypeptides from cheap raw materials so that the synthetic silk could compete with the natural product. Of course, the impending World War II emphasized the urgency. Carothers took two directions: (1) the polymerization of an amino acid, and (2) the polymerization of a diamine with a diacid.

$$(1) \quad n \, H_2N\!-\!R\!-\!\overset{\displaystyle O}{\overset{\|}{C}}\!-\!OH \longrightarrow \left(\!NH\!-\!R\!-\!\overset{\displaystyle O}{\overset{\|}{C}}\!\right)_{\!n} + n \, H_2O$$

$$(2) \quad n \, H_2N\!-\!R\!-\!NH_2 + n \, HO\!-\!\overset{\displaystyle O}{\overset{\|}{C}}\!-\!R'\!-\!\overset{\displaystyle O}{\overset{\|}{C}}\!-\!OH \longrightarrow$$

$$\left(\!NH\!-\!R\!-\!NH\!-\!\overset{\displaystyle O}{\overset{\|}{C}}\!-\!R'\!-\!\overset{\displaystyle O}{\overset{\|}{C}}\!\right)_{\!n} + n \, H_2O$$

In either case, the same type of structure (polypeptide) results. After several years of study two satisfactory products resulted. The best known is nylon-66, so named because it is prepared from starting materials each of which has six carbon atoms.

$$\left(\!\overset{\displaystyle H}{\overset{|}{N}}\!-\!CH_2CH_2CH_2CH_2CH_2CH_2\!-\!\overset{\displaystyle N}{\overset{|}{N}}\!-\!\overset{\displaystyle O}{\overset{\|}{C}}\!-\!CH_2CH_2CH_2CH_2\overset{\displaystyle O}{\overset{\|}{C}}\!\right)_{\!n}$$

nylon-66

Another important commercial product is nylon-6, which is an example of a polyamino acid. It is prepared from the cyclic compound ε-caprolactam. Ultimately, this starting material may be manufactured from the same basic raw materials as are used to make nylon-66 (see discussion following). Only the production of nylon-66 will be discussed here.

The actual manufacture of nylon-66 employs the 1:1 mixture of hexamethylenediamine and adipic acid. Because these compounds are a base and an acid

$$NH_2(CH_2)_6NH_2 + HO\overset{\displaystyle O}{\overset{\|}{C}}\!-\!(CH_2)_4\overset{\displaystyle O}{\overset{\|}{C}}\!-\!OH \longrightarrow$$

$$\underbrace{\overset{\displaystyle +}{N}H_3\!-\!(CH_2)_6\!-\!\overset{\displaystyle +}{N}H_3 + {}^-O\!-\!\overset{\displaystyle O}{\overset{\|}{C}}\!-\!(CH_2)_4\!-\!\overset{\displaystyle O}{\overset{\|}{C}}\!-\!O^-}_{\text{"salt"}}$$

respectively, this 1:1 mixture is a salt. After purifying the salt, it is placed in a reactor and heated to a temperature of about 270°C. When all the moisture and air have been purged out of the reactor, it is sealed and the pressure is allowed to build up to about 250 lb/in.² for a few hours. During this time water is lost on conversion of the ammonium salt to amide

$$R-\overset{O}{\overset{\|}{C}}-\overset{-}{O}\ \overset{+}{NH_3}-R' \longrightarrow R-\overset{O}{\overset{\|}{C}}-NH-R' + H_2O$$

On cooling, the molten polymer is extruded into ribbons which are cut up into flakes or chips for use in the next stage.

The fiber is produced by remelting and extruding the molten polymer into thin filaments, which are spun onto spools. During the melt-spinning process, the filament can be stretched to the desired length to add strength. Normally, stretching is to 3 to 6 times the original length (Chapter 16). The greater the amount of stretching, the greater the strength due to increased orientation of the crystalline regions in the polymer.

In addition to fibers, the remolten polymer can be molded into machine parts such as gears.

29-2. STARTING MATERIALS FOR NYLON MANUFACTURE

Although other starting materials have been used in the past, the most important at present are benzene and butadiene obtained from coal or petroleum. It must be remembered that in commercial manufacturing, even a few cents per pound in production costs can make the difference between success or failure of a product. Thus, the final choice of nylon-66 instead of some other combination may have been made on the basis of costs of raw material, since undoubtedly other diamines and diacids would have yielded equally good polymers.

Benzene is hydrogenated catalytically to produce cyclohexane, which on air oxidation yields a mixture of the alcohol, cyclohexanol, and the ketone, cyclohexanone. This mixture is then oxidized with nitric acid to form adipic acid.

| benzene | cyclo-
hexane | | cyclo-
hexanone | cyclo-
hexanol | adipic acid
+ N_2O + NO |

On reaction with ammonia, adipic acid is converted into the amide, which is then dehydrated to the nitrile. Hydrogenation of the nitrile under high pressure yields hexamethylenediamine (see formulas on page 456).

The method employing butadiene involves chlorination to form the dichloride followed by reaction with sodium cyanide to form the nitrile. Note that a by-product is the very stable sodium chloride. The dicyanide containing a carbon-carbon double bond is then hydrogenated and the resulting adiponitrile may be

$$\underset{\substack{\text{adipic acid}}}{HO\overset{\overset{\displaystyle O}{\|}}{C}(CH_2)_4\overset{\overset{\displaystyle O}{\|}}{C}OH} + \underset{\substack{\text{ammonia}}}{2\,NH_3} \longrightarrow$$

$$\underset{\substack{\text{adipamide}}}{H_2N\overset{\overset{\displaystyle O}{\|}}{C}(CH_2)_4\overset{\overset{\displaystyle O}{\|}}{C}NH_2} \longrightarrow \underset{\substack{\text{adiponitrile}}}{N\equiv C(CH_2)_4C\equiv N} \xrightarrow[\text{pressure}]{H_2,\ NH_3} \underset{\substack{\text{hexamethylenediamine}}}{H_2N(CH_2)_6NH_2}$$

either converted into hexamethylenediamine as above, or converted into adipic acid by treatment with water, the reverse of the above reaction.

$$\underset{\substack{\text{1,3-butadiene}}}{\overset{\displaystyle CH_2}{\underset{\displaystyle CH-CH}{\|}}\overset{\displaystyle CH_2}{\|}} \xrightarrow{Cl_2} \underset{\substack{\text{1,4-dichloro-2-butene}}}{Cl-CH_2 \qquad CH_2-Cl \atop CH=CH} \xrightarrow[\text{CuCl, 80–95°C}]{2\,NaCN}$$

$$\underset{\substack{\text{1,4-dicyano-2-butene}}}{N\equiv C-CH_2 \qquad CH_2-C\equiv N \atop CH=CH} \xrightarrow{H_2} \underset{\substack{\text{adiponitrile}}}{N\equiv C-(CH_2)_4-C\equiv N} \xrightarrow[\text{pressure}]{H_2,\ NH_3}$$

$$\underset{\substack{\text{hexamethylenediamine}}}{H_2N(CH_2)_6NH_2}$$

The reactions described are beautiful examples of the typical ingenuity found in industrial processes. Elegant syntheses are often achieved through the use of extremely inexpensive raw materials: benzene, butadiene, hydrogen, air, ammonia, chlorine, nitric acid, and sodium cyanide.

29-3. SULFUR—A PURE ELEMENT IN THE EARTH

Because of its yellow color, sulfur was long thought to be involved in the formation of gold. Because of the pungent odor of its oxides, it has long been used as a medicinal in purging the body of evil spirits and poisons. Until quite recently, sulfur and molasses was a generally accepted springtime remedy. Its production of pungent gases in the body was taken as evidence that bodily poisons accumulated during the long winter were being released.

The extraction of sulfur from the earth is a relatively unique process which takes advantage of the physical properties of the element. Sulfur is a yellow solid that melts at 114°C. It occurs in Texas, Louisiana, and Mexico in huge deposits 500 to 1500 ft below the earth's surface. The present method of extraction was developed in the 1890s by Herman Frasch. His ingenious process involves melting the sulfur underground and pumping the liquid to the surface where it is stored.

The operation is accomplished (Figure 29–1) by pumping superheated water (180°C) under pressure through the outer part of three concentric pipes. This hot water, well above the melting point of the sulfur, mixes with the liquid sulfur as it forms. Through the inner pipe, compressed hot air is forced into the sulfur, forming a frothy mixture of sulfur, water, steam, and air which is forced up through the middle pipe to the surface. On cooling, the sulfur solidifies and

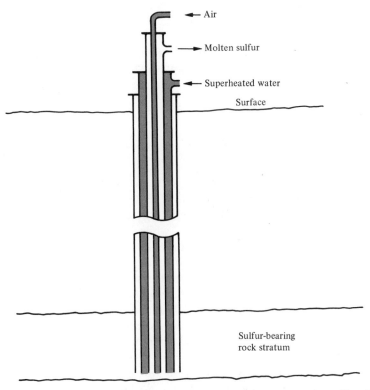

Figure 29-1. Schematic drawing of the Frasch process for extracting sulfur. Air and superheated water are forced down and molten sulfur is forced up.

separates from the water in which it is insoluble. The sulfur thus produced is quite pure.

Sulfuric acid has a long history. It was known to the alchemists of the tenth century who employed it in their various attempts to bring about transmutation.

Sulfuric acid is one of the most important chemical products of our industrial society and leads all other chemicals in the amount produced per year. It has often been said that the annual per capita production of sulfuric acid is a good gauge of the technical prowess of a nation. It is used in the manufacture of fertilizers and pigments, in the iron and steel industry, and very extensively in the production of other chemicals.

The most important method of producing sulfuric acid is the **contact process** developed in 1831. The process involves the catalytic oxidation of sulfur dioxide to the trioxide. First sulfur dioxide is prepared by burning sulfur in air.

$$S + O_2 \longrightarrow SO_2$$

Then the conversion of sulfur dioxide to trioxide is accomplished by the use of a vanadium catalyst at about 550 to 650°C.

$$SO_2 + \tfrac{1}{2}O_2 \xrightarrow{\text{catalyst}} SO_3$$

The contact process is an interesting example of the effect of a catalyst on a chemical equilibrium. The equilibrium constant for the oxidation of sulfur dioxide to the trioxide can be expressed by the equation

$$K_p = \frac{p_{SO_3}}{p_{SO_2} \times p_{O_2}^{\frac{1}{2}}}$$

in which p_{SO_3}, p_{SO_2}, and p_{O_2} are the partial pressures, in atmospheres, of SO_3, SO_2, and O_2, respectively. At a given temperature, the value of K_p is constant, and a catalyst therefore has no effect on the value of K_p but only on the rate at which equilibrium is reached. In terms of an energy diagram, the effect of the catalyst is to reduce the activation energy, E, while it has no effect on the energy of either the reactants or products (Figure 29–2).

Table 29–1 lists various values of the equilibrium constant for the $SO_2 + \tfrac{1}{2}O_2 \rightleftharpoons SO_3$ reaction. Although higher temperatures result in more rapid approaches to equilibrium, they also result in less favorable equilibrium positions; that is, low temperatures favor the formation of SO_3 at equilibrium, but the times required are too long. The use of platinum and, more recently, vanadium catalysts has allowed the use of quite low temperatures and simultaneously achieving high reaction rates.

The reaction is quite exothermic and therefore an elaborate water cooling system is needed. In cooling the reactor, the cooling water is converted into steam

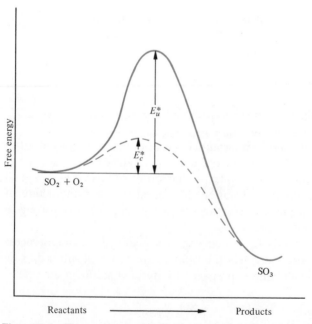

Figure 29–2. Free energy diagram for the conversion of sulfur dioxide to sulfur trioxide. The solid line represents the uncatalyzed reaction. E_u^* and E_c^* are the activation energies for the uncatalyzed and catalyzed reactions, respectively.

Table 29-1. Equilibrium Constants
for the Equilibrium

$SO_2 + \frac{1}{2}O_2 \rightleftharpoons SO_3$

Temperature (°C)	Equilibrium Constant (K_p)
400	397
600	9.35
800	0.915
1000	0.185

and used elsewhere in the plant as a heating source. Thus, a byproduct of the manufacture of sulfuric acid is steam, or rather energy, which is an important raw material in all industries.

The sulfur trioxide thus produced in the contact process reactor is converted into the final product, sulfuric acid, by reaction with water.

$$SO_3 + H_2O \longrightarrow H_2SO_4$$

This reaction, however, is so terribly exothermic that control of the released energy is difficult. The problem is circumvented by using 98% sulfuric acid as the SO_3 absorber. The pyrosulfuric acid ($H_2S_2O_7$) produced is called "fuming" sulfuric acid. This is an extremely reactive substance which, on exposure to the atmosphere, gives off dense white fumes. The absorption of SO_3 by 98% H_2SO_4 is carried out in steel towers packed with steel obstructions to aid mixing. Sulfuric acid of any desired strength can be prepared from the fuming acid by adding the required amount of water.

$$H_2SO_4(SO_3) + H_2O \longrightarrow 2 H_2SO_4$$
fuming sulfuric
acid

Although near 100% sulfuric acid is a very potent chemical, it is not as reactive toward metals as dilute acid. Thus, the highly concentrated ($> 98\%$) acid can be handled in stainless steel equipment. The low reactivity of the concentrated acid toward metals is due to the absence of water and the consequent absence of hydronium ions.

29-5.   THE WINNING OF IRON FROM ITS ORE

The production of metals is the technological cornerstone of any industrialized society. The most important metal by far is iron which, in the form of steel, is the basic structural metal used throughout the world. The entire process involved in locating natural sources of metals, extracting the metal, and rendering it into a commercially useful form is very complex. Geologists, mining engineers, analytical chemists, metallurgical and chemical engineers, along with a host of skilled and semiskilled laborers are required to accomplish the task. Only a broad outline of the steps involved in obtaining iron from its ore will be presented here, however.

These steps, although different in details for various metals, will nevertheless illustrate the chemistry of metallurgical processes.

An **ore** is a natural substance that contains a metal in sufficient quantity and in a form such that production of the metal is economically feasible. The principal iron ores are hematite, Fe_2O_3; limonite, $Fe_2O_3 \cdot nH_2O$; magnetite, Fe_3O_4; and siderite, $FeCO_3$. The first step, called **ore dressing,** is a concentration procedure to eliminate undesirable materials. Crushing and screening to eliminate the fine particles is common. Washing to remove clay bearing particles is sometimes useful. Magnetic separation is quite effective with the strongly magnetic Fe_3O_4, and even weakly magnetic hematite can be separated from worthless materials in this manner if a sufficiently powerful magnet is used. Calcining or heating to drive off moisture and, in the case of carbonate ore, CO_2, is also used. After preparation by one or more of these operations, the ore is ready for **smelting** or reduction of the iron in the ore to the metal.

Smelting of iron ore is accomplished in a **blast furnace** (Figure 29-3) where coke is used to reduce the ore. Such furnaces are sometimes 100 ft high. Coal that has had all volatile components removed from it by heating in the absence of air is known as coke. The hot air blast oxidizes the coke in the bottom of the furnace to carbon monoxide which rises and is further oxidized by the iron oxide to carbon dioxide while the ore is reduced to iron.

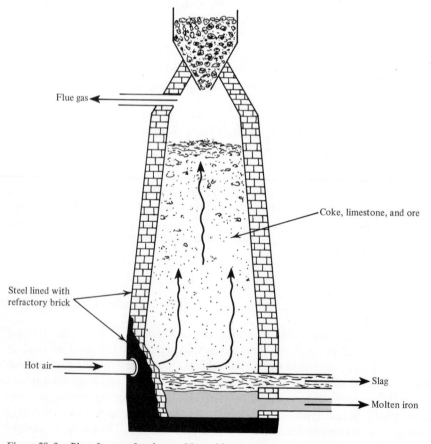

Flue gas

Coke, limestone, and ore

Steel lined with
refractory brick

Hot air

Slag

Molten iron

Figure 29-3. Blast furnace for the smelting of iron ore.

In the lower part of the furnace, the temperature is highest, about 1900°C. A temperature gradient of 1600°C or more exists from top to bottom of the furnace and the chemical reactions vary with the temperature. In the cooler regions near the top, the primary reaction is

$$CO + 3\,Fe_2O_3 \longrightarrow 2\,Fe_3O_4 + CO_2$$

Farther down, where the temperature is higher

$$CO + Fe_3O_4 \longrightarrow 3\,FeO + CO_2$$

and in still hotter regions

$$CO + FeO \longrightarrow Fe + CO_2$$

In the very hottest part, the iron melts and is drawn off from time to time to be either cast in molds as **pig iron** or conveyed directly to a steelmaking plant. As the charge in the furnace moves down, more coke, limestone, and ore are introduced at the top, so that the furnace operates continuously.

Impurities in the ore and the coke ash must be removed. The limestone ($CaCO_3$) serves as a flux that causes these impurities to fuse and form a slag which floats on top of the molten iron and can be drawn off. Chemically, slag is largely calcium aluminum silicate, formed from the reaction of CaO with SiO_2 and Al_2O_3 impurities. The CaO results from the thermal decomposition of the limestone,

$$CaCO_3 \longrightarrow CaO + CO_2$$

Slag is used in large amounts for the manufacture of cement and paving materials. The calcium oxide produced by this reaction serves additionally to remove sulfide impurities through the formation of CaS, which combines with the slag.

Pig iron is far from pure, containing perhaps 4% or more carbon along with lesser amounts of silicon, sulfur, phosphorus, and manganese. Molten pig iron is sometimes simply cast into molds and allowed to solidify to **cast iron.** If the cooling is carried out rapidly, the carbon is in the interstitial compound cementite, Fe_3C, and gives a hard, brittle product. If the solidification is carried out slowly, the carbon separates as particles of graphite throughout the iron and gives a softer and tougher metal.

A material previously of great importance but now largely replaced by steel is **wrought iron.** By melting pig iron with iron oxide and silica in the proper proportions, a purified iron is produced which is then combined with a certain amount of the slag. The product contains 0.1% or less carbon, small amounts of manganese, phosphorus, and sulfur, and up to 2.5% slag. The distribution of the slag through the iron causes the metal to acquire a fibrous crystal structure. The advantage of wrought iron over cast iron is that the former can be worked by rolling, forging, stamping, or drawing into a variety of shapes. Cast iron, on the other hand, is too brittle to be worked.

Iron that contains carbon in variable amounts between 0.1 and 2% and which is hardened by sudden cooling is called steel. Grades of steel are classified according to their carbon content as: soft steel, up to 0.3%; medium steel, 0.3–0.6%; and hard steel with more than 0.6% carbon. Special steels are made by alloying specific metals with the iron, for example, chrome steel.

Purification of pig iron and the introduction of the required amounts of carbon and other substances is largely carried out today in the open-hearth process. Molten iron is placed in a large open furnace along with Fe_2O_3 in such a way that air and combustible gases can be passed over the surface. The carbon is slowly oxidized and eliminated along with various other impurities after which the desired additions are carried out. About 100 tons of material can be processed in 8 to 10 hr in an open-hearth furnace.

A much faster but less easily controlled purification of iron takes place in the Bessemer process. A blast of air is passed through molten pig iron to oxidize the impurities, after which the desired additions of carbon and other substances are made. About 30 tons of charge can be handled in a Bessemer converter, which is a large egg-shaped steel vessel set on trunnions so that it can be tilted to receive the molten iron. It is then turned to a vertical position and air is blown in through holes in the bottom. The carbon and other impurities are quickly oxidized in about 10 to 15 min.

29-6. THE PRODUCTION OF REACTIVE METALS—SODIUM AND ALUMINUM

Many elements are obtained in pure form by electrolysis of their compounds. The process, in effect, supplies electrons from an external electrical source to metal ions and removes electrons from nonmetal ions. Hydrogen and oxygen are produced in this way by allowing an electric current to pass through water containing enough dissolved electrolyte to render the solution a good conductor. The individual half reactions are

Cathode: $\qquad$ $4\,H_2O + 4\,e^- \longrightarrow 2\,H_2 + 4\,OH^-$

Anode: $\qquad$ $6\,H_2O \longrightarrow 4\,H_3O^+ + O_2 + 4\,e^-$

Metals that have reduction potentials less than that of water cannot be prepared by electrolysis from their water solutions. Attempts to do so produce hydrogen instead. These cases include the metals of the first two groups of the periodic table, but for these elements electrolysis of the pure molten salts can be carried out. In the examples that will be discussed—sodium and aluminum—there were also mechanical problems that had to be solved before the otherwise simple principle of electrolysis could be applied.

In the case of sodium and the other alkali metals, the chlorides are used. Sodium chloride is obtained in large quantities in quite pure form in several parts of the world. Its relatively low melting point of 801°C permits electrolysis to be carried out on the molten salt with no major problems, although sodium carbonate is normally added to lower the melting point still further. The major problem that had to be solved in the commercial production of sodium and chlorine was to keep the products separate and away from air or moisture. A carbon anode and an iron cathode are used. The cathode and anode are separated by an iron wall as shown in Figure 29–4.

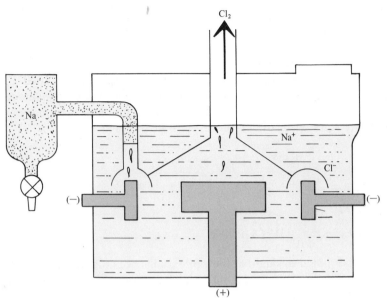

Figure 29-4. Electrolysis of sodium chloride. Sodium metal, due to lesser density, rises into the reservoir. The chlorine gas is trapped by the funnellike separator where it is drawn off in pure form.

The production of aluminum presented a different problem. Aluminum occurs naturally almost exclusively as bauxite, a hydrated form of Al_2O_3. The extremely high melting point of the oxide—over 2000°C—makes the direct electrolysis of the melt impractical.

The solution to this problem was discovered almost simultaneously and independently by two young students in 1886. Both Charles M. Hall, an American, and P. L. T. Héroult, a Frenchman, were 23 years old at the time.

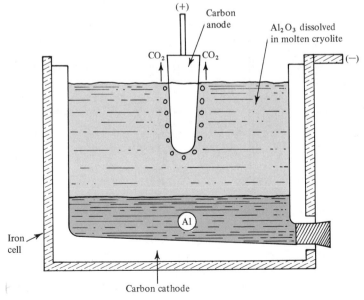

Figure 29-5. Cell for the electrolysis of Al_2O_3 to produce aluminum.

463

The problem was to find a solvent for Al_2O_3 that would melt at a sufficiently low temperature, yet would not interfere chemically to yield undesirable byproducts or impurities.

The solvent chosen was cryolite, Na_3AlF_6, or a mixture of AlF_3 and NaF. This solvent met the requirements and allowed a working temperature of $1000°C$. In the commercial process both anode and cathode are carbon. The carbon anode is consumed through the formation of CO_2. The molten aluminum, being heavier than the electrolyte solution, is drawn off at the bottom (Figure 29–5).

Cathode reaction: $4\,Al^{3+} + 12\,e^- \longrightarrow 4\,Al$

Anode reaction: $3\,C + 6\,O^{2-} \longrightarrow 3\,CO_2 + 12\,e^-$

29–7. THE ELECTROLYTIC REFINING OF COPPER

Copper is widely distributed throughout the world and is even found occasionally in its native or chemically uncombined state. In this form it was used by early man, perhaps ten thousand years ago, to make metal implements. Most of the commercially useful copper deposits, however, are in the form of sulfide ores and, to a lesser extent, oxide ores. Much of the world production of copper is used for electrical purposes such as motors, generators, telephones, radio and television, and power lines. For such purposes, the copper must be free of metallic impurities which would decrease its conductivity.

Most of the impurities remaining in the product of the metallurgical treatment of copper ore are metals such as iron, zinc, nickel, and lead. Small amounts of silver and gold are almost always found in the copper as well. The latter "impurities" are well worth recovering because of their economic value, but the more active metal impurities are worthless and simply need to be removed.

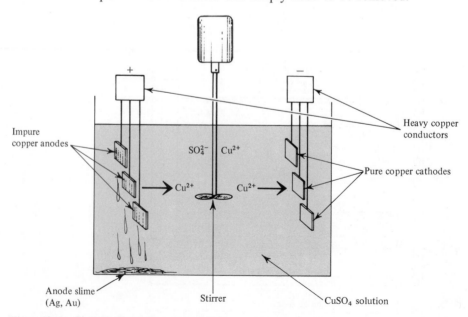

Figure 29–6. Electrolytic cell for copper refining.

An electrolytic process accomplishes both of these desirable ends and gives a product which is better than 99.9% pure copper. Figure 29-6 is a diagram of the electrolytic procedure. Each anode may weigh 500 lb or more. Thin sheets of pure copper serve as the cathodes, and the solution is H_2SO_4 and $CuSO_4$. For every Cu^{2+} ion put into solution at the anode, a Cu^{2+} ion is reduced at the cathode. The solution is stirred to prevent a concentration gradient from arising in the cell, which would increase the internal resistance of the cell. As the electrolysis proceeds, the anodes waste away and the cathodes are coated with pure copper. The more active metal impurities go into solution, but the voltage is regulated so that they are not plated out at the cathodes. The less active metal impurities such as silver and gold simply drop to the bottom of the cell as part of the anode slime from which they are recovered.

29-8. CRYOGENICS—THE ATTAINMENT OF LOW TEMPERATURES

To raise the temperature of a substance requires that energy be added to it. Similarly to lower the temperature, energy must be removed. The simplest way to raise or lower the temperature of a substance usually is to bring it into contact with a hot or cold reservoir. Normally we cool things by immersing them in ice or even in dry ice, if temperatures below zero are required. This method, however, is limited by the second law of thermodynamics; that is, heat flows only from the hotter to the cooler object. Consequently the temperature of a substance cannot be lowered below that of the coolant.

There are, however, other ways of removing energy from a substance. The substance can be made to do work, such as the expansion of a gas where the energy lost by the gas is converted into work. Before proceeding to a discussion of how low temperatures are achieved, perhaps it would be worthwhile to discuss how low temperatures are measured.

The most common type of thermometer is a glass tube that contains a liquid whose density compared with that of glass varies linearly with temperature. Mercury is the most frequently used liquid, although colored alcohol is also used. In this type of thermometer, the length of the column of the liquid in the tube is taken as a direct measure of the temperature.

The problem in using changes in length as the basis for thermometry is in finding two materials, one of which changes much more than the other. A variation of this idea is the bimetal thermometer (Figure 29-7) in which two different metals are bonded together. A change in temperature results in a slightly greater expansion of one metal than the other, with a consequent bending of the bimetal. The angle of bending is then taken as a measure of the temperature. Other thermometers depend upon changes in electrical resistivity or upon thermoelectric power.

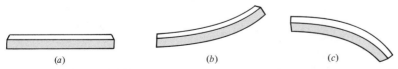

(a)   (b)   (c)

Figure 29-7. A bimetal thermometer (a) at the arbitrary zero of temperature where the metals were bonded to each other, (b) at a higher temperature where the darker metal expands more than the lighter, and (c) at a lower temperature where the darker metal shrinks more than the lighter.

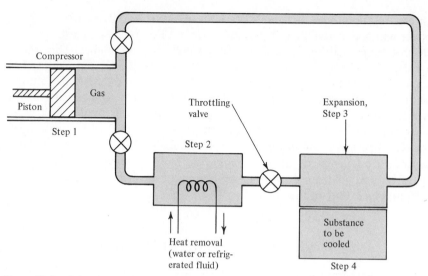

Figure 29-8. Schematic drawing of the essential parts of a cryogenic cooling device.

Low temperature resistance thermometers depend on electrical resistance which varies *inversely* with the temperature; that is, varies with $1/T$. Thus the resistance increases as the temperature drops. Examples of such thermometers are the carbon-resistance and the germanium-resistance thermometers.

Cooling to cryogenic temperatures is fundamentally no different from the process that occurs in the home refrigerator or air conditioner: A permanent gas (one that cannot be liquified at room temperature by pressure alone) is compressed to about 150 atm (2000 lb/in.²) which causes the temperature of the gas to rise (step 1). The heating of the gas is due to the work done on the gas by the compressor. This mechanical work is converted into kinetic energy of the gas molecules. Step 2 consists of cooling the compressed gas to ambient temperature by bringing it in contact with a coil containing running water. An even lower temperature can be obtained if the gas is brought into contact with a refrigerated fluid from a separate refrigerator. In step 3 the cooled compressed gas is allowed to expand. In expansion the gas does work, losing more energy and thereby experiencing a further temperature drop. The cold gas is then brought into contact with the material to be cooled from which it withdraws energy. The warmed gas then returns to the compressor where the cycle is repeated (Figure 29-8).

29-9. IMPORTANT CRYOGENIC LIQUIDS

The elements hydrogen, helium, nitrogen, and oxygen are among the most important cryogenic liquids. All have critical temperatures considerably below room temperature which means that they cannot be liquified at room temperature regardless of the pressure applied. Some important properties of these cryogenic liquids are given in Table 29-2 along with water for comparison.

Oxygen is obtained commercially along with nitrogen by liquifying air. The 13°K difference in boiling point between the two makes separation by distillation fairly easy. Because oxygen is such a vigorous oxidizing agent, "liquid air" is

Table 29-2. Important Properties of Liquid Hydrogen Helium, Nitrogen, and Oxygen

	Molecular Weight	Critical Temperature (°K)	Boiling Point (°K)	Melting Point (°K)	Density at b.p. (g/cm³)
hydrogen	2	32.9	20.3	13.8	0.071
helium	4	5.2	4.2	—	0.125
nitrogen	28	126.1	77.4	63.2	0.80
oxygen	32	154.8	90.2	54.4	1.14
water	18	647	373.2	273.2	1.0

almost never used today. In the past, use of liquid air for cooling purposes was always attended by the danger of an oxygen buildup due to the evaporation of nitrogen. Today, liquid nitrogen is commercially available and cheap, and has almost completely replaced the oxygen-nitrogen mixture as coolant.

Hydrogen is usually produced by the electrolysis of water since it does not occur uncombined in the earth.

Helium is the lowest boiling substance. The isotope, $^3_2$He, is even lower boiling than $^4_2$He by 1°K. It is obtained from some natural gas wells. Although there is as yet no widespread industrial use for helium, its chemical inertness makes it an extremely valuable substance in many areas of scientific research. Table 29–2 lists no melting point for helium. In fact it will become solid only at the lowest temperatures by the application of 25 atm of pressure.

29–10. SYNTHETIC DIAMONDS

Although diamonds are found in Australia, America, and Asia, over 90% of the world's diamonds are mined in Africa. Of these only about 20% are pure enough to be classified as gems. The remainder are used in tools and dies. The largest volume of commercial diamond is used in the form of abrasive powder.

The properties and structures of diamond and graphite were examined in Chapter 15. A knowledge of their structures and of the fact that diamond has a higher density led to the idea that the conversion of graphite into diamond at high temperatures and pressures might be an economic possibility. It was early realized that diamond is the less stable form, and that at 1300°C, it slowly reverts to the more stable graphite.

After several workers had claimed to have made synthetic diamonds from graphite by the use of high temperatures and pressures, F. D. Rossini and R. S. Jessup at the National Bureau of Standards (1938) published an article in which they calculated the graphite ⇌ diamond equilibrium over a range of temperatures and pressures. In 1961, F. P. Bundy and his associates of the General Electric Company extended the data to higher temperatures and pressures. These data are given in Figure 29–9. It can be seen from the low extremity of the curve that diamond is not stable at pressures below about 12 kbars (~11,844 atm).

The first successful laboratory synthesis of diamonds was carried out in the General Electric laboratories in 1955. The process involves heating a mixture of graphite or other form of carbon and a metal catalyst (chromium, manganese, iron, cobalt, nickel, or several others) at temperatures high enough to melt the metal and at a pressure at which diamond is stable.

467

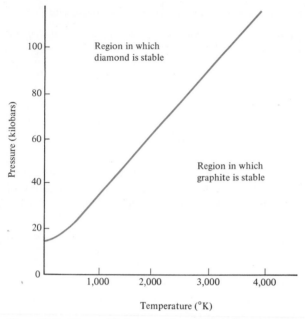

Figure 29-9. The equilibrium graphite ⇌ diamond at various temperatures and pressures. 1 kilobar (k bar) = 987 atm. Any point on the curve represents an equilibrium condition.

Needless to say, one of the major stumbling blocks to a successful process was the design of a suitable apparatus that would withstand the enormous temperatures and pressures required. The apparatus used by the General Electric workers was constructed of tungsten carbide embedded in a stone chamber. The reaction zone was heated electrically. It was found that lighter colors and higher purities were obtained at the higher temperatures.

One of the more important byproducts of this work was the development of techniques for achieving and operating at high temperatures and pressures. Thus, from the new knowledge gained, many new innovations have arisen, especially in the technology associated with space exploration.

PROBLEMS

1. Outline the steps involved in the manufacture of nylon-66. How is this synthetic fiber related to natural silk?

2. Describe the Frasch process for the production of sulfur. What property of sulfur accounts for the feasibility of this process?

3. What is the product when sulfur is burned in air? What further conversion of this product is necessary in the contact process for making sulfuric acid and how is it accomplished?

4. Why is pyrosulfuric acid formed before the addition of water rather than employing a direct reaction between SO_3 and water in making sulfuric acid?

5. Explain the statement: ". . . a byproduct of the manufacture of sulfuric acid is . . . energy".

6. Dilute sulfuric acid readily attacks many metals, yet the concentrated acid is handled in stainless steel equipment with no problems. Explain.

7. What are the principal iron ores?

8. What purposes do the coke and limestone serve in the smelting of iron ore?

9. What is the difference between pig iron and wrought iron?

10. Explain the differences among soft, medium, and hard steels.

11. Describe the open-hearth and Bessemer processes for making steel.

12. Why is it not possible to produce sodium or aluminum from their ores in the same way iron is produced from its ore?

13. Write anode and cathode half reactions for the electrolysis of NaCl.

14. Why is the electrolytic production of aluminum more complicated than the production of sodium? Write the anode and cathode half reactions for the aluminum electrolysis.

15. Why must the anode and cathode compartments be physically separated in the electrolysis of NaCl but not in the electrolysis of Al_2O_3?

16. Calculate the production of purified copper in pounds per day in an electrolytic cell using a current of 2.0×10^3 amperes (amp).

17. Calculate the amount of material consumed at the anodes of the cell in problem 16 if the copper was originally 92% pure. Also calculate the amount of gold recovered if the original impure copper was 0.1% gold.

18. Describe the operation of bimetal thermometers and resistance thermometers. Why is a mercury thermometer useless at very low temperatures?

19. Why has liquid nitrogen replaced "liquid air" for cooling purposes?

20. Explain how the alternate compression and expansion of a gas can produce refrigeration. Is this process a violation of the second law of thermodynamics which states that heat will not flow spontaneously from a cooler body to a hotter body? Explain.

21. At ordinary conditions of temperature and pressure, the energetically more stable form of carbon is graphite (Figure 29-9). Why, then, will diamond under ordinary conditions not spontaneously convert to graphite? Under what conditions will diamond convert to graphite?

22. Why has the price of gem diamonds not been reduced by the introduction of synthetic diamonds?

CHEMICALS THAT AFFECT LIFE

This chapter treats some chemical systems that affect life. Of the great variety of such systems available for study, only a few have been chosen. The ones selected offer a wide variety of chemical phenomena and are easily understood by students. They are a sulfa drug, steroids, compounds that cause cancer, plant growth regulators, drugs that affect the mind, and finally a classification of poisons and their modes of action. It is intended that this survey will provide an introduction to the interface between chemistry and the medical sciences.

30-1. THE FUNCTION OF A SULFA DRUG

Certain bacteria in a laboratory-controlled culture are found to grow well when a small amount of *p*-aminobenzoic acid is present. If a small amount of sulfanila-mide is added to the culture, growth stops. Addition of more *p*-aminobenzoic acid causes growth to resume, and further addition of sulfanilamide stops growth. The development of sulfanilamide and other "sulfa" drugs afforded a great step forward in the late 1930s toward the control of infectious diseases. The explanation of the action of this chemical is an interesting example of the importance of stereochemistry.

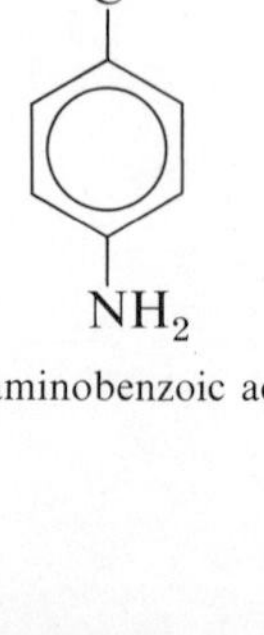

p-aminobenzoic acid

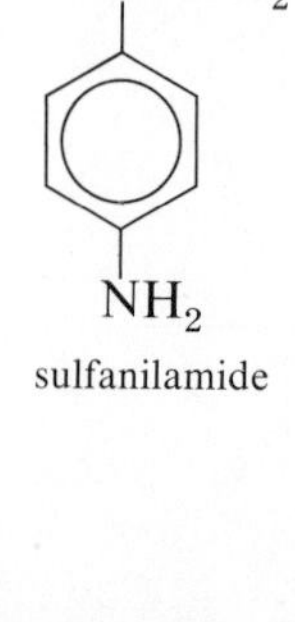

sulfanilamide

Models will show that the sulfanilamide molecule is quite similar in size and shape to the *p*-aminobenzoic acid molecule and can therefore occupy its sites in the enzyme thus preventing growth activity.

30–2. THE STEROIDS

Steroids form a group of compounds of varying functions that are found in plant and animal tissue. They are white, waxy, solids. Although their functions vary greatly, nearly all possess the same basic molecular skeleton of three six-membered rings and one five-membered ring. The generalized carbon skeleton of the steroids is

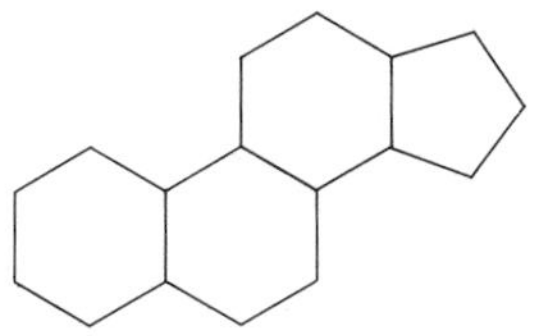

Cholesterol was first isolated in 1770. It is produced in copious amounts in gall stones and is found deposited along the walls of blood vessels in victims of hardening of the arteries. It also occurs in all tissues of the body in greater or lesser extent. Its name is derived from *chole,* the Greek word for bile. By 1928, Adolf Windaus in Germany, the pioneer in steroid research, had, with Heinrich Wieland, proposed a tentative structure for cholesterol for which they shared the Nobel prize. By the following year their structure was shown to be incorrect. The correct structure was finally proposed by Windaus and Wieland and confirmed by X-ray analysis in 1932.

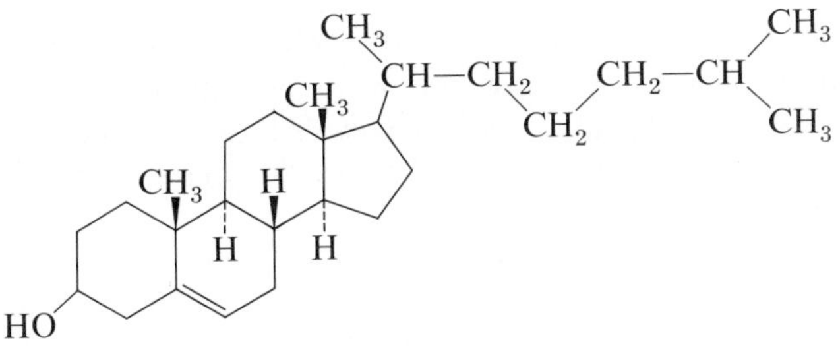

cholesterol

Fish liver oils contain a substance (vitamin D) that prevents rickets. Certain foods, when irradiated with ultraviolet light, are able to counteract a deficiency of vitamin D. Windaus, in his studies of this vitamin, found that ergosterol from yeast could, on irradiation with ultraviolet light, be converted into a substance that was also active in preventing rickets. This compound was named vitamin D_2. In addition to vitamin D_2, which is effective against rickets, several other ineffective isomers were also obtained. Windaus and his associates found several other variants that were also potent against rickets.

471

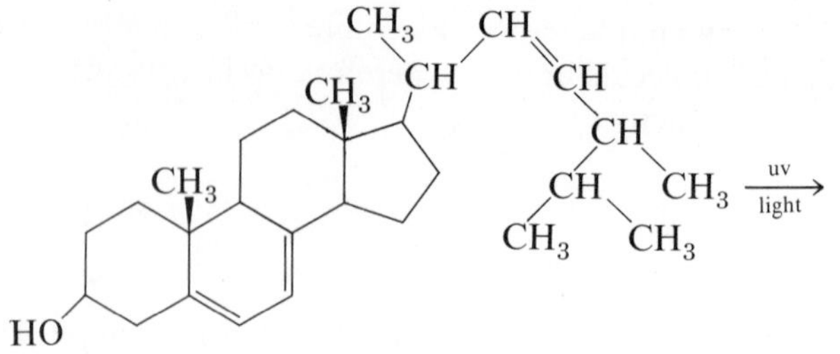

ergosterol

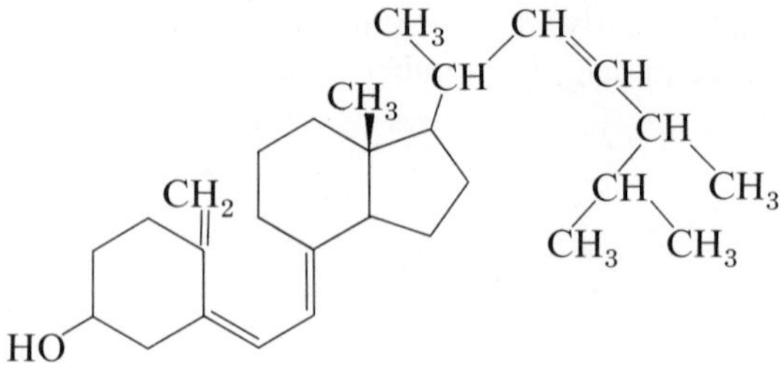

vitamin D₂

One of these was vitamin D₃, which was shown in 1936 to be identical to the natural vitamin D obtained from fish liver oils.

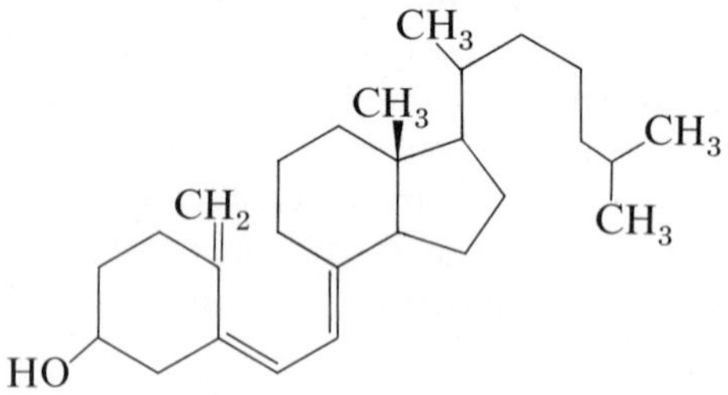

vitamin D (vitamin D₃)

The sex hormones are generated by the gonads (testes or ovaries) and are responsible for the development of sex characteristics and for the sexual responses of males and females. Female sex hormones are called **estrogens;** male sex hormones are called **androgens.**

In 1927, a hormone was isolated from the urine of pregnant women. It had the property of producing sexual heat in laboratory animals. Two years later, the compound, estrone, was isolated from this preparation and was shown to be a steroid by the fact that dehydration yielded the same product as the similar treatment of cholic acid, a known steroid obtained from beef bile. Further research showed that estrone is not the true estrogenic hormone, but instead it is a product of the transformation of the true hormone.

In one of the great feats of chemical science, about 12 mg of the true estrogen, **estradiol,** was isolated from four tons of hog ovaries! It was also prepared from estrone thus establishing their relationship (see formulas on page 473).

Estrogens perform several functions during the 28-day female menstrual cycle. At the beginning of the cycle, the pituitary gland in the brain produces a hormone called **follicle stimulating hormone** (FSH), which proceeds to the ovaries where

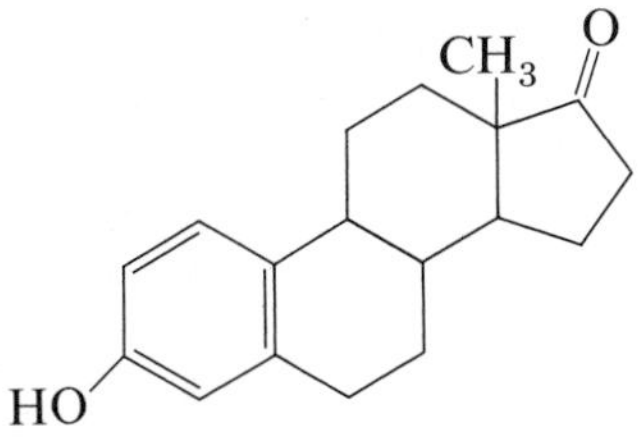

estrone

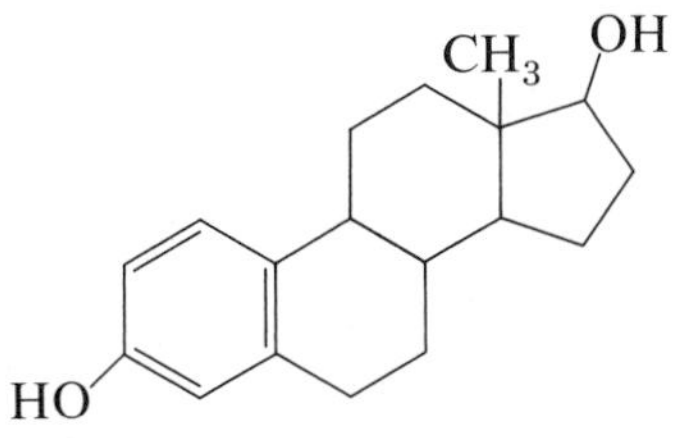

estradiol

it causes a few of the many hundred thousand immature egg follicles to mature. One of these eggs is released during ovulation (fourteenth day) by a hormone called the **lutinizing hormone** (LH) produced in the pituitary. Just prior to ovulation, the ovaries produce estrogen which proceeds to the pituitary where it retards the production of FSH. It also proceeds to the uterus where it stimulates the development of the endometrium—the spongy lining tissue in which the fertilized egg is implanted.

Another female sex hormone, progesterone, is involved in the preparation for and maintenance of pregnancy. This class of hormones is called progestogens.

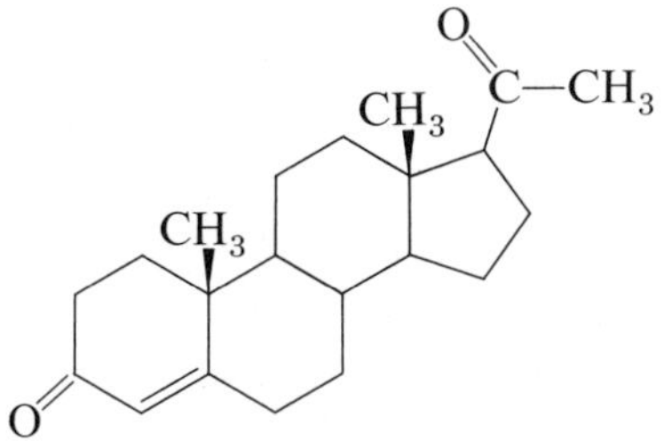

progesterone

Although the activity of progesterone is found only in a few steroids, many compounds are known that exhibit estrogenic activity. An example, which is not even a steroid but is geometrically related, is stilbestrol.

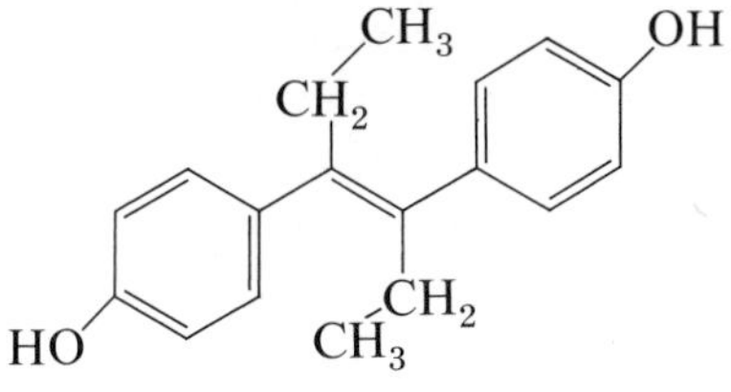

stilbestrol

Progestogens also perform several functions in the female menstrual cycle. Progesterone is produced in the corpus luteum. After ovulation, the progesterone goes to the uterus where it participates along with the estrogen in the preparation of the uterus if fertilization occurs. Immediately after ovulation progesterone is produced by the placenta and moved to the pituitary where it stops LH production thereby preventing overlapping pregnancies. If fertilization has not occurred, the corpus luteum deteriorates; there is no hormone production at all, and menstruation ensues.

473

The first male sex hormone to be isolated was androsterone, which was later shown to be a metabolic product of testosterone, the true male sex hormone.

androsterone   testosterone

Androsterone was isolated from urine. Testosterone was first isolated from steer testes. The steroidal structure of androsterone was demonstrated by the oxidation of cholesterol to androsterone.

Cortisone is used in the treatment of rheumatoid arthritis and in the reduction of inflammation in skin diseases. It is one of a group of steroids called corticosteroids which have been isolated from the adrenal cortex of various animals. These substances control electrolyte balances and the formation of carbohydrates from proteins within the organism.

cortisone

From the examples discussed, one can see that very small variations in molecular structure result in dramatic changes in physiological function. The sensitivity of the body to slight structural differences is essential, otherwise many entirely different structures would have to be synthesized naturally. The striking structural similarities among these compounds, then, is a necessary condition that allows a great efficiency in their natural production from similar, if not the same, starting compounds.

30-3. CARCINOGENIC HYDROCARBONS

A cancerous tumor may be defined as an unregulated growth of cells without obvious cause from previously normal cells.

An interesting story about the effect of accidental discoveries on research concerns the coal-tar industry just after the turn of the century. Coal-tar is obtained by heating coal in the absence of air under conditions that allow the removal of volatile substances. These volatile substances include benzene, toluene, and other commercially useful materials. The remaining tar is called coal-tar, and it is a very complex mixture of polynuclear hydrocarbons (hydrocarbons composed of varying numbers of benzene rings fused to each other). These compounds are to be expected from the decomposition of coal if one recalls that coal is actually an impure form of graphite which is a vast network of benzene rings.

Early coal-tar workers began, after several years, to exhibit an abnormally high incidence of skin cancer. The alarming rate of skin cancer prompted experiments to establish if coal-tar was the cause. The prolonged application of coal-tar materials to laboratory animals was indeed found to lead to such cancerous growths. At that time, 1915, all the known compounds produced from coal-tar were tested for carcinogenic activity with no success. The next step was to fractionate the coal-tar and test each fraction for carcinogenic activity. The active fraction was then refractionated and the process repeated. It was observed that the active fraction always exhibited a fluorescence spectrum with emissions at 4000, 4180, and 4400 Å. These absorption lines were thus taken to be diagnostic for the active compound.

By 1933, the hitherto unknown compound, 3,4-benzpyrene was isolated and shown to be the active carcinogen.

3,4-benzpyrene

It is a light yellow crystalline substance whose structure was ultimately proved by synthesis.

Another very active carcinogen was discovered in the same year during studies on bile acid. In structure studies of the steroids, one of the proofs commonly used was the removal of hydrogen to form the completely benzenoid structures that are readily identified. Such treatment of cholic acid (obtained from beef bile) or of cholesterol led to the compound methylcholanthrene, the most potent carcinogen known.

cholic acid   methylcholanthrene

The easy conversion of these two abundant components of our bodies into a potent carcinogen strongly suggests that spontaneous cancers may indeed originate from the abnormal metabolism of steroids in our bodies.

30-4. CARCINOGENIC ACTIVITY AND STRUCTURE

Although hydrocarbons like the ones discussed in Section 30–3 may not cause spontaneous cancers, their study has continued in the hopes of correlating structure with carcinogenic activity.

Since methylcholanthrene is probably the most potent carcinogen known, it has been used as a *model* compound. Thus, many of its isomers and analogs have been synthesized and their carcinogenic activities measured. A summary of a series of such experiments in which the following compounds were prepared and tested is outlined.

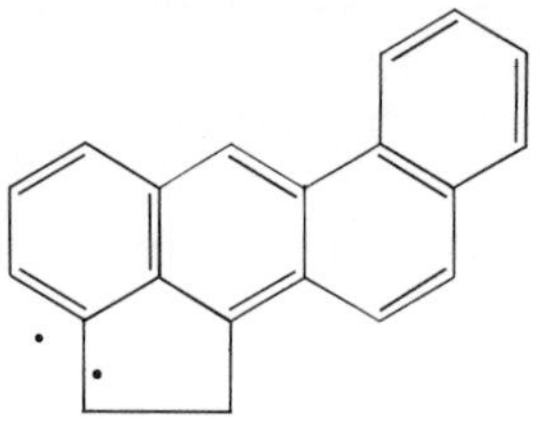

I

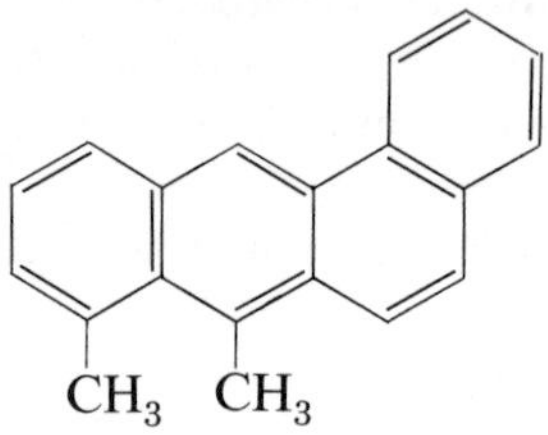

II

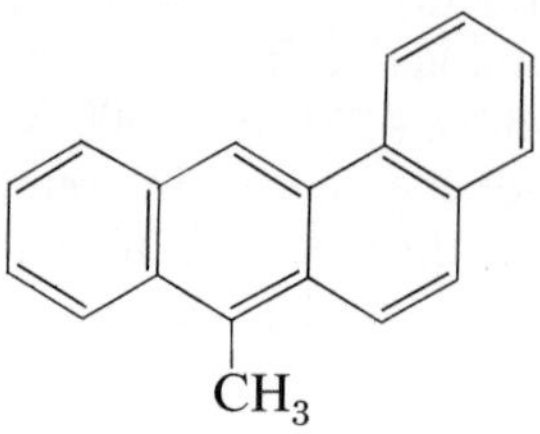

III

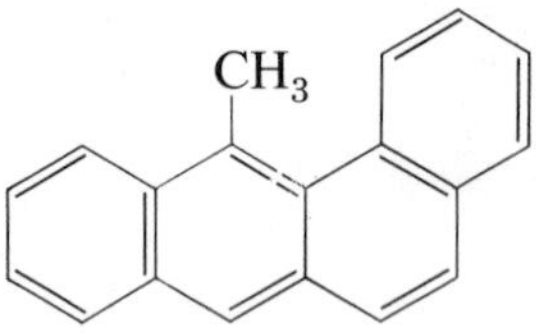

IV

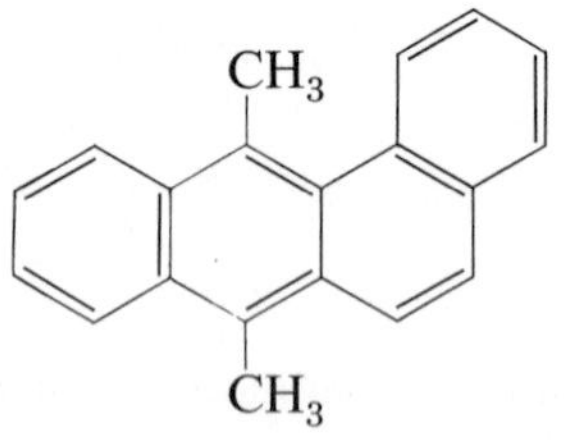

V

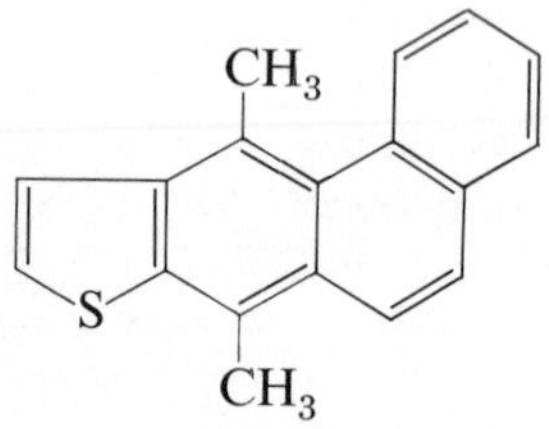

VI

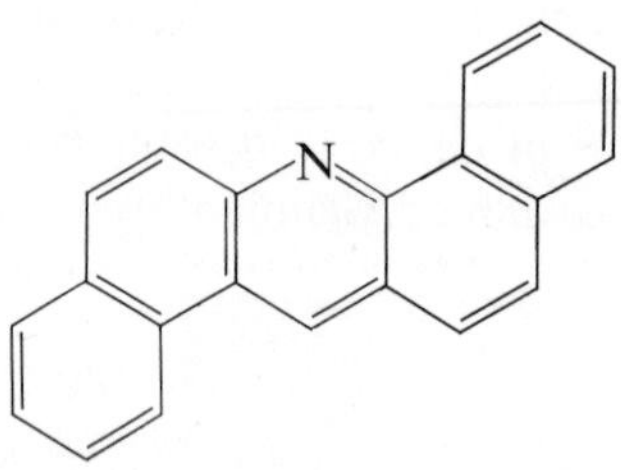

VII

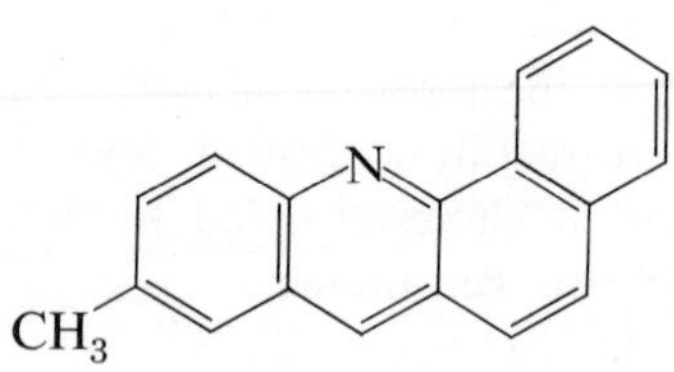

VIII

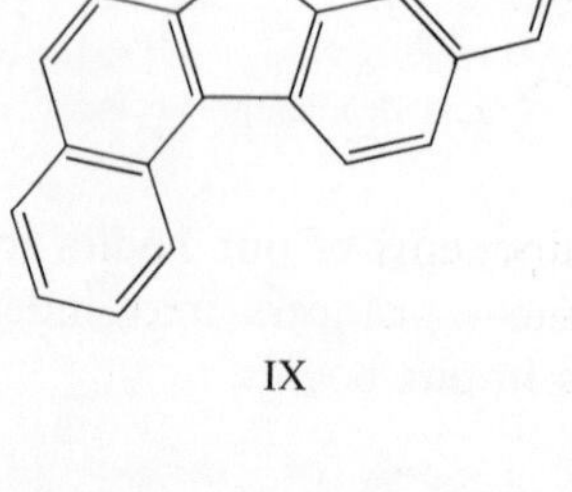

IX

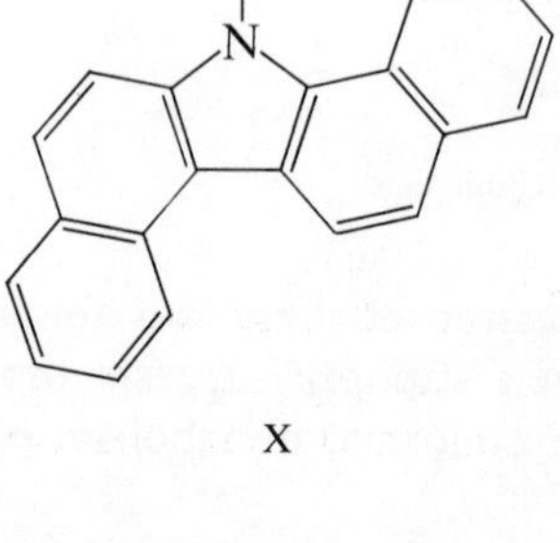

X

The methyl group in methylcholanthrene does not contribute to its carcinogenic activity as evidenced by the equal activity of cholanthrene (I). Similarly, the five-membered ring is also unimportant because II and III exhibit roughly equal activity. In fact, the essential feature besides the benzene ring system itself is the presence of a methyl group in the position shown in III. All the monomethyl derivatives of cholanthrene were synthesized and tested. The most active was III and the second most active was IV. Synthesis of V was therefore undertaken, and it exhibited the predicted high activity. To determine the nature of the ring system, various derivatives were prepared in which atoms other than carbon were present in the ring (VI, VII, VIII, and X). Although VI, VII, and VIII were highly carcinogenic, with the latter comparable to methylcholanthrene, IX and X were found to be only slightly active. These results indicate that the actual atoms present are relatively unimportant as long as the overall shape of the molecule is not altered. Evidently, the essential feature of the ring system is planarity and linearity of the three central rings.

(Required coplanar rings are shown in blue.)

The latter condition is met in VI, VII, and VIII, but not in IX and X. Note also that 3,4-benzpyrene is comparable to compound IV and hence obeys the above generalizations.

Another line of investigation has arisen from the correlation of chemical reactivity with carcinogenic activity. Thus, the most potent carcinogens are also the most reactive in substitution reactions (Chapter 25). Moreover, such substitution reactions occur in that part of the molecule which is most important for carcinogenic activity; namely, the carbon atom in the ring that is attached to the methyl group in III. Although a causal relationship has not been established by these studies, they suggest that carcinogenic action may be associated with reaction of the carcinogen with an enzyme that controls cell growth.

30–5. PLANT GROWTH REGULATORS

Much of our knowledge of substances that affect plant growth derives from studies of **auxins.** Auxins are plant hormones that are synthesized in the apical bud and young leaves and from there are transported to the growing region of stems to aid stem elongation.

A typical experiment involving such substances gives the following results. Sections from the growing region of seedling oat plants were placed in a solution of sugar, mineral salts, and indole acetic acid (IAA). IAA is an auxin that occurs naturally in plant materials. A control sample was placed in a solution containing only sugar and mineral salts. While the control sample showed only very slight growth, the sample in the IAA solution showed a very rapid growth to greater than normal size.

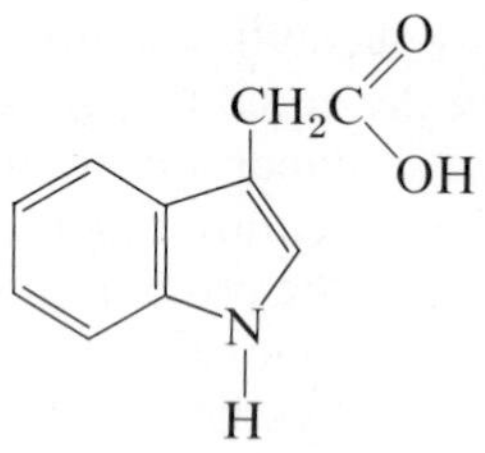

IAA

IAA influences plant growth in many ways by causing a great variety of activities in cells and tissues. It is synthesized in the plant by an enzymatic reaction from the amino acid tryptophan.

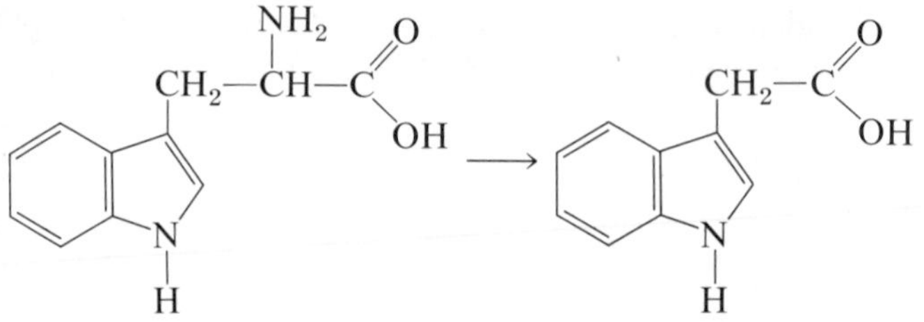

tryptophan   IAA

Naphthalene acetic acid also acts like IAA although it is not naturally occurring.

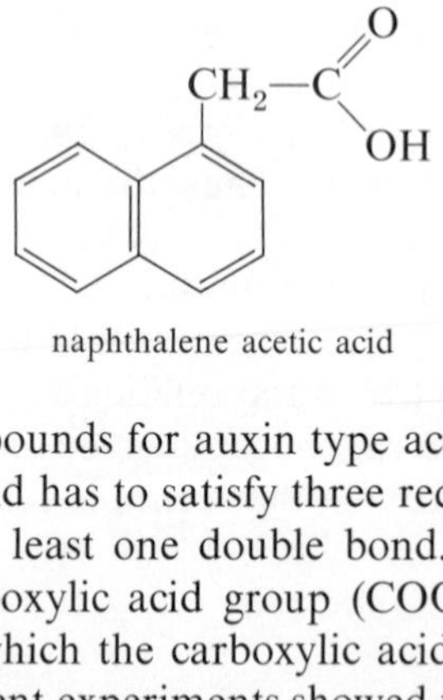

naphthalene acetic acid

The study of many compounds for auxin type activity yielded the conclusion that to be active, a compound has to satisfy three requirements: (1) It must have a ring system containing at least one double bond. (2) It must have a carbon chain terminating in a carboxylic acid group (COOH). (3) It must be able to assume a configuration in which the carboxylic acid group is suitably arranged relative to the ring. Subsequent experiments showed that to be active as an auxin, the ring must have a hydrogen atom of suitable reactivity on the carbon atom adjacent to the carbon chain. This was shown by the relative activities of phenoxyacetic acid (POA) and 2,4-dichlorophenoxyacetic acid (2,4-D)

In the case of POA, the ring hydrogen atoms are not sufficiently reactive. Placement of chlorine atoms in the ring to form 2,4-D apparently activates the remaining hydrogen atom. That the adjacent hydrogen is the one involved in auxin action is demonstrated by the inactivity of the third compound shown in which this hydrogen atom has been replaced by any of several other atoms or groups.

The theory to explain these observations proposes a simultaneous two-point attachment of the auxin molecule to the plant protein (Figure 30-1). Thus,

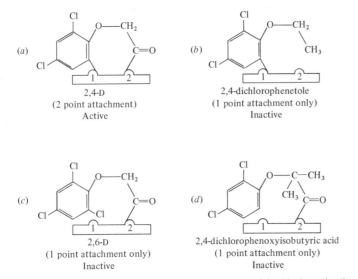

inability to form either attachment renders the molecule completely inactive. In Figure 30–1, compounds *b*, *c*, and *d* are rendered inactive in three different ways. In *b* the carboxylic acid group is absent. In *c*, the adjacent ring hydrogen atom is absent. In *d*, two-point attachment is prevented by the bulky methyl groups that block close approach of the ring to the attachment site.

Figure 30–1. Examples of two point attachment and its blockage in (*b*), (*c*), and (*d*). Sites 1 and 2 represent appropriate reactive sites in the protein molecules.

Auxins are active only at low concentrations. This fact is also in agreement with the two-point attachment theory because, at high enough concentrations, two-point attachment is prevented by crowding of molecules into adjacent sites, as shown in Figure 30–2. Since the concentration of IAA in the plant is regulated, such growth inhibition does not occur. This safety mechanism does not exist with 2,4-D, however, which can thus accumulate and inhibit growth. In this way 2,4-D is used as a herbicide. This is one of many examples in which fundamental knowledge can be applied in either of two directions—in this case, either to aid growth or to inhibit growth.

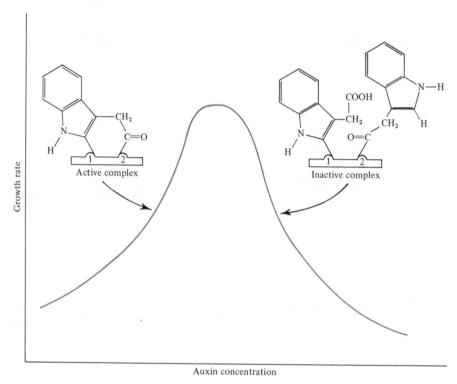

Figure 30-2. Effect of auxin concentration on growth rate. At higher concentrations growth is inhibited by the inability to form two-point attachments due to crowding.

30-6. DRUGS THAT AFFECT THE MIND

Alkaloids □ Analgesics are substances that relieve pain. The most important narcotic analgesics are derived from the opium poppy—*Papaver somniferum*—which is native to parts of Asia. Opium is prepared from the dried juice of the unripened seed pod. Opium is a mixture of several compounds that are called **alkaloids** because they contain amino nitrogen atoms and are thus basic, or "alkalilike". The major component of opium is morphine (Figure 30-3), which has been known since 1805.

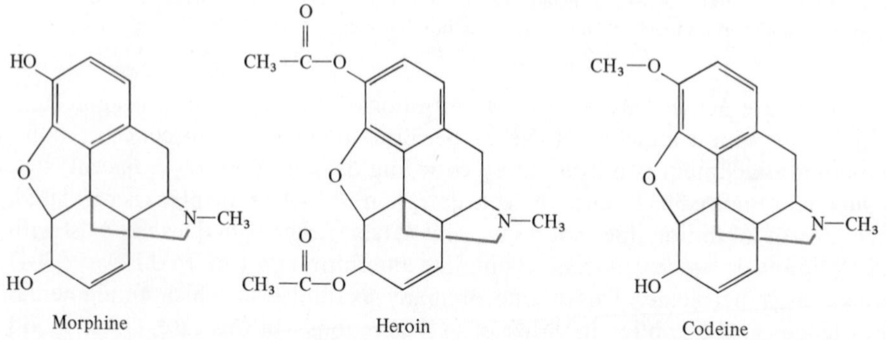

Morphine   Heroin   Codeine

Figure 30-3. Morphine and some derivatives which exhibit narcotic analgesic activity.

The opium alkaloids are extremely effective in their analgesic activity. Unfortunately, however, they are also extremely addicting, and these two properties have not been successfully divorced in spite of much research.

Two terms must be defined in connection with descriptions of drug action: "**tolerance** is said to develop when the response to the same dose of a drug decreases with repeated use. **Physical dependence** is a physiological state of adaptation to a drug, normally following the development of tolerance, which results in a characteristic set of withdrawal symptoms (often called the abstinence syndrome) when administration of the drug is stopped".*

Morphine has five asymmetric carbon atoms and is therefore one of $2^5 = 32$ isomers. Interestingly, the mirror image molecule exhibits none of the physiological properties of morphine.

Codeine and heroin are simply derivatives of morphine prepared by bringing about reactions of one or both of the hydroxy groups. Heroin, the favorite drug among chronic drug addicts, is several times more potent than morphine.

In therapeutic doses, the morphine alkaloids cause depression of respiration and cardiovascular activity, constriction of the pupils of the eyes, a slight elevation of body temperature, and frequently disturbance of the gastrointestinal system. In higher doses, the above reactions are increased, and sufficiently high doses lead to coma, shock, and respiratory failure that produce death. The addicting property of these drugs is severe. Upon withdrawal, the chronic user experiences painful physiological responses that can be dangerous.

A significant fact is the cross-tolerance and cross-dependence of these drugs. This fact has been useful in treating addicts in easing their withdrawal syndrome. An example of a useful drug is methadone which, while itself a narcotic, eliminates the heroin withdrawal syndrome at doses that do not produce euphoria. Addicts are therefore kept on methadone while they undergo treatment for detoxification.

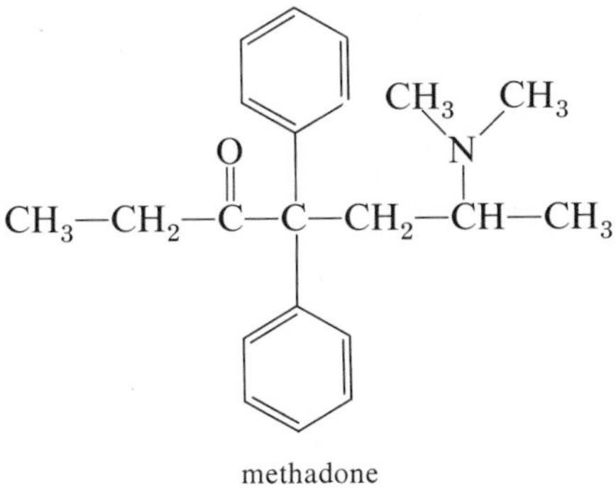

methadone

Amphetamines □ Along with the barbiturates discussed below, the amphetamines (Figure 30-4) are synthetic drugs. They are used medically to treat obesity, mild depression, and narcolepsy (a tendency to fall asleep at any time). The amphetamines are powerful central nervous system stimulants, much more so than the naturally occurring epinephrine. The effects of amphetamines in normal therapeutic doses are similar to those of epinephrine—dilation of the pupils of the eyes, increased blood glucose, elevated pulse rate and blood pressure, in-

*Canadian Commission of Inquiry into the Non-Medical Use of Drugs

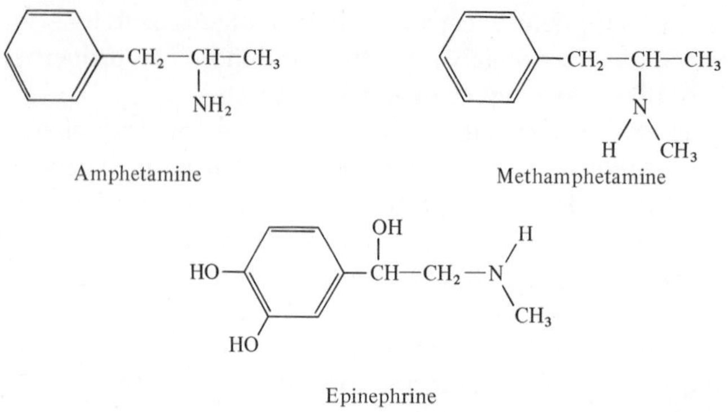

Amphetamine

Methamphetamine

Epinephrine

Figure 30-4. Synthetic amphetamines and their similarity to naturally occurring epinephrine.

creased respiration rate and decreased appetite. The psychological effects vary with the individual, but generally include a feeling of increased alertness, a decrease in fatigue, and an elevation of mood. Sometimes the individual also experiences irritation, restlessness, chest pains, headache, nausea, and cardiac arrhythmias.

Physical dependence on amphetamines has been demonstrated by the onset of atypical electroencephalogram patterns during sleep, lethargy, and depression on sudden withdrawal.

Barbiturates ☐ Barbiturates (Figure 30-5) are the major ingredients in sleeping pills. They are useful in the treatment of insomnia and anxiety. Some derivatives, like phenobarbital, are also useful as anticonvulsants. They thus produce effects opposite to the amphetamines. The two are often used in combination, one to counteract the undesirable effects of the other.

Tolerance to the sedative effects of barbiturates has been amply demonstrated.

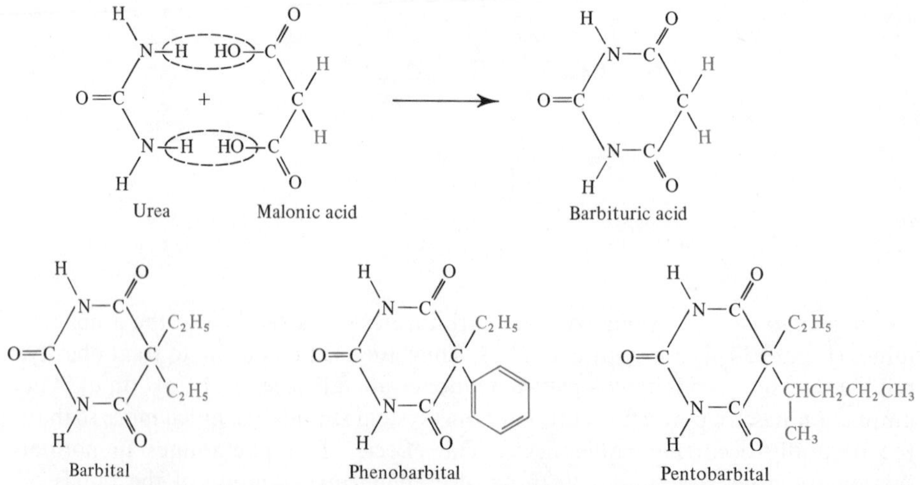

Urea   Malonic acid   Barbituric acid

Barbital   Phenobarbital   Pentobarbital

Figure 30-5. The synthesis of barbituric acid from urea and malonic acid. Barbital, phenobarbital, and pentobarbital are derivatives produced by substituting appropriate groups for hydrogen atoms in the malonic acid (shown in color). Barbituric acid itself is not physiologically active.

However, the lethal dose remains essentially constant; hence, as tolerance increases, the margin of safety decreases. Thus accidental poisoning frequently occurs at doses that no longer provide sedation. Although barbiturates were first thought to be nonaddictive, physical dependence has been adequately demonstrated. Sudden withdrawal in habitual users produces dramatic changes in the electroencephalogram, insomnia, tremor, muscle spasm, delerium, convulsion, and even death.

Caffeine is found in coffee beans and tea leaves, and is a stimulant. It is closely related to two other alkaloids which have similar properties—theophylline and theobromine. It is interesting that these alkaloids are structurally related to the barbiturates which exhibit the opposite physiological effects.

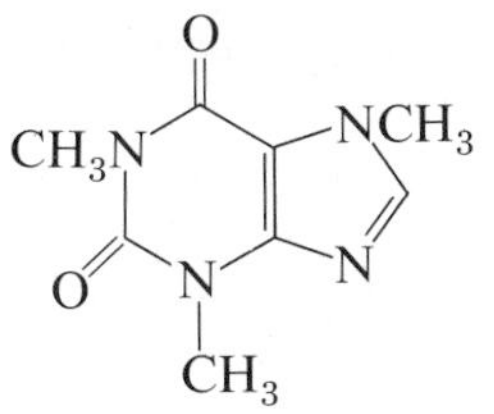

caffeine 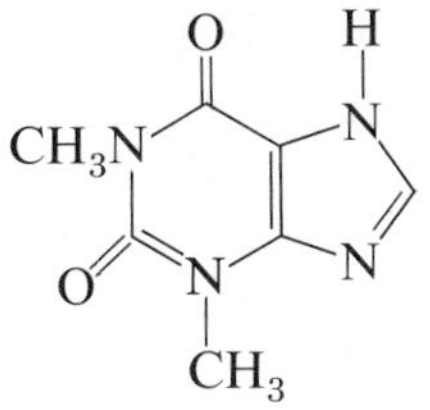   theophylline 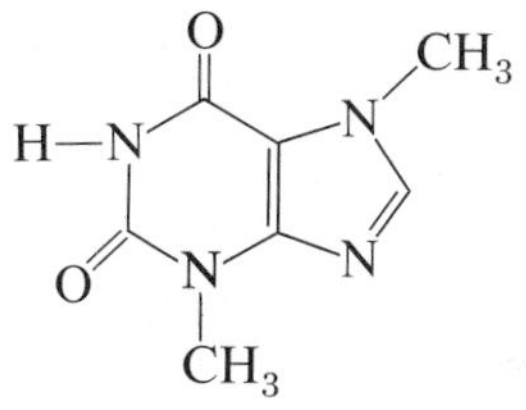   theobromine

Hallucinogenic Drugs □ Hallucinogenic drugs are substances that produce sensory perceptions that have no basis in physical reality. The hallucinations may involve any or all of the five senses. Scientific interest in these drugs stems from their ability to produce experimental psychoses. The first drug of this kind known to man was mescaline (Figure 30-6) which was used in America in ancient times. It is extracted from the peyote cactus. The pure alkaloid was first isolated in 1898. In addition to the typical colorful geometric hallucinations and fluid distortions of objects, tremors, anxiety, exaggerated reflexes, and electroencephalogram abnormalities have been reported.

The structural similarity between mescaline (Figure 30-6) and epinephrine (Figure 30-4) is striking. Such similarities between compounds that produce such different effects are common and often provide clues to brain action.

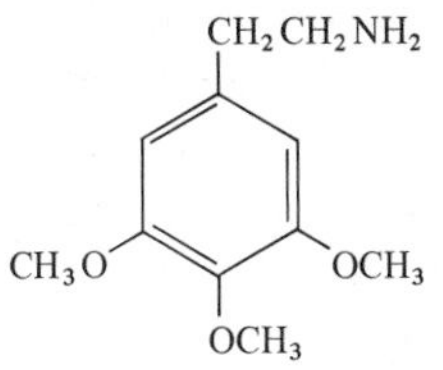

Mescaline 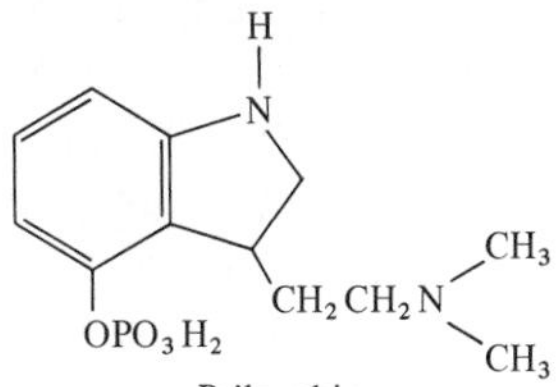   Psilocybin

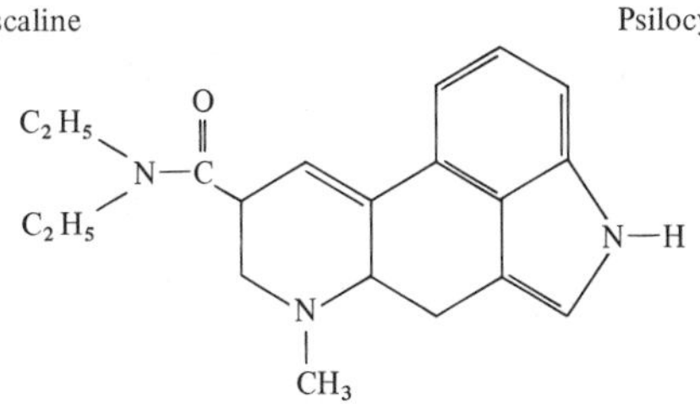

D-Lysergic acid diethylamide (LSD)

Figure 30-6. Some hallucinogenic drugs.

Lysergic acid is one of the alkaloids present in ergot—a fungus that grows typically on some grains. In the Middle Ages, an epidemic periodically occurred due to eating bread made from diseased grain. The disease, now known as ergot poisoning, was called St. Anthony's Fire because people believed a visit to the saint's shrine would cure the illness. The symptoms were vomiting, high temperature, and convulsions. Today, ergot alkaloids (all derivatives of lysergic acid) are useful clinically for their uterotonic activity.

Ergot alkaloid chemistry was extensively studied by Arthur Stall and Albert Hofmann in the first half of the twentieth century. In their studies they prepared many derivatives of lysergic acid amide, which varied in the nature of the two alkyl groups attached to the amide nitrogen atom. The diethyl derivative, commonly known as LSD, was prepared by them in 1938, but they failed to note anything of interest until 5 years later; Hofmann accidentally ingested some and discovered the extreme hallucinogenic property.

Although LSD affects both motor and nervous functions in the body—lack of ability to coordinate voluntary movements, dilation of the pupils of the eyes, drop in blood pressure, increased pulse rate, elevated body temperature, and changes in the encephalogram—the most striking effects are on the psychic functions. These include excitation, mood changes—euphoria to depression—disturbances of perception, hallucinations, depersonalization, and a schizophrenic state or withdrawal from reality.

The mechanism by which LSD functions to bring about these changes is not known. However, it is known that the psychic alterations persist after the elimination of LSD from the body. Possible, but as yet unproved, side effects include deformations in the newborn of mothers who have taken the drug during pregnancy. Physical dependence does not seem to develop; however, tolerance to LSD develops rapidly, and cross-tolerance between LSD and mescaline or other lysergic acid derivatives may develop.

Marijuana □ Although marijuana has been used by man for thousands of years, scientific data on its effects and mode of action are still quite scarce. Even the structure of the active compound (Figure 30-7) was unknown until 1964.

Although tetrahydrocannabinol (THC) acts on both the central nervous and cardiovascular systems, the precise effects are only vaguely known. The drug elevates and depresses mood, often simultaneously. The user usually experiences a pleasant dreamlike state where perception is altered. He often claims profound intellectual and emotional experiences but is seldom able to articulate them intelligibly. Time appears to pass more slowly than normal. Recall of long forgotten details is often reported, while well learned facts often cannot be remembered. Hallucinations are frequent with higher doses.

Among the physiological reactions to marijuana are a dry mouth, light headed-

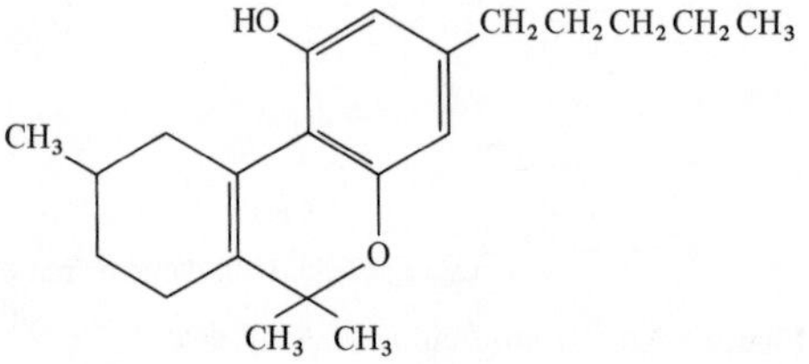

Figure 30-7. Δ^9-3,4-*trans*-tetrahydrocannabinol, believed to be the active component of marijuana.

ness, an increased appetite; with high doses, encephalogram changes, increased heartbeat, red eyes, and dilation of the pupils of the eyes are common.

The mode of action of THC in the brain is as yet unknown. Experiments with monkeys suggest that, at higher doses, marijuana may be physically addictive as defined because tolerance to its effects is developed and withdrawal symptoms develop on suddenly stopping usage. At lower doses habituation occurs through psychological dependence.

Although many claims have been made, both in favor of and against the use of marijuana, most are tentative. With the great amount of interest in this drug, definitive data will undoubtedly be forthcoming in the near future.

Ethyl Alcohol ☐ Ethyl alcohol can be classed as a food because of its caloric content of 7 cal/g in metabolism. Because it contains no essential nutrients, however, malnutrition frequently accompanies alcoholism.

The most notable effect of ethyl alcohol is as a drug that affects the brain and therefore produces behavioral changes. The effects of ethyl alcohol are related to the concentration in the blood. Concentrations up to about 0.05% usually produce slight sedation. Between 0.05 and 0.15% a lack of coordination is usually observed. Intoxication is obvious at concentrations of between 0.15 and 0.20%. Unconciousness usually occurs at concentrations around 0.30 to 0.40%, and death can occur at a concentration of 0.5%. For purposes of legal determinations, especially involving automobile accidents, sobriety is generally defined as a blood alcohol level below 0.05%.

Alcohol exhibits both properties usually associated with physical addiction—tolerance and withdrawal symptoms. Although the biochemical reasons for tolerance to alcohol have not been explained, withdrawal symptoms have been related to reduced magnesium levels (hypomagnesemia) and increased alkalinity of the blood.

Ethyl alcohol is infinitely soluble in water and is therefore rapidly absorbed in the mouth, stomach and small intestines. Detoxification of alcohol occurs in the liver in two steps. In the first it is converted to acetaldehyde by the action of the enzyme alcohol dehydrogenase (ADH). Concurrently, nicotinamide adenine dinucleotide (NAD) is reduced. The reduced form is referred to as NADH.

$$NAD + CH_3CH_2OH \xrightarrow{\text{ADH}} CH_3\overset{\displaystyle H}{\underset{\displaystyle |}{C}}{=}O + NADH$$

In the second step, acetaldehyde is converted into carbon dioxide and water by another enzyme, acetaldehyde dehydrogenase.

$$NAD + CH_3\overset{\displaystyle H}{\underset{\displaystyle |}{-C}}{=}O \xrightarrow{\substack{\text{acetaldehyde} \\ \text{dehydrogenase}}} CO_2 + H_2O + NADH$$

The reduction of NAD to NADH occurs with several metabolic changes, the major one of which is the production of fat. Consequently one of the most common conditions among alcoholics is a fatty liver. Cirrhosis of the liver is a frequent disease among alcoholics. In this disease, destruction of liver cells occurs with the buildup of an excessive amount of connective tissue.

485

30-7. TOXICOLOGY

Toxicology is the study of poisons, their effects, their mechanism of action, and modes of treatment. Broadly defined, a poison is a substance which, by chemical action, can kill or injure living organisms. Almost all chemicals that exhibit physiological effects are poisons at sufficient doses.

Of the many substances that are recognized as poisonous, most can be placed in one of three broad categories.

1. Agents that act so rapidly that no perceptible effects are observed. Death usually occurs as a result of anoxia (oxygen deprivation) and usually follows circulatory collapse. Carbon monoxide and cyanide are examples of such agents.

2. Agents that produce damage on entry into or contact with the body. This group includes corrosive chemicals like the acids and bases, as well as irritant gases like mustard gas.

3. Systemic poisons affect specific organs or systems such as the liver, kidneys, or nervous system. Examples in this group are arsenic oxide which causes fatal injury to the blood vessels; heavy metals that inhibit enzymatic activity such as compounds of lead, mercury, thallium, phosphorus, manganese, cadmium, and chromium; the nerve gases and hydrogen sulfide, which act as central nervous system depressants; and chlorinated hydrocarbons, which attack the liver, heart, and kidneys.

Table 30–1 gives some examples of poisons according to their toxicity.

Table 30–1. Toxicity Scale of Some Common Substances

Toxicity Rating	Examples	Probable Lethal dose (mg/kg)*
(1) practically nontoxic	glycerine, water, graphite, lanolin	>15,000
(2) slightly toxic	ethyl alcohol, lysol, castor oil, soaps	5,000–15,000
(3) moderately toxic	methyl alcohol, kerosene, ether	500–5000
(4) very toxic	tobacco, aspirin, boric acid, phenol, carbon tetrachloride	50–500
(5) extremely toxic	morphine, mercury dichloride	5–50
(6) supertoxic	potassium cyanide, heroin, atropine	<5
(7) supremely toxic	botulinus toxin, some snake venoms	<0.5

*Milligrams of poison per kilogram of body weight.

PROBLEMS

1. Define or explain the following terms: (a) analgesic, (b) narcotic, (c) tolerance (of drugs), (d) physical dependence (of drugs), (e) habituation.

2. Barbiturates and caffein have very similar structures but opposite physiological properties. What inferences can you draw from this fact regarding their modes of action?

3. Can your answer to problem 2 be interpreted in terms similar to the two-point attachment theory of plant auxins, or to the action of sulfa drugs? Explain.

4. The D-vitamins, sex hormones, cortisone, and other steroids have quite similar structures. What biological utility can you ascribe to the striking similarity of such physiologically different substances?

5. The compounds discussed in Section 30–4 showed that the presence of atoms other than carbon has no apparent effect on carcinogenic activity, whereas geometrical shape does. Interpret this generalization in terms of a probable mode of action.

6. Contrast the terms physical dependence and habituation.

7. What is suggested by the fact that the mirror image isomer of morphine is physiologically inactive?

8. List several common poisons and classify each of them into one of the three groups described in Section 30–7. Include in your list the following: LSD, amphetamines, barbiturates, heroin, caffein, 2,4-D (toxicity to plants), 3,4-benzpyrene, sulfanilamide (toxicity to bacteria).

9. Cholic acid is a major component of bile, which is secreted by the liver and stored in the gall bladder for use when needed in the duodenum for the digestion of fats. Suggest a possible explanation for the occurrence of large amounts of cholesterol in gall stones.

10. What does the discussion in Section 30–4 suggest about the role of stereochemistry on physiological action of chemical substances?

11. Explain the order of auxin type activity of the three compounds, phenoxyacetic acid (POA), 2,4-dichloroacetic acid (2,4-D), and 2,4,6-trichlorophenoxyacetic acid (Section 30–5).

THE ORIGIN OF LIFE

31-1. COMPLEXITY OF LIFE

The study of atomic and nuclear structure should have suggested to the student some of the sublime complexities of the smallest particles and of the delicate ingenuity which man has brought to this work. The variability of molecular structure and properties arising from the linking together of atoms has been a major theme of this book. We have also looked at some aspects of chemistry applied to its most sublime and complex branch—the substances of life.

Before beginning our exploration of the origin of life, it must be impressed upon the student what a living substance is. The simplest form of life imaginable must be capable of responding to its environment, of growing, and of reproducing itself. The latter requirement is valid unless we are to consider life to be merely a spurious phenomenon. Life as we know it must sustain itself and its kind.

If one considers only these requirements of living substances, he is apt to lose sight of the enormous chemistry involved in them. The cell, commonly considered to be the simplest self-sustaining living unit, can be thought of as analogous to a large chemical factory in which many raw materials are introduced and converted into many products and in which a strict control is constantly maintained over quality, power requirements, and changing market demands. In addition to this, management of the factory has its counterpart in the living cell so that the cell has all the necessary information to carry out its multifarious processes.

In order for this vast network of very specific operations to be carried out, the molecular complexity of the operating units of the cell must be truly great. Moreover, in order for chemical reactions to occur, several requirements must be met:

1. Most of the reactions must occur in water solution to provide the necessary transport of products from one place to another, as well as to allow the

ionic reactions that must occur. It may be noted in passing that many biochemical reactions, especially neurological processes, are believed to be electrochemical in nature.

2. Moreover, the complexity and specificity of reactions requires that very large molecules be employed. In addition, the needed plasticity and fluidity within the living organism require that most of the substances involved be polymeric. Only polymeric substances like nucleic acids and polypeptides are capable of the complexity and variability necessary to provide the many functions needed in living substances.

3. The reactions must all take place at a temperature not far from the ambient temperature and within the liquid range of water.

4. Finally, acidity control within living systems must be rigidly maintained to allow the necessary structures and reactions to exist among the myriad molecules present.

It can thus be seen that the requirements for life are rather rigid and place very serious restraints on the planet that can support it.

31–2. THEORIES OF THE ORIGIN OF LIFE

Given some appreciation of the tremendous complexity of the simplest living system, it is interesting to speculate on the possible origins of life. An answer to this question has, of course, been one of man's great yearnings and has enriched his religious life.

Essentially two explanations have competed for attention: **divine creation** and **spontaneous generation.** According to the latter theory, life was thought to generate spontaneously and continuously throughout time. The evidence was obvious—anywhere that one found decayed food, one would soon find all sorts of living creatures, which seemed to arise from the rotting mass. The theory, at least in its nineteenth century version, was finally laid to rest by Louis Pasteur, although recently evidence has begun to accumulate in support of spontaneous generation in the prebiotic earth, as will be discussed below. Although the demise of the theory of spontaneous generation was an advance in our knowledge of the world, it left an intellectual vacuum.

In the early part of the twentieth century, a new idea was proposed by the Russian biochemist A. I. Oparin. His speculations were grounded in verifiable chemical facts and paved the way for a whole new line of experimentation. The basis for his ideas rested on the probable nature of the earth and its atmosphere in its early formative period. The prebiotic earth was probably largely rocky with a good deal of liquid surface water and an atmosphere consisting of methane, ammonia, water vapor, and hydrogen. The relatively higher ambient temperature resulted in a greater amount of vaporization and condensation of water and hence probably resulted in perennial rain and electrical storms over the earth's surface. In addition, the very intense ultraviolet rays of the sun were not filtered out by atmospheric oxygen as they are today. Hence, an intense ultraviolet radiation bombarded the earth's surface. Moreover, the amount of radioactive isotopes in the early earth was much higher than it is now, and the resulting high background radiation could provide another large and relatively constant supply of energy.

Given all these conditions—an atmosphere of completely hydrogenated carbon, nitrogen, and oxygen, and several sources of energy; namely, electrical discharges, ultraviolet and nuclear radiation—the synthesis of many simple organic molecules could occur (Figure 31-1).

Laboratory experiments employing such energy sources within synthetic atmospheres of methane, ammonia, water vapor, and hydrogen have been conducted by several eminent scientists who were prodded by Oparin's ideas. One of the earliest was Dr. Harold Urey, who studied the effect of electric sparking on such a gas mixture. After one week of continuous operation, the contents of the apparatus were found to contain small amounts of several simple amino acids and sugars essential to the formation of proteins.

Many other such experiments have been conducted in the last several years, and the evidence strongly supports the possible formation of many simple molecules (Figure 31-2) in the primordial earth.

The demonstration of the synthesis of such simple molecules under the conditions described is a truly great landmark in understanding possible life origins and serves to show that such processes in the prebiotic earth are, in principle, possible. Moreover, means for the conversion of these simple molecules to the extremely complex bioorganic substances used by living cells have been suggested. However, a nearly insurmountable problem arises in trying to bridge the gap between complex molecules, like polypeptides and polynucleotides, and the organization of a living cell.

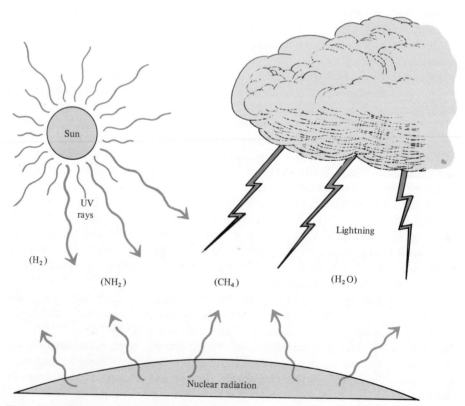

Figure 31-1. Some simple chemistry on the surface of the prebiotic earth.

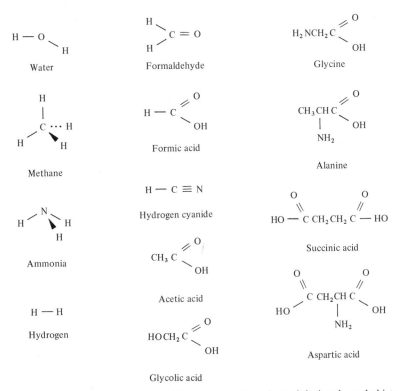

Figure 31-2. Simple organic molecules that could have formed during the early history of the earth. These molecules have been observed to form during experiments with synthetic primitive atmospheres of methane, ammonia, water vapor, and hydrogen.

Thus we have here once again evidence for the impact of a new idea on the way that man views his world. Oparin's theory led the way for many practical and simple experiments that have resulted in a new view of the origin of life on the earth. Whether the time available—2 billion years (the period when the earth had cooled sufficiently to sustain life)—was long enough to render probable a low probability event like life remains to be determined.

PROBLEMS

1. Write balanced equations for the conversion of a mixture of $CH_4 + NH_3 + H_2O + H_2$ into
   (a) CH_2O (d) $HCOOH$
   (b) HCN (e) $HOCH_2COOH$
   (c) CH_3COOH (f) H_2NCH_2COOH

2. List the features that must be present for a substance to be called alive.

3. List the chemical requirements for life to exist.

4. How does Oparin's theory of spontaneous generation differ from the theory that Pasteur overthrew?

5. Why was solar ultraviolet radiation more intense on the early earth than now?

6. Why was radioactivity more intense on the early earth than now?

7. What conditions on the surface of the early earth make plausible Oparin's theory of the chemical origin of life.

8. Evaluate the validity of the assumption that water is an essential requirement of life. Suggest other liquids that might serve in place of water.

NUCLEAR TECHNOLOGY

The first experimental proof that nuclear fission could be controlled and carried out on a large scale came on December 2, 1942, in a converted squash court under the west stands of Stagg Athletic Field at the University of Chicago. There, a group of about 40 scientists under the leadership of Enrico Fermi caused the first nuclear reactor to become self-sustaining.

Unfortunately, nuclear energy was first used in warfare. The awesome death toll of the two atomic bombs dropped on Hiroshima and Nagasaki plus the deformities that appeared in the offspring of some of the survivors has left a scar on mankind which has produced a lasting fear of the power of the atom. It is good that we continue to fear the misuse of atomic energy as well as other engines of war. However, we truly live in the early stages of the Atomic Age in which the proper use of nuclear processes has already begun to affect our lives. The production of power, the diagnosis and treatment of disease, and the chemical analysis of substances are some of the technological aspects of nuclear reactions that are considered in this chapter. Radiation hazards will also be considered briefly.

32-1. NUCLEAR REACTORS

The first nuclear reactor was called a "pile" because it was literally a pile of graphite bricks containing holes into which uranium oxides and uranium metal were placed (Figure 32–1). The graphite served as the matrix to hold the atomic fuel and also as a material to slow down the neutrons liberated in the fission process. Because so little fissionable $_{92}^{235}U$ is contained in natural uranium, the probability of a neutron striking a $_{92}^{238}U$ nucleus rather than the $_{92}^{235}U$ nucleus is great. However, if the neutrons are slowed sufficiently, they will not be captured by the $_{92}^{238}U$ nucleus as easily and will continue on their path until they strike

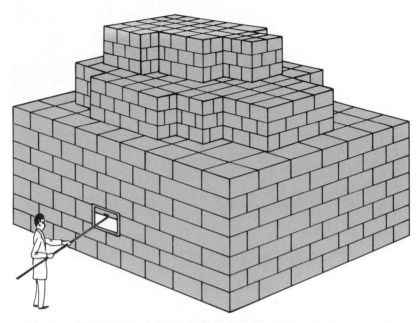

Figure 32-1. The first nuclear reactor built by Fermi and his coworkers was made of a "pile" of graphite bricks within which the atomic fuel was placed.

a $^{235}_{92}$U nucleus and cause fission. To control the number of free neutrons, cadmium rods (which readily absorb neutrons) were placed in the pile at intervals and were withdrawn to increase the neutron bombardment of the uranium or pushed in to shut down the process.

Modern reactors usually use $^{235}_{92}$U-enriched uranium or plutonium as the fuel. Slowing of the neutrons is achieved using a moderator, which is sometimes graphite as in the Fermi model, or heavy water, D_2O, or occasionally ordinary water. Automatic systems of control rods guard against a runaway reactor. Constructed of thick concrete and steel walls, a modern reactor resembles the original 1942 model in about the same way that a modern jet airplane resembles a World War I vintage airplane.

The energy released by a reactor is removed by pumping a coolant through pipes in the reactor. At present, electric power is obtained by pumping the hot coolant through heat exchangers that convert water to steam which is used to drive conventional turbines to generate the electricity (Figure 32–2). Water is often used as the coolant, but since the most efficient reactors operate at a very high temperature, a high boiling coolant is desirable to avoid excessive pressures in the pipes within the reactor. Liquid sodium has been found to be a good substance for this purpose, although a fairly complex heat exchange system is necessary to isolate the sodium which is rendered highly radioactive in the reactor. A recent development that holds promise for simplifying the conversion of the reactor energy into electric power is the use of an inert gas such as helium as coolant. The possibility of using the hot gas directly to drive turbines without the necessity of the generation of steam is economically attractive.

The fuel of conventional reactors must be replenished from time to time. However, reactors of the **breeder** type have been developed which produce more

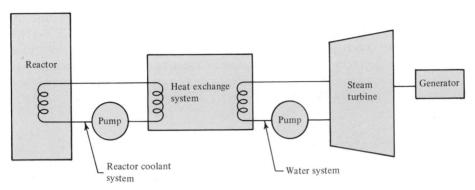

Figure 32-2. Schematic diagram of use of a nuclear reactor for the generation of electric power.

atomic fuel than they consume. The possible utilization of large quantities of low grade uranium ore has stimulated efforts to perfect this type of reactor. Under the right conditions, some of the neutrons liberated by the fission of $^{235}_{92}U$ are captured by $^{238}_{92}U$ nuclides. The conversion of nonfissionable $^{238}_{92}U$ into fissionable $^{239}_{94}Pu$ by neutron capture has been discussed in Section 28–7. In this manner the primary fuel, $^{235}_{92}U$, produces slightly more new fuel, $^{239}_{94}Pu$, than is consumed. Another system operates on the conversion of nonfissionable $^{232}_{90}Th$ into fissionable $^{233}_{92}U$.

$$^{1}_{0}n + {}^{232}_{90}Th \longrightarrow {}^{233}_{90}Th$$

$$^{233}_{90}Th \xrightarrow[t_{1/2} = 22\ min]{\beta} {}^{233}_{91}Pa$$

$$^{233}_{91}Pa \xrightarrow[t_{1/2} = 27\ days]{\beta} {}^{233}_{92}U$$

The efficiency of a breeder reactor is measured by the **doubling time**—the time required to produce twice as much fissionable material as was originally present in the reactor. After the doubling time, the breeder will have produced enough material to refuel itself as well as fuel another reactor of the same size. Efficient breeder reactors have doubling times of 7 to 10 years. At present, there are no commercial breeder reactors in operation.

By the early 1970s nuclear energy plants numbered only a few dozen and accounted for only a few per cent of the total electric power used in the United States. The initial capital outlay is very high for nuclear facilities, and a long lead time of 6 or 7 years from planning a plant until its final construction has slowed the expansion of nuclear power plants below that predicted in the early 1960s. An additional problem has been the thermal pollution of rivers and bays into which waste heat is discharged. At the present time, nuclear generating plants discharge 25 to 30% more waste heat than conventional plants per kilowatt-hour of electricity generated. This excess heat must be disposed of by means of cooling towers, holding basins, or other means to prevent damage to the marine ecology of the area. A plus factor in connection with pollution, however, is the fact that air contamination from a properly designed nuclear reactor is slight, whereas the burning of fossil fuels has been a major source of air pollution.

The Federal Power Commission has estimated that electric generating capacity in the United States will rise to nearly 2000 million kw by the year 2000. The capacity by 1973 was 367 million kw. Any approach to this tremendous increase in energy requirements will surely require different sources from those man has depended upon over the centuries. Nuclear energy may well provide some if not most of these needs in the future.

32–2. MEDICAL USES OF RADIOISOTOPES

As the name implies, radioisotopes are radioactive nuclides. In terms of its ordinary usage, a radioisotope is a species produced artificially by exposure of a stable isotope of a particular element to some sort of radiation. Nuclear reactors provide excellent sources of neutrons, but accelerators are also employed to bombard the stable nuclei with protons or other charged particles to produce radioisotopes. Some of these substances have found important medical applications both in the diagnosis and treatment of disease. For use in the body, a radioisotope should have a fairly short half-life and the daughter element should be harmless. A short half-life means that there will be a significant number of disintegrations per second so that very little of the radioisotope need be used. A short half-life also assures that the body will not be exposed to radiation for a very long period.

An extremely useful nuclide for study of the red blood cells is $^{51}_{24}\text{Cr}$. In the form of the chromate ion, this radioisotope becomes attached to the red cells. The rate of blood flow from the heart, the lifetime of a patient's red cells, and total blood volume are determined by techniques using $^{51}_{24}\text{Cr}$. Ingestion of small amounts of $^{131}_{53}\text{I}$ as iodide ion is used to measure the rate of iodine uptake by the thyroid gland. By measuring the radiation coming from the neck area as a function of time after swallowing a small amount of $\text{Na}^{131}_{53}\text{I}$, the thyroid gland activity can be determined.

The circulatory system can be studied by injecting some sodium chloride containing $^{24}_{11}\text{Na}$ into a vein. By placing counters at various points around the body, the rate of the patient's blood circulation is easily determined. The phosphate ion is, of course, found in normal blood. In many tumors, however, the phosphate concentration is several times higher than that in healthy cells. Consequently, phosphate enriched with $^{32}_{15}\text{P}$ offers a method of locating certain types of cancerous tissues.

Total body water is determined by administering a small amount of tritiated water, T_2O ($\text{T} = {}^3_1\text{H}$). After about six hours, when it has mixed uniformly with all the water in the body, a urine sample is analyzed for radioactivity. From the degree of dilution that has occurred, the total water content can be calculated.

The treatment of cancer has been aided by the introduction of $^{60}_{27}\text{Co}$ as a radiation source. The β particles released by this nuclide are usually shielded out and the intense γ radiation accompanying the decay is employed in the treatment. The $^{60}_{27}\text{Co}$ is easily made from ordinary $^{59}_{27}\text{Co}$ by neutron capture and has largely supplanted radium in cancer treatment. A small amount of $^{60}_{27}\text{Co}$ is more radioactive than the world's supply of radium.

Table 32–1 lists the radioisotopes that have been discussed in this section and gives some of their decay characteristics. The nuclide $^{137}_{55}\text{Cs}$ is also used as a high-intensity radiation source.

Table 32–1. Some Medically Useful Radioisotopes

Radioisotope	Mode of Decay	Half-life
$^{51}_{24}\text{Cr}$	$^{0}_{+1}e$ and K capture, γ	27 days
$^{131}_{53}\text{I}$	β, γ	8 days
$^{24}_{11}\text{Na}$	β, γ	15 hr
$^{32}_{15}\text{P}$	β	14 days
$^{3}_{1}\text{H(T)}$	β	12 years
$^{60}_{27}\text{Co}$	β, γ	5.3 years
$^{137}_{55}\text{Cs}$	β, γ	37 years

32–3. NEUTRON ACTIVATION ANALYSIS

The qualitative and quantitative determination of trace amounts of elements has been greatly facilitated by certain nuclear reactions. First discovered in 1936 by Georg von Hevesy, neutron activation analysis involves irradiation of the sample by neutrons. Some of the atoms comprising the sample absorb neutrons and thereby become radioactive. The subsequent decay of these unstable nuclides produces various particles along with γ rays. Detection of these γ rays with a scintillation counter reveals their exact energies and intensities, which can be directly related to the identity and amount of the decaying nuclei. From this information, the atoms in the original nonradioactive sample can be determined. Figure 32–3 is a schematic representation of the method.

The γ-ray spectrum obtained is compared with known spectra to identify the nuclear source of the γ rays. The half-lives of the radioactive atoms can also be determined and used to identify them. The advantages of neutron activation analysis over other analytical methods are several: (1) the limits of detection are quite low for atoms that readily absorb neutrons, for example, as low as 10^{-12} g

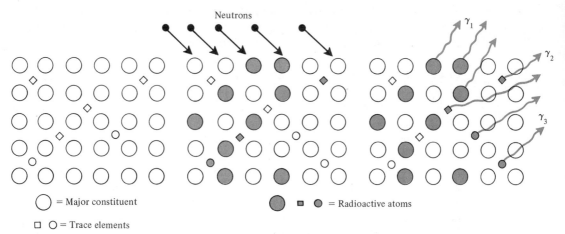

$\bigcirc$ = Major constituent $\bullet$ $\blacksquare$ $\bullet$ = Radioactive atoms

$\square$ $\bigcirc$ = Trace elements

Figure 32–3. Activation analysis involves the irradiation of a sample with neutrons thus making some of the various atoms in it radioactive. Analysis of the γ-rays produced when the unstable nuclei decay reveals the identity of the atoms.

under certain conditions, so that only a very small sample is necessary, (2) the sample is not destroyed in the analysis, (3) a number of elements can usually be identified simultaneously. There are, however, certain limitations inherent in the method: (1) not all nuclides readily absorb neutrons, although use of sources that produce a high density of neutrons plus long irradiation periods can sometimes overcome this problem, (2) elemental analysis only is obtained and molecular structure plays no role in the process, (3) the method is generally more expensive than many other analytical procedures because a reactor may be needed to serve as the neutron source.

Some historians have speculated that the death of Napoleon some 6 years after his exile to St. Helena may not have been from natural causes. Hair that had been shaved from Napoleon's head shortly after he died in 1821 was subjected to neutron activation analysis in 1961. The concentration of arsenic was found to be much higher than the minute amount found in normal hair. A diagnosis of probable chronic arsenic poisoning was thus made 140 years after death!

Modern criminology has received a great assist from activation analysis because tiny specks of dirt, small particles of paint, and other seemingly trivial pieces of evidence can be analyzed to identify their sources or can be compared with traces of materials found on a suspect. Wipings from the hands of a person who recently fired a gun can be analyzed to reveal traces of metals such as antimony, barium, and copper, which are found in gunpowder. By noting the ratio of these trace metals, the brand of ammunition can be identified because the relative amounts of the metals vary from one type to another.

Small amounts of pesticides can be detected in foods, even after processing, by analyzing for elements in the food that could only arise from pesticides. Terrestrial, lunar, and meteoric rocks can be analyzed for their trace components by neutron activation to learn more about the makeup and possible origin of these substances. In short, neutron activation analysis is an excellent tool for the determination of minute amounts of most of the elements.

32-4. RADIATION HAZARDS

The total activity of radioactive sources is measured in the unit **curie,** (c). 1 c is equal to 3.7×10^{10} disintegrations per second. The **millicurie** (mc) is used more often and is equal to 3.7×10^7 disintegrations per second. These units are used mainly in scientific work connected with the activities of radioisotopes. In many cases, the *effects* of radiation are of more interest than the total amount of radiation. Several units are used to express radiation effects. Chief among these is the **roentgen** (r). One roentgen is the quantity of X or γ radiation that will produce 2×10^9 ion pairs in 1 cm^3 of air at STP. One roentgen (1 r) leads to the absorption of approximately 83 ergs/g of dry air. A more useful unit for some purposes is the **rad** (radiation absorbed dose). One rad is the amount of any type of radiation leading to the absorption of 100 ergs/g of any material. The actual amount of radiation necessary to equal 1 rad will vary a little according to the absorbing material. Finally, the **rem** (roentgen equivalent, man) is defined as the amount of any type of radiation that will produce the same biological damage in tissue as that resulting from the absorption of 1 r by the tissue.

The hazards of exposure to radiation vary greatly with respect both to the intensity and to total area of the body exposed. For example, 10^4 r can be applied to a small area of the body without too much lasting effect, but 500 r applied at one time to the entire body will result in only a 50–50 chance of survival. The time of the exposure is also important. If 500 r is spread out over the lifetime of an individual, its effects would probably not be noticeable.

Man is exposed to radiation from various sources, among which are: (1) radiation from the natural environment, or background radiation, from cosmic rays and naturally occuring radioactive materials; (2) fallout from the atmospheric testing of nuclear weapons; (3) radioactive waste from industrial and scientific research activities; and (4) medical examinations and treatments employing X rays, γ rays, or nuclear particles.

For the average person, by far the largest doses arise from natural sources. This radioactivity varies from place to place, but amounts to about 10 r or less over the lifetime of the average person in the United States. The Nuclear Test Ban Treaty, signed in 1963 by the major nuclear powers, ended atmospheric testing, which had noticeably raised the level of worldwide radioactivity. Sources (3) and (4) are at present extremely low level because waste material is stored underground if necessary until activity levels are practically nil, and medical uses of radioactive sources are limited to cases of extreme necessity. X-ray examination is used (or should be) sparingly, and modern instruments expose no more of the body than is necessary. A dental X ray exposes the patient to about 0.005 r over a small region.

Hopefully, as more nations develop a nuclear capability, they will refrain from testing weapons, especially in the atmosphere where radioactive fallout is spread worldwide for years after the detonation. The most dangerous radioactivity arises from nuclides that are produced in relatively large amounts, have long half-lives, are efficiently transferred to man, and are retained for long periods in the body. The three nuclides that meet most or all of these criteria are $^{131}_{53}$I, $^{90}_{38}$Sr, and $^{137}_{55}$Cs. The nuclide potentially the most dangerous because of the large amounts produced in nuclear explosions is $^{131}_{53}$I. Fortunately, its half-life of 8 days reduces the greatest danger to a few weeks. The other two nuclides, however, have quite long half-lives, $^{90}_{38}$Sr has $t_{\frac{1}{2}} = 28$ years and $^{137}_{55}$Cs has $t_{\frac{1}{2}} = 37$ years. Moreover these isotopes are chemically similar to necessary elements in the body, Ca and K, respectively. All three nuclides reach the body through contaminated plant materials and milk from cows that feed on such plants.

It would be dangerous to assume that radiation is harmless as long as it is lower than naturally occurring radioactivity. Radiation causes ionization in cells which results in breaking chemical bonds and consequent destruction of tissue. All scientists agree that any amount of radiation, no matter how small, produces some *irreversible* alterations within biological cells. The difference of opinion arises over the question of what constitutes a *safe* limit of exposure. This question of a safe limit cannot be answered with certainty at this time. The effect of radiation on an individual is called the **somatic** effect whereas effects transmitted to his descendants are **genetic** effects. It is argued by many people that levels of radiation far below those required to produce an observable somatic effect may cause serious genetic effects generations later. Although the truth or falsity of this view cannot be established at this time, most scientists would prefer to err in the direction of safety. Consequently, it is best to avoid any *unnecessary* exposure to any form of radiation.

PROBLEMS

1. List some advantages and disadvantages in the use of water as the coolant for a nuclear reactor. Do the same for sodium.

2. How does the operation of a breeder reactor differ from that of an ordinary reactor?

3. Why is it necessary to have different measures of radiation such as the curie and the roentgen?

4. Thirty years after nuclear fission was carried out on a controlled basis, few nuclear power plants were in operation. Give some reasons for this lag. Discuss the probable future of nuclear power on a commercial basis. Suggest some reasons why nuclear-powered submarines, in contrast with land-based power plants, were produced on a fairly large scale in the 1960s.

5. If a breeder reactor with a doubling time of 10 years operates for 50 years, how many new reactors could it fuel, assuming each new fuel charge were placed in an identical breeder reactor?

6. Why is $^{14}_{6}C$ rarely used as a radioisotope for medical purposes?

7. Give some medical uses of $^{24}_{11}Na$, $^{32}_{15}P$, $^{60}_{27}Co$, $^{51}_{24}Cr$, $^{131}_{53}I$, and $^{3}_{1}H$.

8. Discuss some advantages and disadvantages of neutron activation as an analytical method.

9. Suggest techniques using radioisotopes for the following: (a) monitoring the flow of liquids and gases in pipelines; (b) determination of optimum amounts of trace elements (for example, B, Co, Cu) in plants; (c) monitoring the thickness of sheet metal.

10. Explain why 10 r is a serious radiation dose under certain circumstances but not under others.

11. Discuss the possible sources of hazardous radiation and consider means of control. In your discussion point out why scientific questions are sometimes inseparable from political and social questions.

12. Discuss some of the difficulties inherent in attempting to set "safe" levels of radiation exposure.

ENERGY AS A NATURAL RESOURCE

In his quest for the underlying cause of the many forms of animation that characterize the universe, man was eventually led to the concept of energy. But is this an answer? What is the nature of energy, its source, its end? Is energy in a state of perpetual creation? Or is the universe destined to expire one day?

Newton defined energy as a property of masses in motion. Since his day, however, the definition of energy has evolved to include the quantum, and finally Einstein's relativistic ideas, which led to the identification of energy with substance ($E = mc^2$).

But what of cosmic energy? In what forms does energy occur in the universe? The most obvious are gravitational, thermal, light, nuclear, and chemical. As inhabitants of our planet we have until recently been overwhelmingly concerned with chemical energy, the least significant in cosmic terms. By far, the predominant form of energy in the universe is gravitational. Every mass in space possesses gravitational energy that can be converted into heat or light when the masses fall toward one another. Of course, every star possesses its brilliance as the result of the release of nuclear energy arising primarily from the fusion of light atoms. But these nuclear reactions in stars were undoubtedly initiated by the tremendous heat produced during the gravitational compression of cosmic hydrogen gas. The energy released by the sun in its continual nuclear reaction accounts for a daily weight loss of 3.6×10^{11} tons! In this way much of the material universe is constantly disintegrating into energy radiations, most of which will continue their journey through space forever. The rapidity and magnitude of this degradation of matter into energy are truly staggering.

Certainly one of man's most pressing problems will continue to be finding means to tap the tremendous sources of energy in the universe as it proceeds in its voyage from order into chaos.

33-2. INTERCONVERSIONS OF ENERGY

Although the various energy forms can be interconverted, the interconversions have varying degrees of efficiencies. Energy forms can be rated in terms of entropy such that those forms of lower entropy can readily be degraded into forms of higher entropy, but those of higher entropy can never be completely converted into forms of lower entropy. This ranking of energy forms, therefore, is based on the inherent order or organization of the energy, and this statement is a restatement of the second law of thermodynamics. In other words, the statement specifies the direction of spontaneous energy flow.

In this ranking, gravitational energy is the most highly ordered and has an entropy of zero associated with it. For this reason, the conversion of gravitational energy into electrical energy through a hydroelectric generator can have an efficiency near 100%. The various cosmic energy forms can be ranked in the following rough order of increasing entropy: gravitational < rotational < orbital < nuclear < internal heat of stars < sunlight < chemical < terrestrial waste heat < cosmic radiation. The latter appears to be the ultimate "heat sink."

Man has found many ways to utilize and interconvert the various energy forms. In many examples he seems to have discovered ways of reversing the flow of entropy and thus of violating the second law of thermodynamics. One example is refrigeration (Section 29-8). The violation is more apparent than real, however. Mechanical motion in the piston is used to transfer heat from the system to be cooled into the heat sink thus producing an ordering of the system locally. The overall result, however, is the conversion of mechanical motion of low entropy to waste heat of high entropy in the sink, a process that results in a net increase in entropy.

33-3. EVOLUTION OF THE USE OF ENERGY

Presumably early man first used energy in the form of kinetic energy through the use of sticks and stones. One of his early discoveries was fire, which should be regarded as the first great technological breakthrough. One can only speculate about the discovery, but certainly (like the recent discovery of nuclear energy) fire provided a concentrated and immediate energy source whose uses would reveal themselves as man experimented with it. It affected man's diet, his ability to make tools and to invent new ones, his warfare, his comfort, his adaptability to new and harsher climates, and in many other ways his very survival and social evolution. Fire has even found its way into man's innermost feelings:

> ". . . Nature has perhaps no better offering than fire. The house is closed and it is night outside and lonely; yet how much closer we are to nature than the countryside itself, Platero, here at this window opened on the plutonic cavern! Fire is the universe within our houses. Red and ceaseless as the blood from a wound, it warms us and gives us strength, recalling all our earthly memories."

> *Platero and I*, by Juan Ramon Jimenez

Probably not until the discovery of nuclear energy was there another occurrence of such profound importance to man.

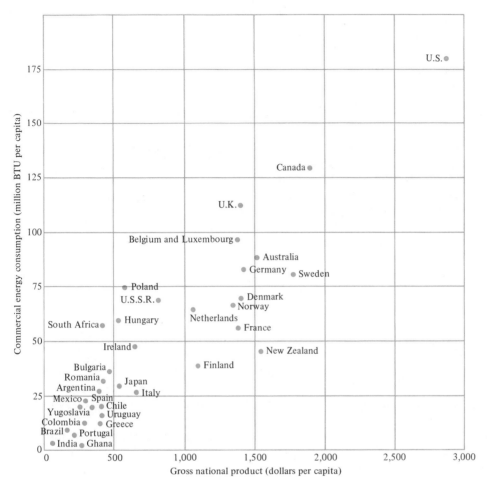

Figure 33-1. [From "The Flow of Energy in an Industrial Society, by Earl Cook. Copyright ©
September 1971 by Scientific American, Inc. All rights reserved.]

Another great discovery in the utilization of energy was the use of the wind
in sailing ships. Only a few moments reflection will suggest the far reaching effect
of this discovery upon man. Other important sources of energy in man's history
are the waterwheel and, much later, the windmill.*

The first mobile power source was the steam engine, a relatively modern
invention. In contrast to the waterwheel and the windmill, the steam engine
converts the more chaotic heat energy into the more ordered mechanical energy.
The efficiency of steam power, however, is correspondingly lower. The more
modern power sources, of course, are the internal combustion engine, the various
turbines, the electric dynamo, the jet engine, and the nuclear powered engines.
All of these suffer from low efficiencies because they all convert energy of a higher
entropy to energy of a lower entropy.

That the ability to utilize energy is directly related to a nation's material
prosperity is indisputable. Figure 33-1 shows the correlation between per capita

*Lest the student of expert hindsight fail to appreciate the intellectual achievements of such devices,
the wheel, which may seem so simple and so obvious, was not discovered by the very advanced
and cultured Maya civilization.

503

consumption of energy and gross national product. Whether or not there is a correlation with the absolute quality of life is another matter. Perhaps an area of future research will be the utilization of energy for the production of human spiritual and emotional well-being.

33-4. EFFICIENCY OF ENERGY CONVERSION

Efficiencies of energy conversion are related to the direction and magnitude of the entropy changes involved. The theoretical limit of efficiency of any heat engine, that is, an engine operating in a cycle that converts heat into mechanical work, was derived by Nicolas Carnot in 1824. His equation relates the efficiency to the difference between the temperatures of the entering steam (or other fluid) and that of the water that leaves the engine to be returned for reheating to steam. The efficiency is expressed as

$$\text{per cent efficiency} = \frac{T_1 - T_2}{T_1} \times 100$$

where T_1 and T_2 are the absolute temperatures of the entering steam and exiting water, respectively. At the turn of the century, typical efficiencies of steam power plants were around 5%. This is the overall efficiency of the conversion of energy in the fuel to electricity. At present, efficiencies of about 33% are typical. This increased efficiency is due largely to the use of higher temperature steam. In 1900, the typical steam temperature was 260°C. Today, temperatures of 540°C are achieved.

Nuclear power plants at present operate at lower efficiency than do fossil fuel steam plants because nuclear plants cannot be designed to operate at high enough temperatures. Typical operating steam temperatures in present nuclear power plants are around 350°C.

Efficiencies of various energy converters are tabulated in Figure 33–2. Interestingly enough, the 100,000,000 automobiles in the United States have an estimated power capacity of over 95% of all prime movers (defined as engines that convert fuel to mechanical energy), and account for over 16% of the fossil fuel consumed in the United States by all forms of transportation. Attempts to alleviate this situation by converting to electrically powered automobiles must take into account the amount of fuel required at the central power stations that would be needed to charge batteries for 100,000,000 automobiles.

33-5. MAJOR ENERGY SOURCES

The principal available energy source on earth has been the fossil fuels—petroleum and coal. These materials represent chemical energy whose conversion into mechanical or electrical energy is quite inefficient. A second energy source whose availability was quite timely was nuclear energy. It was timely because of the apparent need to find a substitute for the fossil fuels that are destined for depletion in the not too distant future. Of the possible nuclear sources, only fission of uranium and similar nuclides has become practical. At present only

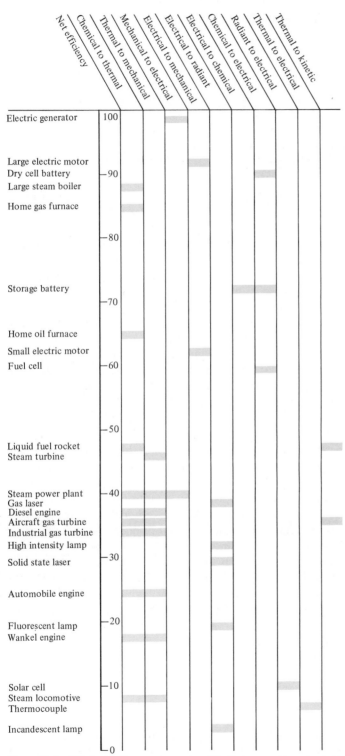

Figure 33-2. Efficiency of energy converters. The figures are percentages and represent the best values attainable with present technology. [From the Conversions of Energy by Claude M. Summers. Copyright September 1971 by Scientific American, Inc. All rights reserved.]

Table 33-1. Breakdown of Fate of Solar Radiation

Fate	Amount of Energy (watts)	Per cent of Total
Scattered back as short wave radiation	$52,000 \times 10^{12}$	30
Absorbed by atmosphere and surface and converted into heat	$81,000 \times 10^{12}$	46.7
Absorbed in evaporation, convection, precipitation, etc., of hydrologic cycle	$40,000 \times 10^{12}$	23.1
Convection and circulation of atmosphere and ocean	370×10^{12}	0.2
Absorbed by chlorophyll in green plants	40×10^{12}	<0.1
Total	$173,410 \times 10^{12}$	

about 5% of the electrical energy output in the United States originates from nuclear plants. However, several hundred nuclear power plants are either under construction or being planned. By the year 2000, it is estimated that nuclear power will account for 50% of our electrical power.

An almost limitless supply of nuclear power is available to us if or when it becomes technologically and economically practical. That power supply is nuclear fusion of deuterium using the water of the oceans.

But perhaps the prime energy source is the sun—the ultimate origin of all the earth's energy. The total solar radiation that strikes the earth's surface is 1.73×10^{17} watts.* A breakdown of the ways in which solar energy acts on the earth is described in Table 33-1.

Various schemes have been proposed for harnessing this tremendous energy source—5000 times the energy input to the earth of all other energy sources combined. These suggestions have included, among others, huge reflecting mirrors that concentrate the rays onto an absorbing surface, and solar batteries that convert solar energy directly into electrical energy.

The main attraction of solar energy, besides its quantity, is that it would add no additional heat load to the earth's biosphere. Such an energy source is called an **invariant** energy system. That is, the same amount of energy strikes the earth regardless of how it is used or converted.

Still another potential energy source is the wind. Windmills have, of course, been in use for centuries. The major problem with either wind or solar energy is in energy storage because both of these sources vary with time. One proposal has been to use the wind's energy intermittently to drive a water electrolysis plant for the production of hydrogen. The hydrogen could then be stored and burned as required. But even under conditions of high efficiency this possibility does not appear to be a significant source of power.

33-6. EPILOG

The history of man's material progress has been the history of the exploitation of energy. Man's initial problems in harnessing the energy around him revolved

*1 watt = work done at the rate of one joule/sec, or 1 amp (1 coulomb/sec) acting under a pressure of 1 volt.

around the first law of thermodynamics. As he progressed, however, he was led again and again to the inexorable movement toward chaos and randomness that is described in the second law. Like man's own mortality, the second law implies a beginning as well as an end. But man seems to face both of these cataclysms with the same perplexing optimism, a fact that attests to his inherent nobility and immortality.

PROBLEMS

1. List all the energy sources that have been used by man throughout history and the approximate time when they were discovered.

2. Make a schematic drawing of a refrigerating unit (Section 29-8) showing qualitatively the energy flows and entropy changes.

3. Arrange the processes below according to expected efficiency. Give your reasons for assigning each to its position.
   (a) water wheel   (d) automobile
   (b) electric motor   (e) jet engine
   (c) steam engine

4. Calculate the maximum theoretical efficiency of a fossil fuel steam plant operating at an initial steam temperature of 540°C and a final outlet water temperature of 40°C. Compare this with a nuclear powered steam plant whose maximum initial steam temperature is 350°C and whose exit water temperature is 40°C.

5. Calculate the energy available in the oceans by the utilization of the deuterium-deuterium reaction

$$\textstyle ^2_1H + ^2_1H \longrightarrow \, ^4_2He + \gamma$$

if the energy released per deuterium atom consumed is 7.94×10^{-13} joule; and if the natural isotopic abundance of deuterium is 0.15% of the total hydrogen present. The total volume of the oceans is 1.5 billion km^3. Assume that only 1% of the total deuterium in the oceans is consumed.

6. Explain why gravitational energy is converted into heat energy when hydrogen in space coalesces to form a star.

7. Using the fact that the energy released by the sun accounts for a daily weight loss of 3.6×10^{11} tons, calculate the amount of energy released.

8. The second law of thermodynamics is not violated when a refrigerator transfers cool air from inside the unit to the warmer surroundings making the unit cooler and the surroundings warmer. Explain why this is so.

9. Why would the conversion from internal combustion to electric engines in automobiles not necessarily alleviate the problem of air pollution?

10. Why is it said that the commercial use of solar energy should result in less *thermal* pollution than the use of nuclear or fossil fuel energy?

11. What are the major fundamental problems in the use of solar, tidal, or wind energy?

APPENDIXES

MATHEMATICAL MANIPULATIONS

A-1. EXPRESSING LARGE AND SMALL NUMBERS: USING POWERS OF 10

It is troublesome to make calculations using numbers such as 623,400,000 and 0.000000561. Also, it may be incorrect to express a very large number as is the first number above, since nine significant figures are implied. Scientific notation makes use of powers of 10 in order to express all numbers as one digit to the left of the decimal point. For numbers larger than 1, each place the decimal point is moved to the *left* becomes one *positive* power of 10. Thus, 623,400,000 becomes 6.234×10^8. For numbers smaller than 1, each place the decimal point is moved to the *right* becomes one *negative* power of 10. Thus, 0.000000561 becomes 5.61×10^{-7}.

■ Examples

1. 2,479,000 is 2.479×10^6
2. 100,042,600 is 1.000426×10^8
3. 0.0240069 is 2.40069×10^{-2}
4. 0.000501 is 5.01×10^{-4}
5. 2,495 is 2.495×10^3
6. 584 is 5.84×10^2

Note: (1) Any number may be considered as multiplied by 10^0 as it stands since $10^0 = 1$. Thus 1.51 is the same as 1.51×10^0. (2) Normally, the power of 10 notation is not used if a number lies between 10 and 100; thus 2.24×10^1 would be written simply 22.4.

☐ *Problems*

1. Express the following numbers using scientific notation:
(a) 637,000,000   (c) 0.00042
(b) 62.1   (d) 0.0901

2. Change the following scientific notation numbers to ordinary numbers:
(a) 5×10^6 (c) 8.70×10^2
(b) 4.06×10^{-3} (d) 7.15984×10^3

A-2.   SIGNIFICANT FIGURES

All numbers following the first *nonzero* digit in a number are significant.

■ **Examples:**

1. 2.479×10^6 has four significant figures.
2. 1.045426×10^8 has seven significant figures.
3. 0.0240 has three significant figures.
4. 0.00240 has three significant figures.
5. 0.002401 has four significant figures.
6. 0.0024 has two significant figures.

In rounding off to a desired number of significant figures, the digit past the last desired digit determines the rounding procedure. If the digit to be dropped is between 0 and 4 inclusive, it is simply dropped. If the digit to be dropped is between 5 and 9 inclusive, the last retained digit is increased by one.

Suppose that in all of the above examples, it is desired to retain three significant figures. We would then have

1. 2.48×10^6 4. 2.40×10^{-3}
2. 1.05×10^8 5. 2.40×10^{-3}
3. 2.40×10^{-2} 6. Cannot have three significant figures.

Suppose that only two significant figures were desired in the above examples. We would then have

1. 2.5×10^6 4. 2.4×10^{-3}
2. 1.0×10^8 5. 2.4×10^{-3}
3. 2.4×10^{-2} 6. 2.4×10^{-3}

☐ *Problems*

1. State the number of significant figures in the following numbers:
(a) 6.9800×10^5 (d) 3.070×10^{-4}
(b) 72   (e) 0.001
(c) 0.30

2. Round the following numbers to two significant figures in each case:
(a) 3.48
(d) 0.00255
(b) 7.541×10^{15}
(e) 4.70×10^{-6}
(c) 6,000

A-3. ADDITION AND SUBTRACTION

In order to add or subtract numbers expressed in scientific notation, the powers of 10 must be the same. The final answer is expressed to the same number of decimal places as that of the term with the least number of decimal places.

■ **Examples**

1. $156.2 + 6.31 \times 10^3 + 50.62 \times 10^{-1}$
   Make all powers of 10 the same:

$$0.1562 \times 10^3$$
$$6.31 \times 10^3$$
$$\underline{0.005062 \times 10^3}$$
$$6.471262 \times 10^3 \text{ which rounds to } 6.47 \times 10^3$$

It can be seen that numbers with many decimal places beyond the term with the least number of places need not be retained in the calculation. Usually only one or two places beyond the least number of decimal places are used in the calculation.

2. $5.7 \times 10^{-2} + 4.64 + 0.02807 + 7.82 \times 10^{-4}$

$$5.7 \ \times 10^{-2}$$
$$464. \ \ \times 10^{-2}$$
$$2.81 \times 10^{-2}$$
$$\underline{0.08 \times 10^{-2}}$$
$$472.59 \times 10^{-2} \text{ which rounds to } 473 \times 10^{-2}, \text{ or simply } 4.73$$

3. $4.631 \times 10^{-2} - 4.509 \times 10^{-5}$

$$4.631 \ \ \times 10^{-2}$$
$$\underline{-0.00451 \times 10^{-2}}$$
$$4.62649 \times 10^{-2} \text{ which rounds to } 4.626 \times 10^{-2}$$

☐ *Problems*

Carry out the following operations:
(a) $6.24 \times 10^2 + 6.80 + 1.220 \times 10^3$
(b) $5.4 \times 10^{-5} + 6.97 \times 10^{-4} + 3.6 \times 10^{-6}$
(c) $7.4 + 9.9 \times 10^{-2} - 4 \times 10^2$

A-4. MULTIPLICATION

In order to multiply numbers expressed in scientific notation, the exponents of 10 are algebraically added and the ordinary numbers are multiplied. The answer is rounded off to give the same number of significant figures as the factor with the least number of significant figures.

■ **Examples**

1. $(6.31 \times 10^4) \times (4.01 \times 10^{-6}) \times (3.2 \times 10^{-2})$
   $= 6.31 \times 4.01 \times 3.2 \times 10^{4-6-2}$
   $= 6.31 \times 4.01 \times 3.2 \times 10^{-4}$
   $= 80.96992 \times 10^{-4}$ which rounds to 81×10^{-4}, or 8.1×10^{-3}
2. $(5 \times 10^3) \times (2.001 \times 10^2) \times (6.0 \times 10^4)$
   $= 5 \times 2.001 \times 6.0 \times 10^{3+2+4}$
   $= 5 \times 2.001 \times 6.0 \times 10^9$
   $= 60.0300 \times 10^9$ which rounds to 6×10^{10}

Note: Many problems in your chemistry course will involve numbers with three significant figures. Use of a slide rule to multiply numbers with three significant figures automatically rounds the answer to three significant figures.

☐ *Problems*

Carry out the following operations:
(a) $5.6 \times 8.0 \times 1.01 \times 10^7$
(b) $9.0 \times 10^{-5} \times (5.401 \times 10^{-2} - 7.81 \times 10^{-3})$
(c) $1.00 \times 10^9 \times 2.00 \times 10^{-4} \times 5.00 \times 10^{-6}$

A-5. DIVISION

In order to divide numbers expressed in scientific notation, the exponent of 10 in the denominator is subtracted from the exponent of 10 in the numerator and the ordinary numbers are divided. The answer is rounded off to give the same number of significant figures as the factor with the least number of significant figures.

■ **Examples**

1. $\dfrac{6.73 \times 10^{-2}}{9.4 \times 10^3}$

   $= \dfrac{6.73}{9.4} \times 10^{-2-3}$

$$= \frac{6.73}{9.4} \times 10^{-5}$$

Now divide the ordinary numbers and carry the division out to give one more significant figure than is to be finally retained. Thus, 0.715×10^{-5} is rounded to 0.72×10^{-5} or, 7.2×10^{-6}

2. $\dfrac{6.01 \times 10^{-4}}{4.00 \times 10^{-3}}$

$$= \frac{6.01}{4.00} \times 10^{-4-(-3)}$$

$$= \frac{6.01}{4.00} \times 10^{-1}$$

$$= 1.502 \times 10^{-1} \text{ which rounds to } 1.50 \times 10^{-1}$$

Note: (1) Many problems in your chemistry course will involve numbers with three significant figures. Use of a slide rule to divide numbers with three significant figures automatically rounds the answer to three significant figures. (2) The order of multiplication or division in a fraction of several factors in the numerator and denominator is unimportant.

$$\frac{6.4 \times 10^2 \times 4.81 \times 10^3 \times 6.1 \times 10^{-2}}{5.07 \times 10^{-2} \times 9.2 \times 10^7}$$

$$= \frac{6.4 \times 4.81 \times 6.1 \times 10^{2+3-2-(-2+7)}}{5.07 \times 9.2}$$

$$= \frac{6.4 \times 4.81 \times 6.1}{5.07 \times 9.2} \times 10^{-2}$$

The factors in the ordinary fraction may be multiplied or divided in any order. However, if a slide rule is used, it is often easier to begin by setting the first number on the left in the numerator on the D-scale, dividing by the first number on the left in the denominator, then without reading the answer, continue by multiplying by the second number in the numerator, dividing by the second number in the denominator, and finally multiplying by the third number in the numerator. Only now is the answer finally read. The proper placing of the decimal is made by estimating the answer. Thus the slide rule process normally is best carried out by dividing, multiplying, dividing, and so on.

$$\frac{6.4 \times 4.81 \times 6.1}{5.07 \times 9.2}$$

☐ *Problems*

Carry out the following operations:

(a) $\dfrac{5.90}{7.61}$

(b) $\dfrac{8.0 \times 7.5 \times 10^2}{4.00 \times 10^5}$

(c) $\dfrac{8.81 \times 10^{-2} \times (3.001 + (7.58 \times 10^{-3}))}{5.04 \times 10^{-4}}$

(d) $\dfrac{748 \times 450 \times 302}{1050 \times 273}$

A-6. UNIT ANALYSIS

Unit analysis is a method of solving problems that is widely used in chemistry and other sciences as well. The method requires that you know the units in which all numbers to be used in solving the problem are expressed as well as the units to be attached to the final answer. Units are then treated as any fraction of numbers would be treated. Many units involve the word "per," for example 50 centimeters per second. "Per" may always be replaced by "divided by" and the units written as a fraction. Thus we could write, 50 cm/sec. The power of unit analysis in problem solving may be seen in the following example where ficticious units are used.

■ Example

How many zaps are there per zip if 1.06 zaps equal 9.34 zoks and 1.00 zok is equal to 5.63 zips?

First, write out all numbers that are equivalent, in fractional form, with units attached:

$$\dfrac{9.34 \text{ zok}}{1.06 \text{ zap}} \qquad \dfrac{5.63 \text{ zip}}{1.00 \text{ zok}}$$

Next, determine the units desired for the answer:

$$\dfrac{\text{zap}}{\text{zip}}$$

Finally, multiply or divide the fractions in such a way that unwanted units will cancel and leave only the desired units. It can be seen that the unit "zok" is unwanted. Also, if the two fractions are multiplied, the "zok" unit will cancel because it is in the numerator of the fraction on the left and in the denominator of the fraction on the right:

$$\dfrac{9.34 \text{ zok}}{1.06 \text{ zap}} \times \dfrac{5.63 \text{ zip}}{1.00 \text{ zok}}$$

However, the units obtained upon multiplying the fractions is seen to be "zip per zap" while "zap per zip" is the desired unit. Therefore, the reciprocal of the above operation is required.

$$\frac{1}{\dfrac{9.34 \text{ zok}}{1.06 \text{ zap}} \times \dfrac{5.63 \text{ zip}}{1.00 \text{ zok}}} = \frac{1.06 \text{ zap}}{9.34 \text{ zok}} \times \frac{1.00 \text{ zok}}{5.63 \text{ zip}}$$

$$= \frac{1.06 \times 1.00 \text{ zap}}{9.34 \times 5.63 \text{ zip}}$$

$$= 2.02 \times 10^{-2} \frac{\text{zap}}{\text{zip}}$$

This rather silly problem should serve to demonstrate what a powerful tool unit analysis can be to the student because problems can be solved even when the units in which the numbers are expressed are unfamiliar. The method does require, however, that units be attached to numbers throughout a calculation.

■ **Examples**

1. If one dozen eggs costs 57 cents, what fraction of an egg is purchased per cent?

$$\frac{12 \text{ eggs}}{57 \text{ cents}} = 0.21 \frac{\text{egg}}{\text{cent}}$$

What would be the cost in cents per egg?

$$\frac{1}{0.21 \dfrac{\text{egg}}{\text{cent}}} = 4.8 \frac{\text{cents}}{\text{egg}}$$

2. How many kilograms of $NaNO_3$ are 50.2 lb of $NaNO_3$? 1.00 lb = 454 g and there are 1.00×10^3 g/kg.

$$1.00 \times 10^3 \frac{\text{g}}{\text{kg}} \qquad 454 \frac{\text{g}}{\text{lb}}$$

$$\frac{50.2 \text{ lb} \times 454 \dfrac{\text{g}}{\text{lb}}}{1.00 \times 10^3 \dfrac{\text{g}}{\text{kg}}} = 22.8 \text{ kg}$$

3. What is the atomic weight of an element if 0.48 g mole has a mass of 0.012 kg?

In order to solve this problem by means of unit analysis, it is necessary to know that the atomic weight of an element is simply the grams per g-mole.

$$\frac{0.012 \text{ kg} \times 1.0 \times 10^3 \frac{g}{kg}}{0.48 \text{ g-mole}} = 25 \frac{g}{\text{g-mole}}$$

☐ *Problems:*

Solve the following using unit analysis:

(a) Dittles cost \$5.00 per gross. Find the cost of $1\frac{1}{2}$ dozen dittles. A gross is a dozen dozen.

(b) 5.2 yips equal 6.8 yaps. There are 4.3 baps per yap. Find the number of baps equal to 10 yips.

(c) The atomic weight of element A is 20. If 1.0 g of this element combines with 0.10 g-mole of another element, X, find the formula of the compound. (The number of g-moles of A per g-mole of B is the same as the atoms of A per atom of B in the formula.)

(d) The speed of light in a vacuum is 3.0×10^{10} cm/sec. If the frequency of a certain radiation is 1.5×10^{15} sec^{-1}, find the wavelength in meters of this radiation. There are 1.0×10^2 cm/m.

A-7. PROPORTIONALITY

Many times two physical quantities or variables are found to be directly proportional. A well known example is a modification of Charles' law, which states that, at constant pressure, the volume and absolute temperature of a gas are directly proportional:

$$V \propto T$$

As one variable gets larger or smaller, so also does the other. A mathematical consequence of direct proportionality between two variables is that they may be set equal to each other if a constant multiplier is attached to one of the variables.

$$V = kT$$

k is called the proportionality constant. If two variables which are directly proportional are plotted on a graph with one variable along the x axis and the other along the y axis, a straight line always results. The line always passes through the origin or zero-point on both axes.

The numerical value of k determines the steepness or slope of the line.

Sometimes it is found that two variables are inversely proportional. A good example of this behavior in nature is summarized by Boyle's law, which states that, at constant temperature, the pressure and volume of a gas are inversely proportional.

$$P \propto \frac{1}{V}$$

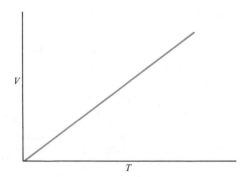

As one variable gets larger, the other gets smaller. A mathematical consequence of inverse proportionality between two variables is that the product of the variables is equal to some proportionality constant.

$$PV = K$$

A plot of two variables which are inversely proportional gives a curve called an hyperbola.

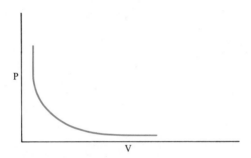

The numerical value of K determines how snug a fit the curve has with respect to the axes.

A final point to be noted is that, if a given variable is proportional to several different variables, then it is also proportional to the product of those other variables.

In the examples above V is directly proportional to T and inversely proportional to P. Therefore

$$V \propto T \times \left(\frac{1}{P}\right)$$

The introduction of a proportionality constant, C, gives

$$PV = CT$$

519

A-8. USEFUL GEOMETRIC RELATIONSHIPS AND DEFINITIONS

1. The circumference of a circle is equal to its diameter multiplied by $\pi(3.14)$. $C = \pi d$. In terms of the radius, $C = 2\pi r$
2. The area of a circle, $A = \pi r^2$
3. Angle size is measured in degrees of arc. $1°$ of arc is $1/360$ of the arc along the circumference of a circle. Thus, any circle has $360°$ of arc.
4. Angle size is also measured in radians. One radian is the arc length of a circle equal to the radius of the circle. Thus $360°$ is equal to 2π radians. One radian is about $57°$.
5. A straight angle is an angle of $180°$ or π radians.
6. A right angle is an angle of $90°$ or $\pi/2$ radians.
7. An obtuse angle is an angle larger than a right angle.
8. An acute angle is an angle smaller than a right angle.
9. The sum of the interior angles of any triangle is $180°$.
10. A right triangle contains a right angle. The side opposite the right angle is called the hypotenuse.

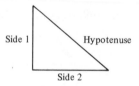

The Pythagorean theorem states:

$$(\text{length of hypotenuse})^2 = (\text{length of side 1})^2 + (\text{length of side 2})^2$$

■ Examples

1. A right triangle has a hypotenuse of 2.5 cm and one side of 1.5 cm. Calculate the length of the other side.

$$(2.5 \text{ cm})^2 = (1.5 \text{ cm})^2 + S^2$$

$$6.25 \text{ cm}^2 = 2.25 \text{ cm}^2 + S^2$$

$$S^2 = 4.00 \text{ cm}^2$$

$$S = 2.0 \text{ cm}$$

2. Express 30 deg in radians.

$$\frac{180 \text{ deg}}{\pi \text{ radians}}$$

$$\frac{30 \text{ deg}}{\dfrac{180 \text{ deg}}{\pi \text{ radians}}} = 0.52 \text{ radian}$$

3. Calculate the radius of a circle with an area of 160 cm.²

$$A = \pi r^2$$

$$r^2 = \frac{A}{\pi}$$

$$= \frac{160 \text{ cm}^2}{3.14}$$

$$r^2 = 51.0 \text{ cm}^2$$
$$r = 7.14 \text{ cm}$$

☐ *Problems*

1. Solve the following:
  (a) The circumference of a circle is 31.4 cm. Find the area of the circle.
  (b) Express $\pi/4$ radians in degrees.
  (c) Calculate the length of the hypotenuse of a right triangle if it has sides of 2.0 m and 5.7 m.
2. Answer the following:
  (a) How many interior acute angle(s) may a triangle have?
  (b) How many interior obtuse angle(s) may a triangle have?
  (c) A right triangle must have two interior angles of what type?

A-9. TRIGONOMETRIC FUNCTIONS

A right triangle has one 90° angle and two acute angles. The sizes of the acute angles are related directly to the length of the sides and the hypotenuse. Each acute angle has a side adjacent to it and a side opposite it. Consider the angle, θ, in the triangle:

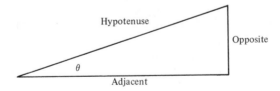

The three most important trigonometric functions are defined

$$\text{sine } \theta \text{ or, } \sin \theta = \frac{\text{length of opposite side}}{\text{length of hypotenuse}}$$

$$\text{cosine } \theta \text{ or, } \cos \theta = \frac{\text{length of adjacent side}}{\text{length of hypotenuse}}$$

$$\text{tangent } \theta \text{ or, } \tan \theta = \frac{\text{length of opposite side}}{\text{length of adjacent side}}$$

The functions for the other acute angle are defined similarly except that the roles of opposite and adjacent sides would be switched. A special right triangle is the 30-60-90 triangle which has the relative lengths of sides and hypotenuse as follows:

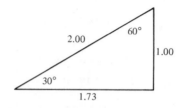

Thus:

$$\sin 30° = 0.500 \qquad \sin 60° = 0.865$$
$$\cos 30° = 0.865 \qquad \cos 60° = 0.500$$
$$\tan 30° = 0.578 \qquad \tan 60° = 1.73$$

Notice that, $\sin 30° = \cos 60°$ and $\sin 60° = \cos 30°$

It is always true that, $\sin \theta = \cos (90 - \theta)$. To obtain the functions for other angles, it is usually easiest to use a slide rule which gives the trigonometric functions or to consult tables in a reference book. It is useful to note that, as an acute angle in the right triangle becomes very small, its adjacent side and the hypotenuse become nearly the same length. Therefore, *for very small angles of just a few degrees*

$$\sin \theta = \tan \theta$$

☐ *Problems*

1. Use the Pythagorean theorem to calculate the lengths of the sides and hypotenuse for a 45-45-90 triangle. Then calculate $\sin 45°$, $\cos 45°$, and $\tan 45°$.

2. Without the use of tables or a slide rule, determine $\sin 2°$ and $\tan 2°$ if $\cos 88° = 0.0349$.

3. From basic definitions determine: $\sin 0°$, $\cos 0°$, $\tan 0°$, $\sin 90°$, $\cos 90°$, $\tan 90°$.

A-10.   QUADRATIC EQUATIONS

A quadratic equation is one in which the unknown quantity is squared. There may or may not also be a first-power unknown term, and there is a constant term in the equation. There are always two solutions to any quadratic equation. The two solutions occasionally are identical, but usually they are different. In chemical problems involving quadratic equations only one answer has physical meaning and the other answer or root of the equation, although mathematically correct, has no physical meaning and is discarded.

If no first-power term is present, a solution is simple:

$$x^2 - 4 = 0$$

or,

$$x^2 = 4$$

and taking the square root of both sides

$$x = \pm 2$$

If the constant term is not a perfect square, as is usually the case in actual problems, a slide rule is the simplest way to obtain square roots. However, *for numbers not much greater than unity,* a simple way to get square roots is to take half of the decimal portion of the number

$$x^2 = 1.16$$
$$x = \pm 1.08$$

because,

$$\frac{0.16}{2} = 0.08$$

If only two significant figures are needed, this method is good for numbers up to about 1.4, thus

$$x^2 = 1.4$$
$$x = \pm 1.2$$

For three significant figures, the 1.16 example above is about as large as the number can be to avoid introducing error.

In general, there are three ways to solve a quadratic equation: factoring, completing the square, and using the quadratic formula. For chemical problems where the first-power term is present, the quadratic formula is the only useful method of solution. For the general case

$$ax^2 + bx + c = 0$$

where a, b, and c represent numbers, either positive or negative, the solutions are

$$x = \frac{-b \pm \sqrt{b^2 - 4ac}}{2a}$$

Often, preliminary simplification of the algebraic expression is necessary before the quadratic formula can be applied; for example, the following quadratic equation cannot be solved as it stands using the formula.

$$\frac{x^2}{1 - x} = 4.0 \times 10^{-3}$$

However, if both sides of the equation are multiplied by $(1 - x)$, there results

$$(1 - x) \times \frac{x^2}{(1 - x)} = (1 - x) \times 4.0 \times 10^{-3}$$

$$x^2 = (1 - x) \times 4.0 \times 10^{-3}$$

Multiplying out the right hand side gives

$$x^2 = 4.0 \times 10^{-3} - (x) \times (4.0 \times 10^{-3})$$

Rearranging to set the left hand side equal to zero gives

$$x^2 + (x) \times (4.0 \times 10^{-3}) - 4.0 \times 10^{-3} = 0$$

The equation can now be solved by application of the quadratic formula with

$$a = 1.0 \quad b = 4.0 \times 10^{-3} \quad \text{and } c = -4.0 \times 10^{-3}$$

■ **Examples**

1. $x^2 - 1.2x = 3.2 \times 10^{-2}$
   First, set everything equal to zero.

$$x^2 - 1.2x - 3.2 \times 10^{-2} = 0$$

Then,

$$x = \frac{-(-1.2) \pm \sqrt{(-1.2)^2 - 4 \times (-)(3.2 \times 10^{-2})}}{2}$$

$$= \frac{1.2 \pm \sqrt{1.44 + 0.13}}{2}$$

$$= \frac{1.2 \pm \sqrt{1.57}}{2}$$

$$= \frac{1.2 \pm 1.3}{2}$$

So that

$$x = \frac{2.5}{2} = 1.3 \quad \text{and} \quad x = \frac{-0.10}{2} = -5.0 \times 10^{-2}$$

Both answers are solutions to the equation, but in a physical problem, only one can have meaning. In most chemical applications, negative roots are discarded.

2. $x^2 + 6.0 \times 10^{-5}x - 2.0 \times 10^{-5} = 0$

$$x = \frac{-6.0 \times 10^{-5} \pm \sqrt{(6.0 \times 10^{-5})^2 - 4(-2.0 \times 10^{-5})}}{2}$$

$$= \frac{-6.0 \times 10^{-5} \pm \sqrt{36 \times 10^{-10} + 8.0 \times 10^{-5}}}{2}$$

$$= \frac{-6.0 \times 10^{-5} \pm \sqrt{8.0 \times 10^{-5}}}{2}$$

In order to take the square root of 10 raised to some power, one simply takes one-half of the power. Therefore, in taking square roots, the power of 10 should be an even number.

$$x = \frac{-6.0 \times 10^{-5} \pm \sqrt{80 \times 10^{-6}}}{2}$$

$$= \frac{-6.0 \times 10^{-5} \pm 8.9 \times 10^{-3}}{2}$$

Thus

$$x = \frac{-0.06 \times 10^{-3} + 8.9 \times 10^{-3}}{2} = 4.4 \times 10^{-3}$$

and

$$x = \frac{-0.06 \times 10^{-3} - 8.9 \times 10^{-3}}{2} = -4.5 \times 10^{-3}$$

Again, in a chemical problem, one answer would be discarded as physically impossible, probably the negative result.

Note: The square root of a negative number is called an imaginary number because there is no way in which a real number can be multiplied by itself to give a negative number. In your work in general chemistry, you will not be concerned with imaginary numbers.

☐ *Problems*

(1) Find the roots of the following quadratic equations:
   (a) $x^2 - 16 = 0$
   (b) $x^2 = 1.8 \times 10^{-5}$
   (c) $x^2 - 1.04 \times 10^{-8} = 0$

(2) Use the quadratic formula to solve the following equation:

$$\frac{x^2}{1.00 - x} = 4.00 \times 10^{-4}$$

A–11. LOGARITHMS

The logarithm of a number is the power to which 10 must be raised to give the number. Thus, $\log 100 = 2$ because $10^2 = 100$. In general if

$$\log N = a$$

then,

$$N = 10^a$$

The number corresponding to a logarithm is called the antilogarithm of the logarithm. Thus, N is the antilogarithm of a. Some useful properties of logarithms are

$$\log (A \times B) = \log A + \log B$$

$$\log A^n = n \log A$$

$$\log \frac{A}{B} = \log A - \log B$$

For most chemical problems, it is sufficiently accurate to use a slide rule to obtain logarithms and antilogarithms. Tables are usually more confusing to use. Remember that any number between 1 and 10 has a logarithm which is 0. . . . . The logarithm of 10 raised to a power is simply the power, so consistent use of scientific notation makes it much easier to express the logarithms of numbers.

■ Examples

1.
$$\log N = 6.32$$
$$N = 10^{6.32}$$
$$= 10^6 \times 10^{0.32}$$

Antilog 0.32 = 2.1. Therefore

$$N = 2.1 \times 10^6$$

2.
$$\log N = -3.41$$
$$N = 10^{-3.41}$$
$$= 10^{-4} \times 10^{0.59}$$

Antilog 0.59 = 3.9. Therefore

$$N = 3.9 \times 10^{-4}$$

3.
$$\log N = -0.064$$
$$N = 10^{-0.064}$$
$$= 10^{-1} \times 10^{0.936}$$

Antilog 0.936 = 8.61. Therefore

$$N = 8.61 \times 10^{-1}$$

4.
$$-0.030 = 0.060 \log \frac{x}{(3.2 \times 10^{-1})^2}$$

$$\log \frac{x}{(3.2 \times 10^{-1})^2} = -0.50$$

$$\frac{x}{(3.2 \times 10^{-1})^2} = 10^{-0.50}$$

$$= 10^{-1} \times 10^{0.50}$$

Antilog $0.50 = 3.2$. Therefore

$$\frac{x}{(3.2 \times 10^{-1})^2} = 3.2 \times 10^{-1}$$

and

$$x = (3.2 \times 10^{-1})^2 \times 3.2 \times 10^{-1}$$
$$= (3.2 \times 10^{-1})^3$$
$$= 3.3 \times 10^{-2}$$

5.

$$N = 4.68 \times 10^{-7}$$

$$\log N = \log (4.68 \times 10^{-7})$$

$$= \log 4.68 + \log 10^{-7}$$

$$= 0.672 + (-7.000)$$

$$= -6.328$$

6.

$$N = 1.04 \times 10^3$$

$$\log N = \log (1.04 \times 10^3)$$

$$= \log 1.04 + \log 10^3$$

$$= 0.017 + 3.000$$

$$= 3.017$$

Note: (1) $\log x / \log y \neq \log x/y$. (2) Although logarithms are often negative (see examples 2, 3, 4, 5 above) there is no such thing as a logarithm of a negative number.

☐ *Problems*

1. If $\log x = 3.42$, find x.
2. If $\log y = -1.22$, find y.
3. Find the logarithms of the following numbers:
   (a) 6.23×10^{-3} (b) 1.05×10^4 (c) 455
4. Solve the following equations:
   (a) $2.63 = 4.03 + \log \dfrac{6.22 \times 10^{-2}}{x}$

   (b) $\log x^2 = 5.14$

ANSWERS TO PROBLEMS

Section A-1

1. (a) 6.37×10^8 (b) 62.1 (c) 4.2×10^{-4} (d) 9.01×10^{-2}
2. (a) $5,000,000$ (b) 0.00406 (c) 870 (d) 7159.84

Section A-2

1. (a) 5   (b) 2   (c) 2   (d) 4   (e) 1
2. (a) 3.5   (b) 7.5×10^{15} (c) 6.0×10^3 (d) 2.6×10^{-3} (e) 4.7×10^{-6}

Section A-3

(a) 1.851×10^3 (b) 7.55×10^{-4} (c) -4×10^2

Section A-4

(a) 4.5×10^8 (b) 4.2×10^{-6} (c) 1.00

Section A-5

(a) 7.75×10^{-1} (b) 1.5×10^{-2} (c) 5.26×10^2 (d) 3.55×10^2

Section A-6

(a) \$0.625   (b) 56 baps   (c) AX_2 (d) 2.0×10^{-7} m

Section A-8

1. (a) 78.5 cm^2 (b) 45°   (c) 6.0 m
2. (a) two or three   (b) one   (c) acute

Section A-9

1. Sides are equal with the hypotenuse 1.41 ($\sqrt{2}$) times the length of a side. $\sin 45° = \cos 45° = 0.707$; $\tan 45° = 1.00$
2. $\sin 2° = \tan 2° = 0.0349$
3. $\sin 0° = 0$, $\cos 0° = 1.00$, $\tan 0° = 0$, $\sin 90° = 1.00$, $\cos 90° = 0$, $\tan 90°$ is infinitely large

Section A-10

1. (a) ± 4.0 (b) $\pm 4.2 \times 10^{-3}$ (c) $\pm 1.02 \times 10^{-4}$
2. $x = -2.02 \times 10^{-2}$, $x = 1.98 \times 10^{-2}$

Section A-11

1. 2.6×10^3
2. 6.0×10^{-2}
3. (a) -2.20 (b) 4.02   (c) 2.66
4. (a) 1.55   (b) 3.71×10^2

NOMENCLATURE

One of the first problems that attends the development of any science is that of nomenclature. In the early stages of development names usually originate from common usage. Such names are often descriptive of the substance or of its use. Examples are "flowers" of sulfur and "milk of magnesia." As the science of chemistry developed into its modern form, with the attendant increase in the number of known substances, such names became less and less useful because they were, like proper names, difficult to associate and difficult to remember.

Eventually the problem became of such major proportions that chemists from all over the world took on the task of descriptive naming as an international project. In 1892, the first international meeting of organic chemists was held to discuss problems of nomenclature. Since that time, the IUPAC (International Union of Pure and Applied Chemistry) has met periodically to extend and revise the system of nomenclature of both organic and inorganic compounds.

B-1. ORGANIC COMPOUNDS

The system is based on a pattern of naming the various families of hydrocarbons and their derivatives. Through this system, all of the million or more known compounds of carbon can be so named that any chemist can, from the name alone, know the exact molecular structure.

Names of saturated hydrocarbons form the basis for naming all of their derivatives, so they will be taken up first (Table B-1). Except for the first four members, the names are derived from the Greek prefixes for the number of carbon atoms. These saturated hydrocarbons are called the paraffins (no affinity) or, according to IUPAC terminology, the alkanes. When one of the hydrogen atoms

Table B-1. Names of the Simple Saturated Hydrocarbons (Alkanes)

CH_4	methane	C_9H_{20}	nonane
C_2H_6	ethane	$C_{10}H_{22}$	decane
C_3H_8	propane	$C_{11}H_{24}$	undecane
C_4H_{10}	butane	$C_{12}H_{26}$	dodecane
C_5H_{12}	pentane	$C_{14}H_{30}$	tetradecane
C_6H_{14}	hexane	$C_{16}H_{34}$	hexadecane
C_7H_{16}	heptane	$C_{18}H_{38}$	octadecane
C_8H_{18}	octane	$C_{20}H_{42}$	eicosane

is replaced by another atom or group, the remaining hydrocarbon group is named by replacing the suffix **-ane** with **-yl.** Examples are as follows:

CH_3Cl methyl chloride

C_4H_9—⬡ butyl benzene

$C_6H_{13}OH$ hexyl alcohol (or hydroxyhexane)

The alkanes all have the empirical formula C_nH_{2n+2} where n is any integer. This is true regardless of branching; for example,

$$CH_3—CH_2—CH_2—CH_3 \quad \text{and} \quad CH_3—\overset{\overset{\displaystyle CH_3}{|}}{CH}—CH_3$$

both have the formula C_4H_{10}. Older names for the above two isomers of C_4H_{10} are n-butane (n = normal) and i-butane (i = iso). These are the so-called common or trivial names. One can readily visualize the problems that would be encountered in trying to name all of the isomers of hexane, C_6H_{12}

$$CH_3CH_2CH_2CH_2CH_2CH_3 \quad CH_3\overset{\overset{\displaystyle CH_3 CH_3}{\diagup}}{CH}CHCH_3 \quad CH_3\overset{\overset{\displaystyle CH_3}{|}}{CH}CH_2CH_2CH_3$$

$$CH_3CH_2\overset{\overset{\displaystyle CH_3}{|}}{CH}CH_2CH_3 \quad CH_3\overset{\overset{\displaystyle CH_3}{|}}{\underset{\underset{\displaystyle CH_3}{|}}{C}}—CH_2CH_3$$

Common or trivial (nondescriptive) names are not available for all these compounds.

The current nomenclature system is a modification of the IUPAC system and is the system employed by the Chemical Abstracts Service. The basic principles are as follows.

Basic Rules for Naming Alkanes □

1. The root or parent name is the name of the longest continuous carbon chain in the molecule. The compound is then named as an alkyl derivative of that parent.

$$CH_3CHCH_3 \qquad \text{methylpropane}$$
$$| $$
$$CH_3$$

$$CH_3CH_3$$
$$| /$$
$$CH_3CHCHCH_3 \qquad \text{2,3-dimethylbutane}$$

2. Indicate by number the carbon atom to which each alkyl group is attached.
3. Begin numbering at the end that has the nearest group.

$$CH_3$$
$$|$$
$$CH_3CH_2CCH_3 \qquad \text{2,2-dimethylbutane, } \textit{not} \text{ 3,3-dimethylbutane}$$
$$|$$
$$CH_3$$

4. If an alkyl group occurs more than once, indicate by a prefix: di-, tri-, tetra-, and so on, the number of times it appears and designate by numbers the position of each.

$$CH_3 \qquad CH_3CH_2CH_3$$
$$| \qquad \quad |$$
$$CH_3CHCH_2CH_2C-CH-CH_2CH_2CH_3 \qquad \text{2,5,5-trimethyl-6-ethylnonane}$$
$$|$$
$$CH_3$$

Basic Rules for Naming Unsaturated Hydrocarbons □

1. Compounds that have a carbon-carbon double bond are called alkenes. Note that the same root name is used and only the suffix is changed from **-ane** to **ene.** If more than one double bond is present, **-ene** is replaced by **-diene, -triene, -tetraene,** and so on.

2. Compounds containing a carbon-carbon triple bond are named **alkynes.** Thus, the suffix **-ane** is replaced by **-yne** to designate the triple bond.

3. In either case, the selection of the longest continuous carbon chain must also contain the multiple bond. For example, the compound

$$CH_2$$
$$\|$$
$$CH_3CH_2-C-CH_2CH_3$$

is named as a derivative of butene (2-ethyl-1-butene) and not of pentane.

4. The position of the double or triple bond is designated by one number. Although the multiple bond is joined to two carbon atoms, use the number of the carbon atom that is nearest the low numbered end of the chain.

$$CH_3CH_2CH{=}CHCH_3 \qquad \text{2-pentene, not 3-pentene}$$

Rules for Naming Cyclic Hydrocarbons ☐
1. The same rules apply to the cyclic hydrocarbons except that the prefix **cyclo-** is used.

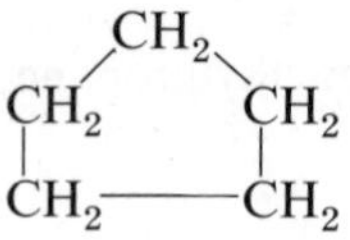

cyclopentane

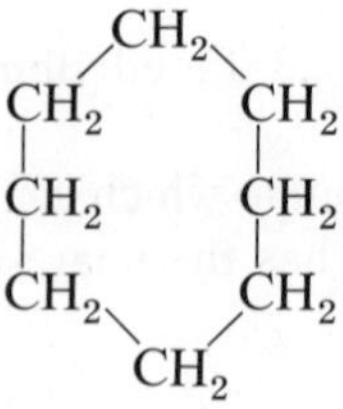

cyclooctane

2. The positions of substituent groups are so designated as to result in the smallest sum of numbers. One substituent is always designated to be at carbon atom number one.

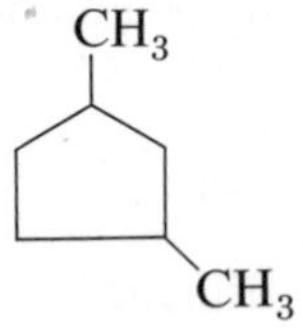

1,3-dimethylcyclopentane, *not*
1,4-dimethylcyclopentane

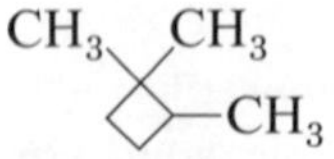

1,1,2-trimethylcyclobutane, *not*
1,2,2-trimethylcyclobutane

1,3-cyclohexadiene

Rules for naming Aromatic Hydrocarbons ☐
1. The rules governing the numbering of cyclic hydrocarbons are applied also to benzene derivatives.

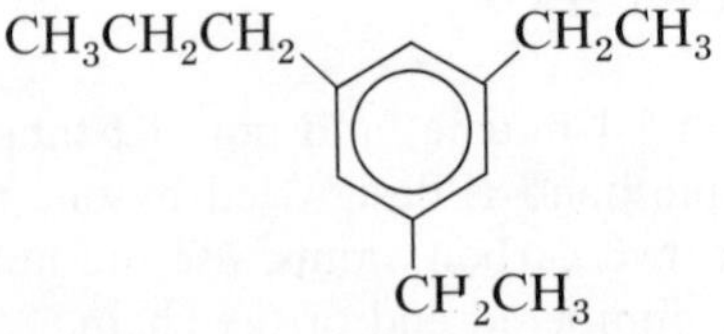

1,3-diethyl-5-propylbenzene

For disubstituted alkyl derivatives of benzene, the prefixes *ortho, meta,* and *para* may also be used in place of 1,2-, 1,3-, and 1,4-dialkyl.

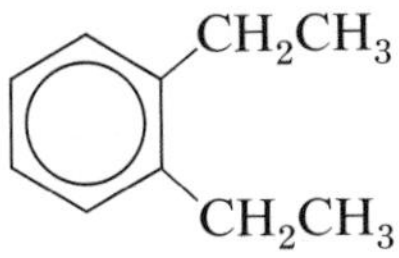

ortho-diethylbenzene

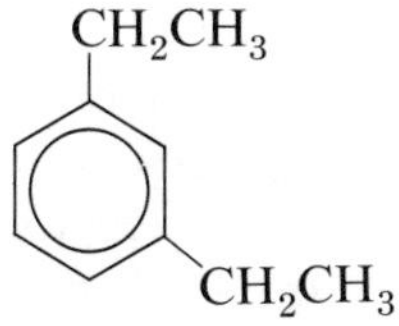

meta-diethylbenzene

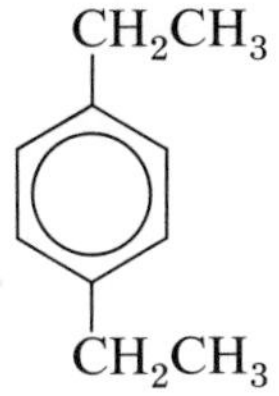

para-diethylbenzene

Rules for Naming Groups other than Alkyl □ The above rules may be applied when groups other than alkyl groups are attached to the parent hydrocarbon. Examples are as follows:

Group	Name of Prefix
$-OH$	hydroxy-
$-NO_2$	nitro-
$-Cl$	chloro-
$-Br$	bromo-
$-I$	iodo-
$-CN$	cyano-
$-NH_2$	amino-
$-NO$	nitroso-
$-SH$	mercapto-

Examples of the use of some of these are

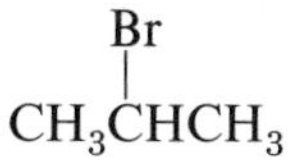

$$\overset{\overset{\textstyle Br}{|}}{CH_3CHCH_3} \quad \text{2-bromopropane}$$

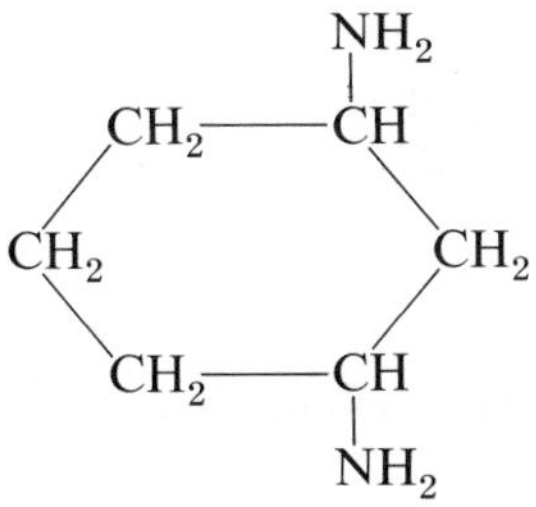

1,3-diaminocyclohexane

533

$$\overset{\displaystyle OH}{\underset{\displaystyle |}{CH_3CHCH_2CH_3}} \qquad \text{2-hydroxybutane}$$

□ *Problems*

1. Write the structure of each of the following compounds.
   (a) 1-amino-4-methylcycloheptane   (e) dinitromethane
   (b) 3-cyclohexyloctane   (f) mercaptoethane
   (c) hexamethylbenzene   (g) 1,1,3-tribromoethane
   (d) 1,2-dicyanoethane   (h) *para*-dicyclopropylbenzene

2. Give an acceptable name for each of the following compounds.

   (a) $\underset{\displaystyle CH_3\overset{\displaystyle |}{CH}-CH_2CH_3}{\overset{\displaystyle CH_2CH_3}{}}$

   (b) $CH_3-\overset{\displaystyle Br}{\overset{\displaystyle |}{CH}}-\overset{\displaystyle Cl}{\overset{\displaystyle |}{CH}}-\overset{\displaystyle OH}{\overset{\displaystyle |}{CH}}-CH_3$

   (c) CH₃ substituted cyclobutane with CH₃ groups

   (d) cyclopentene with CH₃

   (e) benzene with H₂N and SH groups

   (f) $CH_3CH-CH_2-\overset{\displaystyle CH_3}{\overset{\displaystyle |}{CH}}-CH_3$ with $\underset{\displaystyle CH_3 \quad CH_3}{\overset{\displaystyle |}{CH}}$ branch

3. Each of the following names contains an error. Find it and give the correct name.
   (a) 1,4,5-trimethylbenzene   (e) 2,2-methylpropane
   (b) 3-methyl-4-hexene   (f) 1,1,3-chlorobromonitrobutane
   (c) 2-methylcyclohexene   (g) 2-propyloctane
   (d) 3,3-dicyanocyclobutane

B-2. INORGANIC COORDINATION COMPOUNDS

Systems for naming most inorganic compounds are mentioned in Chapter 21. Since students usually encounter little difficulty in naming most inorganic substances, no further discussion will be included here. However, coordination compounds, which involve a central metal atom or ion to which is attached other atoms, molecules, or ions called ligands, do require some special mention because they are fairly complicated species.

Rules for naming these compounds are □
   1. If the substance is a salt, the cation (whether simple or complex) is named first, then the anion.

2. The names of negative ligands end in **-o,** whereas neutral ligands have their ordinary names except for water and ammonia which are called **aquo** and **ammine** (ăm′ēn), respectively.

3. The numbers of ligands of the same kind are indicated by the prefixes di-, tri-, tetra-, penta-, hexa-, and so on, unless the ligand is polydentate in which case bis-, tris-, tetrakis- are used. A unidentate ligand has a single point of attachment to the metal, for example H_2O, NH_3, Cl^-, Br^-, OH^-, CN^-, NO_2^-. A polydentate ligand has two or more points of attachment to the metal, for example

$$\overset{\text{O}\quad\text{O}}{\underset{}{\text{H}_2\text{NCH}_2\text{CH}_2\text{NH}_2, \text{ ethylenediamine, and }\ -\text{O}-\overset{\|}{\text{C}}-\overset{\|}{\text{C}}-\text{O}-,}}$$

$H_2NCH_2CH_2NH_2$, ethylenediamine, and $-O-\overset{O}{\overset{\|}{C}}-\overset{O}{\overset{\|}{C}}-O-$, oxalate ion, are bidentate ligands.

4. Negative ligands are named first, then neutral ligands, then the metal. The reverse order is used in writing the formula.

5. The oxidation state of the metal is indicated by Roman numerals in parentheses.

6. If the entire complex species is an anion, the suffix **-ate** is used with the metal. If the entire complex species is a cation or is neutral, the name of the metal is unchanged.

7. In writing formulas, a complex species (metal and its ligands) is usually enclosed in brackets.

Examples

$Na_3[Fe(CN)_6]$	sodium hexacyanoferrate(III)
$[Co(NH_3)_5Cl]SO_4$	chloropentaamminecobalt(III) sulfate
$[Pt(H_2O)_4Cl_2][PtCl_4]$	dichlorotetraaquoplatinum (IV) tetrachloro-platinate(II)
$[Cr(NH_3)_3Br_3]$	tribromotriamminechromium(III)
$[Cu(en)_2](NO_3)_2$	bis(ethylenediamine)copper(II) nitrate
$[Ni(CO)_4]$	tetracarbonylnickel(0)
$K[Cr(H_2O)_2(C_2O_4)_2]$	potassium bis(oxalato)diaquochromate(III)
$Na_3[Co(NO_2)_6]$	sodium hexanitrocobaltate(III)

☐ *Problems*

1. Name the following compounds.
  (a) $[Pt(NH_3)_2Cl_2]$ **(d)** $[Co(NH_3)_4(en)]Cl_3$
  (b) $Na_4[Fe(CN)_6]$ **(e)** $[Cr(NH_3)_5Cl]_3[Co(NO_2)_6]_2$
  (c) $[Ni(NH_3)_6][Ni(CN)_4]$

2. Write formulas for the following compounds.
  (a) Dichlorotetraamminecobalt(III) sulfate
  (b) Potassium pentachloroammineplatinate(IV)
  (c) Chloroaquobis(ethylenediamine)chromium(III) chloride
  (d) Trinitrotriamminecobalt(III)
  (e) Chloropentaaminechromium(III) tetracyanozincate(II)

ANSWERS TO PROBLEMS

Section B-1

2. (a) 3-methylpentane
   (b) 2-bromo-3-chloro-4-hydroxypentane
   (c) 1,1,3,3-tetramethylcyclobutane
   (d) 1-methylcyclopentadiene-1,3
   (e) 1-amino-3-mercaptobenzene
   (f) 2,3,5-trimethylhexane

3. (a) 1,2,4-trimethylbenzene
   (b) 4-methyl-2-hexene
   (c) 1-methylcyclohexene
   (d) 1,1-dicyanocyclobutane
   (e) 2,2-dimethylpropane
   (f) 1-chloro-1-bromo-3-nitrobutane
   (g) 4-methyldecane

Section B-2

1. (a) Dichlorodiammineplatinum(II)
   (b) Sodium hexacyanoferrate(II)
   (c) Hexaamminenickel(II) tetracyanonickelate(II)
   (d) Ethylenediaminetetraamminecobalt(III) chloride
   (e) Chloropentaamminechromium(III) hexanitrocobaltate(III)

2. (a) $[Co(NH_3)_4Cl_2]_2SO_4$
   (b) $K[Pt(NH_3)Cl_5]$
   (c) $[Cr(en)_2(H_2O)Cl]Cl_2$
   (d) $[Co(NH_3)_3(NO_2)_3]$
   (e) $[Cr(NH_3)_5Cl][Zn(CN)_4]$

QUALITATIVE ANALYSIS FOR COMMON CATIONS AND ANIONS

The acid-base and solubility product principles underlying the separation of a number of metal ions have been mentioned in Section 23–2. This chapter will investigate more fully the practical separation and identification of 21 common cations and 15 of the more common anions. It will be assumed that the unknown is in the form of an aqueous solution. Water insoluble salts can normally be brought sufficiently into solution for anion analysis by prolonged treatment with 2 M Na_2CO_3. About 0.1 g of the insoluble unknown is boiled in 2 ml of 2 M Na_2CO_3 in a small beaker covered with a watch glass for 15–30 min. The water that boils off must be replaced to maintain a constant volume. After removal of any residue, anion tests can often be satisfactorily performed on the solution. The residue, on the other hand, will consist of a mixture of carbonates of the various metal ions originally present in the unknown. Treatment of the residue with 3 M HNO_3 dropwise until effervescence ceases and the solid dissolves usually gives a solution upon which cation analysis can be carried out.

C-1. PRELIMINARY EXAMINATION

A preliminary examination of the unknown should always be carried out before beginning any chemical analysis. The color of the solution should be noted and recorded. Although a given color does not guarantee the presence of a certain ion, because other substances or some combination of colors could be responsible, the *lack* of a color sometimes rules out certain ions. For example, a colorless solution does not contain a highly colored species such as chromate ion, CrO_4^{2-}. If the unknown is a solid, examination with a magnifying glass is often very helpful in getting some idea of the number of different salts present.

The approximate pH of the unknown solution should be determined using commercial "pH paper." This test is quite helpful in suggesting the presence or absence of certain ions. A low pH, for example, rules out the presence of strongly basic ions such as OH^- and CO_3^{2-}. A high pH will rule out a cation such as Fe^{3+}, which gives acidic solutions. Fe^{3+} can exist only as highly insoluble $Fe(OH)_3$ in basic solutions.

Table C-1.   Common Cations and Anions Which Are Colored in Solution

Blue	Green	Yellow	Orange	Pink
Cu^{2+}	Cr^{3+}	Fe^{3+} [due to presence of $Fe(H_2O)_5OH^{2+}$]	$Cr_2O_7^{2-}$	Co^{2+}
	Ni^{2+}	CrO_4^{2-}		Mn^{2+} (pale)
	Fe^{2+} (pale)	$Fe(CN)_6^{3-}$		
		$Fe(CN)_6^{4-}$		

A neutral solution can contain only those ions that either have no acid or base properties, for example, Na^+, K^+, Ba^{2+}, or are exceeding weak acids or bases, for example, NH_4^+, Cl^-, NO_3^-, SO_4^{2-}.

C-2.   GENERAL PRINCIPLES IN THE ANALYSIS OF CATIONS AND ANIONS

Two problems confront the analyst when faced with an unknown consisting of a mixture of ions. First, an overall separation of the ions into groups containing a relatively small number of ions must be achieved. The reason for the necessity of this overall separation is that, in testing for a given ion, other ions often interfere. This interference may take the form of two or more ions giving the same response to some instrument. If detection is to be achieved by means of a specific chemical reaction, it is obvious that two ions which give the same or similar results cannot be simultaneously detected. The initial separation serves to segregate the ions into smaller, more manageable, groups that can be treated separately. The second problem is, of course, the unequivocal establishment of the presence or absence of each ion to be tested. A test that is carried out under conditions such that a given result establishes definitely the presence or absence of a single species is called a **confirmation test.**

Many attempts have been made over the years to circumvent the overall separation step and to develop procedures that would allow the direct testing of a complex mixture for a specific ion. Such a method of testing directly for a single substance is called a **spot test.** Although considerable progress has been made in the development of sensitive spot tests, the problem of interference has never been completely solved. Thus there is a need for some sort of initial separation in almost any analytical scheme whether the final detection is to be made by instrumental or chemical means.

C-3.   SEPARATION OF ANIONS INTO GROUPS

Removal of cations other than Na^+, K^+, NH_4^+ is necessary since any other cations may interfere with the anion tests. A basic solution is desirable in carrying out the tests to prevent undesirable redox reactions among the anions. Both of these results can be realized by treatment of the unknown solution with Na_2CO_3 solution. The undesirable cations precipitate as insoluble carbonates or hydroxides and a high pH is established by the presence of excess CO_3^{2-} ions. The CO_3^{2-} ion must, of course, be tested for *before* this treatment. If it is known that only the cations Na^+, K^+, or NH_4^+ are present, the Na_2CO_3 treatment is unnecessary, but the pH of the solution should be raised by the addition of some NaOH solution.

Table C–2 outlines the overall separation scheme for the anions.

Numbers in brackets [1–7] refer to explanatory notes in Table C–2.

[1] Since the CO_3^{2-} ion is added to remove heavy metal cations, it must be tested for on the original unknown solution. Advantage is taken of the basic property of this ion and the fact that carbonic acid, H_2CO_3, is unstable and decomposes to release CO_2 gas.

$$2\, H_3O^+ + CO_3^{2-} \longrightarrow 3\, H_2O + CO_2$$

The purpose of the $KClO_3$ is to oxidize any sulfide or sulfite ions that may be present and prevent their release as gases (H_2S, SO_2) which could be mistaken for CO_2.

[2] It is necessary to remove any cations other than Na^+, K^+, or NH_4^+ because interference with the anion tests could occur. CO_3^{2-} ion accomplishes this task by precipitating any other metal ions as the carbonate, hydroxide, or oxide. If it is known beforehand that no cations besides Na^+, K^+, or NH_4^+ are present, the addition of Na_2CO_3 is unnecessary. No more than 1.5–2 ml of sample should be used. If the unknown is a solid, dissolve a level microspatula-full (0.1 g) in 2 ml of water and carry out the tests on this solution.

[3] Group I anions are those whose calcium salts are insoluble in neutral or basic solution. $CaCO_3$ will precipitate whether or not it was originally present in the unknown if it has been added to remove heavy-metal cations. If a large excess of CO_3^{2-} ions is present, a rather gelatinous precipitate of $CaCO_3$ may form. Heating in a water bath after the addition of the $Ca(OAc)_2$ precipitating reagent sometimes aids in converting the gelatinous precipitate to a more powdery form. Washing the group I residue will remove small amounts of other anions that may contaminate the precipitate, for example, if CrO_4^{2-} is present in the unknown, some $CaCrO_4$ may precipitate. It must be removed in order to avoid interference with the group I confirmation tests.

[4] The anions of group II consist of SO_4^{2-} and CrO_4^{2-}. Their barium salts are insoluble in basic solution. The color of CrO_4^{2-} ion is so intense that a white precipitate for group II confirms the absence of CrO_4^{2-} and the presence of SO_4^{2-}.

[5] The cadmium salts of the group III anions are insoluble in basic solution. Two of these anions are incompatible because of the redox reaction

$$S^{2-} + 2\, Fe(CN)_6^{3-} \longrightarrow S + 2\, Fe(CN)_6^{4-}$$

ferricyanide   ferrocyanide
[hexacyanoferrate(III)]   [hexacyanoferrate(II)]

Consequently, the presence of S^{2-} confirms the absence of ferricyanide and vice versa. The color of the Group III cadmium precipitate gives a great deal of information.

Table C-2. Overall Anion Separation Scheme

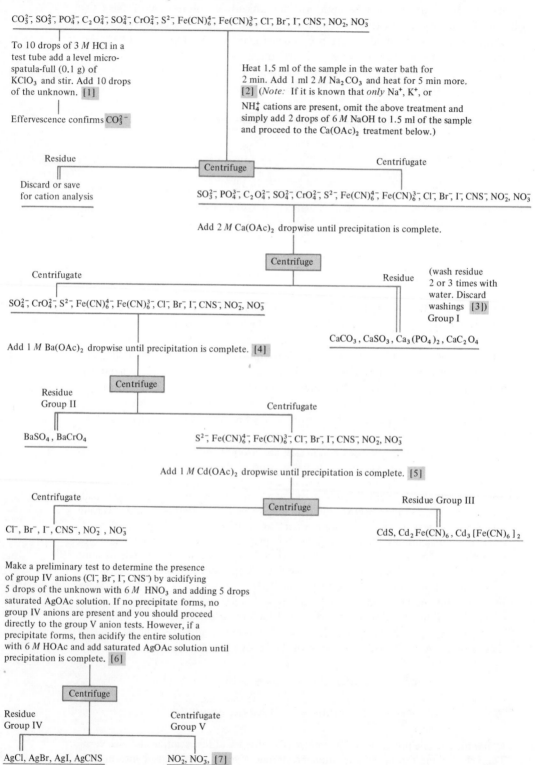

CO_3^{2-}, SO_3^{2-}, PO_4^{3-}, $C_2O_4^{2-}$, SO_4^{2-}, CrO_4^{2-}, S^{2-}, $Fe(CN)_6^{4-}$, $Fe(CN)_6^{3-}$, Cl^-, Br^-, I^-, CNS^-, NO_2^-, NO_3^-

To 10 drops of 3 M HCl in a test tube add a level micro-spatula-full (0.1 g) of $KClO_3$ and stir. Add 10 drops of the unknown. [1]

Effervescence confirms CO_3^{2-}

Heat 1.5 ml of the sample in the water bath for 2 min. Add 1 ml 2 M Na_2CO_3 and heat for 5 min more. [2] (Note: If it is known that only Na^+, K^+, or NH_4^+ cations are present, omit the above treatment and simply add 2 drops of 6 M NaOH to 1.5 ml of the sample and proceed to the $Ca(OAc)_2$ treatment below.)

Residue

Centrifugate

Centrifuge

Discard or save for cation analysis

SO_3^{2-}, PO_4^{3-}, $C_2O_4^{2-}$, SO_4^{2-}, CrO_4^{2-}, S^{2-}, $Fe(CN)_6^{4-}$, $Fe(CN)_6^{3-}$, Cl^-, Br^-, I^-, CNS^-, NO_2^-, NO_3^-

Add 2 M $Ca(OAc)_2$ dropwise until precipitation is complete.

Centrifuge

Centrifugate

Residue

(wash residue 2 or 3 times with water. Discard washings [3]) Group I

SO_4^{2-}, CrO_4^{2-}, S^{2-}, $Fe(CN)_6^{4-}$, $Fe(CN)_6^{3-}$, Cl^-, Br^-, I^-, CNS^-, NO_2^-, NO_3^-

$CaCO_3$, $CaSO_3$, $Ca_3(PO_4)_2$, CaC_2O_4

Add 1 M $Ba(OAc)_2$ dropwise until precipitation is complete. [4]

Centrifuge

Residue Group II

Centrifugate

$BaSO_4$, $BaCrO_4$

S^{2-}, $Fe(CN)_6^{4-}$, $Fe(CN)_6^{3-}$, Cl^-, Br^-, I^-, CNS^-, NO_2^-, NO_3^-

Add 1 M $Cd(OAc)_2$ dropwise until precipitation is complete. [5]

Centrifugate

Centrifuge

Residue Group III

Cl^-, Br^-, I^-, CNS^-, NO_2^-, NO_3^-

CdS, $Cd_2Fe(CN)_6$, $Cd_3[Fe(CN)_6]_2$

Make a preliminary test to determine the presence of group IV anions (Cl^-, Br^-, I^-, CNS^-) by acidifying 5 drops of the unknown with 6 M HNO_3 and adding 5 drops saturated AgOAc solution. If no precipitate forms, no group IV anions are present and you should proceed directly to the group V anion tests. However, if a precipitate forms, then acidify the entire solution with 6 M HOAc and add saturated AgOAc solution until precipitation is complete. [6]

Centrifuge

Residue Group IV

Centrifugate Group V

AgCl, AgBr, AgI, AgCNS

NO_2^-, NO_3^-, [7]

CdS   bright yellow

$Cd_2[Fe(CN)_6]$ cream

$Cd_3[Fe(CN)_6]_2$ orange

[6] The preliminary test is performed to determine whether or not any group IV anions are present. If none is present, the solution can be tested immediately for group V anions. If group IV anions are present, they are precipitated as the silver salts, which are insoluble in acidic or basic solution. It is important, however, to keep the solution acidic by the addition of HOAc (acetic acid) while precipitating group IV to avoid the formation of Ag_2O.

$$2 OH^- + 2 Ag^+ \longrightarrow Ag_2O + H_2O$$

The Ag_2O is brown or black and its presence is confusing. The color of AgI is yellow, so that a totally white precipitate rules out the presence of iodide.

[7] The anions of Group V have no precipitating reagent since almost all salts of these ions are soluble.

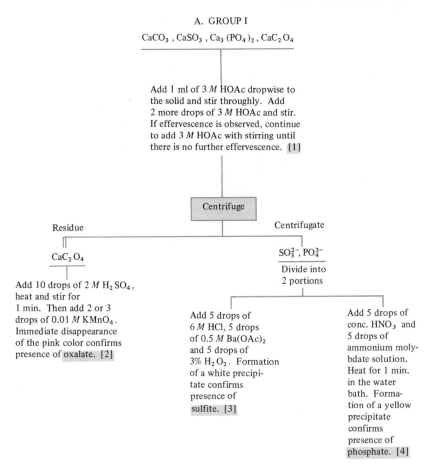

A. GROUP I

$CaCO_3$, $CaSO_3$, $Ca_3(PO_4)_2$, CaC_2O_4

Add 1 ml of 3 M HOAc dropwise to the solid and stir throughly. Add 2 more drops of 3 M HOAc and stir. If effervescence is observed, continue to add 3 M HOAc with stirring until there is no further effervescence. [1]

Centrifuge

Residue

CaC_2O_4

Add 10 drops of 2 M H_2SO_4, heat and stir for 1 min. Then add 2 or 3 drops of 0.01 M $KMnO_4$. Immediate disappearance of the pink color confirms presence of oxalate. [2]

Centrifugate

SO_3^{2-}, PO_4^{3-}

Divide into 2 portions

Add 5 drops of 6 M HCl, 5 drops of 0.5 M $Ba(OAc)_2$ and 5 drops of 3% H_2O_2. Formation of a white precipitate confirms presence of sulfite. [3]

Add 5 drops of conc. HNO_3 and 5 drops of ammonium molybdate solution. Heat for 1 min. in the water bath. Formation of a yellow precipitate confirms presence of phosphate. [4]

541

Numbers in brackets [1–4] refer to the flow chart on page 541.

[1] The acetic acid will destroy any $CaCO_3$ by the reaction

$$CaCO_3 + 2\,HOAc \longrightarrow Ca^{2+} + 2\,OAc^- + H_2O + CO_2$$

$CaSO_3$ and $Ca_3(PO_4)_2$ will dissolve due to the reactions

$$CaSO_3 + 2\,HOAc \longrightarrow Ca^{2+} + H_2SO_3 + 2\,OAc^-$$

$$Ca_3(PO_4)_2 + 4\,HOAc \longrightarrow 3\,Ca^{2+} + 2\,H_2PO_4^- + 4\,OAc^-$$

CaC_2O_4 is insoluble in $3\,M$ HOAc. Absence of any solid after the HOAc addition confirms the absence of oxalate. The mere presence of a solid, however, does not confirm its identity as oxalate.

[2] CaC_2O_4 dissolves in $2\,M$ H_2SO_4 through the formation of soluble oxalic acid.

$$CaC_2O_4 + 2\,H_3O^+ \longrightarrow Ca^{2+} + H_2C_2O_4 + 2\,H_2O$$

The purple MnO_4^- ion is reduced by oxalic acid to nearly colorless Mn^{2+} ion.

$$5\,H_2C_2O_4 + 2\,MnO_4^- + 6\,H_3O^+ \longrightarrow 2\,Mn^{2+} + 14\,H_2O + 10\,CO_2$$

[3] The H_2O_2 oxidizes H_2SO_3 to SO_4^{2-} and $BaSO_4$ precipitates.

$$H_2SO_3 + H_2O_2 + H_2O \longrightarrow 2\,H_3O^+ + SO_4^{2-}$$

$$Ba^{2+} + SO_4^{2-} \longrightarrow BaSO_4$$

[4] The yellow precipitate is ammonium phosphomolybdate.

$$12\,MoO_4^{2-} + H_2PO_4^- + 3\,NH_4^+ + 22\,H_3O^+ \longrightarrow (NH_4)_3PO_4 \cdot 12\,MoO_3 + 34\,H_2O$$
$$\text{(yellow)}$$

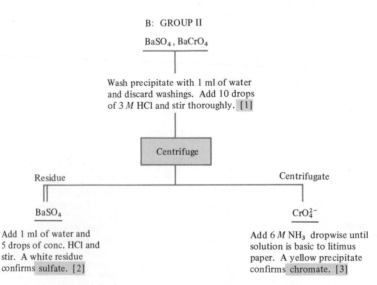

B: GROUP II

$BaSO_4$, $BaCrO_4$

Wash precipitate with 1 ml of water and discard washings. Add 10 drops of $3\,M$ HCl and stir thoroughly. [1]

Centrifuge

Residue

$BaSO_4$

Add 1 ml of water and 5 drops of conc. HCl and stir. A white residue confirms sulfate. [2]

Centrifugate

CrO_4^{2-}

Add $6\,M$ NH_3 dropwise until solution is basic to litimus paper. A yellow precipitate confirms chromate. [3]

[1] $BaSO_4$ is insoluble in acid solution, but $BaCrO_4$ dissolves according to the reaction

$$BaCrO_4 + H_3O^+ \longrightarrow Ba^{2+} + HCrO_4^- + H_2O$$

[2] A white residue after the first treatment with 3 M HCl is indicative of the presence of SO_4^{2-}. However, it is possible that a small amount of some group I anion could be present at this point. Therefore, excess HCl is added to an aqueous solution. Only $BaSO_4$ will remain undissolved after this treatment.

[3] A yellow solution after the first treatment with 3 M HCl is indicative of the presence of CrO_4^{2-}. Addition of 6 M NH_3 until the solution is basic allows the reprecipitation of $BaCrO_4$.

$$Ba^{2+} + HCrO_4^- + NH_3 \longrightarrow BaCrO_4 + NH_4^+$$
$$\text{(yellow)}$$

C. GROUP III

$CdS, Cd_2[Fe(CN)_6], Cd_3[Fe(CN)_6]_2$

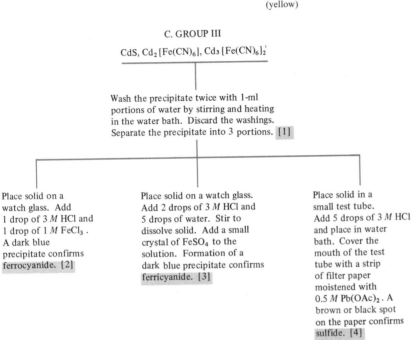

Wash the precipitate twice with 1-ml portions of water by stirring and heating in the water bath. Discard the washings. Separate the precipitate into 3 portions. [1]

Place solid on a watch glass. Add 1 drop of 3 M HCl and 1 drop of 1 M $FeCl_3$. A dark blue precipitate confirms ferrocyanide. [2]

Place solid on a watch glass. Add 2 drops of 3 M HCl and 5 drops of water. Stir to dissolve solid. Add a small crystal of $FeSO_4$ to the solution. Formation of a dark blue precipitate confirms ferricyanide. [3]

Place solid in a small test tube. Add 5 drops of 3 M HCl and place in water bath. Cover the mouth of the test tube with a strip of filter paper moistened with 0.5 M $Pb(OAc)_2$. A brown or black spot on the paper confirms sulfide. [4]

[1] It is possible that some $Cd(CNS)_2$ may be present. CNS^- ion (from group IV) must be removed since it will interfere with the tests for ferrocyanide and ferricyanide.

[2] The dark blue precipitate results from the reaction between ferrocyanide ion and ferric ion. This material has variously been called Prussian blue and Turnbull's blue. The main point is that the presence of iron in the +2 and +3 states is required to give the blue precipitate, which approximates the formula

$$Fe[FeFe(CN)_6]_2$$
$$\text{(blue)}$$

[3] The blue precipitate is presumably the same as that produced in the test for ferrocyanide. In this case, however, ferrous ion is supplied by the test reagent to react with any ferricyanide that may be present.

[4] Cadmium sulfide reacts with the hydrochloric acid to release hydrogen sulfide gas.

$$CdS + 2 H_3O^+ \longrightarrow Cd^{2+} + H_2S + 2 H_2O$$

The hydrogen sulfide then reacts with the lead acetate to produce black lead sulfide.

$$H_2S + Pb(OAc)_2 \longrightarrow PbS + 2 HOAc$$
$$\text{(black)}$$

D. GROUP IV

AgCl, AgBr, AgI, AgCNS

Wash the precipitate with 1-ml portions of water containing 3 drops of 6 M HNO$_3$ until the washings no longer give a precipitate with 1 drop of 3 M HCl. Discard all washings. Finally, wash the group IV residue with 1 ml of water and discard the washings. [1]

Treat the residue with a solution made by mixing 2 drops of 3 M NH$_3$, 2 drops of 3 M KNO$_3$, and 2 drops of 0.1 M AgNO$_3$ with 14 drops of water. Stir thoroughly. [2]

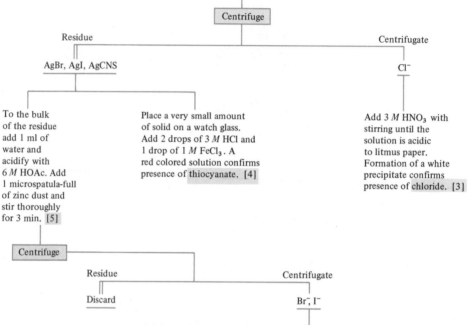

Centrifuge

Residue

AgBr, AgI, AgCNS

Centrifugate

Cl$^-$

To the bulk of the residue add 1 ml of water and acidify with 6 M HOAc. Add 1 microspatula-full of zinc dust and stir thoroughly for 3 min. [5]

Place a very small amount of solid on a watch glass. Add 2 drops of 3 M HCl and 1 drop of 1 M FeCl$_3$. A red colored solution confirms presence of thiocyanate. [4]

Add 3 M HNO$_3$ with stirring until the solution is acidic to litmus paper. Formation of a white precipitate confirms presence of chloride. [3]

Centrifuge

Residue

Discard

Centrifugate

Br$^-$, I$^-$

Test with litmus paper to make certain solution is acidic. If it is not, acidify with a few drops of 6 M HOAc. Add 10 drops of CCl$_4$ and a few small crystals of NaNO$_2$. Dilute with water to a total volume of 2 ml and shake vigorously. The characteristic violet color of I$_2$ in the CCl$_4$ layer (lower layer) confirms the presence of iodide. [6]

If iodine is present, the water layer (upper layer) must be removed and shaken with additional 10 drop portions of CCl$_4$ and the procedure repeated until the color of I$_2$ is no longer seen. The test for Br$^-$ is then performed on the water layer. (If no iodine was present, the water solution can be tested directly for bromide ion.) [7]

To a few crystals of KClO$_3$ in a test tube is added 10 drops of the aqueous solution to be tested. Add 5 drops of conc. HCl to the solution and stir. Add 10 drops of CCl$_4$ and agitate vigorously. A yellowish-brown color in the CCl$_4$ layer confirms presence of bromide. [8]

[1] The washing of the group IV residue removes excess Ag$^+$ ion. If this ion were not removed, the separation of Cl$^-$ ion by solution of AgCl in NH$_3$ would not be complete.

[2] The AgCl dissolves according to the reaction

$$AgCl + 2\,NH_3 \longrightarrow Ag(NH_3)_2^+ + Cl^-$$

Some AgBr and AgCNS would also dissolve in an NH$_3$ solution; however, the ammoniacal silver nitrate solution retards the solution of AgBr and AgCNS but allows the solution of AgCl. The ammoniacal silver nitrate contains $Ag(NH_3)_2^+$, which operates as a common ion to prevent solution of AgBr and AgCNS but yet allow solution of the less insoluble AgCl.

[3] The acid destroys the $Ag(NH_3)_2^+$ ion and allows the reprecipitation of AgCl.

$$Ag(NH_3)_2^+ + Cl^- + 2\,H_3O^+ \longrightarrow AgCl + 2\,NH_4^+ + 2\,H_2O$$

[4] The AgCNS dissolves in hydrochloric acid

$$AgCNS + Cl^- \longrightarrow AgCl + CNS^-$$

and reaction of CNS^- ion with Fe^{3+} gives the red color.

$$Fe^{3+} + 6\,CNS^- \longrightarrow Fe(CNS)_6^{3-}$$
$$\text{(red)}$$

[5] The zinc reduces the Ag^+ ion to elemental silver and gives a solution of Br^- and I^- ions.

$$Zn + 2\,AgBr \longrightarrow Zn^{2+} + 2\,Br^- + 2\,Ag$$
$$Zn + 2\,AgI \longrightarrow Zn^{2+} + 2\,I^- + 2\,Ag$$

The CNS^- ion is destroyed by the zinc treatment by the reaction.

$$3\,Zn + 2\,AgCNS + 6\,HOAc \longrightarrow 2\,Ag + 3\,Zn^{2+} + 2\,H_2S + 2\,HCN + 6\,OAc^-$$

[6] In an acidic solution, NO_2^- ion forms HNO_2 which oxidizes I^- to I_2 by the reaction

$$2\,HNO_2 + 2\,I^- + 2\,H_3O^+ \longrightarrow I_2 + 2\,NO + 4\,H_2O$$

The I_2 is readily soluble in CCl_4 and can be effectively removed from the aqueous solution by repeated extraction with portions of CCl_4.

[7] If iodine is present, it must be removed by repeated CCl_4 treatments or it will interfere with the bromide test. If no iodine is present, the test for Br^- ion can be carried out on the aqueous solution without extraction by CCl_4.

[8] The reaction between hydrochloric acid and ClO_3^- ion produces Cl_2 which oxidizes any Br^- to Br_2. A brown coloration in the CCl_4 layer is indicative of the presence of Br_2.

$$ClO_3^- + 5\,Cl^- + 6\,H_3O^+ \longrightarrow 3\,Cl_2 + 9\,H_2O$$
$$2\,Br^- + Cl_2 \longrightarrow Br_2 + 2\,Cl^-$$

E. GROUP V

$$NO_2^-, NO_3^-$$

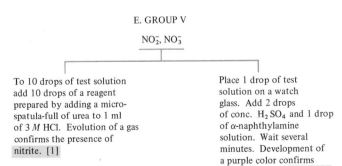

| To 10 drops of test solution add 10 drops of a reagent prepared by adding a micro-spatula-full of urea to 1 ml of 3 M HCl. Evolution of a gas confirms the presence of nitrite. [1] | Place 1 drop of test solution on a watch glass. Add 2 drops of conc. H_2SO_4 and 1 drop of α-naphthylamine solution. Wait several minutes. Development of a purple color confirms the presence of nitrate. [2] |

545

Numbers in brackets [1–2] refer to the flow chart on page 545.

[1] In dilute acid solution, urea reacts with NO_2^- ion to produce CO_2 and N_2 gases. Urea has the structure

$$\underset{\displaystyle H_2N-\overset{\textstyle O}{\overset{\|}{C}}-NH_2}{}$$

$$CO(NH_2)_2 + 2\,NO_2^- + 2\,H_3O^+ \longrightarrow CO_2 + 2\,N_2 + 5\,H_2O$$

[2] In a sulfuric acid solution, the NO_3^- ion oxidizes α-naphthylamine to give a fleeting purple color. The origin of this color is uncertain. α-Naphthylamine has the structure

C-5. SEPARATION OF CATIONS INTO GROUPS

The separation of cations into four groups is accomplished by successive precipitation, first of those metal ions whose chlorides are insoluble followed by insoluble sulfides and finally insoluble carbonates. Reagents are added throughout the procedure as their ammonium salts; consequently, the NH_4^+ ion must be tested for on the original unknown. The sodium and potassium ions form no insoluble compounds with any ordinary anions and will thus fall into a fifth group (soluble group). The Na^+ and K^+ ions are usually detected by the specific colors which their compounds impart to flames.

Table C–3 gives the overall separation of the cations into groups.

Numbers in brackets [1–10] refer to explanatory notes in Table C–3.

[1] Reagents are added throughout the scheme of analysis as ammonium salts, consequently NH_4^+ ion must be tested for on the original unknown. The acid-base reaction between the NH_4^+ ion and OH^- ion produces ammonia, which is easily detected by its odor or by its basic reaction with moist, red litmus paper. Care must be taken not to allow the litmus paper to dip into the basic solution. The litmus paper must be held just above the surface of the solution as it is warmed.

$$NH_4^+ + OH^- \longrightarrow NH_3 + H_2O$$

[2] The group I cations are those whose chlorides are insoluble in acidic solution. All of these are white precipitates.

$$Ag^+ + Cl^- \longrightarrow AgCl$$

$$Hg_2^{2+} + 2\,Cl^- \longrightarrow Hg_2Cl_2$$

$$Pb^{2+} + 2\,Cl^- \longrightarrow PbCl_2$$

If the original solution contains a white precipitate, it may be one of the above substances or possibly an oxychloride of antimony, SbOCl. However, SbOCl will dissolve in the presence of H_3O^+ ions, so that a white precipitate after the addition of the HCl solution can only be a group I compound.

$$SbOCl + 2\,H_3O^+ \longrightarrow Sb^{3+} + Cl^- + 3\,H_2O$$

546

Table C-3. Overall Cation Separation Scheme

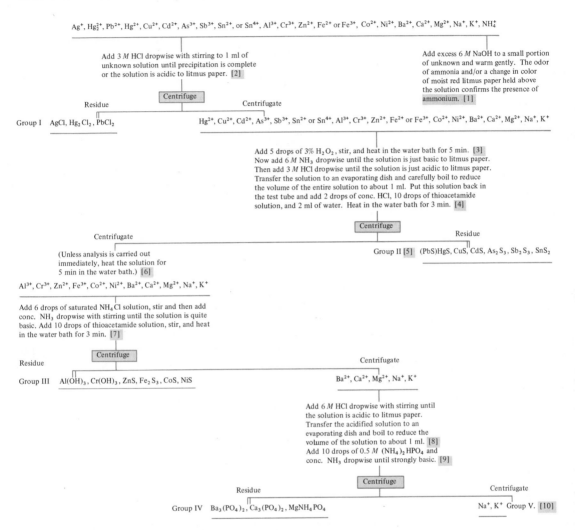

[3] The H_2O_2 is added to oxidize any Sn^{2+} ion to the Sn^{4+} ion prior to precipitation as the sulfide. This is necessary because the separation of the group II sulfides into their subgroups requires tin in the higher oxidation state. Any Fe^{2+} is also oxidized to Fe^{3+} by this treatment.

$$Sn^{2+} + H_2O_2 + 2\,H_3O^+ \longrightarrow Sn^{4+} + 4\,H_2O$$

[4] After the group I precipitation, the solution is quite acidic. It is necessary to adjust the $[H_3O^+]$ of the solution quite precisely. Therefore, the excess acidity is removed by treatment with $6\,M\,NH_3$, and then $3\,M$ HCl is used to adjust the acidity so that the solution is barely on the acid side of neutrality. Adjustment of the solution volume to 1 ml is carried out so that the exact $[H_3O^+]$ can be realized. After addition of the thioacetamide solution and the water, the total volume of the solution is about 3.5 ml (20 drops of an aqueous solution is approximately 1 ml). Therefore, 2 drops of $12\,M$ HCl (conc. HCl is $12\,M$) is diluted to a final volume of about 72 drops.

547

$$\frac{2}{72} \times 12\ M = 0.33\ M$$

The thioacetamide solution produces a saturated solution of H_2S (see Section 23–2), so that $[S^{2-}]$ in this solution can be calculated using the equation

$$[H_3O^+]^2[S^{2-}] = 1.1 \times 10^{-22}$$

Substituting for $[H_3O^+]$

$$(0.33)^2[S^{2-}] = 1.1 \times 10^{-22}$$

$$[S^{2-}] = 1.0 \times 10^{-21}$$

The concentration of metal ions in unknowns normally runs about 0.01 M. The ion product (i.p.) for the sulfide of a dipositive ion is thus seen to be approximately:

$$\text{i.p.} = [M^{2+}][S^{2-}] = [0.01][1 \times 10^{-21}]$$

$$\text{i.p.} = 1 \times 10^{-23}$$

The most soluble sulfide of the group II ions is CdS with $K_{sp} = 1 \times 10^{-28}$. The most insoluble sulfide of the group III ions is ZnS with $K_{sp} = 3 \times 10^{-22}$. It can be seen, therefore that the i.p. exceeds the K_{sp} of CdS by a large factor but does not exceed the K_{sp} of ZnS. However, it should also be obvious that the control of $[H_3O^+]$ is critical. Too low $[H_3O^+]$ will allow the precipitation of ZnS in group II instead of group III, whereas too high $[H_3O^+]$ could result in incomplete precipitation of CdS in group II and cause CdS to appear in group III.

$$Hg^{2+} + H_2S + 2\,H_2O \longrightarrow \underset{\text{(black)}}{HgS} + 2\,H_3O^+$$

$$Cu^{2+} + H_2S + 2\,H_2O \longrightarrow \underset{\text{(black)}}{CuS} + 2\,H_3O^+$$

$$Cd^{2+} + H_2S + 2\,H_2O \longrightarrow \underset{\text{(yellow)}}{CdS} + 2\,H_3O^+$$

Arsenic is probably present as arsenious acid, H_3AsO_3.

$$2\,H_3AsO_3 + 3\,H_2S \longrightarrow \underset{\text{(yellow)}}{As_2S_3} + 6\,H_2O$$

Tin exists as the complex hexachlorostannate ion, $SnCl_6^{2-}$.

$$SnCl_6^{2-} + 2\,H_2S + 4\,H_2O \longrightarrow \underset{\text{(yellow)}}{SnS_2} + 4\,H_3O^+ + 6\,Cl^-$$

Antimony exists in some complex form, possibly $SbCl_4^-$ or perhaps $SbCl_6^{3-}$.

$$2\,SbCl_4^- + 3\,H_2S + 6\,H_2O \longrightarrow \underset{\substack{\text{(freshly pptd.}\\\text{orange)}}}{Sb_2S_3} + 8\,Cl^- + 6\,H_3O^+$$

[5] Although Pb^{2+} ion is a group I cation, it is likely that some lead will appear in group II if it is present in the unknown. The removal of Pb^{2+} ion by precipitation of $PbCl_2$

is not complete in as much as $PbCl_2$ is not nearly as insoluble as the other chlorides of group I.

$$Pb^{2+} + H_2S + 2\,H_2O \longrightarrow PbS + 2\,H_3O^+$$

[6] The purpose of heating the solution is to drive off excess H_2S. If this is not done, some of the sulfide ion may be oxidized by atmospheric O_2 to the SO_4^{2-} ion and cause the precipitation of $BaSO_4$ if Ba^{2+} is in the unknown. If barium is removed from the solution at this point, it will be missed in group IV.

$$H_2S + 2\,O_2 + 2\,H_2O \longrightarrow SO_4^{2-} + 2\,H_3O^+$$
$$Ba^{2+} + SO_4^{2-} \longrightarrow BaSO_4$$

[7] The $[H_3O^+]$ is drastically reduced by the addition of NH_3 so that the $[S^{2-}]$ rises sufficiently to cause precipitation of the group III metal ions. Notice that Al^{3+} and Cr^{3+} do not precipitate as sulfides but rather as hydroxides. The addition of the saturated NH_4Cl solution is necessary to prevent the precipitation of $Mg(OH)_2$ at this point.

$$Mg(OH)_2 + 2\,NH_4^+ \longrightarrow Mg^{2+} + 2\,NH_3 + 2\,H_2O$$

Equations for the precipitation of group III ions can be expressed:

$$Al(H_2O)_6^{3+} + 3\,NH_3 \longrightarrow Al(H_2O)_3(OH)_3 + 3\,NH_4^+$$
(colorless gelatinous
precipitate)

$$Cr(H_2O)_6^{3+} + 3\,NH_3 \longrightarrow Cr(H_2O)_3(OH)_3 + 3\,NH_4^+$$
(green, gelatinous
precipitate)

$$Zn^{2+} + S^{2-} \longrightarrow ZnS$$
(white)

$$2\,Fe^{3+} + 3\,S^2 \longrightarrow Fe_2S_3$$
(black)

$$Co^{2+} + S^{2-} \longrightarrow CoS$$
(black)

$$Ni^{2+} + S^{2-} \longrightarrow NiS$$
(black)

The Zn^{2+}, Co^{2+}, and Ni^{2+} ions are actually present as ammonia complexes after the addition of NH_3, namely, $Zn(NH_3)_4^{2+}$, $Co(NH_3)_6^{2+}$, $Ni(NH_3)_6^{2+}$. The Fe^{3+} ion first forms the brown, gelatinous hydroxide, $Fe(H_2O)_3(OH)_3$ when NH_3 is added. All of these species react with S^{2-} to form their sulfides. The series of equations given thus expresses the *net* reaction between S^{2-} and the metal ions.

Note: The formulations $Al(H_2O)_3(OH)_3$ and $Cr(H_2O)_3(OH)_3$ are expressed in the flow diagram as $Al(OH)_3$ and $Cr(OH)_3$ for simplicity. Although dipositive and tripositive cations are undoubtedly hydrated in aqueous solution, they will henceforth be written as simple ions in order to simplify equation writing.

[8] Boiling the acidified solution removes excess H_2S (see note [6]). This step should be carried out immediately following removal of the group III precipitate. Also, the volume of the solution is likely to have become too large and should be reduced to around 1 ml in order to insure complete precipitation of group IV ions.

[9] The Ca^{2+} and Ba^{2+} ions of group IV precipitate as phosphates from a basic solution.

$$3\,Ba^{2+} + 2\,HPO_4^{2-} + 2\,NH_3 \longrightarrow Ba_3(PO_4)_2 + 2\,NH_4^+$$
$$\text{(gelatinous)}$$

$$3\,Ca^{2+} + 2\,HPO_4^{2-} + 2\,NH_3 \longrightarrow Ca_3(PO_4)_2 + 2\,NH_4^+$$
$$\text{(gelatinous)}$$

The Mg^{2+} ion precipitates as an ammonium phosphate.

$$Mg^{2+} + HPO_4^{2-} + NH_3 \longrightarrow MgNH_4PO_4$$
$$\text{(crystalline)}$$

All these precipitates are white, but the magnesium salt is crystalline, whereas the other two are gelatinous in consistency. If precipitation is not immediate, the presence of Mg^{2+} ion is not ruled out because some time is required for the precipitation of the magnesium ammonium phosphate if Mg^{2+} is present by itself in group IV.

[10] The K^+ and Na^+ ions are not precipitated by any common anions. Consequently, if either or both of these ions are present in the unknown, they will remain in solution after removal of the first four groups. Naturally NH_4^+ ions will also be present since ammonium salts are added throughout the scheme. Flame tests (described later) are the best means of identifying the K^+ and Na^+ ions. (NH_4^+ ion imparts no color to a flame.) Preparation of this group V solution for the flame tests requires the careful evaporation of the solution until a moist (not dry) precipitate remains.

C-6. ANALYSIS OF CATIONS BY GROUPS

Numbers in brackets [1–4] refer to the flow chart on page 551.

[1] The purpose of washing the group I precipitate with hot water is to remove any $PbCl_2$ by solution. Although $PbCl_2$ is relatively insoluble in cold water (0.67 g $PbCl_2$ dissolve in 100 g of water at $0°C$), its solubility is much greater in hot water (3.3 g $PbCl_2$ dissolve in 100 g of water at $100°C$). The two washings, if carried out while the water is kept hot, will remove nearly all of the $PbCl_2$ present.

[2] The yellow precipitate of $PbCrO_4$ confirms the presence of Pb^{2+} ion.

$$Pb^{2+} + CrO_4^{2-} \longrightarrow PbCrO_4$$
$$\text{(yellow)}$$

[3] Addition of the 6 M NH_3 serves two purposes. First, any AgCl will be dissolved and thereby separated from Hg_2Cl_2.

$$AgCl + 2\,NH_3 \longrightarrow Ag(NH_3)_2^+ + Cl^-$$

Second, if Hg_2Cl_2 is present, a disproportionation reaction occurs and the black finely divided elemental mercury that is produced serves to identify mercurous ion.

$$Hg_2Cl_2 + 2\,NH_3 \longrightarrow \underset{\text{(black)}}{Hg} + HgNH_2Cl + NH_4^+ + Cl^-$$

[4] Acidification of the solution destroys any $Ag(NH_3)_2^+$ ion and allows the precipitation of AgCl.

550

$$Ag(NH_3)_2^+ + Cl^- + 2\,H_3O^+ \longrightarrow AgCl + 2\,NH_4^+ + 2\,H_2O$$
$$\text{(white)}$$

A. GROUP I

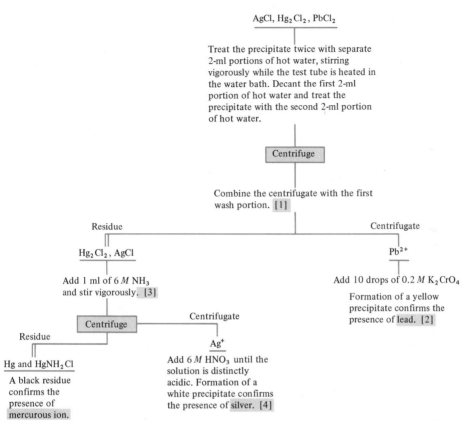

Numbers in brackets [1–13] refer to the flow charts on pages 552–553.

[1] The precipitate is washed to remove excess acidity (HCl). The thioacetamide in the wash solution helps to prevent oxidation of sulfide to sulfate and the NH_4Cl helps to prevent colloidal dispersion of the precipitate.

[2] The sulfides of arsenic, antimony, and tin are rendered soluble in the presence of excess S^{2-} ion in a basic solution. The formulas for these soluble, anionic species are idealized in the scheme. The exact formulas are uncertain. The treatment with NaOH solution is carried out twice to insure a clean separation. Incomplete removal of HCl solution in the washing step [1] will prevent the solubilization of these sulfides.

[3] The washing is necessary to remove excess NaOH solution. HgS is not soluble in $6\,M\ HNO_3$, but some of the HgS may be converted into $Hg(NO_3)_2 \cdot 2\,HgS$ which is white. The sulfides of lead, copper, and cadmium are dissolved by oxidation of the sulfide to sulfur.

$$3\,PbS + 2\,NO_3^- + 8\,H_3O^+ \longrightarrow 3\,Pb^{2+} + 3\,S + 2\,NO + 12\,H_2O$$

$$3\,CuS + 2\,NO_3^- + 8\,H_3O^+ \longrightarrow 3\,Cu^{2+} + 3\,S + 2\,NO + 12\,H_2O$$

$$3\,CdS + 2\,NO_3^- + 8\,H_3O^+ \longrightarrow 3\,Cd^{2+} + 3\,S + 2\,NO + 12\,H_2O$$

The sulfur produced in these reactions will appear as a solid that can be white, black, or yellow. Consequently, the presence of a residue after centrifugation does not necessarily indicate the presence of mercury.

B. Group II

$(PbS), HgS, CuS, CdS, As_2S_3, Sb_2S_3, SnS_2$

Wash the precipitate with 10 drops of water containing 1 drop of thioacetamide solution and 1/4 of a microspatula-full of NH_4Cl. [1] Discard the wash solution. Add 6 drops of 6 M NaOH and 2 drops of thioacetamide solution; stir and heat in the water bath for 3 min. [2]

Centrifuge

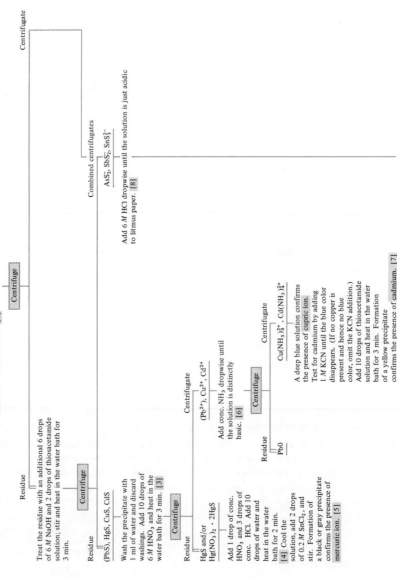

Residue

Treat the residue with an additional 6 drops of 6 M NaOH and 2 drops of thioacetamide solution; stir and heat in the water bath for 3 min.

Centrifuge

Residue

$(PbS), HgS, CuS, CdS$

Wash the precipitate with 1 ml of water and discard washings. Add 10 drops of 6 M HNO_3 and heat in the water bath for 3 min. [3]

Centrifuge

Residue

HgS and/or $Hg(NO_3)_2 \cdot 2HgS$

Add 1 drop of conc. HNO_3 and 3 drops of conc. HCl. Add 10 drops of water and heat in the water bath for 2 min. [4] Cool the solution, add 2 drops of 0.2 M $SnCl_2$, and stir. Formation of a black or gray precipitate confirms the presence of mercuric ion. [5]

Centrifugate

$(Pb^{2+}), Cu^{2+}, Cd^{2+}$

Add conc. NH_3 dropwise until the solution is distinctly basic. [6]

Centrifuge

Residue

PbO

Centrifugate

$Cu(NH_3)_4^{2+}, Cd(NH_3)_4^{2+}$

A deep blue solution confirms the presence of cupric ion. Test for cadmium by adding 1 M KCN until the blue color disappears. (If no copper is present and hence no blue color, omit the KCN addition.) Add 10 drops of thioacetamide solution and heat in the water bath for 3 min. Formation of a yellow precipitate confirms the presence of cadmium. [7]

Combined centrifugates

$AsS_2^-, SbS_2^-, SnS_3^{2-}$

Add 6 M HCl dropwise until the solution is just acidic to litmus paper. [8]

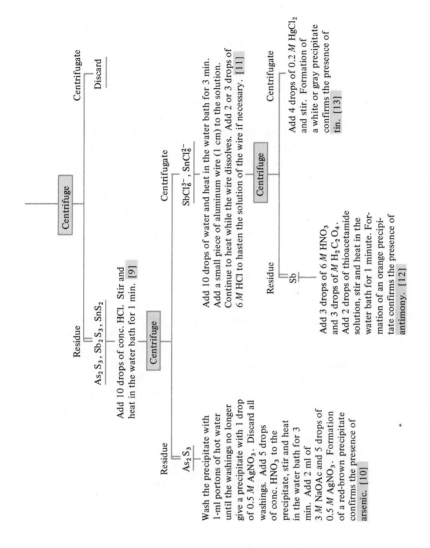

Centrifuge

Residue	Centrifugate
As₂S₃, Sb₂S₃, SnS₂	Discard

As_2S_3, Sb_2S_3, SnS_2

Add 10 drops of conc. HCl. Stir and heat in the water bath for 1 min. [9]

Centrifuge

Residue	Centrifugate
As₂S₃	SbCl₆³⁻, SnCl₆²⁻

As_2S_3

Wash the precipitate with 1-ml portons of hot water until the washings no longer give a precipitate with 1 drop of 0.5 M AgNO₃. Discard all washings. Add 5 drops of conc. HNO₃ to the precipitate, stir and heat in the water bath for 3 min. Add 2 ml of 3 M NaOAc and 5 drops of 0.5 M AgNO₃. Formation of a red-brown precipitate confirms the presence of arsenic. [10]

$SbCl_6^{3-}$, $SnCl_6^{2-}$

Add 10 drops of water and heat in the water bath for 3 min. Add a small piece of aluminum wire (1 cm) to the solution. Continue to heat while the wire dissolves. Add 2 or 3 drops of 6 M HCl to hasten the solution of the wire if necessary. [11]

Centrifuge

Residue	Centrifugate
Sb

Sb

Add 3 drops of 6 M HNO₃ and 3 drops of M H₂C₂O₄. Add 2 drops of thioacetamide solution, stir and heat in the water bath for 1 minute. Formation of an orange precipitate confirms the presence of antimony. [12]

Add 4 drops of 0.2 M HgCl₂ and stir. Formation of a white or gray precipitate confirms the presence of tin. [13]

[4] The HgS and $Hg(NO_3)_2 \cdot 2\,HgS$ are dissolved in aqua regia, formed by the mixture of concentrated HCl and HNO_3.

$$HgS + 2\,NOCl + 2\,Cl^- \longrightarrow HgCl_4^{2-} + 2\,NO + S$$

[5] Stannous ion reduces mercuric ion to mercurous ion which will precipitate in the presence of chloride ion.

$$2\,HgCl_4^{2-} + Sn^{2+} \longrightarrow \underset{\text{(white)}}{Hg_2Cl_2} + SnCl_6^{2-}$$

The Hg_2Cl_2 thus formed is a white precipitate. It is possible for further reduction to metallic mercury to occur in which case a black precipitate will be found which will impart a gray color to the total precipitate.

[6] In the presence of excess NH_3, the complex ions $Cu(NH_3)_4^{2+}$ and $Cd(NH_3)_4^{2+}$ are formed. The deep blue color of the $Cu(NH_3)_4^{2+}$ ion is sufficient to confirm copper. The $Cd(NH_3)_4^{2+}$ ion is colorless.

$$Cu^{2+} + 4\,NH_3 \longrightarrow \underset{\text{(deep blue)}}{Cu(NH_3)_4^{2+}}$$

$$Cd^{2+} + 4\,NH_3 \longrightarrow \underset{\text{(colorless)}}{Cd(NH_3)_4^{2+}}$$

If Pb^{2+} ion is present in the unknown and if it was not completely removed in group I, a white precipitate of PbO will occur at this point. In no case should the presence or absence of a precipitate at this point be taken as conclusive evidence one way or the other regarding lead. However, if Pb^{2+} ion was not detected in the group I analysis and a precipitate does develop at this point, a recheck should be made for group I ions. Any Pb^{2+} ion in solution when the NH_3 is added will precipitate by the reaction

$$Pb^{2+} + 2\,NH_3 + H_2O \longrightarrow PbO + 2\,NH_4^+$$

[7] If copper is absent (no blue color after the NH_3 addition), the thioacetamide can be added directly to give yellow CdS if cadmium is present.

$$Cd(NH_3)_4^{2+} + S^{2-} \longrightarrow \underset{\text{(yellow)}}{CdS} + 4\,NH_3$$

However, if $Cu(NH_3)_4^{2+}$ ions are present, black CuS will precipitate and obscure the test for cadmium. The addition of the KCN solution complexes both ions. The cadmium reaction is straightforward

$$Cd(NH_3)_4^{2+} + 4\,CN^- \longrightarrow Cd(CN)_4^{2-} + 4\,NH_3$$

The copper reaction is more complicated

$$2\,Cu(NH_3)_4^{2+} + 7\,CN^- + 2\,OH^- \longrightarrow 2\,Cu(CN)_3^{2-} + 8\,NH_3 + CNO^- + H_2O$$

The $Cd(CN)_4^{2-}$ ion reacts with sulfide ion, but the $Cu(CN)_3^{2-}$ ion will not.

$$Cd(CN)_4^{2+} + S^{2-} \longrightarrow \underset{\text{(yellow)}}{CdS} + 4\,CN^-$$

[8] Acidification of the solution destroys the complex anionic species and reprecipitates the sulfides of arsenic, antimony, and tin.

[9] In an acidic solution in a high concentration of chloride ion, antimony and tin ions form complexes.

$$Sb_2S_3 + 6 H_3O^+ + 12 Cl^- \longrightarrow 2 SbCl_6^{3-} + 3 H_2S + 6 H_2O$$

$$SnS_2 + 4 H_3O^+ + 6 Cl^- \longrightarrow SnCl_6^{2-} + 2 H_2S + 4 H_2O$$

The As_2S_3 is unaffected.

[10] Washing is necessary to remove Cl^- ions which would interfere with the confirmation test for arsenic. The concentrated HNO_3 oxidizes any As_2S_3 to $H_2AsO_4^-$.

$$As_2S_3 + 10 NO_3^- + 8 H_3O^+ \longrightarrow 2 H_2AsO_4^- + 3 S + 10 NO + 10 H_2O$$

The acetate ion reduces excess acidity so that Ag^+ ion can react with arsenate ion to give the red-brown precipitate of Ag_3AsO_4.

$$H_2AsO_4^- + 2 OAc^- \longrightarrow 2 HOAc + AsO_4^-$$

$$3 Ag^+ + AsO_4^- \longrightarrow Ag_3AsO_4$$
$$\text{(red-brown)}$$

[11] The solution is first heated to expel H_2S. Aluminum wire is then added to reduce antimony ions to metallic antimony and tin(IV) to tin(II).

$$SbCl_4^- + Al \longrightarrow Sb + Al^{3+} + 4 Cl^-$$

$$3 SnCl_6^{2-} + 2 Al \longrightarrow 3 SnCl_4^{2-} + 2 Al^{3+} + 6 Cl^-$$

[12] The metallic Sb dissolves in 6 M HNO_3 in the presence of 1 M $H_2C_2O_4$ (oxalic acid) by the reaction

$$Sb + H_2C_2O_4 + H_3O^+ + NO_3^- \longrightarrow SbO(HC_2O_4) + NO + 2 H_2O$$

The addition of thioacetamide solution then causes the precipitation of the orange Sb_2S_3.

$$2 SbO(HC_2O_4) + 3 H_2S \longrightarrow Sb_2S_3 + 2 H_2C_2O_4 + 2 H_2O$$
$$\text{(orange)}$$

[13] The tin(II) reduces mercuric to mercurous ion, which precipitates in the presence of chloride ion. Some reduction to black, metallic mercury may also occur in which case the white Hg_2Cl_2 will appear gray.

$$SnCl_4^{2-} + 2 Hg^{2+} + 4 Cl^- \longrightarrow SnCl_6^{2-} + Hg_2Cl_2$$

Numbers in brackets [1–9] refer to the flow chart on page 556.

[1] If CoS and/or NiS are present, the precipitate will not completely dissolve. These sulfides are difficult to dissolve in hydrochloric acid once they have been precipitated. Complete solution of the precipitate in 6 M HCl indicates the absence of cobalt and nickel in the unknown. The other solids dissolve by the reactions

$$Al(OH)_3 + 3 H_3O^+ \longrightarrow Al(H_2O)_6^{3+}$$

$$Cr(OH)_3 + 3 H_3O^+ \longrightarrow Cr(H_2O)_6^{3+}$$

C. GROUP III

$AL(OH)_3$, $Cr(OH)_3$, ZnS, Fe_2S_3, CoS, NiS

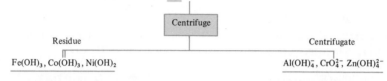

Add 6 M HCl dropwise with stirring until the precipitate has dissolved or until a total of about 10 drops have been added. [1] (If a precipitate remains, add 4 drops of conc. HNO_3.) Stir and heat the solution in the water bath for 3 min. Make the solution basic by adding 6 M NaOH dropwise with stirring. When it is basic to litmus paper, add 5 drops in excess. Add 20 drops of 3% H_2O_2, stir and heat in the water bath for 3 min. [2]

Centrifuge

Residue — $Fe(OH)_3$, $Co(OH)_3$, $Ni(OH)_2$

Centrifugate — $Al(OH)_4^-$, CrO_4^{2-}, $Zn(OH)_4^{2-}$

Wash the precipitate with 1 ml of water and discard the washings. Add 6 M HCl dropwise with stirring until the precipitate dissolves. Add enough water to give a total volume of about 1.5 ml, stir and divide the solution up into three parts. [3]

Add 6 M HCl dropwise with stirring until the solution is acidic to litmus paper. If the volume of the solution is much greater than 2 ml, transfer the solution to an evaporating dish and reduce the volume. Put the solution back in the test tube and add 6 M NH_3 dropwise with stirring until the solution is basic to litmus paper.

Add 3 drops of 2 M NH_4CNS. A deep red color confirms the presence of iron. [4]

Add 1/2 microspatula-full of KNO_2. Heat and stir in the water bath. Formation of a yellow precipitate confirms the presence of cobalt. [5]

Add 6 M NH_3 dropwise until the solution is basic to litmus paper. Add 5 drops of dimethylglyoxime solution and stir. Formation of a bright red precipitate confirms the presence of nickel. [6]

Centrifuge

Centrifugate — CrO_4^{2-}, $Zn(NH_3)_4^{2+}$

Residue — $Al(OH)_3$

Divide solution into two parts.

The presence of a white, gelatinous precipitate at this point confirms the presence of aluminum. [7]

If the solution is yellow, acidify by adding 3 M HOAc dropwise until acidic to litmus paper. Then add 5 drops of 1 M $Pb(NO_3)_2$.

Formation of a yellow precipitate confirms the presence of chromium. [8]

Add 10 drops of thioacetamide solution, stir, and heat in the water bath for 3 min. Formation of a white precipitate which dissolves when the solution is acidified with 6 M HCl confirms the presence of zinc. [9]

$$ZnS + 2 H_3O^+ \longrightarrow Zn^{2+} + H_2S + 2 H_2O$$

$$Fe_2S_3 + 6 H_3O^+ + 6 H_2O \longrightarrow 2 Fe(H_2O)_6^{3+} + 3 H_2S$$

[2] If the precipitate is completely dissolved by the action of 6 M HCl, it is not necessary to add the HNO_3. However, any CoS and NiS must be dissolved by the oxidizing action of HNO_3.

$$CoS + 4 H_3O^+ + 2 NO_3^- \longrightarrow Co^{2+} + 2 NO_2 + S + 6 H_2O$$

$$NiS + 4 H_3O^+ + 2 NO_3^- \longrightarrow Ni^{2+} + 2 NO_2 + S + 6 H_2O$$

Heating the acidified solution is necessary to drive off H_2S; otherwise the sulfides would reprecipitate when the solution is made basic with 6 M NaOH. Any solid present after treatment with the HNO_3 is probably elemental sulfur and should be discarded. The 6 M NaOH causes the precipitation of $Fe(OH)_3$, $Co(OH)_3$, and $Ni(OH)_2$ whereas Al^{3+} and Zn^{2+} ions are converted to soluble, complex anions. The H_2O_2 oxidizes Cr^{3+} ion to CrO_4^{2-} ions.

$$Fe^{3+} + 3\,OH^- \longrightarrow Fe(OH)_3$$

$$Co^{3+} + 3\,OH^- \longrightarrow Co(OH)_3$$

$$Ni^{2+} + 2\,OH^- \longrightarrow Ni(OH)_2$$

$$Al^{3+} + 4\,OH^- \longrightarrow Al(OH)_4^-$$

$$Zn^{2+} + 4\,OH^- \longrightarrow Zn(OH)_4^{2-}$$

$$2\,Cr^{3+} + 3\,H_2O_2 + 10\,OH^- \longrightarrow 2\,CrO_4^{2-} + 8\,H_2O$$

[3] The precipitate is washed to remove any CrO_4^- ions that could interfere with the confirmation tests. The hydroxides of iron and nickel dissolve in HCl in a straightforward manner. The $Co(OH)_3$ undergoes reduction of Co^{3+} to Co^{2+}.

$$Fe(OH)_3 + 3\,H_3O^+ \longrightarrow Fe^{3+} + 6\,H_2O$$

$$Ni(OH)_2 + 2\,H_3O^+ \longrightarrow Ni^{2+} + 4\,H_2O$$

$$2\,Co(OH)_3 + 6\,H_3O^+ + 2\,Cl^- \longrightarrow 2\,Co^{2+} + Cl_2 + 12\,H_2O$$

[4] The CNS^- ion forms a deep red complex ion with ferric ion.

$$Fe^{3+} + 6\,CNS^- \longrightarrow Fe(CNS)_6^{3-}$$
$$\text{(deep red)}$$

Traces of Fe^{3+} ion are often present in the solution. This test is very sensitive, so a faint pink coloration should be disregarded. If iron is absent and cobalt is present, a faint blue color may appear.

[5] The Co^{2+} ion is determined by formation of the complex ion hexanitrocobaltate(III), which precipitates with the potassium ions.

$$3\,K^+ + Co^{2+} + 7\,NO_2^- + 2\,H_3O^+ \longrightarrow K_3[Co(NO_2)_6] + NO + 3\,H_2O$$
$$\text{(yellow)}$$

[6] The addition of dimethylglyoxime to a basic solution containing nickel ions forms the bright red coordination compound, bis(dimethylglyoxime)nickel(II).

$$Ni(NH_3)_6^{2+} + 2\,C_4H_8O_2N_2 \longrightarrow Ni(C_4H_7O_2N_2)_2 + 2\,NH_4^+ + 4\,NH_3$$
$$\text{(bright red)}$$

[7] Acidification of the solution converts $Al(OH)_4^-$ and $Zn(OH)_4^{2-}$ to Al^{3+} and Zn^{2+} ions.

$$Al(OH)_4^- + 4\,H_3O^+ \longrightarrow Al^{3+} + 8\,H_2O$$

$$Zn(OH)_4^{2-} + 4\,H_3O^+ \longrightarrow Zn^{2+} + 8\,H_2O$$

When the 6 M NH_3 is added, aluminum will precipitate as the hydroxide, but zinc remains in solution in the form of a complex ion.

$$Al^{3+} + 3\,NH_3 + 3\,H_2O \longrightarrow Al(OH)_3 + 3\,NH_4^+$$
$$\text{(gelatinous)}$$
$$Zn^{2+} + 4\,NH_3 \longrightarrow Zn(NH_3)_4^{2+}$$

The appearance of gelatinous $Al(OH)_3$ as the $6\,M$ NH_3 is added to the solution confirms the presence of aluminum. The presence of a trace of ill-defined gray solid is probably due to elemental sulfur.

[8] If the solution does not have the characteristic yellow color of CrO_4^-, this ion is absent. However, if the solution is yellow, the presence of CrO_4^- is confirmed by the addition of lead ions.

$$Pb^{2+} + CrO_4^{2-} \longrightarrow PbCrO_4$$
$$\text{(yellow)}$$

[9] The complex ion, $Zn(NH_3)_4^{2+}$, reacts with sulfide ion to precipitate white ZnS.

$$Zn(NH_3)_4^{2+} + S^{2-} \longrightarrow ZnS + 4\,NH_3$$
$$\text{(white)}$$

If CrO_4^- ion is present, it may oxidize some of the S^{2-} ions to elemental sulfur, which can form a milkiness in the solution. However, ZnS is soluble in $6\,M$ HCl whereas S is not. Hence, the formation of a white precipitate which can be partially or completely redissolved in $6\,M$ HCl confirms the presence of zinc.

Numbers in brackets [1–5] refer to the flow chart on page 559.

[1] The phosphates dissolve in acetic acid and barium is removed by addition of CrO_4^{2-} ions.

$$Ba_3(PO_4)_2 + 2\,HOAc \longrightarrow 3\,Ba^{2+} + 2\,HPO_4^{2-} + 2\,OAc^-$$

$$Ca_3(PO_4)_2 + 2\,HOAc \longrightarrow 3\,Ca^{2+} + 2\,HPO_4^{2-} + 2\,OAc^-$$

$$MgNH_4PO_4 + HOAc \longrightarrow Mg^{2+} + NH_4^+ + HPO_4^{2-} + OAc$$

$$Ba^{2+} + CrO_4^{2-} \longrightarrow BaCrO_4$$

[2] Addition of the NH_3 causes the reprecipitation of the phosphates of Ca^{2+} and Mg^{2+} so that the excess chromate ions can be removed. The yellow color of CrO_4^{2-} ions interferes with the confirmation tests for these cations.

$$3\,Ca^{2+} + 2\,HPO_4^{2-} + 2\,NH_3 \longrightarrow Ca_3(PO_4)_2 + 2\,NH_4^+$$
$$Mg^{2+} + HPO_4^{2-} + NH_3 \longrightarrow MgNH_4PO_4$$

[3] The phosphates are again dissolved by the action of acetic acid (see note [1]). The oxalate ion forms a white precipitate of CaC_2O_4 which serves to confirm calcium ion.

$$Ca^{2+} + C_2O_4^{2-} \longrightarrow CaC_2O_4$$
$$\text{(white)}$$

[4] The magnesium ion is confirmed by the formation of $Mg(OH)_2$ which adsorbs the organic dye nitrobenzeneazoresorcinol to give the blue precipitate.

$$Mg^{2+} + 2\,OH^- \longrightarrow Mg(OH)_2$$
$$\text{(blue with}$$
$$\text{dye)}$$

The dye is adsorbed only by a precipitate of $Mg(OH)_2$ under these conditions.

D. GROUP IV

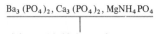

$Ba_3(PO_4)_2$, $Ca_3(PO_4)_2$, $MgNH_4PO_4$

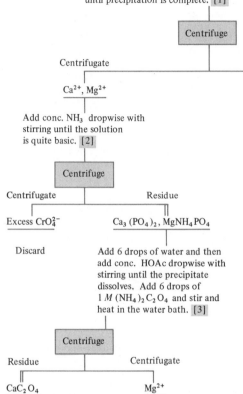

Wash the precipitate with 10 drops of water and discard the washings. Add 10 drops of 6 M HOAc and stir. Then add 1 ml of water. (If undissolved precipitate remains, add 5 more drops of 6 M HOAc.) Add 1 M K_2CrO_4 dropwise until precipitation is complete. [1]

Centrifuge

Centrifugate

Ca^{2+}, Mg^{2+}

Add conc. NH_3 dropwise with stirring until the solution is quite basic. [2]

Centrifuge

Centrifugate

Excess CrO_4^{2-}

Discard

Residue

$Ca_3(PO_4)_2$, $MgNH_4PO_4$

Add 6 drops of water and then add conc. HOAc dropwise with stirring until the precipitate dissolves. Add 6 drops of 1 M $(NH_4)_2C_2O_4$ and stir and heat in the water bath. [3]

Centrifuge

Residue

CaC_2O_4

The presence of a white precipitate at this point confirms the presence of calcium.

Centrifugate

Mg^{2+}

Add 1 drop of nitrobenzeneazoresorcinol solution and 1 ml of 6 M NaOH. Formation of a blue precipitate confirms the presence of magnesium. [4]

Residue

$BaCrO_4$

Add 5 drops of conc. HCl and 10 drops of water; stir. The precipitate may all dissolve; if not, remove most of the solution from any undissolved solid and to the solution add 3 drops of 1 M $(NH_4)_2SO_4$. Formation of a white precipitate confirms the presence of barium. [5]

[5] The yellow precipitate of $BaCrO_4$ is indicative of the presence of barium, but further confirmation that the chromate ion has precipitated barium and not some other ion is achieved by dissolving it in HCl and precipitating $BaSO_4$. Only $BaSO_4$ is insoluble under these conditions.

$$2\, BaCrO_4 + 2\, H_3O^+ \longrightarrow 2\, Ba^{2+} + Cr_2O_7^{2-} + 3\, H_2O$$

$$Ba^{2+} + SO_4^{2-} \longrightarrow BaSO_4$$
$$\text{(white)}$$

Numbers in brackets [1–2] refer to the flow chart on page 560.

[1] Sodium traces are practically always present in any solution to an extent sufficient to give a slight yellow flame. Therefore, only an intense yellow flame should be taken as evidence of the presence of Na^+ in the unknown.

[2] In the absence of Na^+ ions, the violet color imparted by K^+ ions is visible to the naked eye. However, in the presence of Na^+ ions, the bright yellow color obscures

E. GROUP V

Na^+, K^+

Evaporate the solution until a moist (not dry) solid forms. Add 4 drops of conc. HCl and carry out flame tests on this solution by dipping a platinum or nichrome wire into some conc. HCl, then into the test solution and then into a nearly colorless flame. Production of an intense yellow color confirms the presence of sodium. [1]

Production of a violet flame confirms the presence of potassium. [2]

the more delicate violet potassium flame. Consequently, a piece of cobalt glass must be used to screen out the yellow color. Best results are usually obtained by holding the cobalt glass closely up to the eye and viewing the flame at a distance of a foot or more.

PROBLEMS

1. An unknown solution is colorless. Of the following ions, which ones could possibly be present and which ones could not be present in moderate concentration?
   Cu^{2+}, Al^{3+}, Cr^{3+}, Ba^{2+}, Fe^{3+}, Ca^{2+}, Na^+, K^+, NO_3^-, Cl^-

2. An unknown mixture dissolves completely in water to give a colorless solution with a pH of 3. Of the following ions, decide which ones are definitely present, which ones are definitely absent, and which ones are undetermined.
   NH_4^+, Fe^{3+}, Al^{3+}, Pb^{2+}, Mg^{2+}, S^{2-}, CO_3^{2-}, $Cr_2O_7^{2-}$, SO_4^{2-}

3. Outline a flow diagram to show how the following ions could be separated into groups and then specifically identified.
   Ag^+, Hg^{2+}, Cu^{2+}, Cd^{2+}, Zn^{2+}, Co^{2+}, Ba^{2+}, Ca^{2+}

4. Outline a flow diagram to show how the following ions could be separated into groups and then specifically identified.
   CO_3^{2-}, PO_4^{3-}, $C_2O_4^{2-}$, SO_4^{2-}, S^{2-}, Cl^-, Br^-, NO_3^-

5. A certain compound is a white solid that is insoluble in water. Choose the compound from among the following possibilities.
   $Na_2C_2O_4$, $BaCrO_4$, $CrCl_3$, AgI, $CaCO_3$

6. A certain compound is soluble in water and no reaction is observed when a little 3 M HCl is added to the solution. Choose the compound from among the following possibilities.
   $PbCl_2$, Na_2CO_3, $FeCl_3$, $AgCl$, $BaSO_4$

7. List the colors of the following sulfides.
   Ag_2S, HgS, CuS, CdS, As_2S_3, Sb_2S_3, SnS_2, Fe_2S_3, CoS, NiS, ZnS

ANSWERS TO PROBLEMS

1. Absent Cu^{2+}, Cr^{3+}, Fe^{3+}; all others possibly present.
2. Present Al^{3+}, SO_4^{2-}
   Absent Fe^{3+}, Pb^{2+}, S^{2-}, CO_3^{2-}, $Cr_2O_7^{2-}$
   Undetermined NH_4^+, Mg^{2+}
5. $CaCO_3$
6. $FeCl_3$

GROUND-STATE ELECTRON
DISTRIBUTIONS OF THE *ATOMS*

At. No.	Element	1s	2s	2p	3s	3p	3d	4s	4p	4d	4f	5s	5p	5d	5f	6s	6p	6d	7s	7p
1	H	1																		
2	He	2																		
3	Li	2	1																	
4	Be	2	2																	
5	B	2	2	1																
6	C	2	2	2																
7	N	2	2	3																
8	O	2	2	4																
9	F	2	2	5																
10	Ne	2	2	6																
11	Na	2	2	6	1															
12	Mg	2	2	6	2															
13	Al	2	2	6	2	1														
14	Si	2	2	6	2	2														
15	P	2	2	6	2	3														
16	S	2	2	6	2	4														
17	Cl	2	2	6	2	5														
18	Ar	2	2	6	2	6														
19	K	2	2	6	2	6		1												
20	Ca	2	2	6	2	6		2												
21	Sc	2	2	6	2	6	1	2												
22	Ti	2	2	6	2	6	2	2												
23	V	2	2	6	2	6	3	2												
24	Cr	2	2	6	2	6	5	1												
25	Mn	2	2	6	2	6	5	2												
26	Fe	2	2	6	2	6	6	2												
27	Co	2	2	6	2	6	7	2												
28	Ni	2	2	6	2	6	8	2												
29	Cu	2	2	6	2	6	10	1												
30	Zn	2	2	6	2	6	10	2												
31	Ga	2	2	6	2	6	10	2	1											
32	Ge	2	2	6	2	6	10	2	2											
33	As	2	2	6	2	6	10	2	3											
34	Se	2	2	6	2	6	10	2	4											
35	Br	2	2	6	2	6	10	2	5											
36	Kr	2	2	6	2	6	10	2	6											
37	Rb	2	2	6	2	6	10	2	6			1								
38	Sr	2	2	6	2	6	10	2	6			2								
39	Y	2	2	6	2	6	10	2	6	1		2								
40	Zr	2	2	6	2	6	10	2	6	2		2								
41	Nb	2	2	6	2	6	10	2	6	4		1								
42	Mo	2	2	6	2	6	10	2	6	5		1								
43	Tc	2	2	6	2	6	10	2	6	6		1								
44	Ru	2	2	6	2	6	10	2	6	7		1								
45	Rh	2	2	6	2	6	10	2	6	8		1								

At. No.	Element	1s	2s	2p	3s	3p	3d	4s	4p	4d	4f	5s	5p	5d	5f	6s	6p	6d	7s	7p
46	Pd	2	2	6	2	6	10	2	6	10										
47	Ag	2	2	6	2	6	10	2	6	10		1								
48	Cd	2	2	6	2	6	10	2	6	10		2								
49	In	2	2	6	2	6	10	2	6	10		2	1							
50	Sn	2	2	6	2	6	10	2	6	10		2	2							
51	Sb	2	2	6	2	6	10	2	6	10		2	3							
52	Te	2	2	6	2	6	10	2	6	10		2	4							
53	I	2	2	6	2	6	10	2	6	10		2	5							
54	Xe	2	2	6	2	6	10	2	6	10		2	6							
55	Cs	2	2	6	2	6	10	2	6	10		2	6			1				
56	Ba	2	2	6	2	6	10	2	6	10		2	6			2				
57	La	2	2	6	2	6	10	2	6	10		2	6	1		2				
58	Ce	2	2	6	2	6	10	2	6	10	2	2	6			2				
59	Pr	2	2	6	2	6	10	2	6	10	3	2	6			2				
60	Nd	2	2	6	2	6	10	2	6	10	4	2	6			2				
61	Pm	2	2	6	2	6	10	2	6	10	5	2	6			2				
62	Sm	2	2	6	2	6	10	2	6	10	6	2	6			2				
63	Eu	2	2	6	2	6	10	2	6	10	7	2	6			2				
64	Gd	2	2	6	2	6	10	2	6	10	7	2	6	1		2				
65	Tb	2	2	6	2	6	10	2	6	10	9	2	6			2				
66	Dy	2	2	6	2	6	10	2	6	10	10	2	6			2				
67	Ho	2	2	6	2	6	10	2	6	10	11	2	6			2				
68	Er	2	2	6	2	6	10	2	6	10	12	2	6			2				
69	Tm	2	2	6	2	6	10	2	6	10	13	2	6			2				
70	Yb	2	2	6	2	6	10	2	6	10	14	2	6			2				
71	Lu	2	2	6	2	6	10	2	6	10	14	2	6	1		2				
72	Hf	2	2	6	2	6	10	2	6	10	14	2	6	2		2				
73	Ta	2	2	6	2	6	10	2	6	10	14	2	6	3		2				
74	W	2	2	6	2	6	10	2	6	10	14	2	6	4		2				
75	Re	2	2	6	2	6	10	2	6	10	14	2	6	5		2				
76	Os	2	2	6	2	6	10	2	6	10	14	2	6	6		2				
77	Ir	2	2	6	2	6	10	2	6	10	14	2	6	9						
78	Pt	2	2	6	2	6	10	2	6	10	14	2	6	9		1				
79	Au	2	2	6	2	6	10	2	6	10	14	2	6	10		1				
80	Hg	2	2	6	2	6	10	2	6	10	14	2	6	10		2				
81	Tl	2	2	6	2	6	10	2	6	10	14	2	6	10		2	1			
82	Pb	2	2	6	2	6	10	2	6	10	14	2	6	10		2	2			
83	Bi	2	2	6	2	6	10	2	6	10	14	2	6	10		2	3			
84	Po	2	2	6	2	6	10	2	6	10	14	2	6	10		2	4			
85	At	2	2	6	2	6	10	2	6	10	14	2	6	10		2	5			
86	Rn	2	2	6	2	6	10	2	6	10	14	2	6	10		2	6			
87	Fr	2	2	6	2	6	10	2	6	10	14	2	6	10		2	6		1	
88	Ra	2	2	6	2	6	10	2	6	10	14	2	6	10		2	6		2	
89	Ac	2	2	6	2	6	10	2	6	10	14	2	6	10		2	6	1	2	
90	Th	2	2	6	2	6	10	2	6	10	14	2	6	10		2	6	2	2	
91	Pa	2	2	6	2	6	10	2	6	10	14	2	6	10	2	2	6	1	2	

At. No.	Element	1s	2s	2p	3s	3p	3d	4s	4p	4d	4f	5s	5p	5d	5f	6s	6p	6d	7s	7p
92	U	2	2	6	2	6	10	2	6	10	14	2	6	10	3	2	6	1	2	
93	Np	2	2	6	2	6	10	2	6	10	14	2	6	10	4	2	6	1	2	
94	Pu	2	2	6	2	6	10	2	6	10	14	2	6	10	5	2	6	1	2	
95	Am	2	2	6	2	6	10	2	6	10	14	2	6	10	6	2	6	1	2	
96	Cm	2	2	6	2	6	10	2	6	10	14	2	6	10	7	2	6	1	2	
97	Bk	2	2	6	2	6	10	2	6	10	14	2	6	10	8	2	6	1	2	
98	Cf	2	2	6	2	6	10	2	6	10	14	2	6	10	9	2	6	1	2	
99	Es	2	2	6	2	6	10	2	6	10	14	2	6	10	10	2	6	1	2	
100	Fm	2	2	6	2	6	10	2	6	10	14	2	6	10	11	2	6	1	2	
101	Md	2	2	6	2	6	10	2	6	10	14	2	6	10	12	2	6	1	2	
102	No	2	2	6	2	6	10	2	6	10	14	2	6	10	14	2	6		2	
103	Lw	2	2	6	2	6	10	2	6	10	14	2	6	10	14	2	6	1	2	
104	Unnamed	2	2	6	2	6	10	2	6	10	14	2	6	10	14	2	6	2	2	
105	Unnamed	2	2	6	2	6	10	2	6	10	14	2	6	10	14	2	6	3	2	

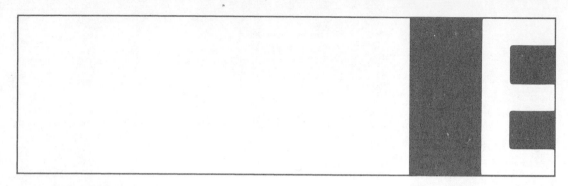

LOGARITHMS

Natural Numbers	0	1	2	3	4	5	6	7	8	9	Proportional Parts								
											1	2	3	4	5	6	7	8	9
10	0000	0043	0086	0128	0170	0212	0253	0294	0334	0374	4	8	12	17	21	25	29	33	37
11	0414	0453	0492	0531	0569	0607	0645	0682	0719	0755	4	8	11	15	19	23	26	30	34
12	0792	0828	0864	0899	0934	0969	1004	1038	1072	1106	3	7	10	14	17	21	24	28	31
13	1139	1173	1206	1239	1271	1303	1335	1367	1399	1430	3	6	10	13	16	19	23	26	29
14	1461	1492	1523	1553	1584	1614	1644	1673	1703	1732	3	6	9	12	15	18	21	24	27
15	1761	1790	1818	1847	1875	1903	1931	1959	1987	2014	3	6	8	11	14	17	20	22	25
16	2041	2068	2095	2122	2148	2175	2201	2227	2253	2279	3	5	8	11	13	16	18	21	24
17	2304	2330	2355	2380	2405	2430	2455	2480	2504	2529	2	5	7	10	12	15	17	20	22
18	2553	2577	2601	2625	2648	2672	2695	2718	2742	2765	2	5	7	9	12	14	16	19	21
19	2788	2810	2833	2856	2878	2900	2923	2945	2967	2989	2	4	7	9	11	13	16	18	20
20	3010	3032	3054	3075	3096	3118	3139	3160	3181	3201	2	4	6	8	11	13	15	17	19
21	3222	3243	3263	3284	3304	3324	3345	3365	3385	3404	2	4	6	8	10	12	14	16	18
22	3424	3444	3464	3483	3502	3522	3541	3560	3579	3598	2	4	6	8	10	12	14	15	17
23	3617	3636	3655	3674	3692	3711	3729	3747	3766	3784	2	4	6	7	9	11	13	15	17
24	3802	3820	3838	3856	3874	3892	3909	3927	3945	3962	2	4	5	7	9	11	12	14	16
25	3979	3997	4014	4031	4048	4065	4082	4099	4116	4133	2	3	5	7	9	10	12	14	15
26	4150	4166	4183	4200	4216	4232	4249	4265	4281	4298	2	3	5	7	8	10	11	13	15
27	4314	4330	4346	4362	4378	4393	4409	4425	4440	4456	2	3	5	6	8	9	11	13	14
28	4472	4487	4502	4518	4533	4548	4564	4579	4594	4609	2	3	5	6	8	9	11	12	14
29	4624	4639	4654	4669	4683	4698	4713	4728	4742	4757	1	3	4	6	7	9	10	12	13
30	4771	4786	4800	4814	4829	4843	4857	4871	4886	4900	1	3	4	6	7	9	10	11	13
31	4914	4928	4942	4955	4969	4983	4997	5011	5024	5038	1	3	4	6	7	8	10	11	12
32	5051	5065	5079	5092	5105	5119	5132	5145	5159	5172	1	3	4	5	7	8	9	11	12
33	5185	5198	5211	5224	5237	5250	5263	5276	5289	5302	1	3	4	5	6	8	9	10	12
34	5315	5328	5340	5353	5366	5378	5391	5403	5416	5428	1	3	4	5	6	8	9	10	11
35	5441	5453	5465	5478	5490	5502	5514	5527	5539	5551	1	2	4	5	6	7	9	10	11
36	5563	5575	5587	5599	5611	5623	5635	5647	5658	5670	1	2	4	5	6	7	8	10	11
37	5682	5694	5705	5717	5729	5740	5752	5763	5775	5786	1	2	3	5	6	7	8	9	10
38	5798	5809	5821	5832	5843	5855	5866	5877	5888	5899	1	2	3	5	6	7	8	9	10
39	5911	5922	5933	5944	5955	5966	5977	5988	5999	6010	1	2	3	4	5	7	8	9	10
40	6021	6031	6042	6053	6064	6075	6085	6096	6107	6117	1	2	3	4	5	6	8	9	10
41	6128	6138	6149	6160	6170	6180	6191	6201	6212	6222	1	2	3	4	5	6	7	8	9
42	6232	6243	6253	6263	6274	6284	6294	6304	6314	6325	1	2	3	4	5	6	7	8	9
43	6335	6345	6355	6365	6375	6385	6395	6405	6415	6425	1	2	3	4	5	6	7	8	9
44	6435	6444	6454	6464	6474	6484	6493	6503	6513	6522	1	2	3	4	5	6	7	8	9
45	6532	6542	6551	6561	6571	6580	6590	6599	6609	6618	1	2	3	4	5	6	7	8	9
46	6628	6637	6646	6656	6665	6675	6684	6693	6702	6712	1	2	3	4	5	6	7	7	8
47	6721	6730	6739	6749	6758	6767	6776	6785	6794	6803	1	2	3	4	5	5	6	7	8
48	6812	6821	6830	6839	6848	6857	6866	6875	6884	6893	1	2	3	4	4	5	6	7	8
49	6902	6911	6920	6928	6937	6946	6955	6964	6972	6981	1	2	3	4	4	5	6	7	8

Natural Numbers	0	1	2	3	4	5	6	7	8	9	Proportional Parts								
											1	2	3	4	5	6	7	8	9
50	6990	6998	7007	7016	7024	7033	7042	7050	7059	7067	1	2	3	3	4	5	6	7	8
51	7076	7084	7093	7101	7110	7118	7126	7135	7143	7152	1	2	3	3	4	5	6	7	8
52	7160	7168	7177	7185	7193	7202	7210	7218	7226	7235	1	2	2	3	4	5	6	7	7
53	7243	7251	7259	7267	7275	7284	7292	7300	7308	7316	1	2	2	3	4	5	6	6	7
54	7324	7332	7340	7348	7356	7364	7372	7380	7388	7396	1	2	2	3	4	5	6	6	7
55	7404	7412	7419	7427	7435	7443	7451	7459	7466	7474	1	2	2	3	4	5	5	6	7
56	7482	7490	7497	7505	7513	7520	7528	7536	7543	7551	1	2	2	3	4	5	5	6	7
57	7559	7566	7574	7582	7589	7597	7604	7612	7619	7627	1	2	2	3	4	5	5	6	7
58	7634	7642	7649	7657	7664	7672	7679	7686	7694	7701	1	1	2	3	4	4	5	6	7
59	7709	7716	7723	7731	7738	7745	7752	7760	7767	7774	1	1	2	3	4	4	5	6	7
60	7782	7789	7796	7803	7810	7818	7825	7832	7839	7846	1	1	2	3	4	4	5	6	6
61	7853	7860	7868	7875	7882	7889	7896	7903	7910	7917	1	1	2	3	4	4	5	6	6
62	7924	7931	7938	7945	7952	7959	7966	7973	7980	7987	1	1	2	3	3	4	5	6	6
63	7993	8000	8007	8014	8021	8028	8035	8041	8048	8055	1	1	2	3	3	4	5	5	6
64	8062	8069	8075	8082	8089	8096	8102	8109	8116	8122	1	1	2	3	3	4	5	5	6
65	8129	8136	8142	8149	8156	8162	8169	8176	8182	8189	1	1	2	3	3	4	5	5	6
66	8195	8202	8209	8215	8222	8228	8235	8241	8248	8254	1	1	2	3	3	4	5	5	6
67	8261	8267	8274	8280	8287	8293	8299	8306	8312	8319	1	1	2	3	3	4	5	5	6
68	8325	8331	8338	8344	8351	8357	8363	8370	8376	8382	1	1	2	3	3	4	4	5	6
69	8388	8395	8401	8407	8414	8420	8426	8432	8439	8445	1	1	2	2	3	4	4	5	6
70	8451	8457	8463	8470	8476	8482	8488	8494	8500	8506	1	1	2	2	3	4	4	5	6
71	8513	8519	8525	8531	8537	8543	8549	8555	8561	8567	1	1	2	2	3	4	4	5	5
72	8573	8579	8585	8591	8597	8603	8609	8615	8621	8627	1	1	2	2	3	4	4	5	5
73	8633	8639	8645	8651	8657	8663	8669	8675	8681	8686	1	1	2	2	3	4	4	5	5
74	8692	8698	8704	8710	8716	8722	8727	8733	8739	8745	1	1	2	2	3	4	4	5	5
75	8751	8756	8762	8768	8774	8779	8785	8791	8797	8802	1	1	2	2	3	3	4	5	5
76	8808	8814	8820	8825	8831	8837	8842	8848	8854	8859	1	1	2	2	3	3	4	5	5
77	8865	8871	8876	8882	8887	8893	8899	8904	8910	8915	1	1	2	2	3	3	4	4	5
78	8921	8927	8932	8938	8943	8949	8954	8960	8965	8971	1	1	2	2	3	3	4	4	5
79	8976	8982	8987	8993	8998	9004	9009	9015	9020	9025	1	1	2	2	3	3	4	4	5
80	9031	9036	9042	9047	9053	9058	9063	9069	9074	9079	1	1	2	2	3	3	4	4	5
81	9085	9090	9096	9101	9106	9112	9117	9122	9128	9133	1	1	2	2	3	3	4	4	5
82	9138	9143	9149	9154	9159	9165	9170	9175	9180	9186	1	1	2	2	3	3	4	4	5
83	9191	9196	9201	9206	9212	9217	9222	9227	9232	9238	1	1	2	2	3	3	4	4	5
84	9243	9248	9253	9258	9263	9269	9274	9279	9284	9289	1	1	2	2	3	3	4	4	5
85	9294	9299	9304	9309	9315	9320	9325	9330	9335	9340	1	1	2	2	3	3	4	4	5
86	9345	9350	9355	9360	9365	9370	9375	9380	9385	9390	1	1	2	2	3	3	4	4	5
87	9395	9400	9405	9410	9415	9420	9425	9430	9435	9440	0	1	1	2	2	3	3	4	4
88	9445	9450	9455	9460	9465	9469	9474	9479	9484	9489	0	1	1	2	2	3	3	4	4
89	9494	9499	9504	9509	9513	9518	9523	9528	9533	9538	0	1	1	2	2	3	3	4	4

Natural Numbers	0	1	2	3	4	5	6	7	8	9	Proportional Parts								
											1	2	3	4	5	6	7	8	9
90	9542	9547	9552	9557	9562	9566	9571	9576	9581	9586	0	1	1	2	2	3	3	4	4
91	9590	9595	9600	9605	9609	9614	9619	9624	9628	9633	0	1	1	2	2	3	3	4	4
92	9638	9643	9647	9652	9657	9661	9666	9671	9675	9680	0	1	1	2	2	3	3	4	4
93	9685	9689	9694	9699	9703	9708	9713	9717	9722	9727	0	1	1	2	2	3	3	4	4
94	9731	9736	9741	9745	9750	9754	9759	9763	9768	9773	0	1	1	2	2	3	3	4	4
95	9777	9782	9786	9791	9795	9800	9805	9809	9814	9818	0	1	1	2	2	3	3	4	4
96	9823	9827	9832	9836	9841	9845	9850	9854	9859	9863	0	1	1	2	2	3	3	4	4
97	9868	9872	9877	9881	9886	9890	9894	9899	9903	9908	0	1	1	2	2	3	3	4	4
98	9912	9917	9921	9926	9930	9934	9939	9943	9948	9952	0	1	1	2	2	3	3	4	4
99	9956	9961	9965	9969	9974	9978	9983	9987	9991	9996	0	1	1	2	2	3	3	3	4

ANSWERS TO SELECTED PROBLEMS

Chapter 2, page 21

9. (a) 83°C; (b) ~86°C
12. 60 g
13. 4.2 ml
15. (a) physical separation of insoluble liquids; (b) dissolve sodium chloride in water; (c) dissolve the phenol in benzene
20. 1.25×10^3 lb mi/hr^2 or 1.13×10^9 g cm^2/sec^2
22. (a) 2.3×10^2 cal
26. $-40°$

Chapter 3, page 35

1. NaCl = 39% Na, 61% Cl; H_2O_2 = 5.9% H, 94.1% O
2. 6.5 g Na combines with 10 g Cl; 0.63 g H combines with 10 g O
4. 92% C, 8% H
8. (a) 1 mole, 6×10^{23} molecules; (d) 0.500 mole, 3.01×10^{23} molecules
11. (a) H = 1; Cl = 71
13. (a) 27.3% C, 72.7% O; (d) 59.9% Ti, 40.1% O
15. N_2O_5

Chapter 4, page 56

5. 574°F = 574°K
7. 2.8 liters
9. 822 torr
11. 28.9
13. 2.50 atm
15. 10 moles
17. 0.57 g/liter
19. (a) 60.0 liters; (b) 75.0 liters
21. 25.3

23. 1.1×10^{23}

25. 61.2 liters

Chapter 5, page 67

5. 3.71×10^{10} ergs or 888 cal

8. 4.04×10^9 collisions/sec

13. 2.0

14. 45.3

16. $273\,^\circ$C

Chapter 6, page 81

2. H_2, $z = 1.04$; CO_2, $z = 0.73$

11. $100\,^\circ$C

13. $-140\,^\circ$C

Chapter 7, page 91

6. (a) O_2; (b) 0.375 mole; (c) 0.750 mole; (d) 16.5 g CO_2 and 13.5 g H_2O; (e) 6.00 g CH_4

9. (a) 0.67 g-mole; (b) 2.0 g-moles; (c) 32 g; (d) 1.0 mole

11. dissolve 2.9 g NaCl in enough water to make 500 ml of solution.

13. 0.0500 mole

15. (a) 4.4 g H_2 in excess; (f) 2.9 g O_2 in excess

Chapter 8, page 105

5. 4.2×10^3 coulombs

7. 1.12 amp

17. (a) 9.65×10^4; (b) 4.83×10^4 coulombs/g; (c) 1.39×10^4

Chapter 9, page 114

11. 75.3% of mass 34.97; 24.7% of mass 36.95

Chapter 10, page 136

11. (a) 4.3×10^{14} sec^{-1} (per second); (b) 6.8×10^7 cm/sec

14. $\lambda = 1.215 \times 10^{-5}$ cm; $v = 2.469 \times 10^{15}$ sec^{-1}

17. (a) $\lambda = 7 \times 10^{-8}$ cm; (b) $\lambda = 2 \times 10^{-36}$ cm

20. (a) $n = 2$, $l = 0$, $m_l = 0$; (b) $n = 2$, $l = 0$, $m_l = 0$, $m_s = \pm\frac{1}{2}$; (c) $n = 3$, $l = 0$, $m_l = 0$, $m_s = \pm\frac{1}{2}$

Chapter 12, page 175

13. (b) $Mg^{2+} + I^-$; (c) $Al^{3+} + F^-$; (i) $K^+ + N^{3-}$

15. (a) tetrahedral; (b) linear; (e) angular; (f) planar trigonal; (g) linear

21. (a) $^-:\ddot{O}{-}C{\equiv}C{-}C{\equiv}O:^+$ (c)

(b)   (d)

Chapter 13, page 197

1. (a) identical; (b) identical; (c) enantiomers; (d) constitutional isomers; (e) diastereomers (cis-trans); (f) identical; (g) identical; (h) enantiomers
10. $H_2O \sim -120°C$; $HF \sim -130°C$; $NH_3 \sim -140°C$
13. yes
15. six

Chapter 14, page 215

5. butane
7. bent, or distorted, bonds
9. (a) C—H bond energy; (b) C—H and C—C bond energy + strain energy from bond angle distortion and ecllipsing of C—H bonds; (c) H—S bond energy

17. cis =

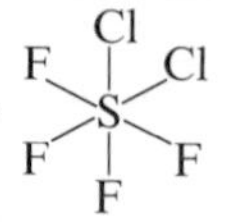

symmetry elements = $2\sigma + 1\ C_2$

Chapter 15, page 252

7. $Al(OH)_3$ is amphoteric, $B(OH)_3$ is not.
23. (a) hydrogen bonding is very strong in HF; HF is a weak acid; (b) oxyacids of F are unknown
31. (b) > (d) ~ (a) > (c) > (e)
33. (a) N_2O_3; (b) N_2O_5; (c) P_4O_6; (d) SO_2; (e) Cl_2O

Chapter 16, page 272

13. hydrogen bonding
21. chiral carbon atoms, and chiral coil of the polymer chain
24. (a) II—1,3, III—3; (b) II—2,4, III—4; (c) II—2,5, III—1,2,4; (d) II—1, III—2,3

Chapter 17, page 285

7. 52.4%
14. Zr_2C, Mo_2C, FeC, Co_2C and CoC, IrC

Chapter 18, page 291

2. (b) VC; (c) 6:6
4. Fe_4C
9. calculated $\Delta H = +38$ kcal/mole

Chapter 19, page 304

10. (a) $[CO_2] = [H_2] = 0.56$; $[CO] = [H_2O] = 0.44$
11. $K = 50$
12. 1%

Chapter 20, page 312

6. $5.0°C$
8. 185

Chapter 21, page 329

12. for 11. (a) $E° = 0.78$ volt, $\Delta G° = -1.5 \times 10^5$ J or -36 kcal, $K = 2.5 \times 10^{26}$
14. for 13. (a) $E° = 1.83$ volt

17. for 15. **(a)** $K = 6.5 \times 10^8$

19. **(a)** 1.80 volt; **(b)** $W = 2.90 \times 10^5$ J or 69.4 kcal; **(c)** $W = 9.67 \times 10^4$ J or 23.1 kcal;
(d) $\Delta G° = -2.90 \times 10^5$ J or -69.4 kcal; **(e)** $K = 7.0 \times 10^{50}$

Chapter 22, page 351

10. **(a)** $HCl + H_2O \longrightarrow H_3O^+ + Cl^-$ **(d)** $HOAc + NH_3 \rightleftharpoons NH_4^+ + OAc^-$
    (b) $HF + OH^- \rightleftharpoons H_2O + F^-$ **(e)** $H_2CO_3 + HSO_3^- \rightleftharpoons H_2SO_3 + HCO_3^-$
    (c) $H_3O^+ + OH^- \rightleftharpoons 2\,H_2O$ **(f)** $NH_4^+ + HPO_4^{2-} \rightleftharpoons NH_3 + H_2PO_4^-$

11. **(a)** H_3Y and HY^{2-}; **(b)** PH_4^+ and PH_2^-; **(c)** HCO_3^- no conjugate base for CO_3^{2-}

12. $K_a = 1.1 \times 10^{-3}$; $k_b = 9.1 \times 10^{-12}$

14. $[OH^-] = 1.3 \times 10^{-3}$

15. for 13. pH $= 4.92$

16. **(b)** pH $= 12.0$, pOH $= 2.00$

17. pH $= 13.0$

18. **(b)** $[H_3O^+] = 1.0 \times 10^{-4}$, $[OH^-] = 1.0 \times 10^{-10}$; **(c)** $[H_3O^+] = 3.2 \times 10^{-12}$, $[OH^-] = 3.2 \times 10^{-3}$

21. $[HPO_4^{2-}] = 0.31$

22. for 16. **(c)** pH $= 1.30$

26. pH $= 1.10$

Chapter 23, page 372

2. **(a)** $3.9 \times 10^{-5}\ M$; **(c)** $6.3 \times 10^{-4}\ M$

3. $K_{sp} = 4.0 \times 10^{-9}$

6. $0.071\ M$

7. CuS

8. pH $= 5.9$

12. **(a)** 20.0 ml; **(b)** pH $= 8.4$

17. **(a)** pH is in the range 5 to 6; **(b)** pH is below 1

18. methyl violet K_a is about 10^{-2}

21. 73.7%

22. 11.1 ml

Chapter 24, page 386

3. **(a)** drop in pressure; **(b)** change in acidity; **(c)** loss of color

5. 1.4×10^{-5} liter/mole sec; 8.4×10^{-1} cm^3/mole min; 2.3×10^{-26} cm^3/molecule sec

8. second order with respect to B

9. rate $= k[B]^2$

13. **(a)** reaction rates; **(b)** equilibria

15. **(a)** is faster because of less orientation requirement, combination of $+$ and $-$ ions, no covalent bonds are broken

17. **(a)** catalyst lowers the height of the energy barrier; **(b)** both rates are increased; **(c)** unaffected

Chapter 25, page 404

2. **(a)** no; **(b)** $364°K$ ($91°C$)

5. second order

7. No. The H—Cl bond energy is too high compared with the Cl—Cl bond energy.

14. **(a)** substitution; **(b)** rearrangement; **(c)** addition; **(d)** substitution; **(e)** oxidation; **(f)** oxidation

15. **(b)** or **(f)**

17. $B =$ [phenyl]$-CH-CH_3$ with Cl; $C =$ [phenyl]$-CHCH_3$ with OH; $D =$ [phenyl]$-CH=CH_2$

$E =$ [phenyl]$-CHCH_2$ with Br Br

Chapter 26, page 414

1. C—C—C—C—C—C   C—C—C—C with C C branches

C—C—C—C—C with C branch   C with C—C—C—C

C—C—C—C—C with C branch   C—C—C—C with C branches

8. **(a)** CH_3COOH is acid to litmus; **(b)** same as **(a)**; **(c)** $CH_3COO—CH_2CH_3$ may be hydrolyzed to $CH_3COOH + C_2H_5OH$; **(d)** aldehyde may be oxidized to carboxylic acid; **(e)** aldehyde is more readily oxidized than alcohol; **(f)** amine is basic to litmus paper; **(g)** same as **(f)**

Chapter 28, page 449

5. **(a)** ^1_1H; **(b)** $^{28}_{13}Al$; **(d)** ^1_0n **6.** **(a)** $^{14}_7N + ^4_2He \longrightarrow ^{17}_8O + ^1_1H$;
(e) $^{107}_{47}Ag + ^1_0n \longrightarrow ^{106}_{47}Ag + 2\,^1_0n$
6. **(g)** $^{24}_{12}Mg + ^2_1H \longrightarrow ^{22}_{11}Na + ^4_2He$
9. $^{10}_6C$
11. $\sim 12 \times 10^3$ years
13. **(a)** mass defect $= 0.00239$ amu; binding energy $= 2.23$ Mev; binding energy per nucleon $= 1.11$ Mev
18. 17.6 Mev

Chapter 29, page 468

16. 125 lb/day
17. 136 lb consumed, 0.136 lb gold

Chapter 32, page 500

5. 31

Chapter 33, page 507

4. fossil plant efficiency $= 61.5\%$, nuclear plant efficiency $= 49.8\%$
5. 5.9×10^{29} J

INDEX

A

Names, Symbols, and Atomic Weights
(weights based on carbon - 12 as 12)
(alphabetical listing)

Name	Symbol	Atomic number	Atomic weight
Actinium	Ac	89	(227)
Aluminum	Al	13	26.9815
Americium	Am	95	(243)
Antimony	Sb	51	121.75*
Argon	Ar	18	39.948*
Arsenic	As	33	74.9216
Astatine	At	85	(210)
Barium	Ba	56	137.34*
Berkelium	Bk	97	(247)
Beryllium	Be	4	9.01218
Bismuth	Bi	83	208.9806
Boron	B	5	10.81
Bromine	Br	35	79.904
Cadmium	Cd	48	112.40
Calcium	Ca	20	40.08
Californium	Cf	98	(249)
Carbon	C	6	12.011
Cerium	Ce	58	140.12
Cesium	Cs	55	132.9055
Chlorine	Cl	17	35.453
Chromium	Cr	24	51.996
Cobalt	Co	27	58.9332
Copper	Cu	29	63.546*
Curium	Cm	96	(245)
Dysprosium	Dy	66	162.50*
Einsteinium	Es	99	(249)
Erbium	Er	68	167.26*
Europium	Eu	63	151.96
Fermium	Fm	100	(255)
Fluorine	F	9	18.9984
Francium	Fr	87	(223)
Gadolinium	Gd	64	157.25*
Gallium	Ga	31	69.72
Germanium	Ge	32	72.59*
Gold	Au	79	196.9665
Hafnium†	Hf	72	178.49*
Hahnium†	Ha†	105	(260)
Helium	He	2	4.00260
Holmium	Ho	67	164.9303
Hydrogen	H	1	1.0080*
Indium	In	49	114.82
Iodine	I	53	126.9045
Iridium	Ir	77	192.22*
Iron	Fe	26	55.847*
Krypton	Kr	36	83.80
Lanthanum	La	57	138.9055*
Lawrencium	Lw	103	(257)
Lead	Pb	82	207.2
Lithium	Li	3	6.941*
Lutetium	Lu	71	174.97
Magnesium	Mg	12	24.305
Manganese	Mn	25	54.9380
Mendelevium	Md	101	(256)
Mercury	Hg	80	200.59*
Molybdenum	Mo	42	95.94*
Neodymium	Nd	60	144.24*
Neon	Ne	10	20.179*
Neptunium	Np	93	237.0482
Nickel	Ni	28	58.71*
Niobium	Nb	41	92.9064
Nitrogen	N	7	14.0067
Nobelium	No	102	(254)
Osmium	Os	76	190.2
Oxygen	O	8	15.9994*
Palladium	Pd	46	106.4
Phosphorus	P	15	30.9738
Platinum	Pt	78	195.09*
Plutonium	Pu	94	(244)
Polonium	Po	84	(210)
Potassium	K	19	39.102*
Praseodymium	Pr	59	140.9077
Promethium	Pm	61	(147)
Protactinium	Pa	91	231.0359
Radium	Ra	88	226.0254
Radon	Rn	86	(222)
Rhenium	Re	75	186.2
Rhodium	Rh	45	102.9055
Rubidium	Rb	37	85.4678*
Ruthenium	Ru	44	101.07*
Rutherfordium†	Rf†	104	(261)
Samarium	Sm	62	150.4
Scandium	Sc	21	44.9559
Selenium	Se	34	78.96*
Silicon	Si	14	28.086*
Silver	Ag	47	107.868
Sodium	Na	11	22.9898
Strontium	Sr	38	87.62
Sulfur	S	16	32.06
Tantalum	Ta	73	180.9479*
Technetium	Tc	43	(99)
Tellurium	Te	52	127.60*
Terbium	Tb	65	158.9254
Thallium	Tl	81	204.37*
Thorium	Th	90	232.0381
Thulium	Tm	69	168.9342
Tin	Sn	50	118.69*
Titanium	Ti	22	47.90*
Tungsten	W	74	183.85*
Uranium	U	92	238.029
Vanadium	V	23	50.9414*
Xenon	Xe	54	131.30
Ytterbium	Yb	70	173.04*
Yttrium	Y	39	88.9059
Zinc	Zn	30	65.37*
Zirconium	Zr	40	91.22

Numbers in parentheses are mass numbers of the most stable isotopes of radioactive elements.

All atomic weights are reliable to ±1 in the last digit except those marked *

*Atomic weights where the reliability is ±3 in the last digit.

SOME FUNDAMENTAL CONSTANTS

Gravitational constant	g	9.8067 m/sec²
Speed of light	c	2.9979×10^{10} cm/sec
Electronic charge	e	1.6021×10^{-19} coulomb
Electronic rest mass	m	9.1091×10^{-28} g
Planck's constant	h	6.6256×10^{-27} erg sec
Faraday constant	$\mathcal{F}$	9.6487×10^{4} coulomb/mole e
Gas constant	R	0.082056 liter atm/mole deg
		8.3143 joule/mole deg
		1.9872 cal/mole deg
Avogrado's number	N	6.0225×10^{23} molecules/mole

†The names and symbols for elements 104 and 105 are in dispute. Kurchatovium, Ku, for 104 and Niels Bohrium for 105 have been proposed by Russian workers, but their evidence of discovery was vague and contained inaccuracies. The names given here are the ones assigned by a group which provided more convincing evidence.